Star Ark
A Living, Self-Sustaining Spaceship

Rachel Armstrong (editor)

Star Ark

A Living, Self-Sustaining Spaceship

Published in association with
Praxis Publishing
Chichester, UK

Editor
Rachel Armstrong, PhD.
Professor of Experimental Architecture
Newcastle University
Newcastle-upon-Tyne, UK

SPRINGER-PRAXIS BOOKS IN SPACE EXPLORATION

Springer Praxis Books
ISBN 978-3-319-31040-4   ISBN 978-3-319-31042-8 (eBook)
DOI 10.1007/978-3-319-31042-8

Library of Congress Control Number: 2016952869

© Springer International Publishing Switzerland 2017
This work is subject to copyright. All rights are reserved by the Publisher, whether the whole or part of the material is concerned, specifically the rights of translation, reprinting, reuse of illustrations, recitation, broadcasting, reproduction on microfilms or in any other physical way, and transmission or information storage and retrieval, electronic adaptation, computer software, or by similar or dissimilar methodology now known or hereafter developed.
The use of general descriptive names, registered names, trademarks, service marks, etc. in this publication does not imply, even in the absence of a specific statement, that such names are exempt from the relevant protective laws and regulations and therefore free for general use.
The publisher, the authors and the editors are safe to assume that the advice and information in this book are believed to be true and accurate at the date of publication. Neither the publisher nor the authors or the editors give a warranty, express or implied, with respect to the material contained herein or for any errors or omissions that may have been made.

Cover design: Jim Wilkie

Printed on acid-free paper

This Springer imprint is published by Springer Nature
The registered company is Springer International Publishing AG Switzerland

Preface

This anthology examines the Interstellar question – i.e. the idea that we may one day live beyond the world we know and settle distant planets. A challenge on this scale requires not only vision, but multiple voices for the acknowledgment of complexity and contradictions. These are inherent in the quest. Taking a multidisciplinary and cultural view of the challenge, the book accordingly seeks to provide a form of cultural catalysis by which an interstellar culture may be seeded (it is, in other words, emphatically *not* a technical manual seeking to offer formal solutions to particular problems). To address such ambitions, the book has been divided into two main sections – Part I and II – in which differing conventions of writing have been deployed.

Part I, written by Rachel Armstrong, proposes a new age of space exploration based on an ecological perspective of the cosmos. It is this that will create the conditions for inhabiting starships and, ultimately, new worlds. Drawing on her leadership of the Persephone Project, this section adopts an experimental, yet testable, and inclusive approach to constructing a livable and self-sustaining starship. Persephone is part of the Icarus Interstellar group's portfolio of work – an international consortium of aerospace engineers aiming to construct a starship research platform in Earth's orbit within the next hundred years. This means a series of Earth-bound experiments are being detailed through a wide range of laboratory types that inform us about how we live with and design ecosystems on this planet – and beyond.

Part II, which is edited by Rachel Armstrong, introduces other voices to explore the Interstellar Question. The editor's aim here has been to create a productive interplay between differing perspectives and disciplinary backgrounds via themed, multi-author chapters. These are organized into, sections, presenting distinct viewpoints for examining the Interstellar Question. Topics include: the interstellar mission (Andreas C. Tziolas, Nathan Morrison, Esther M. Armstrong), space ecology (Michael N. Mautner, Simon Park), (Barbara Imhof, Peter Weiss, Angelo Vermeulen; Astudio – Emma Flynn, Richard Hyams, Christian Kerrigan, Max Rengifo; Susmita Mohanty, Sue Fairburn), space bodies (Kevin Warwick, Arne Hendriks, Rachel Armstrong, Sarah Jane Pell), connecting with the divine and the sacred and becoming cosmically conscious (Steve Fuller, Roberto Chiotti,

Krists Ernstsons), constructing worlds (Jordan Geiger, Mark Morris) and interstellar research methodologies (Rolf Hughes, Rachel Armstrong). The unconventional structure explores how different perspectives must be brought into a productive dialogue when considering the fundamental principles for inhabiting space. If, as a result, the book resembles a Tower of Babel for the space age, this is a design choice that invites us to address our innate diversity. Readers are invited to reflect on what these different perspectives mean for a coherent approach to settling environments far, far beyond the familiar planet we call (for now) "home".

Contents

List of Contributors xiv

PART I An Ecological View of the Interstellar Question

Chapters

1   The Interstellar Question: an ecological view by Rachel Armstrong............ 2
    1.1   Philosophy of "space"... 2
    1.2   Prototyping the interstellar question 5
    1.3   It begins... 13
    1.4   Mission.. 13
    1.5   Aspirations 13
            1.5.1   Introduction 14
    1.6   Ecocene 15
    1.7   Space ecology 18
    1.8   Who are "we"?.. 19
    1.9   Summary 20
    References.. 20

2   Architecture and space exploration by Rachel Armstrong 21
    2.1   The interstellar challenge 21
    2.2   Far, far away.. 22
    2.3   Architecture as a survival strategy 24
    2.4   Space skyscraper 28
    2.5   What is this place called "space"? 33
    2.6   World-making 37
    2.7   StarshipSPIDER.. 39
    2.8   Architecture and hypercomplexity.. 41
    References.. 44

3 Sustainability and interstellar infrastructure by Rachel Armstrong 45
- 3.1 Pale blue dot 45
- 3.2 What is an ark? 46
- 3.3 Apollo's orphans 48
- 3.4 Interplanetary communications 54
- 3.5 Apollo's progeny 55
- 3.6 Organic backbones 59
- 3.7 Directed panspermia 62
- References 71

4 An ecological approach to interstellar exploration by Rachel Armstrong 74
- 4.1 From Interplanetary to interstellar space exploration 74
- 4.2 Laboratory practice 75
- 4.3 Cultural agendas 78
- 4.4 Persephone 81
- 4.5 Mixology of space 89
- 4.6 What is a soil? 91
- 4.7 Who needs soils? 95
- 4.8 Portfolio of materials for making soils 97
- References 101

5 Experimental architecture: on-world and off-world exploration of possibilities by Rachel Armstrong 103
- 5.1 Gel experiments 103
- 5.2 The Hanging Gardens of Medusa 106
- 5.3 Capsule of Crossed Destinies 111
- 5.4 Hylozoic Ground 115
- 5.5 Future Venice 119
- 5.6 Future Venice II 123
- References 127

6 Building a worldship interior by Rachel Armstrong 129
- 6.1 Architectures of elsewhere 129
- 6.2 Worldship interior 132
- 6.3 Soils as urban infrastructure 134
- 6.4 Ecological design: biosphere 137
- 6.5 Space Nature 143
- 6.6 Manifesto for Persephone's "living architecture" 145
- References 146

7 Designing and engineering the infrastructures for life by Rachel Armstrong 148
- 7.1 Natural computing 148
- 7.2 Dissipative structures 153
- 7.3 Computing with hypercomplexity and uncertainty 156
- References 162

8 Choreography of embodiment by Rachel Armstrong ... 164
- 8.1 Interstellar being ... 164
- 8.2 Ecological being ... 166
- 8.3 Nature of life ... 169
- 8.4 A living city: building with organs ... 171
- 8.5 Anandgram: new ways of living and being ... 175
  - 8.5.1 Reconfiguring the leprous body ... 177
- 8.6 Expanded modes of being ... 180
- References ... 182

9 Constructing lifestyles by Rachel Armstrong ... 184
- 9.1 Transformations ... 184
- 9.2 Establishing new and contested territories ... 190
- 9.3 Dying well ... 194
- 9.4 Interstellar possibilities ... 195
- 9.5 Making moonlight ... 196
- 9.6 Other voices ... 200
- 9.7 New beginnings ... 202
- 9.8 Interstellar synthesis ... 204
- References ... 206

PART II Anthology of Interstellar Culture

Chapters

10 The interstellar mission ... 210
Andreas C. Tziolas, Nathan Morrison, Esther M. Armstrong
- 10.1 Introduction to interstellar exploration by Andreas C. Tziolas ... 210
  - 10.1.1 The worldview of the explorer ... 210
  - 10.1.2 The Interstellar Question (and answers) ... 211
  - 10.1.3 Explore, explore more ... 212
- 10.2 Why and where we should go, boldly by Nathan Morrison ... 213
  - 10.2.1 Why? ... 213
  - 10.2.2 Where? ... 218
  - 10.2.3 Our future among the stars ... 224
- 10.3 Scenographies and space: reading the gendered tyranny of the "Blue Ball" by Esther M. Armstrong ... 224
  - 10.3.1 Earth, we have a problem ... 225
  - 10.3.2 The grip of Gaia ... 227
  - 10.3.3 Theater's depiction of tyranny ... 228
  - 10.3.4 The problem of showing the "real" Blue Ball in theater ... 230
  - 10.3.5 Film and the immersive experience of space ... 231
  - 10.3.6 Blue-Ball "fixed" thinking ... 232
  - 10.3.7 The frame of modernity and the rise of modernism ... 233
  - 10.3.8 Revisiting modernist man: *Moon* and male existentialism ... 235
  - 10.3.9 Earth and Its conveyance of the apex of male solipsism ... 236

	10.3.10	Kubrick and the trope of the Earth-bound domestic	238
	10.3.11	Earth-bound *Gravity*	239
	10.3.12	Representing *Gravity's* visual reality	239
	10.3.13	Dr. Stone, the hysterical cassandra	241
	10.3.14	The call of the Ball	242
	10.3.15	Transcending realism	244
	10.3.16	*Interstellar* and the residue of blue-ball thinking	245
	10.3.17	Creating the reality of space in *interstellar*	246
	10.3.18	Attempting to leave the Blue Ball behind	247
	10.3.19	Gender dynamics and the residue of blue-ball thinking	250
	10.3.20	Conclusion	251
Combined Reference list for Chapter 10			253

11 Space ecology 255
Michael N. Mautner, Simon Park

- 11.1 Saving life itself: life-centered ethics, astroecology, and our cosmological future by Michael N. Mautner ... 255
  - 11.1.1 Biology, ethics, and purpose ... 255
  - 11.1.2 The human role ... 256
  - 11.1.3 The origins of life and ecology ... 256
  - 11.1.4 Biological challenges ... 259
  - 11.1.5 A roadmap to the galaxy ... 259
  - 11.1.6 Astroecology ... 260
  - 11.1.7 Human prospects ... 261
  - 11.1.8 Cosmo-ecology and the long-term future ... 263
  - 11.1.9 Quantitative astroecology ... 264
  - 11.1.10 Conclusions ... 265
- 11.2 Saving life itself: science and strategies for seeding the universe by Michael N. Mautner ... 266
  - 11.2.1 Synopsis ... 266
  - 11.2.2 Seeding the universe: motivations for directed panspermia ... 266
  - 11.2.3 Evaluating potential targets ... 267
  - 11.2.4 Propulsion and navigation ... 268
  - 11.2.5 Capture in target zones ... 269
  - 11.2.6 The microbial payload ... 270
  - 11.2.7 Inducing a biosphere ... 270
  - 11.2.8 Quantitative analysis ... 270
  - 11.2.9 Designing the biological payload ... 273
  - 11.2.10 Conclusions ... 273
- 11.3 Silent running: the bacteriology of spaceflight by Simon F. Park ... 273
  - 11.3.1 Introduction ... 274
  - 11.3.2 Earth: planet of the bacteria ... 275
  - 11.3.3 Bacteria in space ... 278
  - 11.3.4 When alien microbiologies meet ... 282
- Combined Reference list for Chapter 11 ... 283

Contents

12 Space architectures 287
Barbara Imhof, Peter Weiss, Angelo Vermeulen, Emma Flynn,
Richard Hyams, Christian Kerrigan, Max Rengifo, Susmita Mohanty
and Sue Fairburn
- 12.1 The world in one small habitat, by Barbara Imhof,
  Peter Weiss, and Angelo Vermeulen 287
  - 12.1.1 MELiSSA: the agency's perspective 288
  - 12.1.2 MEDUSA: designing for living in outer space—underwater!.. 290
  - 12.1.3 Water Walls life support: the next step in evolution 295
  - 12.1.4 Toward a robust and resilient future in outer space 295
- 12.2 Starship cities: a living architecture, by Astudio: Emma Flynn,
  Richard Hyams, Christian Kerrigan and Max Rengifo 296
  - 12.2.1 Introduction 296
  - 12.2.2 Tomorrow's city 297
  - 12.2.3 Architectural ecologies 304
  - 12.2.4 Near-future architecture 322
- 12.3 Terraforming our cities by Susmita Mohanty,
  Sue Fairburn and Barbara Imhof 323
  - 12.3.1 Introduction 323
  - 12.3.2 Urban atmospheric reconstruction 326
  - 12.3.3 Earth as a living laboratory 330
  - 12.3.4 Case studies 333
  - 12.3.5 The future city 340
- Combined Reference list for Chapter 12 340

13 Space bodies 341
Kevin Warwick, Arne Hendriks, Rachel Armstrong, Sarah Jane Pell
- 13.1 Cyborgs: upgrading humans for a future in space by Kevin Warwick 341
  - 13.1.1 Introduction 342
  - 13.1.2 Robots with biological brains 342
  - 13.1.3 General-purpose brain implants 345
  - 13.1.4 Non-invasive brain–computer interfaces 348
  - 13.1.5 Subdermal magnetic implants 349
  - 13.1.6 Conclusions 350
- 13.2 Shrinking into the universe by Arne Hendriks 352
  - 13.2.1 Introduction 352
  - 13.2.2 Genetics 354
  - 13.2.3 Environment 356
  - 13.2.4 Culture 357
  - 13.2.5 View on the future 359
- 13.3 Human reproduction in space by Rachel Armstrong 359
  - 13.3.1 The imperative for human space exploration 359
  - 13.3.2 Human space migration 360
  - 13.3.3 The survival imperative 361
  - 13.3.4 Spacefaring humans 361

Contents

- 13.3.5 Space colonization 362
- 13.3.6 Sex in space 363
- 13.3.7 Human reproduction in space 365
- 13.3.8 Data from animal experiments 365
- 13.3.9 Implications of animal studies for humans 367
- 13.3.10 The future of human reproduction in space 367
- 13.3.11 Self-preservation 370
- 13.3.12 Augmented reproduction 370
- 13.3.13 Symbiotic environments 370
- 13.3.14 Succession of humans 372
- 13.4 Bodies *in extremis* by Sarah Jane Pell 372
  - 13.4.1 Fluid architectures for human survival in space 372
  - 13.4.2 Wet dreams and politicized bodies *in extremis* 373
  - 13.4.3 Human relationships are intricate 376
- Combined Reference list for Chapter 13 378

14 Connecting with the divine and the sacred, and becoming cosmically conscious 383
Steve Fuller, Roberto Chiotti, Krists Ernstsons
- 14.1 Humanity's lift-off into space: prolegomena to a cosmic transhumanism by Steve Fuller 383
  - 14.1.1 What is transhumanism? 383
  - 14.1.2 Which way is up for the human condition? 386
  - 14.1.3 Who's afraid of the human? A theological defense of ecomodernism 389
  - 14.1.4 Ecomodernism as the politics of theological energeticism 390
  - 14.1.5 Expanding the scope of the human in search of the divine 392
- 14.2 The dream drives the action: toward a functional cosmology for interstellar travel by Roberto Chiotti 394
  - 14.2.1 The big picture: how did we get here and where are we going? 395
  - 14.2.2 The universe story as our new story, our new cosmology 397
- 14.3 Spaceship mind/virtual migration to exoplanets by Krists Ernstsons 401
  - 14.3.1 Introduction 401
  - 14.3.2 Science and light 401
  - 14.3.3 Microseeds and migration 405
- Combined Reference list for Chapter 14 408

15 Constructing worlds 410
Jordan Geiger, Mark Morris
- 15.1 Alive without us by Jordan Geiger 410
  - 15.1.1 Introduction 410
  - 15.1.2 Spacecraft and architecture 411
  - 15.1.3 The Very Large 413
  - 15.1.4 Beyond the Very Large 416

	15.2	The scales of *Ouroboros* by Mark Morris..	427
		15.2.1 Introduction..	427
		15.2.2 World-building..	433
		15.2.3 Architectural iterations...	434
		15.2.4 Conclusion...	442
	Combined Reference list for Chapter 15		445
16	**Interstellar research methodologies**		**446**
	Rolf Hughes, Rachel Armstrong		
	16.1	The art of the impossible: beyond reason in the twenty-first century by Rolf Hughes..	446
		16.1.1 Introduction..	447
		16.1.2 Knitting entanglements	453
		16.1.3 Unraveling the human	453
		16.1.4 Catastrophic reversals	455
		16.1.5 Circus: risk, gravity, equilibrium, disequilibrium	456
	16.2	*The Temptations of the Non-linear Ladder*: meditations on scrying space by Rolf Hughes and Rachel Armstrong ...	459
		16.2.1 On scrying..	459
		16.2.2 Background to *The Temptations of the Non-linear Ladder*.........	460
		16.2.3 Background: "protocell circus"...	461
		16.2.4 Principles..	462
		16.2.5 Themes of *The Temptations of the Non-linear Ladder*	462
		16.2.6 On medaka fish...	464
		16.2.7 On Lazarus fish	466
		16.2.8 Additional texts for inspiration	469
	16.3	"It" by Rolf Hughes	474
	16.4	Legacy of the Interstellar Question by Rachel Armstrong	477
	Combined Reference list for Chapter 16		485
Index...			**486**

List of Contributors

Esther M. Armstrong is Program Director for Theatre and Screen at Wimbledon College of Arts, University of the Arts, London. She lectures in Critical and Contextual Studies across the design disciplines of theater and screen arts. Her Ph.D. thesis interrogated the relationship between the reading of set design and national identity. She has also worked and trained in technical arts at the Royal Academy of Dramatic Arts and broadcast journalism at the London College of Printing. She has an M.A. from Regents Park College at the University of Oxford.

Rachel Armstrong is Professor of Experimental Architecture at Newcastle University. She designs lifelike environments for the built environment using technologies that manipulate the building blocks of life such as synthetic biology and smart chemistry. Armstrong trained as a medical doctor, graduating from the University of Cambridge with First-Class Honors and prizes. She completed her clinical training at the John Radcliffe Medical School at the University of Oxford. She also has qualifications in general practice and a Ph.D. funded by the EPSRC in Architecture from the Bartlett School of Architecture, University College London. Armstrong has worked across many disciplines as a multimedia producer, a science fiction author, and an arts collaborator. She is TWOTY Futurist of the Year 2015 and a 2010 Senior TED Fellow. Rachel was named as one of the top 10 UK innovators by *Director Magazine* in 2012 and featured in the top 10 "Big Ideas, 10 Original Thinkers" for *BBC Focus Magazine*. Her TED book on Living Architecture was #1 Bestseller in Biotechnology on Amazon. Her book, *Vibrant Architecture: Matter as CoDesigner of Living Structures*, explores prospects for transformations of matter from inert configuration into lifelike habitable structures, which prompts a reevaluation of how we think about sustainability in our homes and cities.

Roberto Chiotti, B.E.S., B.ARCH., M.T.S., O.A.A., F.R.A.I.C. is Principal at Larkin Architect Limited and Assistant Professor at the Faculty of Design, OCAD University, Toronto, Canada. He obtained his professional architectural degree in 1978 from the University of Waterloo, Canada, and completed a Master of Theological Studies degree at the University of St. Michael's College, University of Toronto, in 1998 with a specialty in

Theology and Ecology obtained through the Elliott Allen Institute for Theology and Ecology at St. Michael's. He is a founding partner of Larkin Architect Limited, a firm specializing in the design of sacred space, and is also currently Assistant Professor within the Department of Environmental Design and the Sustainability Officer for the Faculty of Design at OCAD University. Chiotti brings "big picture" thinking to the topic of sustainability, insisting that his favorite designer is the universe itself, citing many of the major transformations over its 14 billion years of existence and the dynamic laws of cosmogenesis as valuable resources to help address the ecological challenges facing the human community today. He believes that the journey toward sustainability must be grounded in a cosmology that relinquishes the current anthropocentric worldview in favor of one that acknowledges the needs of Earth as primary and those of the human as derivative.

Krists Ernstsons is Architectural Assistant/RIBA Part 2/AA Diploma. He completed his Part I in Architecture at the University of East London and his Part II Diploma in Architecture at the Architectural Association School of Architecture. His thesis project "Virtual Migration to Exoplanets" was nominated for honors and explored the idea of breaking the speed-of-light boundary through quantum entanglement and migrating to Earth-like planets at the macroscale. The key driver for this study was the idea that humanity eventually will grow out of its cradle called the planet Earth. After his Part I, he worked for Foster+Partners in London, where he prototyped a household unit to revolutionize efficiency and functionality of low-income housing. In 2010, he joined Izolyatsia, a platform for Cultural Initiatives in Ukraine, where he worked on master planning, refurbishment, curating a residency program, and executing projects by internationally renowned artists. He recently joined Hawkins\Brown in London, where he works in the infrastructure sector on Crossrail station projects and others.

Susan Fairburn is a Design Educator and Researcher who works between the boundaries of the body and the environment. Her research experience, gained over 20+ years, uses design as knowledge exchange applied to Social Design and Reciprocities in Design for Extremes. She was cofounder of a social enterprise, Design for Development (DFD 2005–2012), which used the design process as a problem-solving and engagement tool in low-income settings. She continues to work in the areas of social innovation and environmental design at Fibre Design Inc. Educated in Canada, Sue holds degrees in Environmental Physiology and Environmental Design.

Emma Flynn leads Research and Development at Astudio. She is a practicing architect and design researcher, whose work broadly focuses on the future architectural landscape in relation to Nature, exploring environmental responsiveness and resilience in the context of climate change and resource depletion. In 2012, she helped to establish Astudio Research as the separate R&D arm of Astudio in response to the need for increasingly innovative responses to issues of sustainability in the built environment. Existing at the intersection of sustainability, technology, and design, her work is highly collaborative and multidisciplinary, exploring fields such as environmental design, energy management and modeling, emerging technologies, facade system design, modern methods of construction, and future cities. Key research projects include Living Architecture: Demonstrating Resilience to Climate Change and Resource Depletion, Bio-Responsive Facade Systems,

In-Use Environmental Performance Monitoring and Metering Systems, and Waste to Value: Micro Anaerobic Digestion for Urban Communities. She is a tutor in Environmental Design on the B.Sc. Architecture course at the Bartlett School of Architecture, is a module leader of the Adaptive Typologies Think Tank, London School of Architecture, and works with students at Brunel University as part of the EU Co-Innovate program.

Steve Fuller is Auguste Comte Professor of Social Epistemology in the Department of Sociology at the University of Warwick, UK. Originally trained in history and philosophy of science, he is best known for his foundational work in the field of "social epistemology," which is the name of a quarterly journal that he founded in 1987 as well as the first of his more than 20 books. He has recently completed a trilogy relating to the idea of a "post-" or "trans-"human future, all published with Palgrave Macmillan: *Humanity 2.0: What It Means to Be Human Past, Present and Future* (2011), *Preparing for Life in Humanity 2.0* (2012), and (with Veronika Lipinska) *The Proactionary Imperative: A Foundation for Transhumanism* (2014). His latest book is *Knowledge: The Philosophical Quest in History* (Routledge 2015). His works have been translated into over 20 languages. He was awarded a D.Litt. by the University of Warwick in 2007 for sustained lifelong contributions to scholarship. He is also a Fellow of the Royal Society of Arts, the UK Academy of Social Sciences, and the European Academy of Sciences and Arts.

Jordan Geiger is an architect and educator whose work crosses architecture and interaction design, considering implications of human–computer interaction for social and environmental issues. He lectures, exhibits, and publishes internationally on theoretical research and on his projects, which frequently investigate globalization's design problems at many scales. Design research outcomes include buildings, objects, and landscape proposals, but also written and graphic analyses and technological investigations. He is editor of the book *Entr'acte: Performing Publics, Pervasive Media and Architecture*, which explores ephemeral and interstitial formations of publics and of public space with the proliferation of new technologies. Geiger has taught architecture, at the University of Buffalo as member of the department's Center for Architecture and Situated Technologies. He has also taught architecture, urban design, and advanced interdisciplinary studios and seminars at the Academy of Fine Arts in Vienna, at UC Berkeley, and at the California College of the Arts in San Francisco. He holds a Master of Architecture from Columbia University and a Bachelor of Arts in Comparative Literature from UC Berkeley.

Arne Hendriks is an Amsterdam-based artist, exhibition maker, researcher, and historian with a Master of Arts from the University of Amsterdam. He teaches at the Next Nature department of the Technical University in Eindhoven, which is concerned with investigating the implications of nanotechnology. He was the second ambassador of the 13th Dutch Design Week and a curator for the next Alternativa exhibition in Gdansk Poland. His work explores the positive transformative power of creative impulses and the importance of fundamental free scientific research. In his speculative design research, the strange and the familiar continuously swap places to provoke conflicting perspectives, which speak about the radicality of everyday experience and the familiarity of radical interventions. He is a strong believer in the transparency of information and an active participant in the open-design movement. His projects include Instructables Restaurant (the world's first open-source restaurant), Hacking Ikea, the Repair Manifesto, and the Academy of Work. His most recent projects include *The Incredible Shrinking Man*—downsizing the human species to better fit Earth; Fatberg—constructing a floating island of fat to ask what this substance means to us in an age of excess, obesity, and starvation; and 8 Billion City—one city for all.

Rolf Hughes Professor of Artistic Research and former Head of Research at Stockholm University of the Arts (inaugurated 2014). He is a prose poet and disciplinary nomad, actively promoting innovative forms of artistic and transdisciplinary research over the past 20 years. He has been expert advisor for artistic research at the Swedish Research Council, the Norwegian Artistic Research Programme, and the Austrian Programme for Arts-based Research (PEEK); Guest Professor in Design Theory and Practice-Based Research at Konstfack University College of Arts, Crafts and Design (2006–2014); and Senior Professor in Research Design at Sint-Lucas School of Architecture (KU-Leuven, Belgium), where he helped create and develop an international, design-led Ph.D. program (2007–2013). He has also served two terms as Vice President of the international Society for Artistic Research (elected by the SAR membership 2011–2013, unanimously re-elected 2013–2015). Hughes holds a First-Class degree in English and Related Literature (University of York), an M.A. (with Distinction) in Creative Writing, and the first ever Ph.D. in Creative and Critical Writing funded by the British Academy from the University of East Anglia, UK. He is currently exploring the potential contribution of magic and the circus arts to the conception and design of a third-millennium experimental research laboratory. Writing and performing arts remain central to his endeavor to link diverse forms of experience, expertise, and knowledge.

Richard Hyams is a Director of Astudio, whose purpose is to leave the planet better than we found it. He is an architect who focuses on Research and Innovation. Motivated by unique design challenges, he enjoys leading design at and beyond the industry boundaries. Blending high-quality design with research, he challenges the professional status quo and leads teams to deliver highly individual buildings that inspire and delight. His pragmatism for understanding a complex brief and finding elegant yet simple solutions has placed Astudio as one of the most successful rising practices. Hyams has encouraged this approach throughout Astudio, and the practice's innovative lateral thinking often surprises clients. This is adopted on all projects to deliver buildings and spaces which are not thought possible in the tight constraints which project briefs often demand. Astudio was awarded Architectural Practice of the Year after just 6 years in practice.

Barbara Imhof is a space architect, design researcher, and lecturer. She cofounded LIQUIFER Systems Group in 2004, a platform of experts comprising different experts from the field of engineering, science, architecture, and design collaborating on R&D projects. She has a background in architecture, having studied at the Bartlett School UCL, London, and graduated from the University of Applied Arts (Studio Wolf D. Prix). She additionally holds a Master of Science from the International Space University in Strasbourg, France, and a Ph.D. from the Vienna University of Technology (VUT). She taught at the VUT (assistant professor for 8 years), the ETH Zürich, and among others at the Chalmers University in Gothenburg. Barbara combines artistic with scientific education.

She has more than 10 years of experience in project lead functions with EU framework programs, the Austrian Science Fund, and contracts for the European Space Agency. She has been the project leader of architectural projects and academic teaching projects that include the first European habitat simulator, SHEE (Self-deployable Habitat for Extreme Environments), and the biomimetics project GrAB (Growing As Building), translating growth principles from Nature into proto-architecture. Currently, Barbara develops RegoLight, a H2020 EU solar sintering project developing building elements for 3D printed habitats on the Moon.

Christian Kerrigan is an architect at Astudio. He studied architectural technology at Dublin Institute of Technology, architecture at Edinburgh College of Art, and completed his M.Arch. with Distinction at the Bartlett School of Architecture in 2007. He enjoys growing strong links in a variety of industries to push beyond the expectations of architecture. Working through concept computer drawings, installations, R&D, and architectural technology, he has informed the design and realization of a number of projects including Avon & Somerset Constabulary, CI Tower New Malden, RIBA Pylon Competition, Raine's Foundation School, and Skyline of London a future vision. Recently, he has been responsible for the team delivery of a commercial office development for Stanhope at 70 Wilson Street. He is project leader in the design of an amphitheater and Memory Stones project on the Isle of Portland. Kerrigan's research explores the 200 Year Continuum that draws from Nature to innovate across disciplines, which began during his thesis year at the Bartlett School of Architecture, UCL. In 2010, he collaborated with chemist Martin Hanczyc to make the world's first synthetic biology drawing, programming ink to move by itself. He has exhibited his work in the USA, Europe, and the UK, and, in 2010, he was awarded the first digital artist in residence at the Victoria & Albert Museum. He has lectured for Astudio at the Aarhus School of Architecture Master's program and tutored at Chelsea College of Art and Greenwich University. He has spoken at numerous conferences such as Leonardo ISSAT Art papers SIGGRAPH, New Orleans; Generative Art International, Milan; GSK New Contemporary Art season at the Royal Academy of Arts; and First To Bloom, Last To Leave as part of the Human–Nature series at Siobhan Davies Dance studio. Christian's work has also been featured on TED collaborating with Dr. Rachel Armstrong on "Architecture that repairs itself."

Michael Noah Mautner is Research Professor at Virginia Commonwealth University. Born in 1942 in Budapest, Hungary, he was saved from the Holocaust by a courageous woman, Irene Gigor-Horvath. He obtained a B.Sc. from the Hebrew University, an M.Sc. from Georgetown University, and a Ph.D. in Physical Chemistry from Rockefeller University. He is the author of over 180 research papers and book chapters on ion chemistry (also under Meot-Ner) and in astrobiology and space science. He served as Associate Professor at Rockefeller University and as a Research Scientist at the National Institute of Standards and Technology. His research interests in chemistry include ion chemistry with applications to astrochemistry and biophysics; space science, astroecology, and the soil fertilities of asteroid/meteorite materials as biological resources; prospects for human populations in the Solar System; and directed panspermia for seeding new solar systems to secure and expand life. His early experiences showed the value of life, especially when we can either destroy our species or perpetuate life for innumerable eons. This led to the principles of life-centered biotic ethics that value life itself and panbiotic ethics that can motivate us to seed new solar systems with our family of organic gene/protein life (see www.astro-ecology.com, www.panspermia-society.com).

Susmita Mohanty is a spaceship designer and aerospace entrepreneur who lives a renaissance life without boundaries. She was a member of an international crew that lived in isolation on a simulated Martian outpost in the Utah desert (2004). She is the CEO of Earth2Orbit, India's first private start-up and her third venture. Prior to turning entrepreneur, she worked for the Space Station Program at Boeing in California and for a brief period at NASA Johnson. Educated in India, France, and Sweden, Mohanty holds multiple degrees including a Ph.D.

Mark Morris teaches architectural design, history, and theory at Cornell University. He is the winner of an AIA Medal for Excellence in the Study of Architecture. Morris has taught at The Bartlett, University College London, The Architectural Association School of Architecture, and the University of North Carolina–Charlotte. Morris studied architecture at Ohio State University and received his Ph.D. at the Consortium Doctoral Program, University of London, supported by a Royal Institute of British Architects grant. His studios focus on modeling techniques, reflexive composition, and craft. Morris's essays have been featured in *Frieze, Contemporary, Cabinet, AD*, and *Domus*. He is the author of two books: *Models: Architecture and the Miniature* and *Automatic Architecture: Designs from the Fourth Dimension*. Host of the iTunes podcast series "Architecture on Air," Morris co-organized the Preston Thomas Memorial symposium "Architecture of Disbelief." His research focuses on architectural models, scale, and questions of representation. Morris has previously served as coordinator of post-professional programs and director of graduate studies in the field of architecture.

Nathan Morrison is the Chief Executive Officer and Director of Research and Development for Sustainable Now Technologies, Inc. (SNT), a biotech firm located in Southern California that he cofounded in 2009. Nathan additionally designs living interiors for the Icarus Interstellar Project Persephone. His work primarily centers on closing the cycle of human respiration. Nathan is a published author, public speaker, and a strong advocate for carbon capture implementation strategies. He has spent most of the last decade developing innovative bioreactor technologies that utilize algae as a living internal component to capture carbon and produce sustainable hydrocarbons as organic biomass. Examples of his work include the development of the Helix BioReactor in collaboration with the late inventor Steven Shigematsu, the Algae Research Module for the University of Greenwich's School of Architecture and Landscape in London, and SNT's Mark IX BioReactor modules. These bioreactors are now entering the market in the UK to enable academic institutions and new construction projects to comply with the Climate Change Act implemented by Parliament in 2008.

Simon Park is a senior lecturer in the Faculty of Health and Medical Sciences at the University of Surrey, where he teaches Bacteriology and Molecular Biology. As an internationally recognized molecular bacteriologist, he has published over 60 papers in international refereed journals, books, and other periodicals. His wider activities, and practice, are driven by the common misconception that microbiological life is primitive and always detrimental and that through collaborations with artists the real nature of the microbiological world can be revealed. In this context, he has been widely involved in many innovative collaborative projects with artists. Funded collaborations include "Sixty Days of Goodbye Poems of Ophelia" with artist JoWonder (funded by The Wellcome Trust), "Exploring the Invisible" with artist Anne Brodie (funded by the Wellcome Trust), and "Communicating Bacteria" with artist Anna Dumitriu (funded by the Wellcome Trust).

Sarah Jane Pell is an artist-astronaut and Australia Council Fellow (AU), TED Fellow (USA), and Scientist-Astronaut (PoSSUM Class 1601) for the NASA Flight Opportunities Program Experiment 46-S, Noctilucent Cloud Imagery and Tomography Experiment, she has an extensive exhibition and publication record. She incorporates themes of human-aquatic adaptation to other worlds and other extreme-performance interfaces in her work. She established the Aquabatics Research Team initiative (ARTi) 2002–2012 for developing aquatic performance and related underwater technologies. In support of her research, she logged 500+ Occupational Dives and demonstrated prototype re-breather systems. Leonardo LABS awarded her Best PhD "Art & Science," MIT 2007, and she was made Official Aquanaut of the Atlantica Expeditions subsea habitat mission. In 2006, she graduated from the International Space University to lead the NASA-sponsored project "Luna Gaia: Closed Loop Habitat for the Moon." At Singularity University, in 2010, she codeveloped exponential technology pathways for NASA to "Boldly Stay" in space. Appointed Cochair of the European Space Agency Topical Team Art Science (ETTAS) 2011–2014, she published the ESA Arts Initiative as the first author. In April 2015, Pell reached Everest Base Camp (5,364 m) and survived the Nepal earthquakes. She was attempting an independent arts-led expedition to summit Mt. Everest as an experiment in space analog training and communications. She currently works as Simulation Astronaut for Project Moonwalk, EU Undersea Lunar Analogue EVA Simulation Trials, and is Prime Crew for the Project Poseidon: 100-day Undersea Habitat Mission, US.

Max Rengifo is a Director of Astudio who uses a range of techniques, from sketches to advanced 3D tools, that brings depth to every project. By leading design reviews, he is able to guide projects, making sure that a clear design response is achieved from its brief response to its material language. Rengifo led the award-winning Wigan Life Centre scheme, which is recognized as an exemplar urban regeneration project. He has also been responsible for TOKKO MyPlace Youth Space in Luton, working and mentoring a group of young clients during the design process. Prior to this, he was a founding Director of CPR Arquitectos, where his work was recognized internationally for the National Health Institute of Venezuela—Production Laboratories and Administrative buildings project. He is also an invited critic and university lecturer in South America and Europe.

Andreas Tziolas is a cofounder and the President of Icarus Interstellar. He has served as Project Icarus Leader (fusion-based starship study) and is currently Project Lead for Project TinTin (interstellar nanosat mission development team). He holds the positions of Chairman for Research and Chairman for Education at Icarus Interstellar. Tziolas completed his Ph.D. in Gravitation and Cosmology at Baylor University in 2009. His dissertation "Colliding Branes and Formation of Spacetime Singularities in Superstring Theory" has implications for the study of black holes in extra dimensions. He holds an M.Phys.

degree in Physics with Space Sciences and Technology from Leicester University, where he worked on the Life Detection Module aboard the Mars Express/Beagle-2 mission to Mars. Tziolas has also held a variety of research positions including two summer research fellowships at JPL/NASA, where he worked with the Outer Planetary Atmosphere's Group, supporting the Galileo mission to Jupiter as part of the Hubble Wide Fields and Planetary Camera team. He was a Graduate Technologist working on the Large Interferometric Space Antenna (LISA) mission development team at the University of Birmingham in the UK. He currently resides in Anchorage, Alaska, where he has held the position of Chief Scientist for Variance Dynamical Corp., developing next-generation analog electronic sensors for use in real-time spectrometers and high-radiation environments. He is the founder and President of the Anchorage Makerspace and currently a Project Manager at one of Alaska's largest nonprofit foundations.

Angelo Vermeulen is a space systems researcher, biologist, artist, and community architect at the Participatory Systems Initiative, Delft University of Technology. He therefore collaborates as easily with practicing scientists as he is comfortable constructing multimedia installations in galleries or building communities through design and co-creation. In 2009, he initiated SEAD (Space Ecologies Art and Design), an international network of creatives working in art, science, engineering, and advocacy. Its goal is to reshape the future through critical reflection and hands-on experimentation. To achieve this, SEAD develops paradigm-shifting projects in which ecology, technology, and community are integrated in synergistic ways. Biomodd and Seeker are the two most well-known SEAD projects. From 2011 to 2012, Vermeulen was a member of the European Space Agency Topical Team Arts & Science (ETTAS), and, in 2013, he was Crew Commander of the NASA-funded HI-SEAS Mars simulation in Hawaii. His space-related work led him to start research at Delft University of Technology, creating new concepts for starship development. He has been faculty at LUCA School of Visual Arts in Ghent, the University of Applied Arts in Vienna, Parsons, the New School for Design in New York, and the University of the Philippines Open University in Los Baños. In 2012, he was a Michael Kalil Endowment for Smart Design Fellow at Parsons, and, in 2013, he became TED Senior Fellow.

Kevin Warwick is Deputy Vice Chancellor (Research) at Coventry University, England. His main research areas are artificial intelligence, biomedical systems, robotics, and cyborgs. Due to his reputation as a self-experimenter, he is frequently referred to as the world's first Cyborg. He was born in Coventry, UK, and left school to join British Telecom. He took his first degree at Aston University, followed by a Ph.D. and research post at Imperial College London. He held positions at Oxford, Newcastle, Warwick, and Reading Universities before joining Coventry. Warwick is a Chartered Engineer who has published over 600 research papers. His experiments into implant technology led to his being featured as the cover story on the US magazine *Wired*. He achieved the world's first direct electronic communication between two human nervous systems, the basis for thought communication. Another project extended human sensory input to include ultrasonics. He also linked his nervous system with the Internet in order to control a robot hand directly from his neural signals, across the Atlantic Ocean. He has been awarded higher doctorates (D.Sc.) by Imperial College and the Czech Academy of Sciences, Prague. Warwick has been awarded Honorary Doctorates by eight UK universities and from Saints Cyril & Methodius University, Skopje. He received the IEE Senior Achievement Medal, the IET Mountbatten Medal, and the Ellison-Cliffe Medal from the Royal Society of Medicine. In 2000, he presented the Royal Institution Christmas Lectures.

Peter Weiss is the Head of the Space and Innovations Department at COMEX. He has 15 years of experience in the management of industrial and research projects in the field of deep sea and space robotics and EVA simulations. He is Technical Coordinator of the FP7 project MOONWALK and Manager at COMEX for the SHEE project on a novel space habitat design. Weiss earned a B.Sc. (FH München, Germany) in Mechatronics, a M.Sc. (EPF Paris, France) in Robotics, and a Ph.D. on kinetic impactor sampling probes (Hong Kong Polytechnic University, China). Peter worked at the DLR (Oberpfaffenhofen, Germany) on a motor drive of a lightweight robot arm, at the Massachusetts Institute of Technology (Boston, USA) on the development of artificial muscles for self-transforming robotic planetary explorers, at Cybernetix (Marseilles, France) as Project Manager for several European Commission Projects and industrial projects, and at the Hong Kong Polytechnic University on a robotic microgravity sampling device for the PHOBOS-GRUNT mission. During his studies in Hong Kong, he led a team of researchers in the frame of the APOPHIS MISSION DESIGN COMPETITION (Planetary Society, NASA, ESA, ASE, AIAA, USRA), where his team won a prize for their innovative proposal.

Part I
An Ecological View of the Interstellar Question

1

The Interstellar Question: an ecological view

Rachel Armstrong

This chapter discusses a vision and context for the Interstellar Question.

1.1 Philosophy of "space"

For a place that we barely know, we talk about "space" with a fair degree of confidence.

This is not surprising—we are, after all, surrounded by space. We see it every night, gazing down on us through billions upon billions of winking stars. As soon as we could imagine, we've asked ourselves how we might live among these stars. In dreaming of its vast potential, we have answered this question in different ways.

In ancient times, space was a place for immortals, ancestors, and beings with magical powers. With the advent of the Enlightenment as Medieval belief-based systems of thinking were replaced by rationality, Galileo Galilei and Isaac Newton gave us the concept of planetary orbits as mechanical systems, whereby we imagined ourselves orbiting the Sun. Later, with the advent of the Industrial Revolution, flying machines escaped Earth's gravitational pull and enabled us to tread a pathway from the imagination towards the horizons of possibility. Today, powerful vehicles ferry us beyond our planetary domain—living in space has become a reality.

With experiments conducted within the sealed environment of our first orbital house, the International Space Station (ISS), we have started to discover just a little about what it means to dwell in a non-terrestrial habitat.

But this is not the limit of our experience of space, as we have extended our senses through the visualization systems of robots into the Solar System. We now receive images that we can layer on to these models of space. They are not "reality" as such, but ideas, images, and proposals overlaid on concepts, which are not (and cannot be) mediated directly through our senses. As wonderful as it may be to "see" the bitterly harsh terrains

---

Rachel Armstrong, PhD. (✉)
Professor of Experimental Architecture, Newcastle University, Newcastle-upon-Tyne, UK

of Mars through the eyes of rovers and New Horizon's survey of Pluto, it's an experience that is very different from actually trying to live there.

So how do we know what we know about space? On what basis do we construct our ideas about the world, the cosmos, and the way in which its perceived reality unfolds? Crucially, how can we maintain a critical understanding of the assumptions that we accept as being truths, so that we can continue to evaluate, develop, and even inhabit them?

Such questions underpin the concepts that construct our philosophical understanding of our own world as well as other spaces. Can we, for example, expect the same principles of material interaction to operate in non-terrestrial locations as they do here on Earth? As we know, the way we think shapes our horizons of expectation and underpins how we imagine unfolding events. Ultimately, a philosophy of space enables us to create innovative living experiences that relate terrestrial and alien landscapes.

This reflection on contemporary conceptions of "space" is the subject of this book. It is examined, explored, and exemplified through a project called Persephone, which is part of the Icarus Interstellar portfolio of projects that propose to construct a starship research platform in orbit within 100 years.

This book is therefore a creative engagement with a broad range of ideas that seek to influence the way we think about space—as an evolution of scientific ideas, for example, and as technologies that may shape our modes of living or experiencing. The concern is less with degrees of virtuosity in describing non-terrestrial environments, or the construction of megastructures, and more with curating those values and ideas that may inform how we live together.

And, yet, to do this critically requires us to confront, dissolve, and potentially resynthesize some of our most deeply held preconceptions. In embarking on extended, unbounded, and strategic disorientations of our senses—and the kinds of knowledge that we take for granted—we should seek to open up new potentialities within the unknown and unexplored reaches of space. This would ask us to connect with our deepest desires and ambitions, to engage with our intuitions, so that we can revisit our aspirations and dreams. Who are we? And what is the purpose of our existence?

Ultimately, these visions may help us to conjure up new ways of working and constructing alternative modes of existence, which open up the possibility of inhabiting terrains that are currently inaccessible to us.

That a philosophical approach to the future of space exploration and colonization should stray into the realm of fiction is hardly surprising—our first starship has not yet been built. This book therefore proposes to create a space for a diverse range of narratives and ideas to shape our conversations about human space exploration. It is also anticipated that readers will differ in their reactions to the multiple ideas and perspectives shared in this book, in which conversation—not consensus—is sought. Of course, when we eventually construct a habitat, we must find modes of working that enable us to reach agreement on what we are doing—but that is not the same as expecting everyone's perspective to be homogenous—nor to conform to any particular form of knowledge, or creed.

Indeed, the current publication should be read as an experiment into different modes of thinking concerning topics that are so vast, complex, heterogeneous, or shaped by so many contingencies that they require us to work with our differences to the extent that we can ultimately co-imagine experiences in those terrains best described as "unknowns."

Such a synthesis of ideas is well encapsulated by an ancient parable from the Far East, which describes the consternation that a group of learned men experience when encountering a specific natural phenomenon for the first time. All observers have an incomplete view of the subject owing to a preferred perspective. Since they are all holding a different part of the elephant, they describe the nature of the beast depending upon where they touch, likening it to a rope, a snake, a tree, a spear, a wall, or a fan. Ultimately, their observations are all correct in some ways and completely wrong in others:

"It was six men of Indostan
To learning much inclined,
Who went to see the Elephant
(Though all of them were blind),
That each by observation
Might satisfy his mind.

The *First* approached the Elephant,
And happening to fall
Against his broad and sturdy side,
At once began to bawl:
'God bless me! but the Elephant
Is very like a WALL!'

The *Second*, feeling of the tusk,
Cried, 'Ho, what have we here,
So very round and smooth and sharp?
To me'tis mighty clear
This wonder of an Elephant
Is very like a SPEAR!'

The *Third* approached the animal,
And happening to take
The squirming trunk within his hands,
Thus boldly up and spake:
'I see,' quoth he, 'the Elephant
Is very like a SNAKE!'

The *Fourth* reached out an eager hand,
And felt about the knee
'What most this wondrous beast is like
Is mighty plain,' quoth he:
''Tis clear enough the Elephant
Is very like a TREE!'

The *Fifth*, who chanced to touch the ear,
Said: 'E'en the blindest man
Can tell what this resembles most;
Deny the fact who can,
This marvel of an Elephant
Is very like a FAN!'

The *Sixth* no sooner had begun
About the beast to grope,

Than seizing on the swinging tail
That fell within his scope,
'I see,' quoth he, 'the Elephant
Is very like a ROPE!'

And so these men of Indostan
Disputed loud and long,
Each in his own opinion
Exceeding stiff and strong,
Though each was partly in the right,
And all were in the wrong!" (John Godfrey Saxe (1816–87), *The Blind Men and the Elephant*)"

This book similarly seeks to remind us that—as citizens of a late modern era—there are certain latent assumptions we make that are the equivalent of the "elephants in the room." The essays that follow accordingly try to re-engage with, and re-describe, some of the bewildering elements of this thing we call space, while acknowledging that this is only one start. The thing we call space remains—thankfully—both largely unchartered and inaccessible to our current imaginations.

1.2 Prototyping the Interstellar Question

The Interstellar Question—whether humankind will ever colonize the stars—is not about business as usual in a place that is just a very long way away.

It is also not a typical question that can be broken down into a finite set of soluble steps. It is more of a conundrum, or manifold that harbors many overlapping and related concepts, which change with context and over time. These are often dissected using the sharp instruments of modern analysis and lead to a mismatch between the nature of the question and the instruments of its validation. Consequently, the Interstellar Question invites much skepticism.

An example of high modern criticism—which positions science and technology as central to human endeavor—is Nick Beckstead's blog from the Future of Humanity Institute in Oxford. He reviewed the starship question in relation to existential risk—the study of human extinction-level risks that may emerge from technological advances, in which this organization is invested. He is interested in rejecting the claim that there is a reasonable chance of colonizing space in the future. If humanity can actually escape the threat of extinction, then it is less pressing to take the Future of Humanity concerns so seriously. During his survey, he makes some insightful comments about interstellar exploration (Beckstead 2014).

Beckstead notes that most authors writing about the Interstellar Question already believe it is possible, which results in bias, as a foregone conclusion leads to a lack of critical debate among experts. He claims that starship advocates tend not to go into great detail with their proposals and also fail to engage with relevant counterarguments.

Beckstead identifies six phases of development whose technological integrity can challenge the starship question, where potential advances in artificial intelligence (AI), robotics, manufacturing, and propulsion technology could overcome the obstacles implicit in each of these stages. Further analysis could anticipate the relevant, necessary advances in

6 The Interstellar Question: an ecological view

AI, robotics, manufacturing, and propulsion technology as an investable road map that would take the project toward realization:

1. colonize (interplanetary) space;
2. resources to build a civilization into a starship;
3. propulsion technology;
4. enough provisions to keep a civilization intact during the voyage;
5. slow the starship down at target location (exoplanet);
6. build a civilization at the target location.

Interestingly, Beckstead observes there are no major voices insisting on the fallacy of the enterprise. Notably, the question has had high-profile coverage by commentators such as Stephen Hawking. If persuasive counterarguments existed, then the publicity surrounding the debate would have already brought them to the fore. Prominent critics like Charles Stross, whom Beckstead suggests makes the strongest counterarguments, also pulls his punches from completely killing the idea. He concedes that the interstellar ambition will eventually be accomplished with the relevant advanced AI and robotics.

Beckstead's observations are typical of a dominant discourse. He places a definable boundary around the challenge and takes an Occam's razor to the Interstellar Question, breaking it down into solvable steps. The interpretation becomes something like: Do we have the industrial capabilities to successfully settle an exoplanet today? The current answer to this highly reduced question is "no." We are simply not set up to do this.

Given that this is the Future of Humanity Institute making the analysis, it is also possible to add a secondary question—that if it is not possible today, then when is success likely? The response to this contingency is "we don't know," although there is a spectrum of opinions on when and how this may be possible.

Beckstead adopts a deterministic viewpoint that assumes the answers to the question are calculable, or finite. Of course, this seems like a good idea, as nobody wants to invest significantly in a project that is doomed from the outset. Yet, the Interstellar Question depends upon so many contingencies that are unknowable in the anticipated timescales involved that it cannot possibly address or control for all variables. For example, it does not allow for radical breakthroughs in scientific research and development, or the discovery of characteristics of space that could significantly alter the chances success of an outcome, such as discovering an Earth-like planet in the Centaurus constellation, in which Alpha is the brightest (binary) star.

According to Beckstead, the feasibility of the enterprise rests on our capacity to make appropriate technical advances and their implied existential risks of extinction, where humans and machines are the only players in the outcome. Yet, the characteristics of space itself may also contribute to being able to travel to another star system, such as the discovery of wormholes, which are theoretical tunnels through the fabric of space-time that could potentially allow rapid travel between widely separated points—from one galaxy to another, for example, as depicted in Christopher Nolan's 2014 movie release *Interstellar*.

A current understanding of technological advances is assumed. This is the approach that Gerard O'Neill took in designing off-world colonies. Again, there is nothing wrong with this perspective, as it lends credibility to the venture by making direct reference to the latest findings. However, it does not acknowledge what Kevin Kelly describes as concurrent landscapes of technical evolution. These create the context in which a breakthrough technology suddenly seems to appear (Kelly 2010). Specifically, the motorcar revolution would not have been possible without simultaneous advances in petrochemical engineering or the construction of highways. Of course, we have the opportunity to address these developments in the second part of the question and consider how, for example, we might first become an interplanetary civilization but, until the "radical" technology emerges, we cannot be sure, in the present, what the relevance of specific "landscape" innovations may be.

Whether we will one day live among the stars is a very old question indeed. It is not just simply about technical accomplishment. It is highly contingent on who we are, what we aspire to, and also on unfolding events such as economic investment, natural disaster, or political agendas. Nor can it be addressed by a single discipline. In addition to the practical feasibility of technical challenges, the Interstellar Question provokes many existential challenges from the nature of humanity to our understanding of the cosmos. Over the millennia, a variety of solutions to the Interstellar Question have been proposed. Currently, we prioritize technical approaches. Yet, the variety of ideas that we share about our future among the stars suggests the protean nature of this question and invites us to be more diverse, collaborative, and expansive in our expectations.

While the Future of Humanity analysis interprets the Interstellar Question through the lens of high modernism, this is not the only possible way in which the challenge can be read.

The Interstellar Question embraces many kinds of concepts, whose causalities are not fixed, which exist at different times and scales.

Perhaps an analogy can be made to quantum physics, where a fundamental split between the world we see around us and its quantum underpinnings can be observed during the course of experiments. For example, the two-slit experiment demonstrates the paradox that light and matter can display the characteristics of both classically defined waves and particles (Al-Khalili and McFadden 2014, pp. 105–118):

> "Quantum mechanics does in fact provide us with a perfectly logical explanation of this phenomenon; but it is only an explanation of what we observe—the result of an experiment—not what is going on when we're not looking. But since all we have to go on is what we can see and measure, maybe it makes no sense to ask for more. How can we assess the legitimacy or truth of an account of a phenomenon that we can never, even in principle, check? As soon as we try, we alter the outcome." (Al-Khalili and McFadden 2014, p. 115)

Stuart Kauffman refers to this kind of question as being one in which "phase space is changing in ways that we can't pre-state" (Kauffman 2012). In other words, the Interstellar Question is so large, complex, and contingent that it is actually evolving. We may call this "black-sky thinking."

So, since we are actually living in a high modern era, how can we open up our range of perspectives?

Space is a screen into which our ideas are projected. Since the Space Age, we have held a largely industrial view of our civilization, which shapes our expectations. Existing space pioneers cite activities such as mining and communications enterprises as being justifications for brave new forays. Yet, we are living in very exciting and confusing times, in which we are moving away from an industrial age of development and toward an ecological view of reality. The growing realization that our species is not simply an isolated body, but deeply entangled in and dependent on its ecosystems, is changing the way in which we live on Earth and, by implication, the way we address the Interstellar Question.

An ecological perspective no longer finds it acceptable to simplify the challenge into a dance of mutual survival between human and machine. Rather, the issues at stake must first be understood through a reading of the cosmos as an ecosystem and working through multiple, overlapping perspectives and includes science, technology, the arts, and humanities.

This requires us to be able to coherently embrace a range of overlapping and sometimes contradictory sets of ideas. It is not about exchanging one dominant model of reality for another. Rather, it is about finding opportunity in the spaces that lie between these differing worldviews.

The literature that begins to describe this perspective exists in science-fiction novels such as Kim Stanley Robinson's 2009 *Green Mars* (2009) (terraforming) and *Aurora* (2016) (worldship). Within these thought experiments, we are still shaping the kinds of questions that are relevant to these third-millennium challenges. They are no longer concerned with the atomic control of matter, but with synthesizing life and increasing the complexity of environments. These perspectives help us survive even the most extreme environments.

Indeed, we have not fully addressed these challenges on our own planet: notably, we cannot build ecosystems from scratch, despite millennia of gardening them, so the research and discoveries that we are currently making will certainly reflect back on our expectations of non-terrestrial civilizations and what they may need to settle new territories. The tools, methods, and materials that will enable us to do so do not yet exist, but they are not hypothetical. They are still emerging, and we do not yet understand all the possible choices they have to offer. These findings will not just affect our ideas about why and how we might travel into interstellar space, but also how we can persist and potentially thrive in ecologically barren terrains, including our own planet.

When considered from an ecological perspective, the Interstellar Question is no longer a challenge posed for human and machine, but invokes new ideas about the broader community and set of conditions for life.

Yet, there is another twist to the Interstellar Question.

To date, the challenge of making grand proposals for starship blueprints has largely been addressed by enthusiasts and theorists, such as Freeman Dyson, and professionals, such as Alan Bond and Tony Martin. These generally explore the physics of the cosmos, knowledge of engineering practices, and projections for future possible forms of existence. Yet, even from the most rational perspectives, these designs make significant assumptions that may take them from concept to reality, such as the propulsion systems available, the existence of certain technologies, or that an industrially based interplanetary society has already been established.

Since possible future events shape the outcomes of the Interstellar Question in unpredictable ways, science fiction has served as a critical literary prototyping instrument for a range of possibilities. Proposals range from the classical *Starship Enterprise* to the organic vessel *Lexx*, and even advanced intelligences, as depicted in Ian M. Banks's "The Culture." However, it is now possible to go beyond thought experiments and actually explore a range of physical systems by interrogating ideas through small-scale prototypes that help us to address the Interstellar Question in new ways.

A quiet revolution in manufacturing has been taking place over the last 10 years or so. Networks of special interest groups, makers, and educators have created an experimental platform that enables the production of low-cost prototypes.

These prototypes are regarded in an engineering context not just as first versions of a final product, but also as design tools such as diegetic objects, which are used to discuss the impacts of technological developments (Sterling 2011), and also models that represent the character of ideas.

When an idea is translated into a physical form, the assumptions, applications, and limitations can be questioned to a degree that is not possible through philosophy or thought experiment alone. This first Icarus Interstellar Starship Congress Hackathon formally recognizes the possibilities of these new research fields. It opens up a new degree of criticality that does not rely on specialist rhetoric and data projections for validation. The production of prototypes encourages critical discussion and invites participation from a much broader audience than that regarded as a self-selecting community of enthusiasts.

But what is the nature of these prototypes? How do they actually relate to the production of a megastructure?

Like the starship question, the production of prototypes and their relevance are massively open and highly contingent ventures. Like Saxe's elephant, it depends on which kinds of questions we are asking and what aspects of the challenge they propose to address. Certainly, at this stage, prototypes are likely to explore concepts that start to help us to see where our current efforts may best be placed. In other words, prototyping discusses an open, bottom-up approach to the production of an interstellar culture, as much as it proposes the development of a vessel that will take colonists to the stars.

While a bottom-up developmental approach is not a substitute for visionary blueprints, it is conducive to the history of many engineering developments. Both top-down and bottom-up approaches are needed to evolve the construction road maps for implementable innovations. Yet, exactly where these synergies may currently lie is not known. Although these methods may at first appear to be quite separate, the space of possibilities between them can converge around specific questions, projects, and enterprises.

Since bottom-up projects are also multidisciplinary and open, they require different kinds of evaluation systems. Prototyping the Interstellar Question is also in keeping with an academic landscape that values open publishing and knowledge transfer. They are authenticated by their ability to help us understand the challenge in different ways. For example, these prototypes may reveal blind spots in our thinking, advocate the need for particular infrastructures, broaden the range of support for the field, or even inspire people to become interested in the challenge. They also build critical mass and may lead to the development of new questions, methods, tools, and production platforms.

While these "prototype" developments may not actually be starships, their existence bears relevance to the starship question—particularly in the context of an ecological culture in which it is essential to be able to accommodate many different worldviews, skills, and stakeholders. Maker platforms advocate the emergence of new kinds of laboratories and the creation of a range of approaches that are not typical of classical research analyses, but generate relevant information through practice-led research. Indeed, some groups are testing ideas—for example, in our homes and cities—with the aim of improving living conditions within our sprawling urban environments, which may eventually help to meet the needs of colonists.

The first Interstellar Hackathon, held at the second Starship Congress at Drexel University, Philadelphia on September 4–5, 2015 invited an experimental approach to the Interstellar Question. "Hackathon"

> "… has become a byword for a caffeine-fuelled creative carnival where entrepreneurs of various stripes get together and engage in problem-solving at a feverish pace to come up with a piece of new software, a new device, a new business, or even a new industry. It's also fertile ground for recruiting talent, for networking, and in some cases, going home with prize money." (Mookerjee 2015)

The goal is to make the information accessible to everyone who is interested in participating, but also to make significant progress in research and ideas, with tangible actions that can be taken. With the participation of an intentional community of shared-interest communities temporarily problem-solving together, a range of experiments was conducted using low-cost manufacturing platforms and knowledge transfer to examine aspects of interstellar exploration such as the design of a space terrarium, biomass recycling for starships, interstellar CubeSat design, and solar-powered transmission to interstellar probes. These highly distributed, small-scale, multidisciplinary approaches point toward the production of prototypes that engage with the development of new questions, methods, tools, and production platforms and opened up the Interstellar Question to a broader community.

A successful approach to prototyping developments in emerging design and engineering practices has been the iGEM (International Genetically Engineered Machine) competition. This annual, worldwide, synthetic biology event was launched in January 2012 by the iGEM Foundation, which was spun out of MIT as an independent non-profit organization and is aimed at undergraduate university students, as well as high-school and graduate students. Multidisciplinary teams work all summer long to build genetically engineered systems using standard biological parts called biobricks. iGEM teams create sophisticated projects that contribute positively to their communities and the world, taking the field of synthetic biology forward.

So, what kind of experimentation is possible in unconventional laboratories that do not have access to the budgets, resources, specialist knowledge, and kudos of academia or national institutions?

When faced with massive challenges with phase spaces that cannot be fully searched by the tools of modern synthesis, we actually need to be much more specific about the kinds of questions we ask of them—in other words, to identify which aspect of Saxe's elephant we are holding.

Of course, Douglas Adams famously toyed with the paradox of questions and answers in his 1995 book, *The Hitchhiker's Guide to the Galaxy*, in which a species of hyper-intelligent white mice invent a powerful computer called Deep Thought. They ask it for the ultimate *answer* to "Life, the Universe, and Everything." It takes Deep Thought seven and a half million years to compute the answer: "forty-two."

Since the question had a massive phase space, it was way too general for the answer to have any real meaning for the mice. So, they build a new computer called Earth to design the ultimate *question* of "Life, the Universe, and Everything." This takes another 10 million years, during which time Earth is destroyed by Vogons.

Question-making is therefore a very important task for an open, engaged community that generates prototypes for which our scientific research communities and formal institutions can provide meaningful answers. The political importance of openness, civic engagement, and impact has changed the research landscape, where science and technology have to be socially invested in the kinds of questions to which they respond.

The discipline of experimental architecture can provide examples of these kinds of engaged experiments that keep the Interstellar Question open, and its audience engaged and critically thinking about its evolution.

As I am a professor of experimental architecture at Newcastle University and project leader for Persephone, my work explores the impact of new technologies on our lives and how we may design for these possibilities. These research interests come together through the Persephone project, which proposes how the living interior of a starship—an inhabitable world for prospective colonists—may evolve. Persephone reruns the story of human civilizations, starting at our technological relationship with the soils. The project asks how life may be different if we do not kill the earth from which our nurturing ecosystems spring but develop a new relationship with them, harnessing its capacity to generate useful work. For example, an acre of fertile soil performs one horsepower of work daily—so how do we begin to develop these ideas to change the performance of our habitats and cities?

So, for example, in the chemistry outreach department, the kinds of questions that we are shaping include: "What is a soil?" This is explored with Masters of Architecture students who are looking at the movement of minerals through activated gel scaffoldings to examine the characteristics of non-equilibrium materials by reading the temporal and spatial patterns that they produce.

We are also thinking about the questions that will enable us to scale these kinds of operations to architectural dimensions. The Hylozoic Ground installation collaboration with architect Philip Beesley, which was shown at the Venice 2010 Architecture Biennale, explored how programmable chemistries were converged with a cybernetic scaffolding to produce a technological soil. Stepping inside the structure can be likened to being inside a giant nose. The spaces in the gallery are the sinuses and the active chemistries are like mucus glands that can smell and taste through the changing carbon dioxide changes released during respiration. The glands produce tiny crystal sculptures the size of a little fingertip and, in this way, produce a collective record of visitor presence.

The Future Venice project is about growing a chemically programmable, soil-like structure to the dimensions of a city. The possibility of creating an artificial reef around the wooden foundations was explored by applying a programmable form of chemistry with the capacity to respond to constantly changing conditions in the Venice lagoon. The final accreted structure aims to spread the point load of the city over a broader base and slow the rate of sinking.

In Future Venice II, the production of soils from problematic and abundant materials in the Venice lagoon is investigated. How may they be used to form a new island for the city, as well as potentially reinvest the materials for use by future generations? By sinking the algae and microplastics into the earth, we may not only end up with a new land mass, but potentially invest in the production of a resource that may be transformed by natural and geological forces into a useful material—a process similar to making chalk, which is made by the natural transformation of the tiny skeletons of sea creatures.

12 The Interstellar Question: an ecological view

The Hanging Gardens of Medusa is a tiered stratospheric laboratory garden that was launched with Nebula Sciences. It asked questions that promoted further thought and conversation about the kinds of technical infrastructures that are possible and necessary to support life in extreme environments. The gardens also examine the potential emergence of synthetic life forms from lively chemistries that may help us to colonize possible new worlds.

Of course, there are also other benefits of question shaping through prototyping. The multidisciplinary nature of the production platforms can draw together paradoxes: such as imagination and pragmatism, life and technology, art and science.

For example, Frederik de Wilde's starship *SPIDER* is a concept craft that is more than an inert hull moving through interstellar space; it is a body that talks about the strangeness of matter between the visible and dark universes. In harvesting scant matter like photonic winds and buckyball carbon atoms, it synthesizes cosmic scaffoldings that may become the infrastructure for future space ecologies.

With the advent of the first Interstellar Hackathon, an experimental approach to the Interstellar Question may help us to obtain increasingly valuable answers to the Interstellar Question, provided we first carefully construct the questions. We cannot just assume that there is implicit value in the generation of engineering blueprints and physics equations. Nor is the investment of a prototyping approach competitive with or an alternative to traditional top-down engineering enterprises. To actually realize something as complex and evolving as the Interstellar Question requires the choreographed orchestration of a whole range of different approaches, some of which are engaged in this publication as a series of expert contributions.

The experimentally based prototyping endeavor that is emerging within the interstellar community transforms its grand quest from being solely concerned with a colossal megastructure project that is so unfeasibly large that it exceeds the capacity of existing technology and its infrastructures to be realized before these approaches become obsolete into an open, critical, collaborative, ongoing enterprise between diverse, shared-interest communities that can take small steps toward a constantly evolving set of possibilities. Taking a bottom-up approach to the Interstellar Question turns the question into a catalyst for human creativity and diversity, in which we all have the opportunity to contribute to the world's most pressing questions.

Prototyping the Interstellar Question does not pretend to be classical science, nor a substitute for formal engineering practices. It is an adjunct—an expanded portfolio of possibilities and an inspiration for these kinds of endeavors. At worst, prototyping activities will help us to examine the challenge across many different disciplines and practices in variably testable ways. At best, it will continue to inspire and compel many generations to come. These are the kinds of leaps of imagination that our civilization needs to make if one day we are to learn to live among the stars. Indeed, the question is likely to stay as open to interpretation as it ever was. However, this book aims to take a fresh look at the challenge given, with particular emphasis on experimentation and ecology.

We may even find that, if we are more critical of our engagement with the Interstellar Question, we may save ourselves 17.5 million years of misplaced effort—unlike Adams's white mice—in pursing poorly constructed questions.

The argument is therefore organized into two sections: Parts I and II. Part I establishes the viewpoint from which the Interstellar Question is being addressed and Part II is an anthology that introduces a range of expert voices, all of whom are responding to the implicit challenges according to their own perspectives. This approach aims to accentuate the mutable nature of the starship question and celebrate its enduring relevance to science, arts, technology, design, and culture. It does not propose to stand as an unassailable text, but to embody a continuation of an ancient ongoing conversation about the nature of humanity, our greater role within the cosmos, and how such responsibilities may one day be realized.

1.3 It begins

"Contact light."

These are the first words spoken from the surface of the Moon, by Buzz Aldrin on July 20, 1969, when Apollo 11 landed. Over six hours later, Neil Armstrong stepped onto the lunar surface and uttered the immortal line, "That's one small step for [a] man, one giant leap for mankind."

In these moments, the doors to the vast expanses of space and its colonization were flung wide open.

1.4 Mission

"Historical precedents … have no meaning in the stars, for interstellar space is not life-supporting. Where the early pioneers opening up a new continent could invariably count on living off the land, the space pioneers have nothing with which to feed themselves. Nor are there oases where they can rest awhile and recuperate in the perpetual night of galactic space.

Hence every star ship must be not only a vehicle it must be a world of its own, a cradle, a home, a repository of memories and, for many, a place of burial. Star ships when they are perfected, will be no more than super-sophisticated robot homes that run themselves." (Strong 1965, p. 124)

This book builds on the incredible developments of space exploration during the late twentieth century and, uniquely, takes an experimental approach to prototyping the living interior of a worldship. Its flagship—Project Persephone—represents a fully artificial form of Nature that exceeds a mechanical model of existence, or "robot home," in which a community of starfarers can potentially persist indefinitely. It provides an evolving, collaborative, theoretical, and practical model that helps us to interrogate our current relationships with and responsibilities to our home planet. It also invites us to reflect on the future of humanity in an ecological era of space exploration—an epoch termed the Ecocene.

1.5 Aspirations

"Mankind was born on Earth. It was never meant to die here." (Interstellar poster, Christopher Nolan 2014)

We are relentless explorers, sometimes curious to our detriment, willing to push back boundaries—simply, to paraphrase George Mallory, because they are there. Yet, the possibility of inhabiting other worlds is an ancient idea. It spans back to the dawn of human history where we formed a deep relationship with the stars. We imagined these bodies as gods and heavenly places, but so far we have not been able to reach and experience them directly. As Voyager 1 begins its journey beyond our Solar System, we can already consider ourselves to be an interstellar culture, our imaginations having long preceded us into inhabiting worlds beyond our own. In our collective mind's eye, we are already an interstellar species and it is simply a matter of time before we take our place among the stars with the gods and heroes that have preceded us.

To turn our dreams into reality, epic projects called starships were conceived to take us to our interstellar destination.

1.5.1 Introduction

Robert H. Goddard first proposed the idea of space arks in 1918 when he wrote *The Last Migration*, envisaging the need for an "interstellar ark" to help humanity escape the death of the Sun. The sleeping crew would awaken centuries later to find themselves in another star system. This sleeper ship concept is different to a generational starship, or worldship, that carries its own environment. In 1928, Konstantin E. Tsiolkovsky first described such artificial worlds in *The Future of Earth and Mankind* (Gilster 2013), in which he described a space colony that traveled for centuries through space on a vessel called "Noah's Ark."

There are two main approaches to reaching the stars—the fast and the slow routes.

1.5.1.1 Fast route

The fast route to interstellar space entails constructing a vessel with incredible powers of propulsion reaching a significant fraction of the speed of light—for example, 0.1 % c, and aiming it toward a specified destination. Everything on this ship is cargo, including its human crew. There is no need for sustained provisions, as it is just a holding space. It is assumed that the quality of life that we expect on Earth will be suspended for the journey and ultimately resumed at the destination. A common proposal to achieve this is for humans to be cryogenically frozen, or put into some kind of deep suspended experience. This process not only saves resources, but also reduces tedium for the passengers. While certain life forms demonstrate incredible abilities to shut down their metabolic activity—like certain species of amphibians, marine life, insects, and reptiles—few can perform the trick as effectively as *Rana sylvatica*. This species of wood frog can be found from the state of Georgia, up through Canada, and into the Arctic Circle. These tiny amphibians can survive for weeks with only two-thirds of their body fluids, which are rich in solutes that protect against the physical effects of freezing. During the winter, they go into full cardiac arrest and become solid—effectively turning into "frogsicles." In the spring, they thaw out and are fully revived (Costanzo et al. 2013). Yet, the science of cryogenics is speculative and controversial when applied to human subjects and we still have a lot to learn about the science of physiological resurrection from various forms of life suspension. However, from a cultural perspective, we are familiar with the idea of cryosleep. Famously, the character Ellen Ripley from the *Alien* film series travels through space in the *Nostromo* in a state of stasis. Notably, the interior of this starship is barren, industrial, and reminiscent of a laboratory. It is much more like a construction site than a biologically enriching environment—without a green plant in sight. Yet, this is not the kind of starship experience under consideration in this book, as the issues faced are mainly to do with propulsion technology and have been covered in a number of books, such as James Strong's pioneering 1965 *Flight to the Stars* that discusses the possibility of such a venture for the first time, Eugene Mallove and Greg Matloff's 1989 *The Starflight Handbook: A Pioneer's Guide to Interstellar Travel*, and Kelvin F. Long's 2011 *Deep Space Propulsion: A Roadmap to Interstellar Flight*.

1.5.1.2 Slow route

The slow route to the stars focuses on the construction of "arks" that are primarily designed to support lively communities. These spacecraft travel vast distances slower than the speed of light (less than 10%) and may take several lifetimes to reach a target destination, which also earns them the title "generational starships." Owing to the very long time periods involved, they provide an interior world to sustain colonists during their journey and are also therefore known as "worldships." During this period, the colonists live complete and fulfilling lives entirely within their starship environment, which supplies them with all their needs. Perhaps the most elegantly depicted example is explored in Arthur C. Clarke's *Rendezvous with Rama*.

A first detailed practical consideration of establishing a human space colony was proposed by Gerard O'Neill in the 1970s that operates through modern structures characteristic of industrial urbanization, namely mines, factories, farms, homes, and markets, which are designed to support populations of around 10,000. Marshall T. Savage and Clarke's *The Millennial Project: Colonizing the Galaxy in Eight Easy Steps* (1992) gives a series of concrete stages for interstellar colonization based on an agricultural system that is underpinned by the exponential growth of blue-green algae. This is first developed as an industry in the terrestrial environment and subsequently exported, first as miniature ecologies and gradually as an interplanetary form of aquaculture that underpins a non-terrestrial economy.

A characteristic of the Interstellar Question is that it is a giant canvas for the projection of contemporary issues and concerns. Owing to the era in which we live, these tend to be various expressions of a high modern, industrial era, underpinned by an industrial-scale agriculture that looks to science, engineering, and computer-controlled machines for its realization.

1.6 Ecocene

This particular publication adopts an ecological viewpoint of space exploration, using the Ecocene as a cultural lens through which a range of assumptions about contemporary challenges is examined and alternatives proposed. The Ecocene is preferred to the widely used term, the Anthropocene—described in 2000 by scientists Erwin Stoermer and Paul Crutzen to indicate that human impact is so extensive that it can be thought of as a new geological epoch (Crutzen and Stoermer 2000). Since these effects are relatively recent in the 4.5-billion-year history of the planet, this idea is somewhat scientifically controversial, as the time period is incredibly short when compared with sustained events like ice ages and mass extinctions. While the mythos of the Anthropocene largely views human impacts as negative and inevitably resulting in our extinction, the Ecocene looks for alternative forms of social and cultural organization than modern industrial practices that invite our ongoing success as a species (see Part I, Chap. 4).

The Ecocene therefore refers to a set of ideas and practices that are starting to emerge in a post-industrial world in which we are beginning to understand the importance of the environment as integral to our ongoing survival. Our twenty-first-century megacities swell and loom around us. They house more than 10 million inhabitants and stretch over hundreds of square kilometers. Increasingly, our everyday encounters with matter appear more paradoxical and uneasy. Drinking water is used to flush excrement. Fertile soils are scorched by intense agricultural and geo-engineering practices that have permanently reconfigured Earth's biological trajectory. Above us, our skies are full of invisible toxins, our industrial landscapes full of "technofossils"—inert materials like concretes, or black carbon—and beyond us stretch oceans of plastic. Yet, we are countering these insidious events by exploring ways of moving from an industrial age toward an ecological era of thinking and practice. This is not new; it has been happening over the last 80 years. Marshall Savage's consideration of algae replacing the foodstock of O'Neill's high-yield varieties of grains, such as wheat-, rice-, and corn-based agriculture reflects this transition.

The Ecocene is not a new hegemony. It is not simply about biomimicry—copying Nature's forms and functions—or the greening of things. It is not as simple as substituting an object-centered view of reality and supplanting it with process, complexity, networks, and nonlinearity. It embraces many different approaches and worldviews that are overlapping for the first time. It involves constructing a framework for understanding a world in continual flux that is navigated by many overlapping models of thought, which require different ways of attributing value to natural systems than, for example, modern economics, which centers on resource scarcity and ignores qualitative criteria such as creativity (Papazian 2013). The impacts of these convergences are thriving owing to the advent of the Internet. The intersecting ideas that shape these conversations also bring about paradoxes in our experience of the environment and therefore influence the way in which we work to solve these complex challenges.

In this context, space travel in the Ecocene does not only imply the setting against which human and technological activity is conventionally foregrounded. In the Ecocene, space explorers and colonists do not stand on the surface of the Moon—as if on the end of a pier and look back at our pale blue dot with fond nostalgia—but look forward in the direction of travel of the starship is traveling and out towards the unfamiliar terrains of the cosmos.

At the heart of the Ecocene is the appreciation that the world is in constant flux and that the matter from which it is formed is lively. Responding to grand global challenges, such as climate change, increasing population density, and the sustainability of cities, architects, engineers, scientists, and designers have been looking for new ways of working with a whole range of strategies to counter the net effects of global-scale, intensive industrial practices that are effectively reverse-terraforming our planet. This ranges from alternative food resources from farming insects to various forms of aquaculture, as well as notions of "circular economy," in which, by design, material flows of biological nutrients re-enter the biosphere safely and technical nutrients are circulated without entering the natural environment. Over the last 30 years or so, new toolsets have become available to enable us to develop alternative platforms with the kinds of plasticity and robustness that may help us to look at these challenges and work with them in new ways.

The Ecocene invokes ecological ideas of location and materiality and notions of identity. In that way, the starting point of interstellar colonization adopted in this book is very different to these modern and technically focused accounts. It establishes a fundamentally philosophical approach to thinking about how life itself arose on our planet through ecopoiesis, biogenesis, succession, and evolution. It then considers the contexts in which these fundamental organizational systems responsible for the sustained liveliness of matter can be ensured, so that it is realistic to anticipate the ongoing survival of inhabitants and their capacity for change over prolonged periods, which are currently predicted to be hundreds, perhaps thousands, of years.

Indeed, space colonization is not possible without an ecological view of humanity beyond this planet. While it may be argued that O'Neill and Savage's solutions are "good enough" for space colonization, we also know that our own attempts to sustain closed ecosystems on our own life-bearing planet, such as BIOS-3 (an algae-based foodstock) and, more famously, Biosphere 2 (grain foodstocks), have demonstrated that the closed-ecosystem design is far from a done deal, with much still to be discovered about their effective and prolonged operation.

Indeed, even on the International Space Station (ISS), the world's first house in space, which has been permanently occupied since the millennium, we have little experience of living off this planet. To date, the longest single human spaceflight record is held by Valeri Polyakov, who stayed aboard the Mir space station for just over 14 months (437 days, 18 h). Although this is an impressive display of human endurance, it is an incredibly short time when considering our requirements for a slow worldship; this is insignificant. With the prospect of colonies on Mars in the next decade, the issue of sustaining human activity beyond the easy reaches of supplies or a quick return home becomes a challenging one.

Moreover, industrial societies have particular challenges to deal with regarding the prospect of long-term thriving. It is essential to identify a set of ideas that enable us to live viably in places that are much less richly supplied with resources than on our home planet. If we do not change the thinking before we propose the solutions, we will soon discover that we are faced with a "groundhog day" of modern development with climbing carbon dioxide levels, pollution, and destruction of natural resources. While our thinking into these issues is indeed advancing, it is far from providing the kinds of infrastructures that would enable us to generate ecopoiesis on a barren terrain.

In this manner, this book develops some of the themes proposed by earlier authors who, as Marshall Savage observes, cannot individually hope to write out the entire course of human development over the next millennium. It also sets out to challenge some assumptions about our current view of space colonization, which make several fundamental assumptions.

The first is that we can begin to colonize space by applying knowledge and technology that originate from a particular stage of Earth's history to address all our challenges, which is dominated by the operations of machines and computers.

The second is to assume that matter behaves in a spaceship in exactly the same way as it does on Earth.

The third is that the Interstellar Question is a monoculture, rather than a diversity of approaches.

This book establishes two particular sets of ideas in relation to the Interstellar Question: how we might understand space ecology—and who is "we"?

1.7 Space ecology

What is a space ecosystem and what are its external and internal environments? Are these landscapes inextricably separated and how may an appreciation of these materialities help us to survive and thrive away from our home planet? Are we, in effect, bringing a new kind of Nature with us?

A real-world project that I am leading for Icarus Interstellar is introduced to further explore these questions. Project Persephone is a starship laboratory that examines how the first terrains may be constructed for ongoing settlement. It is not directly concerned with the construction of the vessel, but rather establishes a laboratory approach that engages with the fundamental conditions for sustained life in a barren landscape. Persephone invites questions about the kind of world we might wish to inhabit and construct from scratch. It develops its ideas by making prototypes that may become part of the living interior of the worldship and also enable further discussion about a range of tactics through which its habitable spaces may evolve. These support living systems while promoting diversity and change.

The method embodied in this experimental practice is "black-sky thinking," which explores solution spaces beyond the conventions of current models and deals with many unknowns, such as continual flux, hypercomplexity (complexity, heterogeneity, multiplicity), uncertainty, the combination of several disciplines, values, community, and forms of knowledge.

Persephone's experiments are "messy," being prepared to deal with conditions of hypercomplexity that exceed their meaningful simplification into individual agents, or forms of representation. They are massive, heterogeneous, constantly changing, and, therefore, exceed the remit of traditional laboratories that are best suited to small-scale, highly controlled environments in which deterministic outcomes are sought and where simulations are meaningful. Since Persephone engages with the process of worlding—the construction of a world—its experimental platform works alongside many diverse agents such as the liveliness of environments and non-human actors. Persephone also understands the value of context in experiment. While traditional science relies on controls to set standards, Persephone's black-sky challenges are such that they are inaccessible to single forms of analysis; therefore, controls are comparators and modes of discussion, not absolutes. Applying determinism and extreme forms of "realism" to the Interstellar Question is akin to observing that a particular feature on a unicorn does not seem believable. Rather,

Persephone appreciates the value within hypercomplex and contingent questions as they enable us to explore synthetic approaches to the production of knowledge, using instruments that establish possible pathways, not final solutions. Alan Turing pioneered this kind of approach in dealing with uncertainty, with the idea of solving hypercomplex challenges using biased observers. Appreciating that some phenomena are so difficult to define that they cannot be meaningfully reduced to symbols, he applied the principles of an "imitation game" to the question of whether machines could "think." This approach became known as the "Turing Test," which set the benchmark for comparison and construction as valid forms of experimental inquiry.

However, Persephone's experimental approach does not propose to invalidate classical and conventional methodologies; rather, it seeks to expand our portfolio for problem-solving and find ways of dealing with multiple ideas. It adopts a range of methods when addressing extremely complex changes that are concerned with enrichment, diversity, and mutability.

Necessarily, then, the kinds of experiments developed in Persephone's laboratories are about exploring through making prototypes using a constructivist approach. Of course, standard benchmarks for measurement or defining systems cannot be established in this manner, as bias must be eliminated from the system. However, such approaches do allow us to propose ways forward, even when many unknowns remain, so that we can explore possibilities iteratively using design-led explorations. At some point in this process, we may even reach an appropriate point at which more empirical methods may be used.

An extension of this experimental approach is encapsulated in the diverse range of author contributions in this book. Experts from many disciplines discuss various aspects of space habitation. Each of these perspectives takes a different position in relation to the Interstellar Question and provides a different lens through which to observe questions about identity, technology, values, and culture. Collectively, their ideas constitute a kind of instrument that resists hegemony within the worldship. Kevin Warwick, Andreas Tziolas, Arne Hendriks, Rachel Armstrong, Steve Fuller, and Krists Ernstsons discuss how technology and environment may influence our being, while Esther Armstrong, Nathan Morrison, and Roberto Chiotti explore epic space narratives. Michael Mautner, Simon Park, and Sarah Jane Pell observe our new relationships with life, while Rolf Hughes, Mark Morris, and Jordan Geiger explore how alternative ideas of space and physicality may emerge. The relevance of the Interstellar Question to contemporary terrestrial challenges is considered by Astudio architects (Emma Flynn, Richard Hyams, Christian Kerrigan, and Max Rengifo) and Susmita Mohanty, Sue Fairburn, Peter Weiss, Angelo Vermeulen, and Barbara Imhof, who propose how designing starships may help us to live better in cities today.

1.8 Who are "we"?

The second set of questions constitutes the "we" that is being perpetuated beyond our own world in another star system. This raises the essential question of directed panspermia as integral to the Interstellar Question, whereby establishing life on non-terrestrial habitats is considered in a responsible and humane manner.

1.9 Summary

However, this book does not propose final solutions, conclusions, or a blueprint for human space exploration in the Ecocene, but aims to open up new conversations about settling non-terrestrial terrains by thinking through the principles of ecopoiesis—the science of how inert environments become lively. It also asks us to consider the much longer-term consequences of these proposals, where they may be taking us, and how this new knowledge may be reapplied back home on Earth.

REFERENCES

J. Al-Khalili, J. McFadden, *Life on the Edge: The Coming of Age of Quantum Biology* (Crown Publishers, New York, 2014)

N. Beckstead, Will we eventually be able to colonize the stars? Notes from a preliminary review. (Future of Humanity Institute, 2014), June 22, www.fhi.ox.ac.uk/will-we-eventually-be-able-to-colonize-other-stars-notes-from-a-preliminary-review/. Accessed 20 Aug 2015

J.P. Costanzo, M.C.F. Amaral, A.J. Rosendale, R.L. Lee, Hibernation physiology, freezing adaptation and extreme freeze tolerance in a northern population of the wood frog. J. Exp. Biol. **216**, 3461–3473 (2013)

P.J. Crutzen, E.F. Stoermer, The "Anthropocene". Glob. Change Newslett. **41**, 17–18 (2000)

P. Gilster, Robert Goddard's interstellar migration. Centauri Dreams (2013), May 6, www.centauri-dreams.org/?p=27453. Accessed 18 Mar 2016

S. Kauffman, Evolving by magic. Arup Thoughts (2012), May 17, http://thoughts.arup.com/post/details/200/evolving-by-magic. Accessed 21 Aug 2015

K. Kelly, *What Technology Wants* (Viking, New York, 2010)

K.F. Long, *Deep Space Propulsion: A Roadmap to Interstellar Flight (Astronomers' Universe)* (Springer, New York, 2011)

E.F. Mallove, G.L. Matloff, *The Starflight Handbook: A Pioneer's Guide to Interstellar Travel* (Wiley, New York, 1989)

A. Mookerjee, "Interstellar hackathon" to chart our path to the stars. Discovery News (2015), July 27, http://news.discovery.com/space/private-spaceflight/interstellar-hackathon-to-chard-our-path-to-the-stars-150727.htm. Accessed 28 Aug 2015

A. Papazian, Our financial imagination and the cosmos. Social Science Research Network (2013), Nov 18, http://dx.doi.org/10.2139/ssrn.2388052. Accessed 20 Aug 2015

K.S. Robinson, *Green Mars* (Bantam Books, New York, 2009)

K.S. Robinson, *Aurora* (Orbit, London, 2016)

B. Sterling, "Architecture fiction: Rachel Armstrong": beyond the beyond. Wired (2011), Dec 21, www.wired.com/beyond_the_beyond/2011/12/architecture-fiction-rachel-armstrong/. Accessed 17 Mar 2016

J. Strong, *Flight to the Stars: An Inquiry into the Feasibility of Interstellar Flight* (Temple Press Books, London, 1965)

2

Architecture and space exploration

Rachel Armstrong

This chapter proposes that the production of different kinds of architecture will be key to how we explore, inhabit, and ultimately colonize space.

2.1 The interstellar challenge

Interstellar travel is vast—with respect to not only the scale of the distances involved, but also the various enterprises needed to embark on such a venture.

This book uses the Interstellar Question to establish the nature of the possibilities and approaches, which Stephen Hawking proposes is not an option or mere intellectual vanity. He asserts that establishing human colonies beyond this Solar System is essential for the long-term survival of our species.

Our home planet may support us for around another billion years, if we have not evolved into something else by then, before our own Sun collapses and becomes a red dwarf star. When this occurs, our world will be so hot that it will make the current concerns of a four-degree temperature rise associated with the current predictions for climate change look like we have been splashing around in Charles Darwin's warm little evolutionary pool.

Yet, the prospect of leaving our "pale blue dot" and venturing into the great unknown evades a single solution and cannot currently be accomplished within a single generation. The first passenger-carrying starship has not been built and the first planet has not been settled. Indeed, those of us living right now are unlikely to be booking our places on an interstellar shuttle to the Centaurus constellation. Yet, the starship question remains an essential set of ideas that helps us to consider our ongoing survival as a species. It confronts us with questions that have no easy answers but ask us to actively reflect on how we live and operate in the world.

Perhaps skepticism can be partly related to the lack of real initiatives that have stemmed from inventive proposals regarding the settlement of the nearest planetary body—our own

---

Rachel Armstrong, PhD. (✉)
Professor of Experimental Architecture, Newcastle University, Newcastle-upon-Tyne, UK

Moon. There has been some incredibly creative thinking about this subject, which begins as early as Bishop John Wilkins's ideas of the Moon in 1638. Wilkins conceived of this body as being more than a shining disk, as a world with a landscape like Earth—while the father of the space age Konstantin Tsiolkovsky criticized Jules Verne's method of reaching the Moon by firing passengers from a cannon that would inevitably kill them due to the force of acceleration. Yet, despite landing a range of objects on the Moon since the Soviet Union's Luna 2 mission on September 13, 1959—the first object to reach its surface—there has still been no formal implementation of a Moon base. Apollo 11's historic 1969 mission heralded a new era in space exploration with the first manned vehicle to land on the Moon and the first human footprint on its surface. Although this was followed by six further manned US landings, the last of which was in 1972, we have still barely begun to consider the practical implications of sustaining life in an ongoing manner beyond our terrestrial sphere.

However, living beyond Earth's nurturing environment is not fiction. The International Space Station (ISS) has been continually occupied since 2000. This home in the sky may be considered the greatest human outpost and ongoing living experiment in Earth's orbit, and sets a tone of possibility for an interplanetary society and ultimately an interstellar culture.

The Interstellar Question bestows license to speculate on long-term possibilities and our ambitions as a species. Since the dawn of humankind, we have embarked on epic adventures using the most cutting-edge transport technology as we have migrated to all regions of our globe. Each traversed boundary—from the Silk Road in the East, to the "discovery" of the Americas in the West—was at some point a seemingly impossible challenge until feats of courage and endurance proved otherwise. Taking the first bold steps required many leaps of faith and departures from dominant conventions of thinking.

The Interstellar Question is not an ordinary challenge. Its responses are unlikely to exist entirely within today's realms of understanding and inquiry. It raises more questions than it provides answers to. Yet, we can grow the seeds of possibility by embarking on the exploration of bold new approaches, which is part of the process and broader value to our civilizations. Without exception, the texts within this book are provocative in providing glimpses of forms of existence and realms that we do not currently inhabit. Yet, they all share a common perspective in taking an ecological approach to interstellar space exploration, which provides the context in which our presence beyond Earth can be sustained. These explorations also offer us a chance of survivability on our home planet and may ultimately help us to decide what strategic moves may help us to become more responsible Earth citizens. With the prospect of an ongoing presence in realms beyond our own, we may also establish the conditions through which we may assert our place among the stars.

2.2 Far, far away

"Space" is well named.

Outer space is colossal. Enormous. It is far bigger than the kinds of scales that we are used to imagining.

Yet, somewhere in the unfathomable vastness of the cosmos, there are sites that we might orbit, or even inhabit one day. One of our first challenges is in finding out where they are.

We are expressions of our own planet—a condition that Bruno Latour calls "Earthbound." Therefore, we need a target destination that is just like Earth if we are to have a chance of ongoingness.

Currently, there is no single preferred destination for a progenitor human colony to make its home. However, it is likely that the first space colonies will be traveling in the direction of the Centaurus constellation, which is a mere 4.3 light years away. Settlers will be looking for an Earth-like planet in this star system that is suitable for habitation. Such destinations are described as being within a "Goldilocks zone"—where the conditions for life in a solar system are "just right" and not too extreme to support life. The alternative is to find a "good enough" location for space colonization and transform its environment through a process called terraforming, so that it is similar to our home planet.

Alpha Centauri is trillions of miles away from Earth—nearly 300,000 times the distance from Earth to the Sun. If we traveled to Alpha Centauri using the Space Shuttle, then we would need about 10,000 shuttle main engines just to build up a decent speed at around one-hundredth of the speed of light, or 0.01 c. If Earth were the size of a sand grain, this would be about the width of a hair in contrast to a 10-km distance to Alpha Centauri. Traveling at a maximum speed of about 17,600 mph (about 28,300 km/h), a starship powered by a single Space Shuttle would have taken about 165,000 years to reach Alpha Centauri and at 0.1 c it would take about 40 years.

While no planetary systems to date have been identified as possessing "life," astronomers estimate that there are billions of potentially habitable worlds within our galaxy alone. Over 1060 exoplanets orbiting star systems have been located. Yet, even if suitable planets were discovered, the possibility of establishing life in close-orbiting binary star systems is controversial, owing to severe transient radiation and gravitational disturbances. At least 8.8 billion Earth-sized planets exist that could potentially support life in habitable zones of Sun-like stars in the Milky Way (Borenstein 2013). NASA's Kepler Space Telescope recently discovered an Earth-sized planet in the habitable zone of Kepler-186, where water could potentially form. While the planet appears to have a rocky surface, this does not imply that it can support life. According to the Habitable Exoplanets Catalog, only about 10 Earth-sized planets in the habitable zones have been confirmed (Planetary Habitability Laboratory 2015, *http://phl.upr.edu/projects/habitable-exoplanets-catalog*).

However, it is indeed a challenge to see the whole of the cosmos from our pinpoint terrestrial perspective, and a worldship may simply be launched toward a particular star system and then look for habitable planets as the new solar system is approached. The risk is, of course, that we never find a suitable planet. Colonists will, therefore, continue to travel, perhaps indefinitely, within their worldship system.

Yet, it is likely that many Earth-like planetary candidates will be discovered in the next 100 years. Identifying one of them as a prospective target destination allows colonists to consider what kind of life they are going to need to prepare for during their journey. Since we do not know the sequence of events that gave rise to biogenesis, it will be impossible to run the exact same developmental program on an alien planet. However, we are aware that certain key planetary-scale events made very specific contributions to the livability of our planet during particular epochs. For example, blue-green algae, or cyanobacteria, are known to have transformed the atmosphere of the Archaean geological period on Earth from being choked by toxic gases such as carbon dioxide and methane into one that is rich in life-giving oxygen.

Although we have become competent gardeners of our terrestrial ecosystems, we do not have technical mastery of our environment. Once ecosystems are destroyed, they vanish forever. On a life-bearing planet, others may succeed extinct species but, in a barren desert, there is no residual capacity of life to draw from. To date, we do not know how to turn a dead planet into a lively one. This is the science of ecopoiesis, which is a scientific discipline that will need much greater understanding and technical development if we are to settle on worlds beyond our own. Indeed, our industrial age is reverse-terraforming our planet by reducing the stores of natural resources such as fossil fuels, soils, and forests. We spew pollution as microplastics into our oceans and new carbon dioxide into the atmosphere without meaningfully reintegrating them into our planetary systems. This creates much stress and strain for natural processes and may even nudge complex networks of living things towards "tipping points," when they simply collapse towards equilibrium states. It is, therefore, essential to understand the material environment of space in its broadest context, so that we may consider all reasonable possibilities to increase our chances of survival as an interstellar species.

2.3 Architecture as a survival strategy

At the start of the third millennium, R. Buckminster Fuller's notion of "Spaceship Earth" is not situated in a condition of abundance, but faces the challenge of serious resource constraints through the global uptake of modern industrial processes. These practices are dependent upon natural cycles, which are being progressively disrupted due to the pressure of people on the planet and the speed at which resources are being drained from natural cycles to support them, such as fresh water, fertile soils, and minerals such as phosphorous. Therefore, the issues that face space engineering and architecture in providing habitats that effectively serve as life-support systems are converging around the issue of "sustainability." Sustainable development is a relatively new concept that has been intensified by an awareness of resource constraints, and invites questions about how the current generations can meet their own needs, as well as securing the welfare of future generations.

Even though we are more considerate now than we have been in the production of buildings over the last few decades, we are still building our living spaces using construction processes that are shaped by industrial practices. In the modern epoch, the way in which we make and use buildings is highly wasteful of resources, energy, and opportunities to develop new ways of producing and occupying space. Currently, architecture is responsible for 40 % of our carbon footprint, most of which is produced in the running and maintenance of buildings that require fossil fuels. With a more considered view, we have sought to produce more energy and resource-efficient buildings, harness renewables, and invent materials with low environmental impact. Despite this, the efficiency route to sustainable building practice leaves us with ideals that have an effective net zero impact on our surroundings. This means that, unless everyone can afford to build a zero-impact building, then, no matter how many of these we construct, we will always be in environmental deficit, because of the platform of human development that accompanies industrialization.

The field of architecture is a discipline in which radical breaks in the ways in which we imagine the world and realize our ideas can be achieved. While there is no single definition of architecture, it is not exclusively focused on the production of buildings. Indeed, it can be thought of as a professional practice concerned with the choreography of space and matter. Its practice is developing spatial tactics and material programs. If these processes are going to have a completely different kind of relationship with our living spaces, then we need to consider the production of spaces that are not predefined by existing conventions, commercial interests, and master plans that can be implemented through industrial methods. We will need to imagine new ways of producing habitats that are interrogated through research and development. Architecture is therefore poised to catalyze complete breaks in the way in which we make our environments through a whole range of approaches, such as new kinds of manufacturing, the application of emerging technologies, and understanding cultural contingencies—or any combination of these.

Sir Peter Cook introduced the term "experimental architecture" into the design lexicon in 1970 in a book with a title of that name (Cook 1970). He critiqued the architectural avant-garde as a way of opening up more experimental forms of architecture that resisted an increasing functionalist approach to the development of cities that were leading to uninspired living spaces and bland metropolitan expanses at the expense of creativity and community. Experimental architecture emerged to question modern architecture's clean lines, but also to challenge post-modernism's deconstruction of modernity as a parody of its values by asking architects to be propositional. Experimental architecture replaces a notion of progress through a succession of building styles such as Parametricism, and therefore moves away from predetermined utopias, through an active practice of experimentation.

While Cook's critique was theoretical, American architect Lebbeus Woods championed the idea of experimental architecture as a practical form of research through drawing. He challenged established disciplinary tropes and raised new issues for architectural consideration that were beyond the reach of established knowledge canons, such as massive-scale events like wars and natural disasters (Woods 2012).

Architect Aaron Betsky observes that experimental architecture as Woods and his cohorts saw it was

> "… a replacement for the utopia that had assuaged the guilty conscience of architects who had for decades claimed that their compromises, their economic boxes, and their blandishments to clients were really promissory notes for a utopia they sometimes sketched in their free time. Instead of radiant cities or crystal mountains, mass-produced housing projects or pure forms dissolving in light, experimental architects drew (deliciously, in the case of Woods) skewed versions of the reality around them, exaggerated in form and function and made both seductive and frightening." (Betsky 2015)

My own contribution to the field of experimental architecture is to move this research field into a laboratory space, where models, prototypes, and installations can be realized. In this context, experimental architecture becomes a multidisciplinary design practice that engages with insights arising from the origins of life sciences to develop ways of choreographing matter through space via a system of discursive prototypes (Armstrong 2015).

My research specifically embodies an ecological approach to architectural construction that is predicated on shifting away from the traditional view of architecture as a static, form-giving subject to develop other methods of producing architecture. In this context, architecture is not simply about making a building, but developing strategic approaches to organizing matter in time and space.

Experimental architecture is currently prototyping materials, methods, and emerging technical systems that may help us to address the question of our survivability in the long term. While we are currently making many resource savings and mechanical efficiencies to increase mechanical building performance, the way in which we make buildings right now is underpinned by ancient practices that deal with the organization of inert chunks of matter. To deal with the pressing environmental issues of our time, the profession is taking an active interest in "sustainable" design, which is expressed through a range of diverse approaches. These include the metaphors of biomimicry to the decoration of our living spaces with elements from rural vistas in green roofs and walls, material performance, carbon "counts," and various approaches to improving the mechanical efficiency of systems. Of course, each of these approaches is valid when framed by their internal logic but, when considering a long-term pathway for establishing models of sustainable development, it is worth remembering that the concepts and practices that form sustainable narratives are still very much in evolution.

Experimental architecture asks whether it is possible to radically change some of these ideas, and generate building materials that can positively influence their environments and make them more fertile or become completely self-sustaining environments. In the biospheres that Fuller imagined when, in 1960, he designed a 2-mile geodesic dome over New York City to regulate weather and reduce air pollution, the plan was to reduce cooling costs in summer and heating costs in winter by maintaining the space at a constant temperature.

Yet, there is so much that we need to learn in understanding how we may shape our environmental relations. By constructing habitable ecosystems within our buildings, we may advance our understanding of generating truly sustainable spaces in extreme environments, as well as develop an increasingly effective portfolio of technologies that enable our terrestrial cities to be more resilient and robust in facing likely resource and space shortages. An approach that can develop with top-down (starship to city) and bottom-up (city to starship) approaches in parallel may enable more effective preparation for our journey to the stars, as well as catalyzing necessary ecology-promoting changes within today's cities. These mutually reinforcing approaches to starship design will help us to better understand how to design and engineer a world by developing life-promoting supra- and infrastructures. This may be achieved by prototyping systems that examine the principles through which we may build ecological networks that promote the fertility and livability of our habitats. Indeed, unless we are unable to do this "sustainably" on Earth, our chances of supporting a multi-generational colony within a synthetic environment are extremely slim.

More specifically, we need to live in spaces in ways that do not simply nod to an industrial efficiency indicator such as a carbon count (for the living metabolisms that influence carbon measures are very much drowned out by the concentrated carbon sources derived from fossil fuels), but establish approaches that promote the liveliness of the spaces within and around us and deal with the recalcitrant nature of the biological realm. It is prudent, perhaps, to remember that all cities are founded on soils that provide natural resources and are best understood through their fertility and biodiversity, rather than the square footage of concreted land under which they have been smothered. We also need to consider the basic physical and chemical interactions that shape the flow of matter through our urban spaces and appreciate just how dynamic the world around us is—and how we can facilitate these exchanges in meaningful ways by, for example, filtering polluted air to create accreted materials. These ways of seeing and thinking present us with the kind of baseline set of conditions that may enable us to re-imagine the present and the future of our living spaces, so that we may actively contribute to reconstructing our concept of architectures and that their production is not parasitic on Earth's processes, but maintains, and even promotes, vibrant, thriving cities.

There is much to learn about the practical aspects of building environments that support ecologies. Our current practices assume that our ecologies are stable and a constant, so there is little consideration given to their advancement. Indeed, we are more likely to destroy, cull, or suppress the "unruly" activities of Nature in our cities than promote them, as they erode inert surfaces and thrive uninvited in hidden niches.

Creating the conditions for livability as we travel to the stars may be achieved in a very immediate and practical sense by weaving starship fabrics into our cities. Such materials may also have technological capabilities whose impacts may be measured differently from machines. For example, rather than looking to parameters of efficiency and resource conservation, these systems may be assessed through their promotion of ecological relationships, by increasing fertility and orchestrating the flow and transformation of matter through dynamic, elemental systems. Such approaches may not only help us to increase the livability of our surroundings, but also establish the fundamental building blocks for constructing worlds.

A very immediate set of actions can be taken toward starship development through drawings, models, and prototypes of these proposals, which may be tested by interweaving them with the fabric of our living spaces. Algae bioreactors, anaerobic digesters, bioluminescent light sources, heat-absorbing and emitting substances (e.g. sodium thiosulfate), and even hygroscopic materials (e.g. calcium chloride) are examples of the kinds of systems that may begin to change the operating systems that underpin the way we live.

To develop such habitats to the stage at which they are meaningful to starship design, it will be vital to use the collated data from our many, varied investigations of our living spaces and take this citizen action to the next stage, working alongside architects, scientists, and construction engineers. Lessons learned from the construction, implementation, and use of these systems might inform space-engineering requirements in deciding whether a starship is slow or fast, wet or dry—or, indeed, whether or not it can ethically sustain humans. On Earth, these experiments will help to advance further architectural developments in "sustainability." To monitor progress, it is also essential to develop appropriate metrics that engage with ecological concepts so that they are not simply collapsed back into a system of mechanical values. In other words, we need to measure not only "efficiency," but other livability factors such as biodiversity, fertility, capacity for change, and re-assimilation into other systems.

Even if we do not advance this project toward a collective starship vision, interrogating the way in which we currently live and questioning the expectations we have of our habitats may help us to change our current practices. Perhaps we will no longer accept sterile surfaces that repel life (because it is unruly), but let it in (where we can work alongside its metabolic processes). Maybe we will no longer find it acceptable to allow the advance of iconic architectural bodies that make mere metaphorical references to Nature that—from an existential perspective—simply await their inevitable digestion by natural forces. In their place, our living spaces may actually become ecosystems that are shaped by natural, cellular landscapes that confer our homes with biological resilience and enable them to negotiate their own territory in the constant struggle for survival that we share with all other life forms against the elements.

28 Architecture and space exploration

Architecture is ideally placed to investigate such projects, as its design portfolios engage both biophysical and sociopolitical ones. In the interrogation and reformulation of our habitats, it is also important that we identify new ways of equitably distributing vital infrastructures, which may begin to help us to live our lives more sustainably and productively within our urban environments. From a citywide perspective, each community or urban expanse could be potentially organized so that it shares resources and infrastructures. They may develop production systems that are not only unique to each social grouping, but also help cities to become independently sustainable in terms of their net consumption and production of resources. Citizens will no longer be obligate consumers, but producers who enhance the quality of their living spaces. In this manner, learning how we may live as if we were already on a starship traveling through unchartered space-time could help us to more sustainably live in cities today.

2.4 Space skyscraper

The Interstellar Question requires an experimental approach to open, low-cost, socially engaged, collaborative projects, so that it may be possible to imagine and prototype alternative ways of living as if we were already space-bound. In this way, we may explore how it may be possible to develop next-generation sustainable buildings and dwell within these environments.

Yet, to generate the kind of thinking that leads to radical breaks in our expectations and practices requires us to propose bold new possibilities. In this section, I will discuss a speculative architectural project that aims to directly transpose terrestrial habitats into Earth's orbit and guide them into a developmental platform for the development of a starship framework. From this initial configuration, scaffolding may be assembled and ultimately launched into interstellar space.

Architect Teodor Petrov's space skyscraper design is a series of six modular sustainable rocket buildings. Ingeniously, they may be assembled as starship infrastructure, establishing the foundations for the construction of an orbital spaceship platform. This concept combines the idea of the starship city, with the production of habitable orbital scaffolding.

The space skyscraper is one of a series of six tall buildings, each of which is constructed according to these key principles. While each may differ in details, they share broad principles—construction, terrestrial habitation, orbital launch, and assembly as a starship platform. They are situated in equatorial locations (Beijing, Mumbai, Cairo, Orlando, Mexico City, Honolulu) for ease of departure. However, for the safety of urban populations, these buildings are independent of and at some distance from their major cities. Together, the six space skyscraper buildings comprise a construction platform for the production of the starship chassis that may serve as a construction site for interstellar research, or for structures that will venture into space.

In the first stage of development, the space skyscraper is made from building materials that are extracted from the environment. One example of a space skyscraper is a 260-m iconic building with 63 floors that is located in Beijing Harbor and built from advanced materials. It rivals the tallest completed building in Beijing, the Fortune Plaza office building. It harbors an anticipatory technology that speculates on the warm transpacific conveyor belt, which is the primary mechanism that rapidly transports air pollution from one continent to another. Its goal is to prototype radically new technologies with positive environmental building impacts that may be taken up more widely, especially within China's megacities. As China is a leading source of air pollution, it is hoped that these kinds of technologies will eventually contribute toward stalling extreme environmental conditions such as global warming or even an ice age. These are directed by a prototyping technology that starts construction at the top of the building going downward. The apex is therefore built first. It is a research laboratory and site of synthesis for energy and building components, from which the rest of the building is produced. Materials are harvested using a biofiltration air system as a microspecies of electrostatic precipitator (ESP). This is a charged collection device that removes fine particulates such as dust and smoke from the airstream. Very large airflows may be treated, although a large area (footprint) has typically been required—a large biofilter (>200,000 ACFM (alternating current field measurement)) may occupy as much as or more land than a football field—which is one of the principal drawbacks of the technology. Yet, high-volume flows may be gained from such a tall building and may extract sufficient materials to be compacted into solid structures and form a kind of concrete that has been made from the air (Fig. 2.1).

At the core of the building is a living wall that constitutes a physiology, which may be integrated into a larger ecosystem of material exchanges. The building itself is a

Figure 2.1 Exploded elevation through a space skyscraper. Credit: Teodor Petrov

self-contained, city-like structure, which can produce its own energy and water, and possesses infrastructures that can be redesigned and shaped over the course of its life cycle. Using the action of microbes to clean water and air, as well as removing malodorous compounds and water-soluble volatile organic compounds (VOCs), this structure enables the recycling of essential substances. These processes may be further refined, where necessary, through a series of biofilters and scrubbers. These advanced technologies are also available to a range of industries that include food and animal products, natural gas from wastewater treatment facilities, and pharmaceuticals, which become key to a space skyscraper's economic and political systems. While inhabitants live out the building's lifespan, which for a modern building is about 25 years, although most buildings are over-engineered for purpose and then, owing to the economics of the built environment, are torn down and replaced by a new building that is more "fit for purpose," Skyscraper does not pander to this paradigm. Instead, when it has reached the end of its usefulness, it is launched into orbit. This ambitious project encourages radical reflection on the idea of multi-use and repurposing of building stock. The capacity of the building to enter an orbital trajectory at some future stage in its life cycle informs the design of the living spaces beneath the laboratory levels. These are designed to train inhabitants to live as if they already inhabited an interstellar craft (Fig. 2.2).

The second phase of the space skyscraper is that the structural steel core can be filled with deuterium, which therefore powers the launch of the structure into orbit. Although this is still a speculative propulsion technology, it would reduce mass and also decrease the need for shielding during take-off. The technical details of a space skyscraper's launch have not been formally established and this idea remains conjectural (Fig. 2.3).

Figure 2.2 Laboratory exploded elevation through a space skyscraper laboratory. Credit: Teodor Petrov

Figure 2.3 A space skyscraper taking off. Credit: Teodor Petrov

Once the space skyscraper has reached an orbital trajectory, it is assembled into starship scaffolding through a highly choreographed sequence of assemblies. Each of the skyscraper buildings from different equatorial locations plays a specific role within the

32 Architecture and space exploration

production of this platform. With the help of giant robotic arms, they are individually brought together and secured using scaffolding that is harvested from space junk. This is collected using the giant robotic arms that are able to grab large items that are whizzing past at speeds of up to 17,500 mph. Despite the precarious assembly, the site becomes a place where teams of workers may actually inhabit the skyscraper infrastructure. Owing to its design for terrestrial sustainability, with a carefully orchestrated ecosystems design and physiological infrastructure, the site is already optimally designed for growing crops and sustaining colonies. Ideally, these megastructures would also bring up new groups of assembly workers, thereby cutting the expense of payloads. Maneuvering of the buildings into place takes place via multi-directional engines that are situated in the body of the building. These allow braking and fine positioning of the megastructures to assist in the adoption of their final configuration (Fig. 2.4).

Figure 2.4 View of a space skyscraper from orbit. Credit: Teodor Petrov

While this project is largely speculative and, therefore, a design concept rather than an engineering project, the principle of not tearing down perfectly good buildings into rubble—plus the idea of prototyping and alpha testing space structures for habitation on Earth before they are inhabited in an orbital context—is an interesting notion that warrants much more detailed reflection and technical study, which is beyond the scope of this current publication. However, Skyscraper embodies a way of thinking that takes us out of our comfort zone and challenges us to think longer-term, collectively, ambitiously, and creatively. This is the promise of the starship question—that, in its conceptualization, we may break free from some of the paralyzing assumptions that bind us to our most insidious habits.

2.5 What is this place called "space"?

"This stuff we call 'matter' is just a fraction of what the 'network' is all about …. The true reality is mostly darkness …. There is scarcely any light or matter—that's just the graphic front end for the cosmic code. Most of the cosmic code is Dark Energy and Dark Matter. The stuff we foolishly call 'reality' is the cute friendly part with the kid-colored don't-be-evil Google graphics. The true, actual, cosmic reality is the giant Google network pipes and the huge steel barns full of Google Cloud. It's vast and alien and terrifying." (Sterling n.d.)

Space has amazing properties, many of which are just beginning to be understood. While the cosmos is mostly a vacuum, it also contains "normal" matter, which possesses mass, occupies a volume, and includes a range of substrates, from leptons and quarks to atoms and molecules. However, in astrophysics and cosmology, dark matter does not obey the characteristics of normal matter, which is the luminous substance in which we are immersed. It "shines"—not because it radiates light, but because it interacts electromagnetically and gravitationally with other forms of matter and radiation. It is composed of leptons, such as electrons, and quarks that are the building blocks of protons and neutrons, which can be used to build atoms and molecules. Normal matter exists within a range of materials, which are of great interest to space explorers, as they constitute our known world and create the conditions for life. These particles form the building blocks of material experiences that are expressed in many forms over a vast range of scales in space—from suns, planets, and orbital bodies to asteroids and smaller meteoroids, as well as interplanetary dust and submicroscopic particles of ionizing radiation.

Planets, their moons, and asteroids are the major sources of matter in the universe. The cosmos also contains "antimatter," which is a unique form of normal matter that possesses the same mass but the opposite spin and charge of its counterpart. When matter and antimatter collide, they annihilate each other and create pure energy as gamma rays.

Historically, we have learned much about the composition of the cosmos from our longstanding observation of the skies with telescopes. However, our understanding of our interplanetary landscape has been unfolding rapidly with the advent of robotic space exploration. It is not yet clear what implications these discoveries may have on our deeper survey of space but their revelations create the conditions for the construction of an interplanetary civilization. This will necessarily be shaped by the distribution of solar radiation and the availability of specific resources—particularly water and organic substances—throughout the interplanetary environment. Where we will go and establish colonies to benefit from these materials will depend on the ease of harvesting planetary resources. This is determined by their distance from the worldship and by the depth of the gravitational wells through which the matter must be lifted onto the vessel. The range of available planetary resources can be considered by thinking about the planets in our own Solar System.

Our own Sun is a second- or third-generation star that formed around 4.5 billion years ago, after the initializing event known as the "Big Bang" that took place 13.82 billion years ago from which our current universe, or universes, arose. This means that our early Solar System was an extremely violent place that was not only capable of destroying a couple of suns, but also produced an environment in which young planets frequently collided. From out of this fiery tempest, nine planets arose that characterize today's comparatively peaceful Solar System.

Mercury is the smallest and closest planet to the Sun. The Mariner 10 probe, in 1974 and 1975, offered views of a puckered, strongly magnetic landscape, while the Messenger probe in 2012 made the surprising discovery that, despite planetary temperatures reaching 400 °C, the craters contain ice at the North Pole.

Despite its reputation as sister planet to Earth, Venus is "a vision of Hell," as attested to by Carl Sagan in Episode 4, "Heaven and Hell," of his popular series *Cosmos*. Although it is of a similar size, with some geophysical features that are similar to those on Earth, that is where the similarities end. Venus is choked by a dense atmosphere of noxious clouds of carbon and sulfur dioxide, and is ferociously hot. Its smooth surface reaches temperatures of 450 °C, and it is scarred, as if by smallpox, by thousands of active volcanoes.

Our own planet was produced by the most significant and last of the collisions of the early planets of today's Solar System in an event called the Great Splat, which took place around four and a half billion years ago. Theia, a Mars-sized planet, smashed into primordial Earth, tearing colossal masses of matter from its surface. Then the vagrant body plunged into Earth's fiery core, where it currently remains. The debris from this violent impact began to roll into several developing embryonic plates of expanding molten matter that hugged a much tighter orbit around the healing Earth than the Moon does today. Around 10 million years later, one of these bodies became unstable and caused the Little Splat as it somnambulated into our colossal proto-Moon. Its painful, slow motion impact would have supercharged the photopigments of the first blue-green algae that colonized its surface, as a tsunami of molten rock soared like fireworks over Earth's young horizon. Today, Earth is continually scrutinized by the orbital gaze of satellites that have transformed our understanding of our living planet. They constantly monitor the climate through ice coverage, sea levels, crop growth, pollution, gravitational anomalies, and ozone levels. These new images of our world provoke new relationships that are conjured by the lenses of light, magnetism, gravity, and seismic waves.

We are eternally optimistic about the potential liveliness of Mars, even after Mariner 4 revealed it to be a barren, hostile terrain in 1965, which was later confirmed by the Viking probes in 1976. However, with increasingly sophisticated technology, more subtle details about the story of Mars are unfolding, which, for example, suggest that water flowed on Mars relatively recently, being trapped in hygroscopic soils, and that primitive life forms could still survive below its surface.

Jupiter is the Solar System's largest planet and seems hostile to life. It does not possess a well-defined surface, but has a rocky core that is surrounded by layers of hydrogen and helium. However, several of its moons, such as Ganymede and Europa, which are heated by the tidal effects of Jupiter's powerful gravitational field, have been shown by the Galileo spacecraft in 1995 to possess subsurface oceans that could provide homes for alien life forms. The most recent close observations of Ganymede were made by the New Horizons probe, which recorded topographic and compositional mapping data of Europa and Ganymede on its fly-by of Jupiter in 2007 on the way to Pluto.

Famously, Saturn is surrounded by rings of ice and rock particles, which have been shown by space probes since 2004 such as Cassini to be highly complex structures that appear to form highly structured bands, which orbit Saturn at very high speeds. Saturn's many moons are also potential sites for the discovery of alien life. Titan has lakes of methane, while Enceladus shoots out plumes of water and organic material into space.

Voyager 2 raced past Uranus at a distance of 81,500 km in 1986 and discovered 11 new moons orbiting its powerful magnetic fields. The probe also revealed that one of Uranus's main moons, Miranda, appears to have disintegrated and then reformed in orbit.

In August 1989, Voyager 2 discovered that Neptune has a violently stormy atmosphere of hydrogen, helium, and methane, with winds reaching speeds of almost 1500 mph.

Pluto is a dwarf planet that is extremely cold, at around 375–400 °C below zero. On July 14, 2015, the New Horizons probe collected images and data of the first-ever fly-by of the dwarf planet and its five moons. These observations are already revolutionizing researchers' understanding of Pluto and other bodies in the faraway Kuiper belt beyond Neptune's orbit. The vast majority of New Horizons' data has yet to be relayed back to Earth—and the probe will probably zoom past a second, much smaller Kuiper belt object (KBO) in 2019 as part of an extended mission (Witze 2015).

Yet, planets are not the only sites of valuable materials in the cosmos. Even within the seemingly spartan landscapes of the void, a primitive ecosystem exists. Comets and asteroid bodies transport water across the vastness of space, meteors are impregnated with organic molecules, and the pervasive life-giving electromagnetic spectrum penetrates the universe—being completely absorbed by the vastness of dark energy and dark matter in which our visible world is enmeshed. From these thready exchanges, life's fundamental fabric has come to exist. While we cannot experience those events that have taken place over geological timescales, it is possible to deduce that the spatial entwining of matter and energy is somehow key to the mysterious processes from which life on Earth arose.

Asteroids are of great interest to space colonists, since they have very shallow gravitational wells and may contain significant amounts of substances that support organic life, such as hydrogen, carbon, and nitrogen, as well as other useful minerals. Indeed, they are imagined to be potential vehicles for transporting organic matter through space and may have even seeded Earth with chemistries that gave rise to terrestrial life in a process known as "panspermia." They are exceptionally valuable when they contain significant amounts of water or ice. "Dirty snowballs" or comets are also possible resources for space explorers, as they are made up of dust that is bound with frozen gases and ice. However, they are not suitable for harvesting, as they travel at high velocity and are extremely difficult to capture.

Yet, this visible realm is the palatable interface of the material realm in space, and only a fraction of a much "darker" cosmic reality. According to the Planck mission team, and based on the standard model of cosmology, the total mass–energy of the known universe contains 4.9% ordinary matter, 26.8% dark matter, and 68.3% dark energy (NASA 2013). Unlike normal matter, dark matter is non-luminous. It is a hypothetical substance, first proposed by Jan Oort in 1932 as a way of accounting for "missing mass" in the universe. Its existence and properties are a complete mystery and are inferred from the gravitational effects on visible matter, radiation, and the large-scale structure of the universe. Dark matter cannot be seen directly with telescopes, as it does not respond to the presence of light, although it may emit its own unique kind of gamma ray. This may be a fundamental property of an as-yet uncharacterized type of subatomic particle, whose discovery is one of the major efforts in particle physics today. So, while there is more dark matter than normal matter in the universe, the most abundant substance is actually dark energy, which may be an innate property of space.

Another strange substance is radiation that interacts with matter. It is created by matter, can create matter, and is emitted by matter, but is not actually matter. Radiation is massless and takes up so little volume that it is just too ephemeral to "be" matter. According to Einstein's theory of relativity, mass and energy are equivalent. When particles of radiation collide with sufficiently high energy, mass can be created. Space radiation may, therefore, also be exploited as a resource in space to potentially produce matter, or provide energy for propulsion. Ionizing radiation, such as gamma rays, protons, and neutrons, possesses sufficient energy to remove electrons from the orbits of atoms. This produces charged particles and releases electrons from inner orbits, which destabilize atoms, rendering them highly reactive, so organisms need protection from their effects. However, non-ionizing radiation microwaves, radio waves, and visible light are not toxic to life, as they do not remove electrons from their orbits. Galactic cosmic radiation originates outside the Solar System and is made up of ionized atoms, ranging from a single proton to a uranium nucleus. The rate of flow of these particles is very low. However, since they travel very close to the speed of light, and because some of them are composed of very heavy elements such as iron, they produce intense ionization as they pass through matter.

As our space probes travel farther, with increasingly high-resolution imaging equipment, out beyond our Solar System and into interstellar space, we are likely to encounter unexpected findings like the bright spots of Ceres, a dwarf planet and the largest object in the asteroid belt. It is also possible that we may come across strange fundamental fabrics that can be interrogated back on Earth through their replication in colossal terrestrial instruments such as the Large Hadron Collider, which accelerates particles to the speed of light and observes them fragmenting as they smash into each other. The synergy between field observations and terrestrial experiments creates new knowledge of the universe, with exciting opportunities for design and engineering in interstellar space.

Ongoing questions about navigation, exploration, and the settlement of distant terrains are raised by our encounters with very large or invisible materials—particularly when they are so very far away and we do not have a good grasp of their characteristics—let alone the technical systems to design with them. While we may not fully understand what it means to design across irreconcilable scales of natural experience in an age of elusive Higgs bosons and mysterious dark matter, provoking these questions inspires the human imagination not only to generate new materials, technical systems, and experiences of space—but also to imagine new futures amongst them.

2.6 World-making

"If we are going to live off-world," reported the New World Consultation Committee, "we need our own habitat. With all the things we've evolved to depend on—the right temperature range, specific gravity, light cycles of night and day."

Teams of consultants, visionaries, futurists, engineers, scientists, designers, artists, entrepreneurs, and chancers dreamed along with each other, attended meetings, teleconferences, and workshops. In between these formal events, they babbled ceaselessly in private and public discussions. During dialogues of observation and uncertainty—in the moments before the "object" or "subject" of their discussions were agreed—their roast coffee beans nurtured compost processors, plastic cups were recycled, and waistlines thickened from non-stop formal meals. Surprisingly, they came up with one big idea:

> "Rather than inflating a sheet of metal to become a spinning metal station, wouldn't it be better to re-create our planet by building an artificial world from scratch?"

Then they got back together again to decide what "artificial" really meant. Some of them interpreted the idea as either being a replica of an actual planet—a colossal chunk of matter that was indistinguishable from existing moons, asteroids, and planets. Others decided that this was merely a form of landscape gardening. Instead, they proposed that it would be possible to generate an object that performed the same kinds of functions as Nature but could be improved upon. Therefore, this structure did not need to look anything at all like a real planet.

Of course, they conceded that making a planet was not a new idea. The Chief Systems Engineer observed that John Desmond Bernal had already thought of creating an artificial world first in 1929. It aimed to address human expansion by housing a target population of around 20,000 people within a 16-km-diameter spherical shell.

Then, in 1975, Gerard K. O'Neill developed this idea further in his "Island One" project, which was an artificial satellite with a much smaller Earth-like interior than the Bernal Sphere. The structure was imagined as a solar power station that transmitted solar energy by microwave to Earth. While Island One was 500 m in diameter and could accommodate 10,000 people within an equatorial valley, it was soon upgraded to an Island Two version, which spanned around 1800 m in diameter. Sunshine would be delivered through a complex mirror system, which would also provide enough light for agriculture. Gravity-free manufacturing plants would provide employment for residents who were expected to spend their recreation time in unusual activities such as human-powered flight and zero-gravity sports.

"I don't want to put a downer on the situation," said the Director of Physics, "but we have a hell of a job on our hands here." Inevitable disagreements arose. What should the megastructure be made from? While steel was the obvious choice, the expert advisory board suggested Zylon, as it is seven times stronger than steel and twice as tough as Kevlar. It offered the distinct advantage of possessing an extraordinary strength-to-weight ratio profile that would keep any planet-sized rotating structure from ripping apart. Of additional interest was that the carbon-containing synthetic fiber could also be mined from carbon-rich asteroids.

"An artificial world will never possess the stability of a real planet," grumbled the Astrogeology Director. "Active maintenance will be continually needed for the environment to perform with the right conditions." "It doesn't have to be perfect," asserted the Director of Space Engineering. "We could cheat by packing just a tenth of Earth's mass—say, 700 quintillion tons—into a sphere the size of the Moon."

More discussion followed, as phone calls were missed, important meetings postponed, and tannin-stained teeth lubricated by an overabundance of over-boiled coffee and tea, munched on packets of tasteless biscuits.

"Maybe we could acquire core material by growing the planet through a bottom-up process of construction? You know, like growing a star." While ripples of dissent reverberated through the consultation community, the idea was compelling. Newly forming stars grow by aggregating hydrogen atoms. Gradually, these primordial elements are condensed into a series of elements through a process of nuclear fusion. Over millions of years, these proto-materials have the potential to develop into a world.

"Potentially, super-fusion technology could speed this process up by using magnetic fields to artificially accelerate the evolution of matter," said the Director of Physics. "We'd need the densest known elements like osmium, iridium, and platinum. While these elements can currently only be made in the thermonuclear explosions of supernovae, all we need is a spectacular fusion technology."

"I like the idea of using a mega structure to make a mega structure," observed the Chief Systems Engineer, "one that actually obeys the classical laws of physics."

"Ingots of these materials could be launched piecemeal to the construction site of the artificial planet," continued the physicist. "The construction process itself would generate significant heat. Somewhere in the region of the surface temperature of the Sun." Several astrobiologists tried to interject but were waved down by the Committee Chair.

"After a century of cooling ingots of crustal elements, such as silicon, magnesium, and iron could be layered over the dense core. Then we'd need a period of cooling that would last about 10,000 years, before we could introduce water and begin the first steps of origins of life-making process."

"Unless something radically different happens," insisted the irked astrobiologists, "building your own planet appears to be no quicker, or easier than colonizing a new one." So, they lived in the hollow of an inflated metal structure that slowly leaked breathable air, like an old balloon. Planet Bernal was not an artificial world, but an engineered structure. Yes, it had its glitches, gravitational inconsistencies, and some of the inhabitants went crazy, as their circadian rhythms could not be adjusted by the reflected light of their interiorized world.

But most residents did not remember any other kind of Nature. When you ask them about their day, they generally describe the sunlight as softer now, yellower, and less golden. The light is sleepy, red-eyed, and even bleary on some days and, as night falls, the Bernaldians describe skies of bulging veins stretching into a varicose atmosphere, which slams tightly shut as night falls.

Until tomorrow.

2.7 StarshipSPIDER

> "She's a magnificent, dark figure that grows out of the black spider sea. As she walks she fractures the ground with impenetrable membranes of concrete and tar macadam .... Life starts to fade at the edges of her existence .... A spectre so greedy that she gives nothing back. Not even light is reflected from off her frame. So, she casts no shadows of her own. She is a void in the shape of a woman." (Rachel Armstrong, *Origamy* (in press))

Tension exists between engineering plans and concept vessels in considering the kinds of craft that may take us from our home planet across the Solar System and out into interstellar space. Technically developed plans, such as the Bond/Martin 1984 starship, are engineering proposals that are concerned with the buildability of a structure and its performance, while concept vessels, such as the organic craft that resembles a colossal dragonfly from the science-fiction series *Lexx*, speak to space as an operatic production of human endeavor and experience.

Naturally, frictions exist within the interstellar community between the two approaches. While aerospace engineers wish to be pragmatic, so that potential backers take buildable proposals seriously, architects and designers are compelled to unleash the human imagination and push feasibility beyond its limits. Yet, the approaches are coupled. When we stay too close to where contemporary knowledge resides, we are at risk of constructing something outdated at the time of its completion; on the other hand, fantastical ideas do little to persuade a broader community of the seriousness of the interstellar community and the viability of the proposed project.

Frederik de Wilde's StarshipSPIDER is a concept craft that is inspired by specific geometric forms—the *architectons* of Kazimir Malevich (MOMA 2013). At the turn of the 1920s, Malevich started applying the Supremacist theories, which he had previously developed in painting, to architectonic forms as projections of his cosmic dreams of the *planits*. These were the animated equivalent of floating cities with the names of Greek letters that Malevich called architectons. These utopian, spaceship-like vehicles were set out in great detail and did not relate to any specific scale. Yet, Malevich, who rejected utilitarianism, did not insist that his models should be made into real objects. His ambitions were absolute and free of context, which perhaps embodied a desire to escape the history of the Soviet Union and its specter of "human misfortune." (Fig. 2.5).

Yet, de Wilde's StarshipSPIDER proposes to be more than an inert hull moving through interstellar space. It is a body that talks about the construction of the interstellar medium, its Spartan ecology, and the strangeness of matter. Its absolute blackness links matter with the abyss. It spins its own scaffolding that it uses as a locomotory system to infiltrate unknown territories. As its multiple, intersecting carbon planes voyage through the emptiness of space, it weaves threads formed by photonic winds and buckyballs. Tirelessly constructing cosmic theaters between galaxies, StarshipSPIDER's web snares epic vagrant celestial bodies that invent narratives and strange events as they shake and struggle in its dark cosmic web. Vibrations from these performances resonate throughout the universe, fusing with primal notes from its background noise.

Figure 2.5 Starship SPIDER, a concept ship that produces threads of matter upon which future space ecosystems may be built. Credit: Frederik de Wilde

StarshipSPIDER is pitilessly black. Its artistic roots arise from deep within the satanic mills of the Industrial Revolution that spewed black carbon into Earth's atmosphere and marked the urban environment with its dark scars. It rejects of the depiction of objects as its design ambition and gestures towards pure expression, evoking moods that range from joy to chaos (Peplow 2015).

In an interstellar context, StarshipSPIDER yearns for kinship with both the visible and dark universes, generating scaffoldings that communicate between these incompatible realms, and, in doing so, synthesizes an unlikely, restless space ecology.

The concepts underpinning StarshipSPIDER are a very long way from being buildable, however; the elegance of the project emphasizes the idea of space as an inhabited ecology in which a range of epic non-human inhabitants dwell. It conceives the very purpose of the interstellar craft not just as a transport hull, but also as a discovery vehicle with a metabolism that makes and shapes space as it passes through it. Splitting and repairing the very fabric of the cosmos, the trails of our vessels may now lay down pheromones to attract other vehicles, and synthesize highways as successive craft pass through these spaces. Sites of significance may even emerge along the way. StarshipSPIDER is an attractor of material organization and a catalyst for life, as we have not yet encountered it.

How StarshipSPIDER actually influences spaceship design is as yet unknown. But, to fully grasp what is at stake, let us turn its conceptual design back on our cars, our ships, and our planes. What might happen if our vehicles today actually repaired rather than destroyed environmental relationships and created opportunities for ecodiversity rather than an indelible legacy of biospherical scars?

StarshipSPIDER reminds us of the primitive status of our own legacy of space exploration with our clumsy, industrial hulls and trails of pollution, and asks us to think again about our future relationship with the fabric of space and the assumption that it is not alive.

2.8 Architecture and hypercomplexity

When considered from first principles, rather than from a vernacular association with the production of buildings, architecture is concerned with the choreography of space and matter. It reflects on the relations between substances, processes, and their entanglements with the environment and how these associate with human culture. These relationships deal with the process of living, and at the heart of an architectural approach to these issues is a deep relationship with "life."

In the Middle Ages, the design discourse of life was centered on ideas of the body as a literal form of measurement that was implemented, such as the "foot," "hand," and "inch" (distal phalanx of the thumb). During the Renaissance, the concept of imagination was introduced, enabling architects to design forms inspired by Nature in ways that were more three-dimensional, fluid, and natural than the Gothic designs that preceded them (Cairns and Jacobs 2014). Nevertheless, the materials, tools, and concepts available limited the structures and performance of buildings. In the twentieth century, motion itself became integral to the understanding of structure through cybernetics. For example, Gordon Pask started to build dynamic systems using different self-assembling substrates—crystal growth via his chemical ear project, for example (Bird and Di Paolo 2008)—and Stafford Beer used pond life (*Daphnia*) as the substrate for designing with ecosystems (Beer 1994). However, the technological advancements necessary for taking these fabrics to the next stages of physical and dynamic complexity did not exist in ways that could be accessed by architects/designers.

To sort and order living processes requires a computational platform. Take, for example, the field of molecular biology that aims to better understand how a DNA sequence can be translated into a dynamic, functional protein, so that it can be designed from a linear, binary-compatible code of nucleotide base pairs A–C and T–G. Molecular engineer J. Craig Venter proposes to build systems that "crank out synthetic chromosomes ... [and] ... [make] robot[s] that could build a million chromosomes a day" (Brockman 2007). These codes are compiled and can be read by cellular systems and translated into various polypeptides, which are early forms of proteins. Each of these protein chains is snaked into various topologies such as helixes, loops, coils, and sheets. Modern digital computers are starting to use software programs such as molecular dynamics to help them to predict the folding potential of polypeptides and the design and construction opportunities that lie within them.

The tertiary folding of polypeptides is another question altogether. It remains an extremely difficult and unresolved undertaking, as the solution spaces become astronomically large. These problems can be partially bypassed in "comparative" methods, the most famous of which is Alan Turing's approach to dealing with a hypercomplex system—human intelligence. Turing suggested that, owing to the difficulties in definitions and cultural expectations of certain phenomena, personal experience and subjectivity could significantly collapse the solution space. By introducing bias into the system, an answer could be reached. Such approaches are being used in the synthesis of complex chemistries (Cronin et al. 2006). Yet, this would only ever be a working solution, never a deterministic one. Likewise, in tertiary protein folding modeling, the search space is pruned by the assumption that the protein in question adopts a structure that is close to another homologous protein.

While these challenges may already sound impossibly complex, quaternary relationships, in which several independent polypeptide chains contribute to the three-dimensional configuration of a protein molecule, add further complexity still. Of course, any further interactions of tertiary proteins with sugars when they are biologically packaged in the process of glycosylation are simply beyond the capacity of our digital computers to depict accurately.

The increased processing capacity anticipated with quantum computing (Gershenfeld and Chuang 1998) may go some way toward alleviating the challenges with hypercomplexity. Dave Ackley proposes that we need to spring the "bear traps" of computation if we are going to deal with this particular challenge. He observes the deterministic assumptions that govern-bounded solution spaces, which are typical of digital programs, are actually a property of small systems, so they do not scale to the degree needed to effectively address hypercomplex phenomena:

> "Over sixty years ago, von Neumann recognized the scalability limits of his namesake design. He predicted a future computing model, a 'logic of automata,' in which short runtimes would be preferred and all computational operations would tolerate faults … we continue to delay that future at our peril." (Ackley 2011)

To develop digital computers that can deal with hypercomplexity, new tactics for sorting and ordering reality are needed, as the current formalisms are now all hindrances to continued computational growth. Ackley proposes that seriously scalable computing requires a *robust spatial computer* in which resilience, survivability, and graceful degradation must be inherent not just in the hardware, but also spread upward throughout the computational stack.

Such challenges for increasing, not reducing, the complexity of reality is music to the ears of architects. Constructing architectural "life forms" in an applied context requires an understanding of the dynamic spatial relations between participating bodies. This is an incredibly challenging task to compute and cannot be completely modeled by the tools of modern synthesis. These are experimental tools, and fundamental findings that integrate genetics, paleontology, systematics, and cytology within a new, expanded structure for biological thought that prioritizes mechanism over population-scale phenomena (Rose and Oakley 2007). The current approach is to use "parametric" design that generates a digital model, which allows architects to select a few dominant variables (e.g. sunlight, stress, prevailing wind) and view changes in their impact on a site or structure before choosing a preferred set of parameters. However, these approaches are approximations of reality and there is still much analog work to be done between modeling architecture and implementing these designs as a building on a particular site. The tactics and toolsets employed by architects in spatial thinking are concerned with the notion of material relationships in time or space, and are characteristic of living systems and Nature, whose computational capacities were of great interest to Alan Turing (Turing 1952).

A different approach to computing is required that may address hypercomplexity by evading the need for abstraction and representation, so that it works directly with the generative forces of the natural realm.

While biotechnological developments in the last 30 years or so enable us to consider life's processes as a technology, nobody has succeeded in actually building something from inert ingredients that has the status of being alive. However, around the turn of the third millennium, the idea of buildings that could operatively perform in a manner usually attributable to living systems, such as growth, self-repair, and movement, became a subject for architectural research agendas. Using a combination of materials, technologies, and methods, architects began to explore the possibility that buildings could actually be "living" and possess lifelike properties. For example, an exploration examined by Terreform One examined whether tissue culture could be used in an architectural capacity (Terreform n.d.). Other architects considered how organisms could be incorporated directly into buildings where they could perform a specific set of operations, such as food production, as in ecoLogicStudio's *Algae Canopy*, an algae façade for the Milan Expo 2015 (ecoLogicStudio 2014). In this way, architecture no longer represents the idea of life as an inspirational blueprint for design, as in previous forms of architecture. Rather, it is concerned with the construction of alternative life forms and expressions of the natural realm that interrogate our relationship with the living world, which are not bound by naturalistic conventions.

Various forms of such "natural computing" practices exist (Denning 2007). They are concerned with the choreographies between dynamic substances that can produce a range of configurations depending on a set of variables such as the production of bonds, environmental conditions, and spatial organization. Strategies for coordinating these exchanges do not exclusively invoke science and engineering, but also enfold principles of art and design. This expanded toolset of materials, technologies, and methods is likely to be of increasing value to all disciplines working with hypercomplex systems such as biofilms, embryogenesis, living bodies, and extremely complex molecular configurations.

Yet, the tactics that may address hypercomplex challenges that challenge existing categorical confines through hypercomputation and metadimensionality cannot be expected to hold onto existing ideas about universal solutions, objective perspectives, environmental belligerence, or one-size-fits-all approaches. Rather, since they deal with incredibly large, heterogeneous, complex, and durational challenges, which are implicit in the work of Jordan Geiger (Part II, Chap. 15), Mark Morris (Part II, Chap. 15), Robert Chiotto (Part II, Chap. 14), and Astudio architecture (Part II, Chap. 12), they are likely to have unique, seemingly odd characteristics, such as varying degrees of hyperlocality, subjectivity, environmental responsiveness, heterogeneity, recalcitrance, and unpredictability. In facilitating syntheses between science, technology, culture, and society through notions of parallel processing, multiple functions, peculiarity, transmutations, transfigurations, synthesis, complexification, catalysis, expansion, enrichment, and even paradoxes of existence in which, for example, non-life appears to be "living," the kind of environments and toolsets that characterize the natural world may be produced and explored. Ultimately, they may provide the kind of manufacturing platform through which the construction of our "living" starship interior may begin.

REFERENCES

D.H. Ackley, Beyond efficiency. Commun. ACM (2011), www.cs.unm.edu/~ackley/be-201301131528.pdf. Accessed 17 Aug 2015

R. Armstrong, *Vibrant Architecture: Matter as a Codesigner of Living Structures* (De Gruyter Open, Berlin, 2015)

S. Beer, A progress note on research into a cybernetic analogue of fabric, in *How Many Grapes Went into the Wine: Stafford Beer on the Art and Science of Holistic Management*, ed. by R. Harnden, A. Leonard (Wiley, New York, 1994), p. 29

A. Betsky, Experimental architecture emerged to question postmodernisms jokes. Dezeen (2015), Aug 6, www.dezeen.com/2015/08/06/aaron-betsky-opinion-experimental-architecture-question-postmodernism-jokes/. Accessed 15 Aug 2015

J. Bird, E. Di Paolo, Gordon Pask and his maverick machines, in *The Mechanical Mind in History*, ed. by P. Husbands, O. Holland, M. Wheeler (MIT Press, Cambridge, MA, 2008)

S. Borenstein, Milky way teeming with billions of earth-sized planets. Huffington Post (2013), May 11, www.huffingtonpost.com/2013/11/04/earth-size-planets_n_4215873.html. Accessed 21 Aug 2015

J. Brockman, Craig venter—life: what a concept. Edge.org (2007), Aug 27, https://edge.org/conversation/craig-venter-life-what-a-concept. Accessed 17 Aug 2015

S. Cairns, J.M. Jacobs, *Buildings Must Die: A Perverse View of Architecture* (MIT Press, Cambridge, 2014)

P. Cook, *Experimental Architecture* (Universe Books, New York, 1970)

L. Cronin, N. Krasnogor, B.G. Davis, C. Alexander, N. Robertson, J.H.G. Steinke, S.L.M. Schroeder, A.N. Khlobystov, G. Cooper, P.M. Gardner, P. Siepmann, B.J. Whitaker, D. Marsh, The imitation game—a computational chemical approach to recognizing life. Nat. Biotechnol. **24**, 1203–1206 (2006)

P.J. Denning, Computing is a natural science. Commun. ACM **50**(7), 13–18 (2007)

ecoLogicStudio. Algae canopy (2014), Oct 1, www.ecologicstudio.com/v2/project.php?idcat=3&idsubcat=59&idproj=137. Accessed 17 Aug 2015

N. Gershenfeld, I.L. Chuang, Quantum computing with molecules. Sci. Am. **278**, 66–71 (1998)

A.R. Martin, Worldships: concept, cause, cost, construction and colonization. J. Br. Interplanet. Soc. **37**(6), 242–253 (1984)

MOMA, Kazimir Malevich, inventing abstraction (2013), Dec 23, www.moma.org/interactives/exhibitions/2012/inventingabstraction/?artist=54. Accessed 28 Aug 2015

NASA, Planck mission bring universe into sharp focus (2013), Mar 21, www.nasa.gov/mission_pages/planck/news/planck20130321.html#.VN5lrL7e2S0. Accessed 13 Aug 2015

M. Peplow, The reinvention of black. Nautilius (2015), Aug 13, http://nautil.us/issue/27/dark-matter/the-reinvention-of-black?utm_content=buffer05acc&utm_medium=social&utm_source=facebook.com&utm_campaign=buffer. Accessed 20 Aug 2015

M.R. Rose, T.H. Oakley, The new biology: beyond the modern synthesis. Biol. Direct **2**(30), doi:10.1186/1745-6150-2-30 (2007)

B. Sterling, From beyond the coming age of networked matter (n.d.), www.iftf.org/fanfutures/sterling. Accessed 14 Aug 2015

Terreform, In vitro meat habitat (n.d.), www.terreform.org/projects_habitat_meat.html. Accessed 17 Aug 2015

A.M. Turing, The chemical basis of morphogenesis. Philos. Trans. R. Soc. Lond. B Biol. Sci. **237**(641), 37–72 (1952)

A. Witze, Pluto and Charon come into sharper focus. Nature, News & Comment (2015), July 13, www.nature.com/news/pluto-and-charon-come-into-sharper-focus-1.16971. Accessed 15 Aug 2015

L. Woods, Inevitable architecture. Lebbeus Woods (2012), July 9, http://lebbeuswoods.wordpress.com/2012/07/09/inevitable-architecture/. Accessed 17 Aug 2015

3

Sustainability and interstellar infrastructure

Rachel Armstrong

This chapter explores how the thinking and exploration of ideas and ambitions shape the approaches adopted in space exploration and the construction of habitats.

3.1 Pale blue dot

The propagating Earthrise and Moon-landing images popularized in publications such as Stewart Brand's *Whole Earth Catalog* (n.d.) that converged the emerging Internet and space exploration transformed the privilege of the few astronauts into something that belongs to us all.

Images of our planet against the backdrop of a void not only remind us that we live in space—they are also inspiring and unifying. They provoke a sense of "oneness" that reminds us that we have always been in space, and always will be—whether we leave the planet or not. In a very real sense, all of Earth's life forms are astronauts and members of R. Buckminster Fuller's Spaceship Earth's crew. Such perspectives have precipitated a profound cultural transformation, which began the environmental era.

Consistently, astronauts describe deeply personal, life-changing experiences when witnessing the spectacle of our planet from the vacuum of space (White 1998). When Alan Shepard was asked about how he felt the first time he looked back at Earth while standing on the Moon during the Apollo 14 mission in February 1971, he said he cried. On looking back at Earth from the Moon in July 1969, Neil Armstrong observed how diminished he was in the vastness of space, where "I didn't feel like a giant. I felt very, very small."

Of course, the Moon-landing photographs were only the first of many stirring Earth portraits. Now, astronauts such as Scott Kelly, who is currently spending a year in the International Space Station (ISS), post streams of photographs through their Twitter feeds that constantly provide stunning new views of our incredible planet.

Rachel Armstrong, PhD. (✉)
Professor of Experimental Architecture, Newcastle University, Newcastle-upon-Tyne, UK

Farther away from home, interplanetary space has become a giant pier from which to look back at ourselves and continue to gaze back in awe from where we have come. Messenger beamed the tiny pixels of Earth and the Moon from its Mercurial orbit, while the Curiosity rover stared at Earth in the night sky over the russet landscapes of Mars that are now its home. The Juno spacecraft relayed amazing "starship-like views" of Earth and the Moon as it made a speedy fly-by past our planet on its way to Jupiter, as the Cassini probe played peekaboo underneath the structured rings of Saturn.

As we map our little blue dot into images relayed from the robotic probes that lead us on into interstellar space, our Earth-boundedness takes center stage. Yet, with Pluto in its rear-view mirror, NASA's New Horizon's probe will now journey on through a region of space called the Kuiper belt. This is made of icy bodies that formed the original building blocks of the Solar System—a terrain that we've never seen up close before. Moreover, Voyager 2, which was launched in 1977, sped past Neptune and Uranus before it began its journey into the heliosphere. It is now about 16 billion kilometers from Earth in the outer part of the solar bubble called the heliosheath and is still transmitting data revealing the very limits of our conception. In these realms, the image of our world is a colorless speck as depicted by Voyager 1 in 2012, when it was 11 billion miles from home. Indeed, robotic imaging techniques are critical in advancing our understanding of deep space. Against proliferating blackness, our planet seemingly teeters as a fragile oasis, its blue oceans persuasively beckoning space travelers back home, like a forget-me-not.

In describing the "pale blue dot" image of Earth in *Time* magazine (January 9, 1995), taken by the Voyager 1 spacecraft six billion kilometers away in 1990, Carl Sagan observed that "there is perhaps no better a demonstration of the folly of human conceits than this distant image of our tiny world." Through this inwards-looking gaze, Earth becomes an existential selfie. Yet, so far, we've always been looking "home."

In her essay, Esther Armstrong examines the paralysis induced by the powerful iconography of our planet. She wonders whether the scenography against which our existence is cast is actually holding us in a state in which radical breaks in our identity have become impossible (see Part II, Chapter 10). Undoubtedly, in the Ecocene, our perspectives will inevitably shift again as we will look outward with a new gaze and sense of belonging in the cosmos. We will begin to identify as the spaceship crew on not just one brilliant planet, but on many. And, we will cry again at the profundity at such a realization as we make the transition from one era to another, like Ed White, who, at the conclusion of the first US spacewalk during the Gemini 4 mission on June 3, 1965, sorrowfully declared, "I'm coming back in … and it's the saddest moment of my life."

3.2 What is an ark?

> "All flesh that moved on the earth perished, birds and cattle and beasts and every swarming thing that swarms upon the earth, and all mankind; of all that was on the dry land, all in whose nostrils was the breath of the spirit of life, died. Thus He blotted out every living thing that was upon the face of the land, from man to animals to creeping things and to birds of the sky, and they were blotted out from the earth; and only Noah was left, together with those that were with him in the ark." (Genesis 7:21–23)

One of the earliest descriptions of an artificial environment that could support life is the Biblical tale of Noah's "ark." This survival vessel is described in the Old Testament, where God proposed to destroy humankind by a massive flood. However, Noah had earned favor and was given forewarning of cataclysmic events. He therefore prepared for a new world by designing an ark to preserve life as he knew it. Anticipating mass extinction, Noah paired each known life form—as if they were human married couples—and shepherded them into the giant vessel two by two. Being favored, Noah was exempt from these rules and could bring his entire family. On the vessel, animals and humans lived together for 40 days and 40 nights, which in Biblical terms appears to indicate "a very long time." Fortunately, supplies were adequate. There is no evidence that the carnivores polished off their native prey, or that herbivores destroyed all the plants. Noah's expectation was that, when divine retribution had ceased, the world could be restored to just the way it was before the environmental catastrophe. A dove carrying an olive branch indicated that dry land had been reclaimed, since water levels were dropping lower than the trees and the resolution of the flood was close. Finally, the ark rested on the top of Mount Ararat and Noah, his family, and the animals could alight and make their way in the world anew.

The notion of an ark therefore has special significance. It suggests a time of waiting, which is endured by travelers in the hope of a better future. It also suggests that the life forms on the vessel are escaping a terrible event, for which provisions need to be carefully calculated, so that life can return to normal again.

It is hard to imagine what life might be like if we did things very differently, or lived in extremely unfamiliar places. Radical change is stressful, as it requires us to completely change our way of thinking and to deal with new situations. While an ark implies that its construction follows devastating ecocide, it is more likely that our starship will be constructed to accommodate the next wave of human expansion. Indeed, it hails from a time of plenty, in an age of bold exploration and in challenging frontiers. It is an ark of discovery, carrying a message of bounty and of "life," which dares to venture beyond the Solar System to places that are an extremely long way away.

While our ambitions to reach the stars may be ancient, our steps toward achieving this goal in a technological capacity are very recent and brief in comparison. Interstellar capabilities are most likely pragmatically achieved from a coordinated long-term vision by a prosperous Solar-System-wide civilization that has already established a significant interplanetary presence. In other words, rather than representing an act of desperation, the construction of an interstellar ark represents coordinated human expansion through developments. We are currently witnessing a new wave of space entrepreneurs that demand advances in a whole range of commercializable practices like mining, manufacturing, and terraforming techniques. Advances in these technologies may help us to thrive within interplanetary space and so eventually venture into other star systems.

Our ark, in which Persephone will provide a "living" environment, takes a slow route to its terminus. The journey itself is likely to last more than one human generation, affording a permanent home to its residents as the only world they know. Consequently, it requires a set of particular conditions for the survival of its colony—particularly with the precondition that the ark is "wet" and provides a plentiful source of water to sustain active life (Bond and Martin 1984).

Yet, to leave and thrive in unfamiliar terrains requires new customs and habits. Our expectations of a certain quality and way of living are deeply rooted in our psyche and ey are fundamental to our identity and are therefore hard to unlearn. With the ...nd of investment, interstellar colonists may not consider themselves travelers, but ...ready living at their permanent destination. In the same way as Earth's diurnal rhythms and seasons betray our constant motion around the Sun, the ark's residents may also be quite oblivious that they are traveling at all. Yet, unlike Earth, our ark is not moving in ellipses, but through a linear trajectory. Perhaps this unidirectional trajectory will shape the experiences and expectations of the colonists so that the humans reaching their interstellar destination may be quite unlike their ancestors.

Of course, the human imagination can take us to places to which we have never been, and even open up spaces that do not yet exist. Indeed, existence aboard a starship entertains the possibility of evolution—or even the potential extinction of its residents. In other words, this interstellar ark is an open system, which is capable of change and constitutes a complete, evolving world that changes with time, internal context, and, of course, its position in space. Arthur C. Clarke imagined Rama as a star ark, taking the form of a cylindrical alien starship, initially mistaken for an asteroid near Venus. The story depicts how a manned solar survey vehicle is sent to investigate and discovers that it has an interior world with a breathable atmosphere. The ship's interior is composed of enormous urban geometric structures, a cylindrical sea, which stretches around Rama's central circumference, and colossal cones, which are part of its propulsion system. Once a phenomenon such as Rama is conjured into our imaginations, we can potentially use our intellectual and physical dexterity to actualize the possibilities. By making drawings, building models, and even developing prototype experiences that allow us to reflect on new possibilities for existence, we can prepare for change.

Nonetheless, our ideas are still grounded in a reality reinforced by multiple contingencies and sets of expectations, so it is challenging to deal with the truly unknown or "black sky" venture. Indeed, we are much more comfortable reflecting on different versions of the present in which only a few variables are changed. This book helps us to navigate and potentially go beyond those attitudes, to not only imagine life aboard a starship, but also propose and shape new futures for ourselves within our Solar System, with ultimate (beneficial) impacts here on Earth, too.

3.3 Apollo's orphans

"We should ask, critically and with appeal to the numbers, whether the best site for a growing advancing industrial society is Earth, the Moon, Mars, some other planet, or somewhere else entirely. Surprisingly, the answer will be inescapable—the best site is 'somewhere else entirely'." (Gerard O'Neill, *The High Frontier*, 1977)

At the very start of the Space Age, it seemed entirely plausible to live beyond Earth's surface. The Moon landings not only raised our hopes regarding a new age of human potential, but there were also persuasive economic grounds for pursuing life in space. For example, the lunar surface could be mined for titanium, and the anticipated growth in commerce and employment opportunities for lunar workers was also expected to catalyze a tourist industry.

Princeton physicist Gerard O'Neill proposed a near-term scheme for settling space using existing technologies of the time. He created a scheme to house human settlers in suburban greenbelts that stretched around massive toroidal vessels situated at L5 in the Moon's orbit—a point at which, owing to a cancelling-out of gravitational forces between the Moon and Earth, objects remain still in space. Thousands of visitors could reach these stations using space elevators or skyhooks, or arrive in the Space Shuttle, which was proposed to ferry regular payloads between Earth and orbital space up to 50 times a year (O'Neill and Dyson 2001). Yet, such potential was never realized.

Space exploration has traditionally been carried out by government agencies such as NASA, the Russian Federal Space Agency, and the European Space Agency (ESA). Yet, space expenditure is a very small proportion of gross domestic product (GDP) in those countries with the biggest space programs. Based on 2004 figures, the US spends 0.14% of GDP, Russia 0.06%, Japan 0.05%, Europe 0.03%, India 0.03%, and China 0.02% (Ask an astronomer 1997–2015).

However, once the Moon landings had been achieved, the Cold War incentives that funded them dwindled. NASA budgets plummeted from 5% of the US federal budget to less than 0.5% (Gamble 2014). Compounding the financial challenges, the *Challenger* (1986) and *Columbia* (2003) disasters almost brought the conquest of our near frontier to its knees. Already feeling betrayed by President Nixon's decision to end the NASA Apollo Moon program and profoundly disappointed by the lean years of nominal space activity that followed, the children who grew up in the hope of fulfilling the promise of the Apollo missions became its heartbroken orphans.

However, born to an age of accelerating technological advances and being empowered by the exponentially escalating power of computers and their microminiaturization, Apollo's orphans took matters into their own hands. Brandishing a remarkable entrepreneurial spirit, telecommunications and space investor Walt Anderson and space advocate Rick Tumlinson requisitioned the Russian Mir space station to set up MirCorp, an outpost for the first phase of a trillion-dollar space business. The project was highly controversial, as it created a roadblock to the planned International Space Station (ISS). This ultimate start-up company proposed to establish asteroid mining, gravity-free laboratories, space hotels, and research facilities. It was unlike anything the universe had ever seen. However, it was short-lived. Russia caved in to pressure to decommission the space station from orbit. Yet, these Apollo's orphans had already successfully demonstrated their effectiveness in managing a manned space station and staked outrageous claims for what they regarded as their birthright.

So began an ambitious new commercial space revolution, in which the visions of Apollo's orphans turned the notion of space habitation by ordinary citizens from speculation into reality.

This was an incredible reversal of fortunes for the space industry, particularly since private individuals going into space was just an emerging idea at the turn of the third millennium. Less than two decades later, Anousheh Ansari, Denis Tito, Mark Shuttleworth, Richard Garriot, Guy Laliberte, and Charles Simonyi had all journeyed beyond Earth's atmosphere in a Soyuz spacecraft and stayed on the ISS.

While national space agencies continue to develop the infrastructures for colonizing space habitation at a snail's pace, market forces are increasingly disrupting established practices through the provision of services, space tourism, and harvesting resources. Commercial space firms are providing investors with an opportunity to shape a brand new space industry. Today's industry is currently worth a staggering US$300 billion (£193 billion). Most of that revenue fuels the development of satellites and launchers. The future of the near-term space industry is yet to be imagined. Our world has already been transformed by space, where large commercial organizations recognize that satellite technology is big business. Yet, beyond the secure markets, the rise in private individual investment is helping to fund some of the more unusual space ventures, such as space mining and tourism. In the past decade, space ventures have been commercialized in much more adventurous investment to the tune of around US$13 billion (£8.3 billion) by companies such as Space X and Virgin Galactic.

Elon Musk's Space X (Space Exploration Technologies Corporation) successfully completed the first private cargo mission to the ISS for NASA (Hennigan 2014). Its next step is to fly up to seven people in the cone-shaped capsule called Dragon to space per year, starting from 2015 to 2017. Further developments for a booming space-tourism industry were promised by Richard Branson's Virgin Galactic, which had signed up 700 people, including businessmen, Hollywood stars, and Branson's own family, for private forays into space (Travelmail Reporter 2014). However, on November 1, 2014, Branson's dream ended with the tragic crash of SpaceShipTwo over California's Mojave Desert. The craft broke up at an altitude of about 45,000 ft, shortly after separating from its mother ship, WhiteKnightTwo. Michael Tyner Alsbury, a project engineer and test pilot at Scaled Composites, a Northrop Grumman Corp subsidiary that built and designed the spacecraft for Virgin Galactic, was killed outright. Alsbury was flying for the ninth time aboard the craft, having been the co-pilot on the vehicle's first rocket-powered test flight on April 29, 2013. Another crewmember, Peter Siebold, was injured in the crash. However, Branson is not the only Apollo's orphan trailblazing the space-tourism industry. New destinations for spacefarers are also under construction and the news in private space exploration is not all bad; on November 24, 2015, Blue Origin landed the New Shepard space vehicle—and set a landmark in space exploration in successfully demonstrating the possibility of reusable rockets, which will bring down the cost of space travel.

Additionally, the world's first space elevator has also just been approved by the US Patent and Trademark Office, which granted Ontario-based Thoth Technology the rights to a pneumatically pressurized structure for location on a planetary surface. The elevator would be more than 20 times the height of the tallest building in the world, the Burj Khalifa in Dubai. The freestanding structure would enable spacecraft to be launched with 30% less fuel than rockets. It would also provide additional services such as renewable energy, communications, and tourism (Withnall 2015).

Beyond providing services to the ISS, the prospect of commercial vehicles is rendering near space a luxury market. Holidays to the Moon, lunar mining, and trips to Mars appear to be within the reach of ordinary people. Indeed, NASA envisages a Moon base within the next couple of decades, which will be co-funded by an increasing number of private investors (O'Kane 2015).

Bigelow launched the first prototypes of inflatable space hotels in the Genesis I mission in 2006 and Genesis II in 2007. These inflatable modules aim to greatly exceed the usable space of the ISS (Bigelow 2014). There are ongoing discussions about the US making a return to the Moon to exploit mineral rewards (Millward 2014). Indeed, Norman Foster is working with ESA, anticipating such a prospect by creating a process for developing lunar bases by spraying an infrastructure consisting of a modular tube and an inflatable dome with regolith. These dwellings propose to house four people and offer protection from meteorites, gamma radiation, and high-temperature fluctuations (Armstrong 2014).

Google and NASA are venturing even farther into interplanetary space by proposing to set up systems to mine the treasures of asteroids (Riseborough and Biesheuvel 2012). Planetary Resources, an asteroid mining company founded by Eric Anderson and Peter Diamandis that lays claim to making the first object 3D printed from asteroid dust (Medlock 2016), views the vastness of resources in the Solar System as leading to huge material returns on Earth. They believe that there is enough economically valuable material in the asteroid belt to "redefine wealth" on our planet in true accelerationist style; the idea that either the prevailing system of capitalism or certain technosocial processes that historically characterized it should be intensified to generate radical social change (Cruddas 2015). Deep Space Industries is another American, privately held company with global operation founded by Rick Tumlinson and others, with interests in space technology and resources that are specializing in asteroid mining operations. It currently has three spacecraft and patent-pending microgravity manufacturing technologies under development. The commercial ambitions of these private space prospectors also propose to support our species with access to extraterrestrial supplies as our own finite resources begin to dwindle.

Private corporations are also leading in the race for interplanetary space with bold new visions, such as the Mars One mission, which was founded in 2011 by Bas Lansdorp and Arno Wielders. Their goal is to achieve a successful unmanned mission in 2018, with a view to subsequently sending crews of four to establish a permanent human settlement on Mars, starting in 2024 (Mars One n.d.), while Buzz Aldrin comments that "Mars has been flown by, orbited, smacked into, radar examined, and rocketed onto, as well as bounced upon, rolled over, shoveled, drilled into, baked and even blasted. Still to come: Mars being stepped on." Elon Musk is certainly not to be dissuaded or outdone in the art of visionary space enterprise. As CEO and chief designer of Space X, he proposes to establish commercial missions to Mars in 2025 by providing communications infrastructure and cargo supplies for national space agencies making forays to Mars. Indeed, speaking in May 2013 about the possibility of humans eventually settling on Mars, Musk boldly observed that "It's a fixer-upper of a planet but we could make it work." Controversially, Musk's proposal to terraform Mars requires heating it up by pulsing nuclear fusion explosions over the poles to release frozen carbon dioxide. This would thicken the atmosphere and create a livable greenhouse effect. He has even publicly described his ultimate goal as being buried on the red planet (Harlow 2013). Although there are no guarantees on the returns, the potential gains appear to be significant enough for angel Apollo's orphan investors to take the risk of establishing their portfolio of interests in the evolving space markets, which will soon be worth 10 % of the world's economy (Cruddas 2015).

If the spirit of enterprise alone is not enough to catalyze renewed vigor for space colonization, Cold War tensions between Russia and the US over Ukraine, which generated the political drivers for the original momentum of the Moon landings, appear to have built up again. These renewed hostilities may further contribute to an increased interest in securing and occupying orbital territories, which only increases the possibility of new enterprise.

Other global competitive tensions also may further accelerate both private and national interests in developing inroads into the blossoming space industry. With the unprecedented rise of economic power in the East over the last 30 years, patterns of world trade have been transformed and lifted millions out of absolute poverty. With the prospect of a shift to the East in world power relations, momentum for the advancement of human presence in space is being garnered. China is now the third country to independently send humans into space, with plans to develop a permanent Chinese space station in 2020. To date, Japan and India have not developed a manned spaceflight program, yet they demonstrate sufficient technological capacity, underpinned by the necessary economic power and population density, to catalyze developments in space colonization.

While they may have been heartbroken by the failure of the Apollo program to deliver on its vision of space exploration and settlement, Apollo's orphans have generated a whole from which a new range of discoveries may begin, and are setting a comenda for the exploration and settlement of space.

otives also exist for pushing back the frontiers of space and include:

- *to "do good"* — these values are tightly coupled with the heroic human spirit and notions of altruism that may ultimately be perceived as primarily benefiting humankind, such as supplementing planetary resources and the justifications of self-sacrifice, as an integral aspect of the pioneering spirit to become the first settlers of a hostile planet, expressed by Mars One candidates; however, Michael Mautner (see Part II, Chapter 11) takes an expansive view of a greater purpose through the foray into the cosmos — that we are part of a much broader community of life whose cosmic dissemination is in itself a worthy project;
- *compulsive risk-taking* — pioneers, daredevils, and explorers such as Felix Baumgartner, who jumped from the edge of space on October 9, 2012 to become the first human to break the sound barrier outside a vehicle, are irresistibly compelled to push themselves beyond their limits; such adventures are unlike heroism in that there is no declared benefit to humankind at stake, simply the urge to take risks for their own sake;
- *rational arguments* that address perceived needs (such as scientific knowledge) and develop new technologies with terrestrial impacts (often framed by economic benefits); these ideas are tightly coupled with notions of the Anthropocene, in which humankind has conquered, colonized, and exploited new worlds to increase its power and wealth;
- *survivalist strategies* for the preservation of humanity that engage with notions of population- and ecological-scale collapse;
- *belief systems* — while no religious "sect" has publicly proposed space missions, there are a number of ideologies such as the Raelians that believe that their ancestors are of extraterrestrial origins; indeed, in the science-fiction movie *Prometheus*, the motivation of the interstellar travelers is initially a faith-based system.

The extreme living conditions and time spans that colonizing space demands require a radical rethink of our approach, which has largely been shaped by ideas originating from the 1970s. So we need to be prepared, not with twentieth-century approaches, but by developing the capacities of space exploration with the advances that characterize this new millennium and propose new visions for the future of our species. A handful of organizations are already researching the possibilities as a kind of staging post toward this transition. For example, Icarus Interstellar proposes to construct a crewed interstellar ship in Earth's orbit within a century, while the Initiative for Interstellar Studies ambitiously proposes to develop a range of robotic interstellar probes for missions within the next 30 years. Owing to the vast distances involved that require phase shifts in the kinds of technologies used for space exploration, we will need to start immediately in making preparations for this prospect.

Yet, preparing for the next phase of human development is not the only reason to construct such a project. We urgently need to challenge the global developmental conventions that are holding us in an environmental gridlock here on Earth, where, effectively, our industrial practices are reverse-terraforming our planet. By considering our priorities in sustainably inhabiting the most extreme environments, we may completely rethink the foundations for human development in ways that promote — not destroy — the ecosystems on a life-bearing planet.

Apollo's orphans 53

If we are to believe that it is possible to live beyond our planetary confines, then we must consider the fabric of space not as a sterile site or industrial resource fit only for plundering, but as a living, potentially fertile system that can be better adapted to human existence. We will need to understand our habitats as dynamic ecosystems so that we can construct the kinds of infrastructures we need to colonize the hostile terrains beyond Earth's hospitable climes. Seen from orbit, this sustenance is afforded by a physical membrane—an extremely thin film of air, water, and light—which appears as a mere horizon of possibility against the void of space. Astronauts can see these membranes as a thin blue layer surrounding the planet, whose presence, retained by Earth's gravity, stretches up into the atmosphere to about 300 miles (480 km) thick. However, most of the functional atmosphere is within 10 miles (16 km) of our planet's surface (Fig. 3.1).

"ASTRONAUT TWITTER QUOTE
Scott Kelly @StationCDRKelly Jun 18
Day 82. The sun sets on another day aboard @Space_Station. Goodnight! #YearInSpace."

Figure 3.1 Star Ark membranes—online photograph taken during Scott Kelly's year in space from the ISS in 2015. Credit: Scott Kelly

How we convey this umbilical cord along with us as our mother Earth's blood—or whether we sever this tether and transcend it altogether in a new existence by breathing our own air—depends on whether we can establish an ecological infrastructure in which we propose to dwell.

This book therefore develops the principles for the construction of a living habitat within a worldship—a generation starship that contains its own planetary-scale, sealed environment that supports human colonists as it travels across great distances between stars at a speed much slower than that of light. Such a comparative experiment and dialog between worlds—our home planet and alien terrains—create a context in which we must think, explore, imagine, design, engineer, and act differently. The goal is to generate new platforms for human development and even make the first steps toward realizing our potential as a fully interstellar species.

3.4 Interplanetary communications

"The future is the interstellar Internet: fault-tolerant communication for exploration of the galaxy …. We will need the interplanetary backbone to build a synthetic aperture receiving system to receive signals from such galaxy-roving spacecraft." (Vint Cerf, vice president and chief Internet evangelist of Google, Google+2012)

Chris Hadfield became an overnight online sensation through his video diaries from the International Space Station (ISS). These collapsed the former distinction between the terrestrial social sphere and orbital living, bringing the experience of being on a spacecraft right into our living rooms. Through a compelling demonstration of ordinary events, such as brushing his teeth, wringing out a wet towel, or crying, Hadfield explained the science behind why things were different in reduced gravity, and inspired a new generation of space pioneers and creatives, for which extraterrestrial locations have become part of the society of the spectacle. While this method of communications currently extends to the ISS, a communications protocol that establishes an interplanetary information highway does not exist. Yet, it is essential in establishing interplanetary settlement, economy, and culture.

However, work is in progress. National space agencies, such as the European Space Agency (ESA) and NASA's Jet Propulsion Laboratory (JPL), are already laying the foundations for an interplanetary communications network (IPCN), which is made up of a network of regional internets that share fundamental characteristics. These regions are on a planetary scale rather than within, say, a country or continent. These are likely to catalyze widespread changes in the exploration and use of interplanetary space in a similar manner to the way in which the World Wide Web transformed terrestrial society. Fostering diverse exchanges between Earth and new territories will provide enterprise infrastructure for bold Apollo's orphan pioneers. While still in their planning stage, these communications systems are likely to create opportunities for robotic services, entertainment, and research opportunities that are of increasing interest to a space-hungry, image-consuming public. For example, scientists at Carnegie Mellon University have developed a lunar rover that could be controlled remotely by anyone with an Oculus Rift virtual-reality headset (Dawson 2014).

It is likely that, within the near future, the spectacle of space will be vastly extended through an IPCN at much faster rates than are currently possible. NASA's JPL and Space X are already proposing to transform terrestrial networks through tests being carried out with laser communications technology, which has the potential to speed up transmissions between the ISS and Mission Control by a factor of 200 (Ungerleider 2014).

Indeed, Musk is proposing to build a second Internet in space to connect people on Mars to the terrestrial Internet, which I will call "MuskNet." It will be, in the next 5 years, a vast network of communication satellites that will exceed anything that has been proposed to date. This information architecture will speed up the general flow of data on the Internet, delivering high-speed, low-cost Internet services to more than three billion people who still have poor online access. Hundreds of satellites in much closer geosynchronous orbital locations will be used at altitudes of up to 22,000 miles, which would cut down lag by reducing the distance for electromagnetic signals to travel. Potentially, MuskNet could rival fiber-optic cables on land and bring online access to remote and poor regions, extending all the way to Mars, where Musk plans to set up a colony and found a city in the coming decades.

While Musk's proposal to settle non-terrestrial environments may seem an outlandish enterprise, even for one of the visionary Apollo's orphans, establishing an IPCN is likely to increase our familiarity with space habitation through the increasing forays of robots into interplanetary space and our prospective ability to interact with them.

Indeed, the remote extension of humans into interplanetary space and our robot interactions through vastly improved communications networks is likely to transform developments in unlikely ways and greatly increase the potential traffic into interplanetary space. These highways are a major precondition for our development toward creating the necessary infrastructure to build vessels designed for deep-space exploration and taking the first steps to becoming an interstellar species.

3.5 Apollo's progeny

"**Philae Lander** @Philae2014 Jun 14
Hello @ESA_Rosetta! I'm awake! How long have I been asleep? #Lifeonacomet"

While our ambitions for interstellar programs may ultimately be concerned with human prosperity, the pioneers of this terrain are undoubtedly non-human. Since October 4, 1957, when the Soviet Union launched the first artificial satellite, Sputnik 1, we have been living in a solar system that is colonized by robots—a community that includes artificial satellites, landers, and rovers.

These agencies have become increasingly autonomous so that, now, the Mars exploration rovers are highly autonomous and can work independently for extended periods of time. Robots are not just workhorses for human ambitions. They are extending our senses into the Solar System through their embedded communications systems, gathering data so that we can make new stories about our interplanetary landscapes. Trailblazing generations of rovers have already helped us to feel much closer to the lunar and Martian surfaces. Adventurous robotic probes such as Cassini–Huygens have even taken us into inhospitable planetary systems, like Saturn. Not only do these robots provide data and mediatized experiences of interplanetary environments, but they are also constructing spaces for organic life forms and human habitation.

Take, for example, the Foster + Partners whose project with the European Space Agency (ESA) proposes to build habitats on the Moon using inflatables and 3D printing of regolith that is sprayed over the soft scaffoldings (Foster+Partners 2013). Robots are likely to work together to produce environments that may well nurture houseplants before humans arrive or produce swarms of greenhouses at Lagrange points. NASA is teaming up with Google to send self-sustaining habitats for vegetation that will provide air, food, and water. They will also act as "canaries"—living systems that can be read through the damage they sustain on the lunar surface to provide information about the viability of human habitation before colonists arrive. The plant habitat will be delivered to the Moon via the Moon Express, a lunar lander that is part of the Google Lunar X-prize, a competition to create a robotic spacecraft that can fly to and land on the Moon (Plafke 2013). Other possibilities for robots generating the infrastructures for life in non-terrestrial environments include pioneer AstroGardening robots that are likely to generate the first steps toward habitable sealed biomes (Preston 2013). By gradually potentiating the survival of bacteria and other microorganisms, which may even form their control system, or "brain" (Heyde and Ruder 2015), increasingly diverse communities of life may be established. With the first waves of humans settling non-terrestrial environments for prolonged periods, we are also likely to see robot assistants such as NASA's Robonaut series of dexterous humanoid robots that work alongside people to reduce risks associated with construction and discovery (NASA n.d.). Robots will therefore help us to prototype the first terraforming systems for the Moon, asteroids, and Mars.

Despite our increasing familiarity with interplanetary landscapes as seen through robot vision, these realms and those beyond them are not a continuum of terrestrial plains, and the modes for establishing our presence in alien environments are still being explored. Indeed, robots have changed the very nature of space discovery, since they are much better suited to the extraterrestrial environment than human bodies—they last longer, pose fewer ethical dilemmas when faced with the prospect of one-way missions, and even possess a mysterious ability to come back from the dead.

Philae Lander @Philae2014 Jun 14
Hello Earth! Can you hear me? #WakeUpPhilae"

Philae, a 100-kg robot probe—packed with nearly a dozen instruments for a groundbreaking mission to learn more about the origins of life on Earth—had been dropped onto the surface of 67P/Churyumov-Gerasimenko comet in November 2014, whose material archives are thought to hold secrets written in dust and ice about the formation of our Solar System around 4.6 billion years ago. It reached its target after a 10-year trek by its mother ship, *Rosetta*. However, it landed badly and bounced on the comet's surface, ending up in a dark recess. Its solar battery ran flat after only 60 h, providing 2 min' worth of signals and 40 s of data. However, the comet's position in relation to the Sun changed and so the solar panels were charged again after 7 months of silence. Happily, Philae reestablished communication with *Rosetta* and managed to discover several organic molecules, including four that had never before detected on a comet. However, by July 9, 2015, all went quiet again for Philae's Mission Control engineers at the German Aerospace Centre in Darmstadt. They sent a command to the fridge-sized robot that aimed to boot up its flywheel and shake dust from its solar panels so that it could better align with the Sun but, so far, there has been no response. As the comet and its companion robot speed some

300 million kilometers from the Sun by January 2016, temperatures will fall below −51 °C and Philae will no longer be operational.

Yet, no robot has endeared the public imagination more than Yutu, named after a jade rabbit that always stands on his hind feet in the shade of a cinnamon tree, whose startling resurrection transfigured the rover into the Chinese mythological character it was named after. Yutu lives on the Moon, making magic herbal potions for the immortals, and belongs to Chang-O, the Moon goddess who was banished to the barren rock 4000 years ago, when she stole an immortality pill from her husband.

Yutu—the rover—is a 140-kg, six-wheeled exploration vehicle. It is equipped with cutting-edge radars that allow it to study the Moon's crust and multiple cameras that allow it to photograph its surroundings and beam them back to Earth. It landed on the lunar surface on December 14, 2013, on the Mare Imbrium. This was an incredible moment for the Chinese authorities, as it was the first soft touchdown on the Moon since the Soviet Union's Luna 24 mission in 1976. The rover traveled in a zigzag fashion for about 115 m, sending back its first photographs but, sadly, ended its explorations in January 2014, following a technical glitch that left it paralyzed.

> "**Jade Rabbit** @JadeRabbitRover 30 Jan 2014
> If only … I could … move."

Six weeks into its mission, on February 12, 2014, Yutu was declared officially "dead" and was "mourned" by Chinese social media users. It entered into hibernation mode for about 50 days during the long period of lunar night, when there was no sunlight to charge its solar batteries. Fifty days later, Yutu the mythological rabbit came back to "life." Although is still functioning, it cannot move and is no longer sending any new data, but it has become a living legend. It is no longer a mere technical accomplishment, but embodies the lunar poetry of immortals.

> "**Jade Rabbit** @JadeRabbitRover 18 Apr 2014
> Don't forget me, world, don't forget me. #JadeRabbitRover"

Robots are part of our story of becoming an interplanetary culture. Indeed, terrestrial beings are no longer the only inhabitants of our Solar System. Developments in the field of robotics are extending their capabilities to the extent that they are beginning to demonstrate a certain degree of self-awareness (Millward 2015). Kevin Warwick discusses some of the machine–human couplings that can extend our capabilities in the process of becoming an interplanetary culture in Part II, Chap. 13. Increasingly, interplanetary pioneers and colonies are increasingly likely to be made up of groups of non-human and remotely human-operated beings that demonstrate varying degrees of autonomy from human agencies. Rovers have already settled and ended their existences on the Red Planet way before Elon Musk may follow suit. Satellites wander as mechanical George Mallorys into the harsh environments of interplanetary space—"because it's there"—and because they can.

Mars itself is already "inhabited" by a colony of terrestrial robots. The activities of the US-launched surface dwellers Opportunity, which landed in 2004, and Curiosity, which landed in 2012, are relayed back to Earth by five functional orbital satellites, namely Mars Odyssey, launched on April 7, 2001 by the US; Mars Express, launched on June 2, 2003 by ESA; Mars Reconnaissance Orbiter, launched on August 12, 2005 by the US; Mars Orbiter Mission (MOM) or *Mangalyaan*, launched on November 5, 2013 by India; and Mars Atmosphere and Volatile Evolution (MAVEN), launched on November 18, 2013 by the US. Further interplanetary colonization will be made possible through the proposed interplanetary network that will enable an information relay between robot explorers that are foraging for data, images, and natural resources. So, while entrepreneurs such as Elon Musk are playing a personal role in pushing back the boundaries of space exploration with developments such as MuskNet, in the longer term, robotic missions are likely to be the mainstay of national space agency endeavors in interplanetary space. Perhaps, in the near term, human exploration may even be regarded as little more than an extreme sport for the super-adventurous and super-wealthy.

This first generation of space missions are fly-bys that speed past planets like paparazzi, taking as many photographs as possible before racing on to their next target. Some of them, like Dawn, will orbit a celestial body while others, like the Voyager probes and New Horizons, will disappear into interstellar space. Those probes that go into orbit need to brake at high speed, which places huge demands on fuel and leaves less carrying capacity for instruments. However, such maneuvers have already been successful, as for the Galileo probe, which reached Jupiter in 1995, and Cassini, which went into orbit around Saturn in 2004. In the next decades, we are likely to see even more complex trajectories that enable robotic probes to pull up alongside the subject, as in the case of the Rosetta comet-chasing mission. Achieving that sort of trajectory requires complex choreography, where spacecraft have to skim close to various other planets to steal energy and adjust their orbits through slingshot maneuvers. Owing to this complexity, Martin Lo from the Jet Propulsion Laboratory designed an interplanetary superhighway (IPS) that proposes to enable even deeper and speedier forays into space. Pathways are required that enable the probe to be at the right point in space at the right time, which sometimes means that they have to travel more slowly to maximize their energy efficiency and imply longer journey times. This is a trade-off. The IPS maps out a route where forces between planets and moons effectively cancel each other out and produces a topology of possible flight paths that are interwoven into rope-like tubes through which spacecraft can travel and are being mapped out for the whole Solar System. With this in place, robotic spacecraft could easily travel these pathways through the gravity fields while burning very little fuel. An IPS would also mean that robots could operate at distant locations and function for longer periods in space. Moreover, the IPS establishes a set of building blocks for an extended human presence, which may ease the passage of human colonists into the interplanetary environment.

With a formal transport highway such as the IPS in place, there will be many more robots yet to come. With the deepening of our knowledge of the universe through the information that Voyager 1 and other space probes have relayed back to us, and with more sophisticated hardware on its way, robots will do so much more and help us to learn so much more about our Solar System. It is clear that interplanetary space and the realms beyond are strange terrains that are beyond the settlement experiences of modern terrestrial culture. Indeed, if we are to make a transition toward becoming an interstellar species, we are likely to see our interplanetary environments "inhabited" by increasing numbers of non-human bodies that are becoming part of its ecology and establishing vital infrastructures for us to follow in their wake.

3.6 Organic backbones

Sustaining ourselves completely in the absence of Earth's vitalizing environment is an incredibly difficult transition to make. At the current time, we certainly have not been able to completely cut ties with Earth's nurturing membranes—air, water, and soil. While explorations to date have taken us to the surface of the Moon—a mere 400,000 km away—this distance is still within the umbilical reach of our home planet. As we become an interplanetary and, ultimately, interstellar species, we cannot assume that the conditions of life in which we currently flourish will continue when we cut the placental ties to our home planet.

If we are to live in extreme environments, we will need to produce the conditions and contexts for our survival with which we have very limited experience, owing to the extremely robust conditions on our home planet that enable us to make many assumptions about the way we design and engineer with living systems. We can, for example, assume that the Sun will shine, the rain will fall, the wind will blow, and the soil will regenerate our organic life forms. We cannot assume these conditions will equally apply in non-terrestrial environments.

Yet, if we are to generate the platforms that will enable us to establish ourselves as an interplanetary society, then we will need to understand the operational principles of the ecosystems we encounter, so they may be applied to build appropriate infrastructures, vessels, and environments that will ultimately enable us to venture beyond the Solar System.

The "ecology" implied within interplanetary or, indeed, cosmic space does not refer to the "greening" of things. It invokes the physical conditions that imply the infrastructures of life, which give rise to active fields that precipitate the existence of objects and shapes their relations. These physical conditions may be shaped through the principles of the Ecocene, through the overlapping of multiple technical, cultural, and material practices to provoke complex, material outcomes.

Indeed, the notion of a space ecology takes a much more primordial approach to observing and establishing the materials and infrastructures for life than we may expect to consider in a terrestrial context, since they are more spartanly distributed and extend to events that precede life itself—a process called ecopoiesis.

Notions of ecology have evolved according to our expectations and perceptions about the relations between organisms and their environment. When Ernst Haeckel coined the term in 1886, he was primarily concerned with the relationship between animals. In the mid-twentieth century, Eugene Odum championed Vladimir Vernadsky's notion of biogeosphere as an agent that shaped environmental events. In the late twenty century, a whole range of actors including political, social, and urban development became equated with the term, which reflects developments in systems thinking across a wide number of disciplines (Armstrong 2015).

The principles through which the material systems that underpin the infrastructures of life can be applied using the ideas implicit in the Ecocene to shape our experiences of space ecology, which adopts a fluid, contingent, bottom-up approach. This is quite different from an industrial perspective, which establishes a solution to a perceived problem and implements it through a series of predetermined steps. The Ecocene does not assume that our story in interplanetary space unfolds from the late modern era, or even from the start of the Industrial Revolution. It also does not assume universality in ecosystems, but embraces the contingencies within space. It works with opportunities and the potential for the unknowability of habitats and environments. The Ecocene seeks to produce maps of ideas and approaches that help us read an existing landscape through the process of discovery without predetermining its potential. Within the interplanetary landscape, it seeks to reveal the capacity within multiple, varied sites and work in a flexible, contextualized, robust manner with them, to establish the conditions for synthesis. This approach is not tool-, technology-, or paradigm-specific. An ecological approach to interplanetary settlement invites convergences and a chance to co-produce a potential range of outcomes, like life itself.

Yet, the ecological approaches of the Ecocene do not exclude industrial methods, but include them in an expanded portfolio of technical platforms that take a number of different routes to solve a particular challenge such as using microorganisms, robots, and humans to establish an oasis in a lunar biome. They seek exchange between participating agencies such as planetary bodies, so that we produce extended plateaus of potential that increase the possibility of life. Ecological approaches also look to transform the performance and character of existing environments so that they become more habitable and also embrace the unruly recalcitrance of living systems.

By taking a flexible, robust, networked, and complex approach to the synthesis of infrastructures through understanding the generative fields that shape them, the notion of a space ecology increases the probability of life, but cannot guarantee it. For example, Francis Field and colleagues propose to produce an ocean by the collision of two asteroids at the translunar Lagrange point L2 to serve as a site for colonization and as a resource to fuel future missions (Field et al. 2014). Whether this project works in practice is yet to be seen. Yet, such robust and propositional strategies are necessary when we are still very much in an exploratory phase of interplanetary development, where we need to preserve a great deal of malleability and creativity in our journey toward ongoing survival in hostile environments.

This is not a small task. It has already taken 500 years to make a transition from the Middle Ages to the industrial age to perfect our modern machines. It is possible that the Ecocene will also take several hundred years to reach maturity. Yet, the opportunities for new ways of thinking and working are being constantly revealed by new generations of robotic probes that are streaming images and data back to our observation laboratories. Increasingly, we are becoming aware of the highly diverse, lively environments and terrains within our Solar System, in which we have barely begun to take our first footsteps. As intriguing puzzles about the composition of our interplanetary landscape arise, such as the bright spots on Ceres and the lively character of Pluto, we need to establish approaches toward space exploration that enable us to remain open and responsive to potentially surprising developments.

An ecological view of space exploration is, therefore, one in which the environment of travel is appreciated and engaged through the discovery process. It is not sealed off, but is made semi-permeable to internal ecosystems, so they can keep on evolving. It also invites the potential for meaningful exchanges between new consortia of life and their surroundings. An ecological view of space exploration aims to find ways of connecting the many terrains that generate our interplanetary landscape. It considers what a metabolism at the scale of the Solar System may be, whereby opportunities and challenges within one landscape may connect with needs in another.

We may also consider how to work alongside existing dynamic processes such as asteroid trajectories that may be used as a strategic transport system for the movement of matter throughout the Solar System. Questions about ecopoiesis and terraforming may be explored, whereby material resources, such as frozen water, may be unlocked from planetary sources and used to support or be transformed by life. These are not simply industrial questions that engage with deterministic outcomes and simple causalities. They require more complex forms of engagement that require tuning and continual adjustment, which is more typical of control strategies used in complex, dynamic systems like agriculture, gardening, and cooking, which are characterized by their contingent character.

Our current forays into hostile environments require the construction of protective barriers such as space stations, bathyscaphes, and radiation suits. Current notions about colonizing the Moon and Mars engage with notions of setting up biomes or sealed ecosystems that are, in principle, self-sustaining, such as NASA's Water Walls project, whereby the growth of algae cleans water and provides biomass in closed-loop ecology. However, in practice, ongoing ecosystems are always open. An ecological perspective on space settlement will necessarily require new ideas about how environments are identified, constructed, maintained, kept open to external events, and are creatively evolved. Our experience in managing ecosystems is based on our relationship with the terrestrial environment, which establishes very particular expectations of what an ecosystem is composed of and how it should perform. An ecological view of an ecosystem is much less proscriptive, much looser, and somewhat opportunistic regarding the agreements established between participating agents and bodies. It is always open to change or transformation. Yet, none of this can be sustained for any prolonged period from behind a solid barrier.

In order to persist in spaces beyond those we already know, it is imperative to consider that life and environment are inescapably entangled, so that we can develop a robust problem-solving toolset for interplanetary settlement. Key to off-planet sustainability is the opportunism inherent to in-situ resource utilization (ISRU), which describes the proposed use of resources found or manufactured on other astronomical objects, such as the Moon, Mars, or asteroids, to further the goals of a space mission. However, these approaches also involve 3D printing and robotics, which are still at a relatively low technology readiness level, so there is much uncertainty regarding their specific requirements like energy, materials, or waste management, which are compounded by a lack of operational data.

In this way, the task in developing an Ecocene approach to working with the fabrics of space ecology is not to construct sealed-off environments for human habitation, but to establish meaningful connections between living consortia and their surroundings like the ESA's MELiSSA project—an artificial ecosystem that recycles organic waste back into useable resources. It is conceived of as a microorganisms- and higher-plants-based ecosystem to gain understanding of artificial ecosystems, and produce technology for a future regenerative life-support system for long-term manned space missions such as a lunar base or a mission to Mars. An ecological view of space discovery is semi-permeable to habitable spaces, so that environments remain as open as possible and the ecosystems that support life can be nurtured and keep on evolving. This does not mean, of course, that colonists will be deliberately subjected to lethal doses of radiation or toxic substrates. Rather, it means that the buildings, technologies, communications systems, transport networks, and organisms that necessarily accompany us will actively contribute to our survival. They will become part of a consortium of interdependent agencies that increasingly filter and process our needs, while enabling us to exist within an effectively "open" realm. They will, in effect, become "living" and perform as an integral part of our space ecology. This already occurs naturally at a planetary scale, where Earth's magnetic fields and atmosphere shield us from the damaging effects of ionizing space radiation while allowing solar radiation and matter such as meteorites to reach us.

An ecological perspective of space exploration also accepts that colonized non-terrestrial environments will be transformed by the presence of new consortia, perhaps in an irreversible manner. This is a controversial viewpoint in the current climate of industrial-age thinking, where the presence of humans is inextricably linked with environmental destruction and reductions in biodiversity, and may seem like an anathema that we should prevent at all costs.

However, transformation and change are aspects of evolution and do not have to be a Machiavellian enterprise. Rather, an ecological era of space exploration requires an ethics and methods for constructing bases and habitats in responsible, considered, and integrated ways. It prepares to actively manage our communities and prepare for surprises, so that methods can be adapted accordingly. While it is impossible to guarantee outcomes, or ensure that there will be no negative impacts resulting from space settlements, ongoing management and spreading risk between the various contributing agencies may enable us to make positive transformations to the interplanetary landscape, from which the benefactors are not just human, but the greater community of life.

3.7 Directed panspermia

The migration of humans from Earth, to settle other worlds and environments, is a messy affair. Yet, the interplanetary and interstellar expansion of the human race will be most successful if achieved biologically, through the extension of ecosystems into non-terrestrial environments planet by planet, out into the Kuiper belt and Oort Cloud. The success of such an enterprise may depend on the mutual relationships between artificial and natural systems, exemplified in Freeman Dyson's Astro Chicken thought experiment—a 1-kg self-replicating spacecraft that is a combination of organic and electronic components (Dyson 1979). Organically grown habitats and resilient species, perhaps like Dyson trees, which are genetically engineered plants capable of growing in a comet, also epitomize the ambition for an ecology-centered vision of space exploration that may expand the presence of humans, star system by star system. Yet, if we are to settle non-terrestrial environments, then we must figure out how to live in anarchic and self-propagating contexts, like life itself. This is not a clinically clean, sanitized vision that sterilizes its spacecraft of opportunistic microbial passengers. It is dirtier, contaminated, and full of surprises. However, a balance needs to be struck between the degree of transmission of materials and organisms between interplanetary environments, so that a precautionary approach toward preserving the character of our interstellar environment may be taken to minimize the risk to and from possible indigenous alien life forms. Yet, since it will be impossible to eliminate all risk, acceptable limits will need to be established. We already work this way; for example, water providers must adhere to providing a service in which the number of bacteria is below a threshold of safety in a drinking supply.

When Giuseppe Cocconi and Philip Morrison wrote one of the first formal rational arguments supporting the search for extraterrestrial intelligence, "Searching for Interstellar Communications," published in September 1959, they observed that "The probability of success is difficult to estimate; but if we never search the chance of success is zero."

Similarly, the chances of establishing a non-terrestrial habitat, without exploration and experiment regarding the ecosystems and construction methods to shape them into environments that support life, will also be extremely challenging without actually working with the kinds of living agents that will help us to establish the necessary infrastructures for space colonization.

When we actually observe the challenge of inhabiting environments that are not terrestrial, it implies having to make an extraordinary and evolutionary transition from gaining sustenance from the wet, oxygen-rich environments of our mother planet and learning how to breathe in the Spartan vacuum of space on our own. Therefore, we have to design a new mission—not based on resource consumption, because that does imply designing and engineering the fundamental infrastructures that may enable us to live freely from Earth. The project is more than simply meeting our current set of needs based on modern synthesis calculations, but demands how we can transform barren terrains (such as spaceship interiors and non-terrestrial terrains) into more lively places that not only support, but also potentiate immensely complex ecosystems. This is not something we can currently do. When we design with ecological networks, we garden, rather than produce them. Moreover, achieving these aims is extremely risky, raising issues about the control of these kinds of environments and invoking many ethical challenges as well as physiological ones. Indeed, the science that may allow us to do this is controversial, such as the issue of directed panspermia, and does not formally exist yet. However, these approaches are also not fiction. Bacterial biomes are already being considered as strategies that can help us to mine valuable metals from asteroids (Wall 2015). Indeed, such technical toolsets could profoundly and irreversibly impact other environments during our expansion away from Earth.

The processes that require engagement to establish the interplanetary infrastructure of life are ecopoiesis—the transformation of barren landscapes into dynamic ones and terraforming to produce increasingly Earth-like environments. Currently, these processes are imagined within constrained environments but, for longer durations, these transformations will benefit from being as extensive as possible and, ideally, planetary-wide. Persephone is invested in ecopoiesis, where increasing liveliness within an environment is, on the one hand, an intrinsic property of the natural terrestrial realm, which constantly produces new species, and, on the other hand, is a potential side effect of engineering highly lively environments in non-terrestrial spaces.

Yet, the deliberate transformation of environments has negative connotations on Earth. There is polarized regard for industrial processes that, on the one hand, are sites of great wealth and, on the other hand, are nuclei of environmental destruction. Indeed, during the industrial age, we destroyed environments by ripping out natural resources from soils, forests, and deep ground reserves, and carelessly disposing of our junk and waste products. These cumulative acts of willful neglect have come back to haunt us through dramatic losses of biodiversity and massive plastic deposits in our oceans. While many of these side effects have been the results of neglect and bad management, projects that propose to actively transform environments with remedial ambitions such as geo-engineering are generally held in extremely poor regard. Yet, these are the very processes that will enable us to survive in interplanetary environments.

However, we are not compelled to repeat our mistakes, or re-enact human history in a forward direction like some eternal "groundhog day." This does not mean that we should be casual with our environmental responsibilities, but simply admit that, wherever we go, we will transform sites and have an impact on them. Charles Darwin observed that change is characteristic of life, which is in mutual exchange with its surroundings. The issue is, therefore, not that we produce change and transformation in an environment; it is how we develop the quality of the experiences that we produce. Transforming interplanetary sites so that they may become livelier, creative, and responsive to events may be done sensitively and intelligently, but we need to change our approaches. Yet, to do this, we need to work differently, using the principles of the Ecocene, and taking our first steps into interplanetary space may be an ideal opportunity to see what we are capable of and bring back the benefits of our learning to our terrestrial environments. In dealing with complex environments and living systems, we are always learning and may obtain our best results by fashioning crude instruments, asking questions through them that help construct something that constitutes a positive legacy. However, it is impossible to predict the impacts that calling forth the existence and support for complex, dynamic matter may produce. No matter how well such systems may be designed, there will be no perfect control or outcome. The best that we can hope for are effective approaches that may not be ideal. Yet, to survive within extreme environments in suboptimum conditions may be sufficiently effective to take us through to the next stage of iterative development to refine processes with new insights from the previous set of propositions. In other words, we are choreographing not just the construction of spaces, but the legacy that enables us to colonize unfamiliar territories. This is a story that recounts not a series of perfectly positioned objects and final solutions, but sporadic landmarks whose successes and failures document the cosmic story of life.

It is not possible to overemphasize how important developing an interstellar strategy toward the status and propagation of "life" is. By sending human colonists to interstellar destinations, we are inevitably spreading organisms throughout the cosmos—not just human life, but other organisms that are part of our biome (microbes), our community (pets, domestic animals), and our biosphere (wildlife), too. Humans do not and cannot exist as isolated organisms. We are entangled in complex ecosystems of interdependent species that we take along with us, such as bacteria (NCBI 2009). An existential issue is at stake in directed panspermia that relates to human identity. Over the last 10 years, the discovery of the human biome indicates that we are not pristine bodies, but are outnumbered by bacteria that actually form part of our living system by digesting our food, producing oils that increase our moods, and forming part of our immune system. When these organisms are not working, say after a dose of antibiotics, we feel quite unwell. This observation complicates the idea of sending humans anywhere in space, as we are necessarily also sending consortia of functioning organisms, most of which are not human. Using the idea of consciousness as a qualitative identifier that separates out the human from the non-human and renders us unique is a logical fallacy. Cell for cell, bacteria are more numerous and physically more robust than we are. We are much more likely to perish during an interstellar mission than our bacterial biomes, and we already know that the same bacteria transform into our thanatobiome at the moment of our death. The thanatobiome is the moment at which our once-friendly bacteria invade our bodies to begin the cycle of decomposition. These same bacteria that were once so important to our well-being in life now start to liquefy our corpse and turn it into simple organic compounds, ready for organic recycling. In this survival process, human consciousness makes no contribution to the ultimate outcome of these relationships. There is no loyalty, sentimentality, or conscious regard for our passing. So, as we send people to interstellar destinations, we are also sending bacteria that can operate quite independently from us. Consequently, human space colonization is not a special case of conscious being, but anthropocentrically directed panspermia.

A paradox therefore exists in our expectations between promoting human colonization and the precautionary approach to seeding life in the extraterrestrial environment, which currently constrains this possibility. Our own advancement in the cosmos can only be achieved in partnership with other life forms that will advance with us. Whether we like it or not, the consortia of organisms that help us survive within ecosystems will, at some point, move beyond the barriers and membranes constructed to utilize and contain them. At this point, they will be liberated into the interplanetary and interstellar environments.

Persephone does not propose either to initiate biogenesis on the worldship or to deliberately seed life throughout the universe in the course of its exploration. While the emergence of life is not its primary aim, a point in its various projects could be reached at which biogenesis—or its equivalent—may be highly probable, if not inevitable. Anticipating this possibility, Persephone aims to incorporate them into its philosophy, technology, and culture.

If such an event occurs, it is unlikely to produce an existing terrestrial life form.

Biovigilance and security need to be conscious aspects of the colony's culture. This is not to say that communities need to immediately track down and destroy "deviants" according to some tyrannical anatomic scheme, as in John Wyndham's 1955 science-fiction story, *The Chrysalids*, but to be careful and constant gardeners for novelty in the environments they inhabit.

Persephone's inhabitants will need a clear understanding of ethics and community, and how they may build relationships between a host of different and potentially competing agents by establishing forms of "ecological peace" between them. This can be thought of as the active management of contested resources and territories to avert escalation toward tipping points. Such a task is by no means straightforward. We cannot manage this for our own species without incident here on Earth, let alone propose to perfect the art for many differentiating and potentially new kinds of organisms. However, this must be part of the responsibility of the worldship's citizens, so that environments are managed and unwanted organism "surprises" are minimized. It must also be appreciated that, with a fertile creative space, there will inevitably be unforeseen events. Peace brokers must resist being environmental tyrants that restrict the freedoms of populations in ways that reduce biodiversity on the basis of prejudice or fear. In the long term, this is likely to reduce the robustness and longevity of the worldship community.

Yet, we must design the possibility of radical breaks in continuity into the worldship environment if we are to realistically persist for any significant time away from our planet of origin.

Such transitions are at the heart of ecopoiesis. It is essential that we understand not only how materials become lively, but also how this process can be initiated, designed with, and enhanced. Establishing methodologies that engage with notions of fundamental liveliness are essential in "building" ecosystems, rather than gardening existing ones, and are essential for thriving in the longer term. When a worldship finally arrives at its destination, perhaps hundreds or thousands of years after it has begun its journey, the colonists may need to initiate ecopoiesis and terraforming, as what may have initially appeared to be viable Earth-like conditions could have radically changed. Under these circumstances, colonists are already trained to meet this challenge. More pressingly, from a terrestrial perspective, establishing ways in which environments may be made livelier or constructed from inert materials, so that they indefinitely support "life," may help us to repair the industrially induced scars on our home planet.

In constructing life-promoting soils, the transformational qualities of existing organisms, such as bacteria, fungi, and algae, are likely to be brought into play if these systems are to mature. These agents are likely to be essential for establishing complex ecosystems through the process of biological succession and developing an essential infrastructure of chemical signaling systems that form a sensory network within soil bodies (Hird 2009). Yet, by invoking the participation of multi-cellular organisms in the construction of environments, questions of planetary bioengineering, and even of the possibility of directed panspermia are raised.

The robustness, negotiation, and complex relationships between dynamic agents that are needed to produce mature ecosystems in non-terrestrial environments are dependent on the cooperation between many different bodies—machines, humans, and non-humans. The most intimate and sustained relationships are established through processes with technological qualities that can perform useful work, such as soils. These may be regarded as the most fundamental kind of technology that, over geological time periods, has acquired webs of metabolic relationships. Soils also provide a site for myriads of chemical and biological interactions, and so provide a platform for the sustained growth, decomposition, and recycling of organic matter. Compositionally, soils may be thought of as giant organic bodies made up of many loosely associating components such as clay, sand, organic waste, water, air, bacteria, and microorganisms. Operationally, soils are powerful transformers that organize raw materials into tissues that are the foundation of all organic life.

The construction of soils is entangled with the idea of planetary bioengineering. Such strategies have already been proposed for Mars, to create a system that can thicken the atmosphere to produce life-sustaining infrastructures (Fogg 1993, 1995). Technological alternatives to achieve these goals are not a realistic proposition, as only organic life can generate these outputs.

While the 1967 Outer Space Treaty and 1958 Committee on Space Research (COSPAR), founded by the International Council for Scientific Unions, propose that we must take a precautionary approach to the exploration of new terrains such as Mars (Mars One n.d.; Howell 2015), this presents a paradox. Our very presence in an alien environment contaminates the surroundings—from the metabolites we make to the bacteria that cling to spacefaring vessels—all kinds of environmental changes travel in our wake. While we may be as reasonably considerate as we can, it is likely that the costs of maintaining absolute sterility simply outweigh its perceived benefits on seemingly barren terrains.

Despite very little evidence that there is life on the Red Planet, it is of great historical interest. It seems that, once, Mars was much more conducive to life than today. There are signs that running water once flowed on its surface, that it possessed an atmosphere, and even had tectonic plate activity. Scientists continue to conduct tests to establish its geological story and potentially deduce whether life may once have existed on what is now a hostile environment. Recently, hygroscopic salts have been identified in soils in the northern latitudes of Mars, and widespread chloride-bearing evaporitic deposits have been found in the southern highlands, where water is locked into the ground substrate. These could provide a local and transient source of liquid water that would be available for microorganisms on the surface. In Earth's Atacama Desert, such terrains have become a habitat for hardy microorganisms that can take advantage of the relative humidity in the area. Yet, missions such as Mars One propose to colonize the planet, and the introduction of engineered bacteria to support a human colony, for example, would complicate the story. Despite precautions, we may need to establish whether any bacteria identified in the future are Martian or terrestrial in origin. To read events correctly, it will be important to ensure that the recordings we are picking up are not simply our own contaminants. Already, such controversies are raging, as rovers may be carrying bacteria that survive strict sterilization procedures (Madhusoodanan 2014).

But it would be wrong to indefinitely consider terrestrial bacteria as contaminants to non-terrestrial environments. If we are serious about settling on the Red Planet, then a strategy that enables colonists to work with terraforming bacteria and leave large sections of the planet untouched will need to be attempted. Whether this is technically possible and how much effort we invest in biosecurity are also important issues to consider.

Even more ambitious projects exist to enliven interplanetary and interstellar environments by seeding their lifeless surfaces with microorganisms. Michael Mautner proposes that it is our moral duty to seed life throughout the cosmos using the techniques of directed panspermia (Mautner 2004). Humans have a cosmic purpose to fill the universe with life by protecting and propagating all its forms. Our window of opportunity to challenge the Fermi paradox—the contradiction that exists between the high probability for life elsewhere in the universe, and the lack of evidence for it—may be narrow. He suggests that we should act quickly before we lose the opportunity to promote biogenesis elsewhere in the universe. Mautner's directed panspermia is accompanied by a panbiotic astroethics that takes an expansionist approach toward seeding the universe with life. Proposing that life has purpose, namely self-propagation, and an ethics, he suggests that a set of values can be derived in which life can be viewed as "good" and that which harms life is "evil." Mautner's ethics have resonances with biospheric utilitarianism; specifically, Arne Naess's 1973 notion of deep ecology and Aldo Leopold's *A Sand Country Almanac*, in which humans are "plain members of the biotic community" and land ethic where "a thing is right when it tends to preserve the integrity, stability and beauty of the biotic community. It is wrong when it tends otherwise." Lending weight to his proposition by establishing that no convincing evidence for extraterrestrial civilizations exists, Mautner asserts that directed panspermia is a morally just decision. Seeding nearby solar systems and star-forming zones in interstellar clouds would be the first targets for most effectively expanding the community of cosmic life. He identifies microbial life, such as bacteria and algae, as being the agents capable of achieving this project. Each directed panspermia mission is capable of seeding dozens of new planetary systems with primordial cells, with the potential to evolve within their new environments. The success of Mautner's method is measured through time-integrated biomass, which obtains the greatest impact on cosmic locations before any natural life evolves.

Mautner observes the fundamental conditions of the panspermia hypothesis—the idea that life exists throughout the universe, being distributed by cosmic matter such as meteoroids, asteroids, and comets. These bodies may also have already spontaneously seeded life on Earth. If the hypothesis is correct, then directed panspermia is, therefore, the engineering of a natural process. Mautner offers details on the compounds that can be found on certain kinds of meteorites. They offer a similar kind of fertility to that on Earth and have even grown plants like asparagus in their residues.

In its colonizing ambitions, Persephone therefore also invokes the controversial issue of directed panspermia. If we accept that human colonization is necessary, then we will inevitably be spreading other life forms throughout the cosmos—both from our biomes and also from the ecosystems that support us. While contamination from worldships may be accidental, other proposals are quite deliberate in increasing the life-bearing potential of worlds beyond our own.

Henry Lin and Abraham Loeb propose that life can travel between stars and spread in a characteristic pattern where rapidly spreading clusters of life forms grow and overlap like bubbles. These could be spread through a natural process of panspermia, such as the use of gravitational slingshots of asteroids and comets. Alternatively, intelligent life will travel outward deliberately using its own technologies. Lin and Loeb modeled this process to assume that seeds from one living planet spread outward in all directions. If seeds reach a habitable planet orbiting a neighboring star, it may establish a colony. Over time, the result of this process would be a series of life-bearing oases blooming in the galactic landscape—a kind of epidemic in which the Milky Way could become infected with pockets of life (Lin and Loeb 2015).

While current attitudes place us as the fouler of non-terrestrial spaces, or sites for alien invasion, we may also have to change our perspective on the role that we are playing in the exploration of space. Rather than solely seeking to serve human interests, we may choose to uphold an ecological perspective, such as that proposed by Michael Mautner—that our mission is to enliven our cosmos (Mautner 2004). Without accepting the basic premise that we will inevitably contaminate, enrich, seed, and complexify non-terrestrial environments, we have no future in interplanetary space, let alone among the stars. We therefore have to see the intermingling of species in different territories as part of the process of exploration. These entanglements are part of the great diversity and ecological remixing that we bring. We have to learn to design, engineer, and manage the consequences of these events.

Mautner's rationale for directed panspermia—the strategic sowing of life forms in developing star systems beyond our own (Part II, Chap. 11)—is to fire off a bacterial capsule to reach a forming star system at the earliest stages of the Solar System's formation. This early "infection" of developing planetary systems will ensure that life is more likely in these regions of the universe. However, a challenge is faced if native organisms were to arise from that star system, as the terrestrial strains could provide a threat to the survival of the alien life form. Owing to the scale that this kind of operation would invoke, it is likely to exceed the capacity for human intervention once biogenesis has started. Mautner details the transformative effects of microorganisms and how they have established a baseline for the living things we know on what was effectively a hostile planet during the Archean epoch 2.5 million years ago (MYA). In particular, the metabolisms of ancient species known as cyanobacteria transformed light and carbon dioxide into biomass to produce soils, trap nutrients, and begin to enrich our atmosphere with oxygen. Directed panspermia is contentious, because it challenges some of our deep-seated notions of Nature, human identity, biosecurity, and the cosmic role of humanity, which are fundamental concepts in establishing the ethics, morality, and culture of a worldship civilization.

Yet, panbiotic ethics is not the only approach that could be taken toward establishing a living colony of non-humans in a non-terrestrial environment. Qualitative transformations between microorganisms and their niches could be observed much nearer to home and much later in the evolution of the planet, so their impacts may be studied in advance of scaling biologically engineered approaches to planetary dimensions. The difference between directed panspermia and terraforming using microorganisms is not technical, but procedural, whereby planetary engineering techniques take place much later in the life of a planet, under more controlled and local conditions. Indeed, such approaches are being proposed for terraforming Mars or using bacteria to mine asteroids.

However, the 1967 Outer Space Treaty and COSPAR regard life as something with very clear boundaries that either needs to be conserved and protected or acts as a contaminant. Indeed, the prospects of establishing interplanetary ecosystems are, in principle, hampered by these imperial agreements, which require life-promoting tactics to build soils to support interplanetary colonies or even a multi-generational starship. Indeed, the Outer Space Treaty takes a conservationist, precautionary approach to the deliberate seeding of life in our Solar System. It operates a case-by-case approach, weighing up the risks of contamination with those of scientific interest. Yet, recent studies suggest that, if we really want a sterile cosmos, then we will have to up our game on the standards to achieve it.

Hardy species of bacteria can survive current sterile practices that are applied to robotic probes, and it is likely that we have already seeded a range of species on Mars (Madhusoodanan 2014). Indeed, if we are to establish an interplanetary presence, then it is essential that we create a range of infrastructures to develop robust ecosystems that will support human colonization—particularly through the production of life-giving soils, which also rely on bacteria and other microorganisms to complete composting cycles, so that organic matter is effectively reused within an ecosystem. While the Martian surface is little more than radioactive dust, other interplanetary materials are fundamentally similar to, or exceed the fertility of, terrestrial soils, and may sustain terrestrial life forms (Mautner 2002). Moreover, carbonaceous chondrite asteroids, which are a group of stony, nonmetallic bodies formed from dust and small grains that were present in the early Solar System and contain organic compounds and water, have been proposed as suitable materials from which to harvest resources in space (O'Neill 1974, 1977; O'Leary 1977; Lewis 1993; Sagan 1994; Dyson 2000). Such substances could potentially sustain microbial or plant biomass that support larger human populations (Bowen 1966). Similar carbonaceous chondrite materials on Phobos and Deimos, the moons of Mars, may also usefully serve as reserves for the production of soils.

Indeed, the presence of these life-supporting materials throughout our Solar System suggests that planetary resources can accommodate substantial biological and human activity. Potentially, a balance between promoting life and preventing the unwanted side effects of human settlement needs from contaminating undiscovered planetary ecosystems could initially be attained through the construction of solar-powered biomes with selective permeability to the movement of organisms through their external "membranes." Conceivably, the design principles used to establish the semi-permeability of surgical operating theaters, which require people to move in and out of sterile environments, could be employed in the construction of these spaces—although, given the emerging hardiness of certain bacterial species, and that, potentially, the most tenacious of life forms will be best suited to hostile environments and organisms that can survive in extreme conditions, this necessarily poses a significant biosecurity consideration. Yet, the questions remain, and the risk of working with living systems in non-terrestrial environments becomes a question of whether it is ethical not to conduct these kinds of experiments that will enable us to most effectively transform barren landscapes into living terrains before humans settle them.

Indeed, intermingling is what life "does." It is an established strategy for increasing survival by provoking and developing new toolsets that enable restless change. Preservation and stasis are the realms of very few species, like the coelacanth, which appears in the fossil record around 150 MYA. It was presumed extinct until the 1940s, when several specimens were discovered around the coast of South Africa, being largely unchanged. Indeed, the miracle of our existence is instability and change, not stasis. Moreover, the ecological condition itself engages risk, not security. The challenge is to establish the limits of biosecurity through which tipping points of ecological collapse are reached with sufficient complexity and diversity to promote rich, evolving communities within the ecosystem. Potentially, coordination of this operation invokes notions of ecosurveillance and citizenship, whereby values and decisions related to observations and findings are translated into social action.

The Interstellar Question does not presuppose an affirmative answer about the ultimate feasibility of the human colonization of other planets. However, it appreciates that there are significant anthropocentric benefits to be gained in considering the possibility of building artificial environments and potentially settling non-terrestrial environments in some detail relevant to our long-term survival. Yet, the utilitarian notions and rational arguments for the possibility of traveling to the stars may not exceed the importance of the existential questions that accompany this endeavor. At the very least, the Interstellar Question may help us reflect on contemporary society and identify how we may precipitate changes in our own global civilization so that we may become better stewards of our planet.

In the long term, the reality of our demise is no fantasy. At some point, around 4.5 billion years from now, our own Sun will die. With it, so may life in the universe. Difficult conversations are needed to deal with the possibility of our own extinction. Perhaps we need to adopt new perspectives on our traditional human-centered view of existence and consider alternative, potentially much more inclusive ways of thinking about existence, and our purpose as part of a much broader community of life. Some of these kinds of conversations are started in Part II, and include Mark Morris on the possibility of building our own alternative planets to colonize (Part II, Chapter 15) and Krists Ernstsons, who imagines a future form of humanity that can solve the interstellar "gap" once and for all (Part II, Chapter 14).

REFERENCES

R. Armstrong, Space is an ecology for living in. Archit. Des. **84**(6), 128–133 (2014)

R. Armstrong, *Vibrant Architecture: Matter as a Codesigner of Living Structures* (De Gruyter Open, Berlin, 2015)

Ask an Astronomer, How much money is spent on space exploration (1997–2015), http://curious.astro.cornell.edu/about-us/150-people-in-astronomy/space-exploration-and-astronauts/general-questions/921-how-much-money-is-spent-on-space-exploration-intermediate. Accessed 21 Aug 2015

Bigelow, Mission statement (2014), www.bigelowaerospace.com. Accessed 10 Aug 2014

A. Bond, A.R. Martin, Worldships: an assessment of the engineering feasibility. J. Br. Interplanet. Soc. **37**, 254–266 (1984)

H.J.M. Bowen, *Trace Elements in Biochemistry* (Academic, New York, 1966)

S. Cruddas, Rocketing prices: the investors eyeing the riches of space. BBC.com (2015), Aug 11, www.bbc.com/future/story/20150810-rocketing-prices-the-investors-eyeing-the-riches-of-space?ocid=twfut. Accessed 15 Aug 2015

A. Dawson, They plan to let you drive a space robot across the moon without leaving your home. Mirror Online (2014), Oct 14, www.mirror.co.uk/news/technology-science/technology/plan-you-drive-space-robot-4482930. Accessed 17 July 2015

F. Dyson, *Disturbing the Universe* (Basic Books, New York, 1979)

F. Dyson, *Imagined Worlds: Jerusalem–Harvard Lectures* (Harvard University Press, Cambridge, MA, 2000)

F. Field, J. Goodbun, V. Watson, Space-time and architecture. J. Br. Interplanet. Soc. **67**, 322–331 (2014)

M.J. Fogg, Dynamics of a terraformed Martian biosphere. J. Br. Interplanet. Soc. **46**, 293–304 (1993)

M.J. Fogg, Terraforming: a review for environmentalists. Terraform. Rep. **2**(2), 92–111 (1995)

Foster + Partners, Foster + Partners works with European Space Agency to 3D print structures on the moon. Fosters + Partners (2013), Jan 31, www.fosterandpartners.com/news/archive/2013/01/foster-partners-works-with-european-space-agency-to-3d-print-structures-on-the-moon/. Accessed 18 July 2015

J. Gamble, How do you build a city in space? The Guardian (2014), May 17, www.theguardian.com/cities/2014/may/16/how-build-city-in-space-nasa-elon-musk-spacex. Accessed 17 Aug 2014

J. Harlow, This is major Elon to ground control: bury me on mars. The Sunday Times (2013), Feb 17, www.thesundaytimes.co.uk/sto/newsreview/features/article1215255.ece. Accessed 21 May 2014

W.J. Hennigan, SpaceX cargo mission blasts off to ISS at last, defying latest hurdle. Los Angeles Times (2014), Apr 18, www.latimes.com/business/la-fi-mo-spacex-nasa--launch-20140417-story.html. Accessed 10 Aug 2014

K.C. Heyde, W.C. Ruder, Exploring host–microbiome interactions using an in silico model of biomimetic robots and engineered living cells. Sci. Rep. **5**, Article number: 11988 (2015)

M. Hird, *The Origins of Sociable Life: Evolution after Science Studies* (Palgrave Macmillan, London, 2009)

E. Howell, How can we protect mars from earth germs while searching for life? Astrobiology Magazine (2015), Feb 11, www.space.com/28517-mars-life-earth-germs--protection.html?cmpid=514648. Accessed 13 Feb 2015

S.J. Lewis, *Resources of Near-Earth Space* (University of Arizona Press, Tucson, 1993)

H.W. Lin, A. Loeb, Statistical signatures of panspermia in exoplanet survey. Cornell University Library (2015), Aug 6, arXiv:1507.05614. Accessed 25 Aug 2015

J. Madhusoodanan, Microbial stowaways to mars identified. Nature News & Comment (2014). doi:10.1038/nature.2014.15249

Mars One, The next giant leap for humankind (n.d.), www.mars-one.com. Accessed 10 Aug 2014

M. Mautner, Planetary microcosm bioassays of Martian and carbonaceous chondrite materials: nutrients, electrolyte solutions, and algal and plant responses, planetary bioresources and astroecology. Icarus **158**, 72–86 (2002)

M. Mautner, *Seeding the Universe with Life: Securing Our Cosmological Future* (Legacy Books, Christchurch, 2004)

K. Medlock, This is the first object 3D printed from asteroid dust. Inhabitat (2016), Jan 9, http://inhabitat.com/first-object-3d-printed-from-asteroid-dust-revealed-at-ces/. Accessed 10 Jan 2016

D. Millward, US ready to return to moon. The Telegraph (2014), Feb 3, www.telegraph.co.uk/news/worldnews/northamerica/usa/10614953/US-ready-to-return-to-moon.html. Accessed 10 Aug 2014

D. Millward, Robot passes self-awareness test. The Telegraph (2015), July 18, www.telegraph.co.uk/news/worldnews/northamerica/usa/11748084/Robot-passes-self-awareness-test.html. Accessed 18 July 2015

A. Naess, The shallow and the deep: long-range ecology movements. Inquiry **16**(1–4), 95–100 (1973)

NASA, R2 Robonaut (n.d.), http://robonaut.jsc.nasa.gov/default.asp. Accessed 19 July 2015

NCBI, The NIH human microbiome project (2009), www.ncbi.nlm.nih.gov/pmc/articles/PMC2792171/. Accessed 19 Aug 2015

S. O'Kane, A new NASA-funded study lays out a plan to return humans to the moon. The Verge (2015), July 20, www.theverge.com/2015/7/20/9003419/nasa-moon-plan-permanent-base. Accessed 15 Aug 2015

B.T. O'Leary, Mining the Apollo and Amor asteroids. Science **197**, 363–364 (1977)

G.K. O'Neill, The colonization of space. Phys. Today **27**, 32–38 (1974)

G.K. O'Neill, *The High Frontier* (William Morrow, New York, 1977)

G.K. O'Neill, F. Dyson, *High Frontier: Human Colonies in Space* (Apogee Books, Toronto, 2001)

J. Plafke, NASA teams with Google to grow plants on the moon by 2015 to create habitable environment. ExtremeTech (2013), Dec 3, www.extremetech.com/extreme/171975-nasa-teams-with-google-to-grow-plants-on-the-moon-by-2015-to-create-habitable-environment. Accessed 18 July 2015

L. Preston, How to plant a garden on mars—with a robot (2013), May 14, www.universetoday.com/102151/how-plant-a-garden-on-mars-with-a-robot/. Accessed 18 July 2015

J. Riseborough, T. Biesheuvel, Google-backed asteroid mining venture attracts billionaires. Bloomberg (2012), Aug 7, www.bloomberg.com/news/2012-08-06/google-backed-asteroid-mining-venture-adds-billionaire-investors.html. Accessed 10 May 2014

C. Sagan, *Pale Blue Dot: A Vision of Human Future in Space* (Random House, New York, 1994)

Travelmail Reporter, Sir Richard Branson to develop hypersonic commercial planes that could travel up to 19,000 mph, flying from New York to Tokyo in one hour. Mail Online (2014), May 7, www.dailymail.co.uk/travel/article-2622378/After-Virgin-Galactic-Sir-Richard-Branson-wants-develop-supersonic-planes.html#ixzz31JfmjJxD. Accessed 10 Aug 2014

N. Ungerleider, New laser communication system could speed up communication between astronauts and earthlings, thanks to Space X and NASA. Fast Company (2014), Apr 18, www.fastcompany.com/3029337/most-innovative-companies-2014/new-laser-communication-system-could-speed-up-communication-b. Accessed 22 May 2014

M. Wall, Asteroid miners may get help from metal-munching microbes. Space.com (2015), Jan 29, www.space.com/28320-asteroid-mining-bacteria-microbes.html. Accessed 7 Apr 2015

F. White, *The Overview Effect: Space Exploration and Human Evolution.* (AIAA, Reston, VA, 1998), p. 179 (Originally published in Y. Gagarin, V. Lebedev, *Survival in Space*, Frederick Praeger, New York, 1969)

Whole Earth Catalog, Whole earth catalog family of publications (n.d.), http://wholeearth.com. Accessed 18 Mar 2016

A. Withnall, Canadian company's plans to build a giant lift that is 20 km high and takes astronauts straight into the stratosphere. The Independent (2015), Aug 17, www.independent.co.uk/news/science/a-canadian-company-is-planning-to-build-a-tower-thats-20km-high-and-could-making-flying-to-space-like-taking-a-passenger-jet-10459058.html?icn=puff-2. Accessed 18 Aug 2015

J. Wyndham, *The Chrysalids* (New York Review Books, New York, 1955)

4

An ecological approach to interstellar exploration

Rachel Armstrong

This chapter proposes that a new understanding of "sustainability" will enable us to inhabit extreme environments and spaces that cannot be bridged in a single natural human lifetime.

4.1　From interplanetary to interstellar space exploration

So far, we have explored the context in which the seeds of potential for our transition toward an interstellar species may arise.

Of course, when our first terrestrial pioneers set out on their incredible slow journeys on Earth, they could be resourceful in using the materials around them. When food ran low, they could cast nets over the side of a ship's hull, or use cool surfaces to draw moisture from the air when water was scarce. However, these survival tactics will not work in the void that exists between planetary bodies and celestial systems.

Beyond the incredible distances posed by our interstellar ambitions, perhaps the most daunting challenge is its irreversibility. The starship question is not just about building the fastest and largest vehicle that has ever existed—but in constructing an entire world.

The first habitable artificial environment was described in a philosophical essay, *The World, the Flesh and the Devil*, by John Desmond Bernal (1929, reprinted in 1969). His visionary rotating 16-km-diameter architecture proposed to accommodate a population of around 20,000 colonists. Built from matter harvested from asteroids and regolith with a hard, transparent, thin shell, it proposed to prevent the escape of an internal gaseous atmosphere (Bernal 1969). The Bernal sphere is particularly significant in the construction of artificial worlds, as it established the notion of artificially simulating gravity using centripetal acceleration in a megastructure to exert a gravitational-like force on the inhabitants. In these environments, the gravitational forces are proportional to the radius and increase with growing distance from the axis of rotation. Although, in theory, a gravitational field can be engineered to be constant, in

---

Rachel Armstrong, PhD. (✉)
Professor of Experimental Architecture, Newcastle University, Newcastle-upon-Tyne, UK

practice, torque in the system is likely to produce variations in microgravity. Spherical bodies produce more gravity at their equator than at the poles, so, while a cylindrical design can produce a more consistent surface gravitational force, colonists cannot assume that gravity will always be at an absolute constant and variations in the gravitational field are likely to have to be compensated for. Yet, fluctuations in Earth's gravitational field also occur on Earth itself. Since their effects on biological systems are not formalized, their significance is uncertain.

Conscious of the constructed nature of these worlds, interstellar visionaries have considered the kinds of ecologies that could indefinitely sustain life within these colossal biospheres. Dandridge Cole suggested that hollowed-out rotating asteroids equipped with many of the amenities of planets could act as evolutionary seedbeds, containing evolvable agents that may produce a diverse range of possible life forms or *"macrolife"* (Levitt and Cole 1963). *"Nomadic"* human colonists could, therefore, facilitate a possible expansive union between organic, cybernetic, machine intelligences and disseminate them throughout the cosmos. Gregory Matloff developed the concept of *"slow, or non-nuclear, arks"* (Mallove and Matloff 1989, p. 18), and Eric Jones and Benjamin Finney proposed that nomadic space cities may *"diffuse"* toward nearby stars by hitching a ride on comets, which could result in the possibility of 10,000-year human migrations (Finney and Jones 1985). Indeed, Arthur C. Clarke reflected on the possibility of encounters with biomechanical alien life forms, or *"biots,"* in his 1973 novel, *Rendezvous with Rama* (Clarke 1973). Yet, Cole fully understood the difficulties in creating self-sustaining biospheres in reality. He proposed that experimental environments such as underwater bases would be needed as test beds to practically realize such ideas. In the 1970s, Gerard K. O'Neill considered the Bernal sphere as a buildable structure into for orbital terrestrial communities, and proposed to make them habitable by deploying existing technologies (O'Neill 1981). His projects, Island One and Island Two, were space architectures that were internally lit by polar mirrors, which relayed sunlight into crystal palaces to house agricultural terrains that could sustain tens of thousands of residents living at the equatorial regions. O'Neill proposed that these structures could actually be positioned at the fifth Lagrangian libration point (L5), where a spacecraft can remain in the same position between Earth and the Moon owing to gravitational effects reaching equilibrium. Such concepts laid the groundwork for the tangible design of space megastructures, like worldships, which did not propose to design the life within them, but provide a recognizable context for their existence.

Since the starship question is unlikely to diminish, as the prospect of interplanetary space settlement is realized, it needs to be confronted conceptually. An experimental approach will enable the necessary plasticity to being prototyping our approaches while taking real steps toward the construction of starships. Cumulatively, these investigations may begin to produce a road map of possibilities that can be iteratively developed and explored through relatively low-cost and highly distributed collaborations. Indeed, this project may define our mission for the third millennium.

4.2 Laboratory practice

"At first it was believed that Tlön was a mere chaos, an irresponsible licence of the imagination; now it is known that it is a cosmos and that the intimate laws which govern it have been formulated, at least provisionally." (Jorge Luis Borges, "Tlön, Uqbar, Orbis Tertius"; Borges 2000, p. 32)

In 1624, the father of modern scientific experiment, Francis Bacon, ended the age of superstition that characterized the Middle Ages by designing a city that was shaped by modern technology. Indeed, this New Atlantis provides the framework for modern cities. Even now, nearly 500 years later, we can clearly recognize the kinds of services that Bacon conjured in his vision—communications networks, advanced optics, complex mechanics, desalination, airplanes, submarines, and genetic modification.

Arguably, the most compelling aspect of New Atlantis is how Bacon imagines his scientific method helping us live and work together through an urban environment, where innovation and discovery are fundamentally social.

Perhaps the biggest experimental platform that has generated the most relevance for space exploration has been science fiction. The medium of writing becomes a way of developing thought experiments as scenarios that are extrapolated through storytelling into events that shape the lives of communities. Of course, fictitious characters and interpretations of locations are used to explore possibilities raised by the plot, but, by actually thinking through the complex cycles of innovation, cause, effect, and social transformation, writers like Arthur C. Clarke, Philip K. Dick, and Ursula LeGuin have engaged the public in a consensual thought experiment which commands the content, not just of books, but cinemas (*2001: A Space Odyssey, Star Trek, Star Wars, Gravity, Interstellar*), songs (David Bowie, 'Trentemøller'), theater productions (Tom Stoppard, *Rocky Horror Picture Show*), and social gatherings (Comic Con).

While it may be easy to dismiss these events as being "just fiction," they are not trivial. The process of conceiving and realizing the possibility of worlds beyond our own, through stories and their implementation in graphics, shapes the way we think about contemporary society and our relationship with space. They are forms of experiment. Of course, these are not scientific experiments, but forms of discovery, from which we may learn something new or consider ideas from different perspectives.

Yet, space is real. The scientific community has also developed laboratories that have already helped us to realize seemingly impossible ambitions and ideas.

US officials who were alarmed by the October 4, 1957 launch of the first Soviet satellite, Sputnik 1, and the potential threat to security created a national research platform and leadership agency for space technology. In 1958, they established NASA, which runs on a yearly budget of around US$20 billion (which is about 0.5% of the US$3.5 trillion of the total US GDP).

Additionally, great instruments such as the Large Hadron Collider, the world's largest Swiss watch and cathedral to quantum physics, built by the European Organization for Nuclear Research (CERN) in 2008, which cost €7.5 billion, have provided new information and inspired novel ideas about the nature of reality and the character of space itself.

Today, the economic systems that have supported the industrial age are potentially being reinvented, as world powers are experiencing an age of "new normal." In the current landscape of weak economic growth following the global financial crisis in 2008, widening global differences in economic performance and policies are emerging. These are interlinked with instabilities in geopolitical events, such as the conflicts between Russia and Ukraine as well as terrorist attacks and environmental disasters and are creating a climate of uncertainty. It seems unlikely, therefore, that the kinds of laboratories which are needed to push our thinking into interplanetary and interstellar space will be funded by the same centralized systems that supported the "Space Race."

Yet, the nature of the laboratory is changing. Laboratories are no longer exclusively pristine environments for the acquisition of academic knowledge. In an age of communication networks, shared-interest groups, and maker movements, we are seeing like-minded communities sharing information and ideas and experimenting locally. For example, the DIYbio laboratories that were established in 2008 have provided citizens with places to find out about biotechnological advances and explore what this may mean in everyday contexts—whether this is biosecurity, entrepreneurship, or simply as a creative toolset.

Space exploration is becoming increasingly networked and extended through communications systems, such as the Search for Extra Terrestrial Intelligence (SETI), which has searched the skies for radio-wave signals from alien civilizations since 1960, and has been in operation online as SETI@home since 2012. It is the most successful public participation project of all time, where six million people in 226 countries have taken part in the experiment, each contributing their computer processing time to the search for a signal from intelligent beings in the depths of space. Increasingly, the current International Space Station (ISS) astronauts are using social media to extend their experiences to online audiences. New businesses are being established to fly payloads to the edge of the atmosphere or even launch small objects into orbits using CubeSat technology. As communications systems into interplanetary space strengthen, we will also be interacting with rovers and probes to discover space as a collective project.

These places are under construction and are asking important third-millennium questions that relate to our vision, identity, and purpose for the foreseeable future, and what kind of legacy we will leave our next generations. There are a number of essential questions provoked by the prospect of interstellar exploration. Not least of these is our long-term survivability. While the prospect of the natural extinction of humanity appears to be somewhat far off, if we do not begin to formulate the relevant questions now, we will have squandered our chances of actually making our ancient ambitions to live among the stars possible when the time comes. In this context, we may ask what a laboratory for a starship would look like.

To successfully create new environments in deep space, humanity must incorporate, or "entangle," emergent sociocultural movements that promote a viable future, extending beyond the physical limitations of the terrestrial environment. It is, therefore, essential to establish an actual experimental platform in which new materials, technical systems, and approaches can be tested through the production of prototypes that help us refine our interrogation of the space. However, since Icarus Interstellar's infrastructure will not be constructed for 100 years, its experiments are being conducted in a range of settings on Earth.

These laboratory principles are designed to enable the tape of human civilization from the story of the technology of soils to be replayed through experimental approaches, when people first started to understand that the very ground they stood on possessed riches. These exceeded the capacity to promote the growth of crops, but also produced metal from concentrated seams of exposed minerals and performed the vital role of composting in which organic matter is recycled and reused.

So, what kind of laboratory exemplifies the platform?

Francis Bacon's model for the modern laboratory—pristine, exclusive, precise—enabled the birth of modern science and the industrial age. Bacon's traditional laboratory approaches very complex challenges by reducing them into components that can be tested through measurement and analysis.

Third-millennium needs relate to understanding how to move from an industrial compartmentalized world of objects into one of ecological synthesis. These principles underpin the Ecocene, which demands recomplexification of reality, whereby many overlapping systems are coordinated to generate ways of making and inhabiting highly complex, dynamic, evolving environments.

We therefore need laboratories that operate differently to the classical model.

By situating these laboratories at the heart of a city, as Bacon did with New Atlantis, it becomes possible to use them as sites that explore, evaluate, and propose responses to millennial questions to address challenges that are hypercomplex, heterogeneous, massive, unevenly distributed, and in constant flux. Since these sites actually generate the living spaces for colonists, they cannot be closed or sterile. Rather, they are open and messy, and engage the computational properties of the material realm directly in addressing worldship challenges without obligation to human intent. Such a laboratory may be thought of as producing experiments whose outcomes are perpetual prototypes that were never finished, and constitute the operating system for the production of synthetic ecologies.

4.3 Cultural agendas

"I had in my hands a substantial fragment of the complete history of an unknown planet, with its architecture and its playing cards, its mythological terrors and the sound of its dialects, its emperors and its oceans, its minerals, its birds, and its fishes, its algebra and its fire, its theological and metaphysical arguments, all clearly stated, coherent, without any apparent dogmatic intention or parodic undertone." (Jorge Luis Borges, "Tlön, Uqbar, Orbis Tertius"; Borges 2000, pp. 11–12)

Societal transformation becomes possible when biophysical processes are spatialized and brought into differential contact with communities of actors—which may consist of human and non-human agencies such as robots, biofilms, and crops. If Persephone's worlding process is not founded on industrialization, then alternative visions of culture and social organization may begin to inform Persephone's civilization.

The current, modern industrial age is described in reference to the Anthropocene. To date, we have encountered four industrial ages, the first being shaped by the age of steam that used water and fossil fuels, particularly coal, to create powerful engines. The second followed networks of electricity and harnessed the power of even more concentrated forms of fossil fuel in crude oil and natural gas. The third hooked this energy up to modern computers and wireless to generate a communications and data revolution, and the fourth is set to couple this information into automated environments shaped by cyberphysical systems—an event popularized by Hewlett Packard as the "internet of things."

The Anthropocene is significant, as it has produced two distinct images that pertain to our modern culture and practices. The first of these is the massive scale of fossil-fuel burning, as encapsulated by the Burning Man festival. The second is the iconic view of our pale blue planet, Gaia, as seen from space during the 1968 Apollo 8 mission, which symbolizes our planetary-scale metabolism (Vimeo 2014).

Implicit in the Anthropocene is a discussion about how we make better choices about human development. Specifically, the technical platform we are using is characterized by the tools of modern synthesis, which transcends our engagement with gadgets and works with a range of materials that include genes, cells, organisms, and species. Modern synthesis develops mathematical relationships with these systems and frames them within an algorithmic reality. These realities are bounded (deterministic), finite (small), and fragile. Yet, although they are extremely powerful, the sciences of modern synthesis do not have a complete grip on certain kinds of challenges, including multiple dynamic processes such as microbiological interactions, symbiosis (nested groupings of heterogeneous agents such as lichens and coral reefs), or developmental phenomena (that change with time).

By extrapolating from past events, the Anthropocene confronts us with future scenarios, all of which propose our inevitable extinction. No matter how we appear to play the Anthropocene, we seem to lose out—big time. If we have a "good" Anthropocene, we end up being replaced by hyper-intelligent machines, as predicted by commentators such as Ray Kurtzweil in *The Singularity is Near* (2005). Indeed, Microsystem's Bill Joy, Elon Musk, and scientist Stephen Hawking express concern about the existential risks posed to our planetary stewardship by our lively, intelligent technologies, and even the future of the human race. Yet, if we have a "bad" Anthropocene, then the sixth great extinction occurs, as cautioned by Edward O. Wilson and colleagues.

Since Persephone does not begin its tape of human civilization in the industrial era, it rejects the Anthropocene extinction scenarios. These would be an unethical foundation on which to build a cultural platform that aims to provide a sense of ongoingness for humanity and terrestrial systems—albeit in new contexts. Persephone looks to a different way of eliciting the events within its space where "making" becomes an intrinsic part of a life rather than an allocated occupation—or job—that constrains time, choices, and modes of living. Rather than this restricted notion of a life, inhabitants are otherwise engaged with gardening various fields of activities and material systems, which set up a range of tensions that are shaped through dynamic interactions in which systems of production and exchange are inevitably established, and follow, rather than command, the way the colonists live.

Already, we can see new kinds of possibilities for social and biophysical production emerging through maker movements and artisan practices, with the potential to generate new cultural agendas.

Markus Keyser used a solar-powered, semi-automated low-tech laser called the Sun Cutter to harness the power of the Sun through a traditional manufacturing platform. Using a cam-guided system, Keyser could "laser" cut 2D components by bending the Sun's rays through a glass-ball lens. The crudeness of the fusion between a renewable resource and machine produced objects with a particular aesthetics that Keyser called *"nature craft."* This process uniquely reflected the variations in the environment in the production process, such as solar intensity due to weather fluctuations, which conferred every object with a unique character.

In the desert, the Sun Cutter was able to overlap the field of potentiality between the Sun and sand by melting silica-rich sand using the Sun's heat to produce obsidian.

A 3D printing system was coupled to this sintering process to generate a new kind of production system: selective laser sintering (SLS), which could manufacture glass objects with precision. The system was fully automated using digital computing to become a mature solar-sinter platform, which represents a new solar-powered production tool of great potential to change production processes and environmental impacts in artisan-led manufacturing practices (Kayser 2011).

A kick-starter crowd-funded collaboration between Studio Swine (architect Azusa Murakami and artist Alexander Groves) and Kieran Jones combined artisan skills with an abundance of environmental waste plastics to manufacture open-source bespoke furniture. Responding to the staggering scale of plastic pollution in our oceans that constitutes a plastic soup of planetary-scale proportions, Swine and Jones developed a manufacturing system that could be used on board sea vessels. They harvested plastics from the environment, sorted them according to their color, and broke them into small pieces. These fragments were then melted at 130 °C in a homemade furnace. The molten plastics could be molded. Two flat slabs of heavy metal were used to compress softened plastics to form the seat for a sea chair. Other components were fashioned into a desired shape using bent scraps of aluminum. The sea-chair components were cooled using seawater to solidify the plastic and screwed together (Dezeen 2013b).

Studio Swine has also built a mobile foundry and used it to cast aluminum stools using cooking oil and soft-drink cans that they collected on the streets of São Paulo. By creating a system that enabled the community of independent waste collectors, known as *catadores*, to make saleable products from recycled rubbish, they transformed wanton street matter into metal objects, such as stools and even small-scale architectures (Dezeen 2013a).

These practices establish new ways of making that demand alternative resources, infrastructures, and distribution systems that challenge industrial manufacturing modes. Such examples suggest that the materiality within an emerging worldship environment can profoundly influence the way in which a community and city develop.

This publication does not intend to document every possible alternative social system or mode of production that may inform Persephone's manufacturing systems. It aims to develop a set of examples, ideas, and principles that may help us design useful experiments and prototypes from which informed decisions could be made. At an early stage of development, the design process aims to facilitate discussion and diversity in approaches. At a later stage in the construction of the project, certain approaches may become more formally programmatic as specific resources are allocated to the worldship.

Yet, alternative worldviews are essential for developing a rich toolset of approaches that help design new futures within the context of emerging cultures. They help us to invent, construct value systems, and connect with each other. In developing a broader portfolio of strategies through which we may approach a pathway of common survival, new kinds of *mythoi* must be underpinned by alternative technical systems. These platforms generate the flexibility to differently embody the proposed perspectives. Their alternative forms of organization also help us to access different parts of the physical realm beyond what already exists and enable us to implement alternatives directly.

In particular, the Ecocene calls on the incredible generative and destructive forces of the natural realm that are diverse, multiple, and have the capacity to synchronize our lives with the biosphere. These events take place not only in concert with all different kinds of life, but also through matter and life-bearing elements such as water, rocks, oceans, and the atmosphere. Collectively, they offer something else that is potentially more livable and ongoing than the Anthropocene. They venture beyond the sharing of ideas and propose to bring the worlding process into functionality.

Persephone's project particularly invites us to think more radically about the range of tools that Nature has to offer. In particular, its vast portfolio of biologically active metabolisms could be explored as a dynamic, constantly generative, self-renewing set of instruments with the technical potential to shape events within the community of life. In fact, Timothy Morton, author of *Ecology without Nature* (2007), proposes that we should design with metabolism. But what does this mean in practice, and how can we even start to do this?

In the last 30 years or so, the conditions for viewing life itself as a technology have started to emerge. While living things, such as horses and dogs, have always been harnessed to perform work for people, through advances in biotechnology, we can now work with and understand life's processes at a scale and precision that was previously unthinkable. With the advent of Synthia, for example, a cell with an artificially transplanted genome, we are beginning to redesign the character and capability of biological systems. In effect, we are redesigning the details of the carbon cycle. Yet, given the complexity and number of biologically active metabolisms in the world, this seems like a highly inefficient thing to go about doing. Our mindsets are still influenced by expectations characteristic of the industrial age, which seek standardization, homogeneity, and permanence in their technical operations. Because of these limitations, the Anthropocene does not even equip us with the capacity to keep a life-bearing planet lively. If we are to have a chance of surviving our current environmental challenges, or if we are going to inhabit a worldship environment indefinitely, then we need to start building life-promoting environments. Despite paradoxes that may exist within ecological systems, with their recalcitrance and constantly changing relationships, colonists may find ways of identifying synergies between them so that we may begin to construct new spaces and stories that embody a starship culture.

4.4 Persephone

"A cloak of loose, soft material, held to the earth's hard surface by gravity … that lies between life and lifelessness." (Wallace H. Fuller, *Soils of the Desert Southwest* 1975)

Icarus Interstellar is an international organization dedicated to technical achievement enabling interstellar spaceflight. Volunteer citizen scientists with a broad range of backgrounds, ranging from aerospace engineers to professional scientists, academics, science-fiction writers, artists, and space enthusiasts, perform research related to the Interstellar Question—how we prepare for colonizing bodies beyond our own Solar System.

The nature of the Icarus Interstellar vessel that carries the artificial world has not been specified; it is imagined as being something like the slow, cigar-shaped, wet worldship first proposed by Alan Bond and Tony Martin in 1984 (Martin 1984; Bond and Martin 1984) that possesses its own ocean and will make its way towards a habitable Earth-like planet in the Goldilocks region of the Alpha Centauri system.

Design teams are coordinated around research projects with a central theme, such as Project Icarus, an interstellar probe design study envisioning the design of a mainly fusion-propulsion-based engine, and Project Hyperion, a human-occupied worldship (interstellar migration ship) study (Hein et al. 2012). It was created from members of Project Icarus, initiated in September 2009 in an effort to explore multiple interstellar spacecraft systems simultaneously, and was registered as a 501(c)(3) non-profit organization in Alaska in February 2011.

Persephone is a project and experimental platform that is concerned with the process of "worlding"—a neologism popularized by Martin Heidegger in his 1962 *Being and Time*, which is widely used in cultural studies, philosophy, politics, and technological criticism. It embraces a range of concepts, from origins to boundaries, ethnicities, governance, art, and even consciousness itself. Its laboratories produce prototypes that collectively form a "living interior" and explore the design and construction of the worldship's ecosystems that propose to support colonists for many thousands of years, if not indefinitely.

Culturally, Persephone draws from the mythos of the Anthropocene, using the tools of modern synthesis, as well as agents that can deal with hypercomplexity such as natural computing, which operationally speak to the idea of the Ecocene. Functionally, the living interior is situated within a resource-abundant environment where there is plenty of energy and matter to harvest and nurture the process of life and is therefore a co-producer of events. Persephone identifies the thin, life-bearing membranes that are characteristic of our home planet—air, sunlight, water, and earth—as the fundamental conditions for terrestrial life. The field intersections between these realms provide the creative, life-bearing fabrics from which our society has constructed its habitats. Persephone is, therefore, an instrument that orchestrates the folding of these fields. Its aim is to provoke emergent events that result in the objects and processes that we recognize as characteristic of a lively world and even life itself. Most of these interactions take place within an "open" or limited-resource intake, even when the vast distances and vacuum of interstellar space make it difficult to replenish resources. From a design perspective, Persephone deploys a range of tactics and spatial programs that choreograph events within the environment. Socially, Persephone is considered part of the worldship community and not separate from humans or other organisms. During these early stages of development, however, Persephone primarily aims to open up imaginative spaces within that can be interrogated as potential sites of further exploration that may be worked into experiments and prototypes over the coming decades.

Persephone's primary experiments are with soils that are regarded as the integrating material, which can bring together the planet's membranes in a practical, yet highly creative and life-bearing manner. These highly spatial, hypercomplex material systems prevent the matter in the starship from rapidly reaching equilibrium, or dying. On Earth, the story of soil has allowed the transition of life from the oceans to the land and laid the foundations for life. Soils do this by facilitating the flow of elemental fields into a tapestry of geological-scale exchange that is shaped by many agencies like rivers, and continually changes with time. They are formed by continual weathering and the flow of matter caused by natural cycles, but also through their invasion by microbial life. Living organisms tear apart inert, inorganic infrastructures like rocks to create new surfaces and environmental niches that can be further colonized by living invaders. They generate an ecological circulatory system capable of building, sustaining, and evolving the tissues of the land so that humanity can feed its communities and construct cities around deltas such as the Nile, Tigris, and Euphrates. Life-bearing soils are, therefore, essential for the successful human colonization of space.

Persephone's laboratory, therefore, does not start its experiments in the modern age, or even at the Industrial Revolution, because these points in Earth's history have not invested in the production of soil, but in harvesting its resources.

The worldship laboratory proposes to explore new ways of living by applying the principles of a thought experiment proposed by Stephen Jay Gould who wondered whether we would have recognizable biology if we began biogenesis again—a principle that he called *"replaying the tape of life."*

Starting with the premise that all cities are built on fertile soils, Persephone's "tape" is set to run from the start of human civilization, when we first began to utilize the soil as both a material and technology. Persephone places a laboratory at the heart of its habitat, whereby the artifacts and systems arising from the experiments can be thought of as a city. Yet, Persephone is not a utopia—it is a generative materiality that does not operate according to a deterministic way of thinking in which every aspect of a city can be designed into existence. Indeed, Persephone does not assume that the environment will evolve to produce something recognizably city-like.

The laboratory experiments explore how different cities might be if our construction processes preserved the function of our many different kinds of earth, so that we are able to live in spaces that have an equivalent fecundity to the natural realm. Instead of living in habitats that are made from baked dead soils, Persephone investigates the potential of applying super-smart, complex living fabrics with lifelike characteristics, as construction materials for worldship habitats—or "living" cities.

Persephone's experimental questions set out to establish whether rich soils with the potential to support plants and animals indefinitely can exist in a worldship.

Industrialization poses a particular challenge for a worldship environment, as its principal driver is the harvesting of excess natural resources and trading them against deficits. For this to work best in a worldship context, the colony would need to be in an environment where there was obvious bounty and depletion. For the society to be humane and not set up inequalities to keep a worldship colony trading, it is possible that the culture may be one of miners stopping off on exploratory pathways to harvest significant resources from other planets. These would need a trading community from which other supplies that were scarce in the worldship could be replenished or traded on elsewhere at a profit. While it is possible to assume there will be a market for the exploitation of excesses when settlers are close to Earth, where markets in excesses are established, at much more distant locations, it is unlikely that this will be the case—at least until an interstellar market is established.

In this sense, Persephone takes a different starting point from Gerard O'Neill's view of a modern suburban habitat, whose prosperity is based on an already structurally formalized, industrialized society that is organized around Le Corbusier's utopian principle of human development—that buildings are machines for living in (Le Corbusier 2007).

O'Neill was convinced that a planetary surface was not the correct place for an expanding civilization, and proposed that, instead, we would inhabit giant orbital cylinders within recognizably Earth-like environments. These O'Neill cylinders were pressurized environments that were imagined as inside-out planets, which would be constructed at the

Earth–Moon L4 and L5 Lagrange points and based on the principles of industrialized societies—specifically on resources that were being harvested from the Moon. O'Neill continued to develop ideas about human space colonization based on social principles and technologies typical of the 1980s, as he assumed that humanity would remain unchanged as it expanded into the Solar System.

O'Neill's designs suggested that suburban and agrarian landscapes could be directly transposed en masse from terrestrial environments to spaceships. For example, he imagined that avocado and papaya crops could be produced without ever needing to use insecticide sprays. Yet, these ideas are aesthetic proposals, as there is no mention of what kind of insects would be so abundant that they might become pests (rather than new foodstuffs), or an examination of what pollinator system would ensure the production of fruiting bodies (wind, bees, flies, hummingbirds, lemurs, people using paintbrushes). O'Neill imagined industrial forms of employment and technologically more efficient land-management practices than are possible on Earth. Yet, the augmentation necessary to improve soil productivity beyond Earth's environment is not made explicit. His approach, therefore, was to transpose the image of an industrial society from a terrestrial environment into an entirely different context and—as we do on Earth in gardening practices—assume that biophysical differences could be factored out. Such generalizations are dependent on a particular kind of environmental matrix that is dynamic, hypercomplex, massively robust, resilient, responds to contingency, is fundamentally heterogeneous, and is not one that we are in full control of.

Currently, space habitats such as the 1975 NASA Stanford Torus project (Johnson and Holbrow 1976) and 2012 Project Hyperion propose to support colony sizes of around 10,000 people. These are based on studies of genetic diversity (Takahata 1993). The population densities have also been estimated upward, based on expectations of improved agricultural methods with genetic engineering. Persephone does not currently find it useful to make these numeric predictions until it understands the fertile capacity of the soils. It makes a whole world of difference whether a civilization lives on a rocky desert or a lush pasture. Moreover, we have no evidence that it is possible for humans to have children in non-terrestrial environments. Indeed, the outcome of Persephone's experiments may indicate that worldships, as we currently conceive them, are not fit for human habitation at all.

Persephone proposes to explore alternative models for social organization and to examine and extend a portfolio of alternative frameworks that may underpin human civilization. The beneficiaries of this community will also include the living environment, since it is considered part of the worldship community and not separate from humans or other organisms.

Both design experiments and scientific experiments are used to establish what the possibilities for developing the worldship environment may be. The distinction between the two cultures' approaches is that design experiments aim to produce better questions (which keep our eye on the ball) and scientific experiments seek to develop better answers (which keep our feet on the ground). When challenges are highly complex, we need both approaches. Persephone also proposes to achieve these investigations by engaging researchers over highly distributed networks of exchange to conduct thought experiments, hold events, and carry out real-world experiments as collaborative enterprises.

The transformation of an inert worldship interior into one that can produce soils is not an easy one. It asks many searching questions about the assumptions we make about life, and how it is possible to sustain it by establishing natural cycles of water and mineral flow (e.g. phosphorous, nitrogen, carbon, oxygen) within a closed environment. While it has been possible to generate the illusion of a cloud within a space, such as in Bernauldt Smilde's artwork, whereby the climate of an interior space is controlled and theatrical techniques are used to produce the impression of a cloud, it is important to understand how to precipitate these kinds of events more spontaneously within an environment. We know that it is possible to produce dissipative structures using a whole variety of techniques, such as in the work of John T. Snow at Purdue University in West Lafayette, Indiana, who in the 1970s made a tornado machine (Church et al. 1977). It is a cylinder that is 3×3 m across and is driven by an apical fan which pumps air from the cylinder. At the bottom, a rotating screen stirs the air into rotation. Vents emit swirling white vapor into the cylinder, air ascends, and a vortex quickly forms on the updraft. Currently, though, engineering fields in a spatial context in which complex weather may arise spontaneously outside fixed containers are not yet within our grasp. Yet, we will need to realize these kinds of weather-producing systems if we are to create soils to potentially sustain us indefinitely within a closed space that is a fraction of the volume of Earth's biosphere.

Persephone's fundamental assumption, then, is that, to sustain Earth life, these conditions will need to be produced in some manner. Yet, this is unlikely to be as simple as transplanting the soils that are on Earth into the worldship cylinder, as if we were potting a plant. If this were possible, then it would ease Persephone's journey toward establishing its settlements and cities. However, it is likely that many urban experiments will need to be conducted and studied before our mastery of soil production is sufficiently advanced, so that we understand how fertile soil-making systems can be constructed.

Take, for example, the Cheonggyecheon River in Seoul—a beautiful project that has created park space around what was formerly a stream that had been buried by urban development. The Cheonggyecheon is an 8.4-km creek flowing west to east through downtown Seoul, and then meeting the Jungnancheon, which connects to the Han River and flows into the Yellow Sea. On its reinvigoration, the stream had to be given new life by pumping 120,000 tons of water daily from the Han River, its tributaries, and groundwater from subway stations. Walking along the straightened body of water provides breathtaking views of the sunrise and sunset. Yet, walking along the riverside is sad. The reeds are dying and it feels withered around the edges. The feeling is that succession has not been able to take place and that the landscape has been gardened into existence.

The difference between the original and artificial rivers is that, in natural systems, there is a series of structures and events that follow each other to produce landscapes that enable the next generation of life to inhabit them in increasingly complex and innovative ways. This is called succession. Starting with simple dirts, consortia of microorganisms lay down biofilms along the riverbanks, fungi produce mycorrhizal networks, multicellular plants gain traction and generate root systems, which form chemical communication networks that create relationships with soil creatures such as earthworms, burrowing animals, and insects. Gradually, the structure of the delta silts is complexified and transformed so they are able to establish and maintain multicellular plants and their incredibly robust life cycles. In turn, vegetation converts solar radiation into biomass using carbon dioxide, root nodules fix nitrogen into the ground, and green leaves breathe out oxygen that is necessary for life. Within these complex webs, chemical energy is stored and transmitted through high-energy phosphate bonds, making and breaking metabolisms. Molecule by molecule, the landscapes of life are forged through the most incredible orchestration of bionanotechnology—woven at the molecular scale but experienced at the human scale.

Yet, such environments take time to build in the natural realm because they are not "just" information; they possess mass, can parallel process matter, and there are many possible routes to their realization. It is simply not enough to model them. Even if it were possible to simulate these exchanges, it would be almost impossible to implement them from the data alone. Indeed, the challenges that are faced with modeling such systems exceed the limits of our conventional computing platforms. However, models are useful comparators and Persephone employs both digital and analog methods to construct its propositions and establish ways forward. Yet, as the characteristics of these hypercomplex, dynamic, life-bearing fabrics and their rich details are characterized, it will then be possible to say something about the life-bearing capacity of the space and the kind of colony that Persephone could reasonably expect to support.

Persephone consciously promotes experimental methods to generate prototypes that help establish the fundamental processes that increase the likelihood of these systems to transition from inert to living matter. These orchestrated fusions, which ultimately promote lifelike events, propose the genesis of a unique ecology. A sustainable worldship interior cannot be recapitulated as a function of identified parts, or replicated by copying a pre-existing system. However, working with open, non-equilibrium systems and notions of abundance flies in the face of modern design paradigms, because we assume that our ecologies are closed and their surroundings are at equilibrium—in other words, we are most comfortable constructing worlds of composed of objects in which the environment is a constant.

The biophysical dynamics within a worldship environment must be capable of the massive transfer of matter and energy. While this technology cannot be built out of components, it may, nonetheless, be orchestrated using the technology of soils. Permeable fabrics that are used in association with soils to separate, filter, reinforce, or drain terrains called geotextiles also offer a critical and potentially buildable system, with the capacity to potentiate life-supporting structures through which a range of elemental cycles may be generated and filtered. David Gissen describes the combination of these primal organizing forces as "*sub-natures*" (Gissen 2011), which include environmental forces such as dust, mud, gas, smoke, debris, weeds, and insects as inimical to architecture. Our ability to manage, rather than erase, these expressions of the radical creativity of the material realm establish the limits of civilized life. In short, there is no possibility of "Space Nature" without sub-natures.

It is important to note that soils possess a different character than native regolith, which are the sands and dusts found on non-terrestrial planets. They are formed of minerals that have been physically processed by environmental events. The term "regolith" was introduced to lunar science by astrogeologist Gene Shoemaker, referring to a geologically recognized name for the collective rock wastes on the surface of the land. On Earth, regolith includes earthy material produced by the process of erosion including volcanic ash, glacial drift, alluvial deposits, aeolian deposits, and soils rich in humus. Yet, on the Moon, the primary erosional process is meteoritic impact. Harold Urey introduced the idea of "gardening" for the manner in which bare rock becomes broken and forms a fragmental debris layer that is turned over by meteoric impacts, which thickens with age. Regolith is spread by the ejecta from significant impacts and may have been further eroded by smaller projectiles raining down on these layers. The process yields a continuous seriate distribution of fragment sizes, which does not acquire organic material and therefore is a regolith, not a "soil."

The study of organic processes is therefore critical to Persephone's experimental agenda and how soil construction systems may fully draw from a whole range of environmental events to build, complexify, and nourish its soil systems. Of course, such a process also raises questions of valuing substances and the nature of "waste." Rather than viewing fertility as being the primary quality of value, other processes such as composting, decay, weathering, and sediment flows may be recognized for their contribution to the cycling processes that underpin soil performance and its fundamental capacity for change that links the cycles of life and death.

In the early construction phase, Persephone's orbital scaffolding may even be constructed from space debris, which is a growing challenge in near-Earth orbit. NASA is tracking more than half a million pieces of debris that are about the size of a marble or larger, as they orbit Earth. Of these, 20,000 are larger than a football. Yet, there are millions of particles in Earth's orbit that pose the greatest risk to space missions, as they are untrackable. Space junk can be made of natural (from meteorites) and artificial (man-made) particles. Meteoroids are in orbit about the Sun, while most artificial debris is man-made, no longer useful, and circles Earth. Orbital debris includes non-functional spacecraft, abandoned launch vehicle stages, mission-related debris, and fragmentation debris. Junk can travel up to 28,000 km/h, which is fast enough even for a relatively small piece of orbital debris to damage a satellite or a spacecraft.

There have been a number of suggestions as to how the risk of space junk collisions may be removed. Although harvesting trash would not be straightforward, owing to the very high velocities of space debris, construction workers using robot arms could potentially grab large objects, and giant electromagnetically charged nets could trap some of the smaller junk. This harvesting process constitutes a primitive industrial-style metabolism that reduces the energy and resource demands of the construction of the craft. Clean-up strategies include capture methods (nets, harpoons, robotic arms, tentacles), pushing debris out of space using specially commissioned orbital clean-up satellites and solar-powered technologies such as CubeSail, electrodynamic tethers, and high-altitude balloons that use space debris elimination methods to disturb the paths of low-Earth-orbital debris. More esoterically, slingshot methods that harness the momentum used during clean-up to keep the process going, a network of nanosatellites, and giant foam balls have also been suggested.

Yet, space junk does not need to be tipped back into Earth's atmosphere, where it would burn up in the heat produced by the friction of its descent; it could also be useful as a source of scaffolding, recycled digital systems, and construction materials. Contemporary space agencies only use the best semiconductor devices (transistors, diodes, etc.) in their spacecraft. Titanium and aluminum parts could be harvested and recycled. It may be possible to melt these pieces in orbital furnaces under microgravity and high-vacuum conditions—a process that is not possible on Earth's surface (Fig. 4.1).

Figure 4.1 Persephone's concept of a "caddis fly" larva shell approach to the construction of its semi-permeable shell using recycled space junk. These concept drawings by Jon Morris and Phil Watson explore the possibilities of assimilating space junk and other matter to produce Persephone's initial structural scaffolding. *Credit*: Jon Morris and Phil Watson

If the constructed pieces of recycled metal are not of suitable material for incorporation into the main body of the starship, then they may be of value as a space dust shield that could harvest materials from interplanetary and interstellar space.

These substances are qualitatively different to orbital debris. Rather than discarded machines, asteroids (carbonaceous chondrite asteroids containing 3–22 % water and significant amounts of tarry carbon-rich compounds) or dormant comets that might contain even more water and carbon compounds and S-type and metallic M-type asteroids (that are rich with iron, nickel, cobalt, and platinum group metals) would be the most abundant harvestable substances; see Mautner 2004). Even fractions of the electromagnetic spectrum could be harvested as the worldship body travels through the interstellar medium. These slowly accumulating resources could be processed by an artificial soil-making system that supports the interstellar colony and stacks the odds against its ecosystems grinding to an energetic halt. This connection between the external and internal environments feeds the biophysical processes within the worldship, but also generates a set of resources that are exploited by the colony, which shape their social organization and cultural practices.

Persephone's laboratories take the form a network of coordinated experiments that are performed on mobile platforms and exist as a range of projects on a number of different sites that evolve over time. Collectively, they speak to the overarching questions that guide the explorations—namely of soil synthesis and the formation of ecological systems that also embody the value systems of emerging starship communities. Therefore, a civilization based on a soil-generating culture holds the potential to catalyze humanity forward into interstellar space for exploration through experimentation, discovery, and their application in the exploration of interstellar space.

Over the next 100 years, Persephone will test the validity of its assumptions and push the essential guiding questions to the next stage, as an iterative, constantly evolving process. Gradually, the laboratory experiments propose to reach a new point of significance, at which human activity can influence living systems at the smallest and largest possible scales.

In the longer term, we need to do more than be incredibly creative with existing resources. We must find new forms of energy and ways to make materials in space. One idea is for Persephone to carry its own star—a nuclear fusion reactor that also provides its propulsive power. However, nuclear fusion also happens to be the process by which planets are naturally formed over many billions of years. With the right technical system, it may be possible to use the starship reactor to make new materials by squashing simple atoms like hydrogen together to form more complex elements like nickel, iron, carbon, silicon, or nitrogen. These can then be used in many ways on the worldship.

Despite the significant research needed and likely time frames anticipated before major advances are realized in the field of interstellar exploration, this publication advocates positive steps toward this huge undertaking. By going beyond a purely theoretic approach to exploration beyond our Solar System, the contributors do not need to actually inhabit interstellar space to make drawings, build models, and create prototypes of starships to collectively begin to develop the questions and necessary skills for an ecological approach to space colonization.

Even if we do not see these goals reached in our own lifetimes, it is essential to keep the debate, experiments, and prototypes evolving. Indeed, Persephone's conversations are not just about the construction of a starship; they are about us, the legacy we leave, and about where we are heading—in every sense of the word.

4.5 Mixology of space

"Civilization begins with distillation." (William Faulkner; Martineau 2015)

Persephone may be viewed as a transferrable laboratory practice for the practice of worlding, which produce highly varied environments. Its methods can be applied across a vast range of experimental spaces that invite the participation of all knowledge sets and practices. Participants design questions and provide answers that inform the production of Persephone's living spaces.

Life thrives when hypercomplexity pervades the fundamental conditions of existence. Typically, hypercomplex systems are constantly changing, responding to alterations in their context and radically reinventing themselves. Matter matters in these environments where dynamic operations are choreographed by continual flux within a matrix of elemental infrastructures. Persephone's spaces share many similarities with soils, where its entangled structures are rich with resonances—yet none is identical. The evolutionary potential of hypercomplex systems is characteristic of natural phenomena such as biofilms, embryology, and climate change.

Yet, in exploring the construction of spaces that harbor or produce such fabrics, Persephone does not seek to police their production, but sets out to provoke their existence by empowering constituent agencies to make choices and invent ways of being. Participants provoke the limits of a portfolio of ecological operations in ways that promote the possibility of ongoingness, mutuality, diplomacy, and coexistence.

Persephone's rich environments may be evaluated and appreciated using a range of criteria. Some of these include empirical measurements such as energy efficiency. However, the possibility of considering qualities that cannot be easily evaluated like poetics are also key to its success, which may, for example, inspire long-term involvement in the project.

Persephone challenges the traditional approach to experimentation as a single method carried out as a controlled set of empirically recorded causalities. Instead, it offers a kind of mixology for the production of space that engages a whole range of investments and skill sets to produce a shared experience with common values—albeit this is not homogenized or universal. In this way, the rituals of production of the various environments from which Persephone is composed are as important as its physiological performance and materiality.

Persephone's mixology can be likened to the production of a fine drink. Alcoholic beverages are associated with distinct set of techniques, ingredients, and cultures that create the foundation of a community with common interests. These ideas and practices shape the design and experience of a real place—like a distillery in brewing, or worldship environment. They also enable and enrich imaginary spaces that include the poetics of mixology and the stories told by colonists. Both aspects are equally important for framing the actual operations of production and also to create continuity and meaning across generational timescales.

The production of fine alcoholic drink provides a way of understanding how a worldship may be realized through the orchestration of hypercomplex events—and highlights the importance of what initially appears to be trivial events. For example, in China, fortified wines seek not to create specific flavor profiles, but to embody an experience. Distillers therefore actively contribute to the success of celebrations like weddings or anniversaries. Documentation of the composition of the alcohol in terms of the various proportions of ingredients may give us an indication of its specific chemistry but they do not explain why we care, or conjure stories that offer timeless recollections of events.

In fact, it may not be possible to exactly record every individual parameter that shapes the production of an alcoholic drink. Differences in alcoholic beverages are produced by regionality, owing to the variability in blends of the local yeast and microorganisms that form microconsortia, which are like tiny cities or factories that provide the processing power and materials for the fermentation process. Unique sensations are produced by a combination of ingredients working in combination with the air and local terrains.

Certain Chinese fortified wines such as Strong Aroma are produced by not simply the chemistries mixed within a container, but also what kinds of substances leak into these blends during the fermentation process. Strong Aroma gets its unique character through its maturation in a mud pit where it absorbs added yeast and other microorganisms from the environment. In other words, the actual site contributes to the quality of the environment. Flavor gets absorbed into the mud walls as the alcohol is kept within the pit so that some of the recipes have been in continuous use since the sixteenth century.

Indeed, the sensory experiences associated with fine drinks exceed the capacity of the senses to adequately appreciate, or language to convey them. Mixology, of course, famously uses its own poetics to express shared experiences. Few of these are literal and yet the community understands what is being shared.

A language of contrasts begins to describe the character of a blend (e.g. whisky) whether it is dignified or youthful, soft and firm, big and subtle, dry and sweet, or austere and rich. Specific tones are indicated through highly evocative words like cereal, creamy, dark, grassy, hard, herbal, hot, malty, mouth-coating, nutty, oaky, orangey, citrusy, peaty, peppery, salty, smoky, spicy, woody, and winey. More scientific evocations provoke the chemical composition of the drink, such as estery, spirit, phenolic, warm, or hot.

Mixology helps to weave a pathway of experience that keeps a process and community together over many generations, while allowing the system and constituents to change in response to inevitable alternations in the place, community, or available materials.

While mixologists do not practice an exact science, their art and culture are critical if we are to realize ambitious and epic projects that are invoked by the construction of a worldship interior. We need to draw inspiration from the mixologist's portfolio if we are to excite communities about being part of an interstellar culture. Such a community cares as much about the anticipation of the interstellar quest as it does the actual realization of the pursuit itself—like producing a 100-year-old whisky.

Such long-term visions and commitment are not beyond our capacity to achieve but are a skill set that we have not rehearsed in the modern era in which products are expected to have short time-to-market cycles and economically benefit their inventors. We may learn how to appreciate much more subtle and ephemeral qualities that underpin our existence by considering the mixologist's art, which may help us to better explore hypercomplex phenomena—and lest we merely regard ourselves as disposable construction workers, or starship cargo—when, indeed, we could be emissaries of the cosmos.

4.6 What is a soil?

"If the doors of perception were cleansed every thing would appear to man as it is, infinite.
   For man has closed himself up, till he sees all things thro' narrow chinks of his cavern." (William Blake, *Marriage of Heaven and Hell* 1906)

If we are to believe that it is possible to live beyond our planetary confines, then we must consider the fabric of space not as being a sterile site or an industrial resource fit only for plundering, but as a living, potentially fertile system that can be better adapted to human existence. Yet, to currently survive within hostile spaces, we need to bring our living fabric with us. How we convey our Earth's blood as an infinite umbilical cord that binds us to our origins wherever we travel in the universe—or whether we sever this tether and transcend it altogether in a new existence—depends on how we establish the infrastructure in which we propose to dwell.

The world's first orbital home is already built. At 400 km above Earth's surface, the International Space Station (ISS) has continuously housed between three and six crewmembers for the past 15 years. Teams from the US, Japan, Canada, and Europe stay for about 6 months and live on supplies from Earth. The main question for survival is how and what the inhabitants are going to eat. This is a particular challenge because reduced gravity messes up the astronauts' metabolism and their freeze-dried dinner floats around while it is being reconstituted with water.

Imagine, then, the delight on August 10, 2015, when astronauts were able to snack on freshly harvested, space-grown, red romaine lettuce (see Section 4.7 below). This veggie experiment was the first time that NASA crews could eat what they grew. But it took a lot of effort. Plants were carefully nourished with hydroponic solutions—which are highly refined soils, without the muck, while special lighting made the leaves look tasty. The salad was cleaned with citrus sanitizing wipes, then nibbled raw. Later, it was soused in salad dressing.

Curiously, it transpires that the daily kilo of poop produced by ISS astronauts, rather than being used to cultivate space plants, is simply jettisoned. Admittedly, the spread of illness must be avoided, but astronauts reaching the station are generally healthy. But an effective sewage processing system in a closed space—after 15 years of continual occupancy—would seem to be a sensible idea.

By December 27, Scott Kelly, who is spending a year in space, tweeted on the progress of the lettuce:

"Our plants aren't looking too good. Would be a problem on Mars."

In practical terms, we have a long way to go before our ideas about space ecology even reach the level of organic complexity proposed in Ridley Scott's new film, *The Martian*. Stranded Mark Watney, played by Matt Damon, manages to feed himself with potatoes by mixing his excrement into the red soil.

Yet, the veggie experiment is an incredibly important step in space exploration. It emphasizes not only the importance of food in living away from Earth, but also the degree of difficulty in producing it under such extreme conditions.

This challenge is only going to intensify as we set up bases on the Moon, or Mars, which are too far away to rely on Earth for supplies. Somehow, we will have to figure out how to take our ecosystems with us—and they depend on exchanges between our oceans, air, and soils. But, when it comes to ecological thinking in space, we are way behind the level of sustainable practice that would be considered acceptable on Earth.

However, lest we assume that terrestrial practices can be seamlessly transposed into alien locations, we should remember that there is no "Nature" beyond Earth's surface. And, while we can *garden* ecologies, we don't actually know how *build* them from scratch. Our entire practice of farming depends on a robust set of material exchanges, which we largely take for granted—a largely overlooked assumption that was spotlighted on in the early 1990s by the Biosphere 2 experiment, in which a crew of eight explorers was housed in a three-acre, nine-storey-high, sealed enclosure for 2 years. This prototype for colonizing space, it was claimed, would also benefit society by leading to better understanding of ecosystems. Within this artificial world, five wilderness areas were designed to naturally recycle air, food, and water for the explorers during their isolation—without help from the outside world. However, within weeks of the launch, problems began to surface. There were rumors that supplies were being smuggled inside and that the atmosphere was scrubbed of carbon dioxide, which compromised the experimental findings. Whatever the truth, after about a year, oxygen levels fell steadily, the ocean acidified, internal temperatures rose, and carbon dioxide levels fluctuated. The system clearly began to fail at around 18 months as composting was stopped to save oxygen and the soils collapsed. Explorers suffered personal hardships including suffocation, weight loss, and mental stress (Poynter 2006). Vertebrates and pollinating insects died, while only the cockroaches and ants thrived. Although the experiment was regarded as a media sensation, it had largely failed as a scientific experiment.

If we can't maintain a sealed ecosystem on a life-bearing planet, what chance do we have of building them in space?

One challenge is that the 1967 Outer Space Treaty, which establishes the legal principles for space exploration, is effectively anti-life. It assumes that humans—or our objects, like robots and probes—either contaminate non-terrestrial environments or are contaminated by them.

Yet, life is *all about* contamination! It is *messy*!

The human body is not a discrete, closed structure, but an "ecology" of mutually interdependent cells and overlapping ecosystems. While we might ignore these relationships on our home planet, they are impossible to discount when living in extreme conditions. For example, in the absence of our natural bacteria, we cannot digest certain foodstuffs. And, without green plants, we are starved of oxygen.

So, while the need to avoid contaminating pristine alien terrains may sound sensible, it will not help us to establish life-promoting environments. As we venture into space, we will inevitably bring other organisms with us—microbes, pets, domestic animals, and plants—all of which make up our biosphere. We also cannot assume that non-terrestrial environments can perform the same kind of function as Nature does. We will have to build the fundamental cycles of life and death from scratch, regenerating organic compounds through life-giving composts whose ingredients read like a gourmet cookbook. This is a very different practice from incineration, which is a one-way waste-disposal system.

As we explore the Solar System and beyond, we must actively design and engineer the infrastructures of life. This applies not just to alien worlds, but also to engineered habitats, like biospheres and worldships.

Persephone identifies soil as the fundamental structure with which we can design and engineer these possibilities. While oceans and air are also vital for life, soil is a material system that opens up spaces in which water and gases can interact with matter in the huge surface areas that are offered by soil structures. Yet, contemporary notions of soil are naturalistic. They describe the range of substances that comprise our Earth as a terrestrial material made up of organic matter, minerals, water, and life forms that constitute giant self-organizing bodies. These great mats inhabit the upper layers of Earth as dynamic systems for many thousands of years. They are responsible for cycles of organic life on Earth, continually transforming the fallen bodies. They sink as compost into its substance and rise again through the regenerative powers of microorganisms and soil ecosystems.

Yet, such materiality is unique to Earth's surface. To date, no other material condition has been identified on any other celestial body. While the Moon and Mars are described as having soils, the loose material on the surface of these barren planets is little more than sand and dirt, with no organic matter or water content. Consequently, the structures of these fabrics are homogenous. Although they present a large surface area for catalysis, they are absent of heterogeneous environments typical of terrestrial soils (Fig. 4.2).

Figure 4.2 Rendering of Persephone supersoil that consists of engineered organisms, smart chemistries, and silicon circuitry designed to integrated living and mechanical systems. *Credit*: Jon Morris and Phil Watson

For Persephone's experimental purposes, soils are not merely a reconstruction of their terrestrial materiality, but may be designed and produced in a whole range of contexts—as a Cambrian explosion of possibilities in which the fertility of sites can be augmented by a range of strategies from engineered superbugs to the addition of chemical catalysts. Such highly designed and engineered substrates may be likened to the kinds of modifications that synthetic biology proposes for natural organisms to alter their efficiency or function. These fabricated earths may be thought of as "supersoils" that are designed to perform specific tasks. Their composition and functions are highly contextualized, being shaped by the local performance of gravity, atmosphere, and even weather conditions. Understanding how natural soils are formed, thrive, and how their propagation, connectivity, and metabolic exchanges may be modified is essential not only for life beyond our planet, but—since we are losing over 6.4 billion tons of soil each year—for our long-term survival on Earth also (Fig. 4.3).

Who needs soils? 95

Figure 4.3 Persephone soil, a supersoil species. *Credit*: Jon Morris and Phil Watson

4.7 Who needs soils?

Kjell Lindgren: "That's awesome!"
Scott Kelly: "Tastes good."

International Space Station (ISS) astronauts Scott Kelly, Kjell Lindgren, and Kimiya Yui enjoyed the first fresh food grown on the space station, with a little salad dressing thrown in for flavor, on August 9, 2015. The tasty red romaine lettuce had been cultivated for 33 days from seeds planted by Kelly on July 8. In becoming part of their meal, the salad set a new benchmark in space dining—a developing culinary frontier.

The lettuce was cultivated in "pillow" formations within the experiment Veg-01 facility, a collapsible, expandable unit that emits a pink light that is produced by a combination of red, blue, and green LEDs. About half the lettuce will be consumed, while the rest is returned to Earth for further study. The hydroponics technology is destined for further development in highly productive systems that will provide Martian explorers with the means of growing sustainable food for deep-space missions.

Hydroponics is not new technology. It has been used for centuries to grow plants in soil-free conditions. Around 35 % of all British raspberries, strawberries, blueberries, and other soft summer fruit are no longer raised in soil. For example, in glass houses, strawberry plants grow in coir—the hair-like substance found in the husks of coconuts—in plastic pots held in place on gutters hanging from the ceiling. Water and nutrients, including nitrogen, phosphorus, calcium, potassium, and magnesium, are dripped onto the coir.

Recent advances in hydroponics have enabled a high degree of control and customization of soil-free conditions that give growers more control over nutrient mixtures, temperature, humidity, and growing schedules.

There are many advantages of hydroponics over soil-based nutrient systems. Hydroponics has been successfully used to grow a wide range of plants including tomatoes, strawberries, watermelons, cantaloupes, tomatoes, blueberries, blackberries, raspberries, and grapes. Many hydroponic gardeners also successfully grow more exotic fruits such as kiwi and pineapples. Fresh herbs are also very popular.

Hydroponic systems are popular, as they save an incredible amount of space compared with traditional soil gardening by submerging the roots in a bath of oxygenated nutrient solution, which constrains the naturally spreading roots. Hydroponics speeds up plant growth by providing them with all their nutritional needs in solution, which uses only a small percentage of the water needed to irrigate soil-based systems. A covered, recirculating nutrient reservoir is used to prevent evaporation and seepage. Around 90% of the water used in traditional gardening can be conserved in simply switching to a hydroponic setup. Moreover, no incidental weeds, infections, or pests grow when there is no soil.

Yet, although hydroponics proposes to be "soil-free," the systems functionally act as soils. In other words, the infrastructures that support these plants—the coir, the flow of water and minerals—are effectively the kinds of strategies that natural soils employ to support their native ecosystems but stripped down to a minimal configuration. Moreover, the possibility of spontaneous biodiversity is reduced as not all plants can be grown hydroponically. Hydroponics does not therefore completely replace soil function as they are so highly abstracted. While they may be highly controllable, hydroponics gardens also have to be carefully managed as they are not forgiving and the whole operation still needs supervision so that major trouble can be avoided. They need much more human input than natural soils because they are dependent on the gardener's decisions and input for their very survival, as they are much more susceptible to invasion by external agents. If they become infected with pathogens, they will completely collapse, since there is little resilience or robustness built into the system.

Additionally, organic soils are not just substrates for crop growth. They perform many more functions such as absorbing carbon dioxide and building many complex landscapes that can coexist in the same environment. Although organic earths possess native contaminants of fungal spores, pests, and bacteria, which are a continual concern for every crop grower, they stand a good chance of making a spontaneous recovery after damage and infestations. Also, if they are abandoned, they may become unruly and transform into a weed- and bug-infested mess of rotted fruit, but the plants, in general, can be revived when specific ecological choreographies and spatial systems are restored. When organisms die, they are recomplexified and returned into lifecycles via assimilation through many of the entangled nutrient and elemental cycles that are enabled by sunlight. Organic terrestrial soils are key to those metabolic processes of decay and composting, which actively transform no-longer-viable organic substances into bioavailable compounds that are then assimilated into active ecosystems.

So, while the abstraction of the principles of soil-based operations through the field of hydroponics helps us to understand soil in a context of computationally programmable chemistries and water flow, it only represents part of the portfolio of life-supporting strategies that our terrestrial life-bearing membranes are capable of producing.

It is essential that, in adopting an ecological view of non-terrestrial environments, such as in the construction of a "living" habitat within a worldship, continual cycles of life and death are enabled. These ongoing active exchanges may persistently increase the liveliness of a variety of territories and ultimately enable the propagation of robust, creative, resilient, and biodiverse worlds.

4.8 Portfolio of materials for making soils

A set of materials, methods, and technologies needs to be identified to create the conditions for experiments in soil construction. Potentially, these toolsets may enable the construction of "supersoils." These are geotextiles and complex organic fabrics that augment environmental performance or enable new processes to occur. Similarly to the way in which synthetic biology has transformed the expectations of organisms, so supersoils are earths that have been engineered to meet specific criteria, or exceed natural performance expectations.

Conventional methods of producing soils mimic terrestrial processes by mixing agreed proportions of ingredients together. Existing dirt, decomposing organic matter, mineral particles (coarse sand/perlite), and fertilizer are folded together in cake-making style, to form a well-draining, nutrient-rich mixture.

Yet, gravity and chemical self-assembly, the very processes which organize these substances into the complex spatial structures that are characteristic of organic earths, may not occur in Persephone's environment. For example, subtle alterations in the performance of chemistry and physics beyond Earth's surface may result in essential dysfunction. Water leakage is not wasted in terrestrial soil systems and contributes to sustaining natural cycles as it pools in clay or mineral layers to form aquifers and water tables, from which springs may ultimately flow and carry nutrients from one site to another. Gases are also processed in the multitudinous spaces and catalytic surfaces of soils. Carbon dioxide is absorbed through diffusion—by the chemical processes of microorganisms, inorganic reactions—and re-released through a spectrum of physical and metabolic activities, which include burning peat as fuel and aerobic respiration. In fact, soils possess such an affinity for gases that they are recognized as an essential sink for carbon compounds. As we lose our naturally fertile soils to industrial processes, urban settlement, and intensive farming practices, we are compromising not only a food-producing fabric, but also an essential carbon-fixing resource. Understanding the technical and dynamic characteristics of soils may help us to develop new ways of dealing with some of the biggest challenges that we face on the planet today.

In considering the construction of soils within a worldship environment, it is worth establishing how far it is possible to design these properties and how far they rely on the terrestrial "constants" that we assume, such as diffusion, airflow, or gravity. A more fundamental examination of the principles of construction is therefore proposed by using a range of dynamic, self-organizing chemical systems, which may also be manipulated through their environment. The processes that engage with their design and engineering as supersoils are considered as a kind of computational system that will be discussed in more detail in Part I, Chapter 7.

Conjecturally, soils may be considered in a technological context as a production platform. They seamlessly combine the actions of lively bodies and materials such as dissipative structures, elemental matrices, and natural computing processes. Collectively, these agents produce a system of dynamic complexification in which lifelike systems may be precipitated, sustained, and multiplied. Effectively, then, the work produced by complex, spatialized material systems can be strategically harnessed through the technology of soils to achieve desired effects—namely increasing the fertility of a site. Indeed, the context of producing living fabrics for space soils may be regarded as a fundamentally technical platform that not only generates the conditions for the persistence of life, but also enables it to flourish.

The two main groups of substances that will be experimentally explored, as a starting point, are gels and dynamic droplets.

Gels provide a model system that explores the infrastructural support needed within a complex material to facilitate the free flow of water, matter, and air. While their structure is nowhere near as complex as a cell milieu or terrestrial soils, they create a set of conditions such as diffusion and gradient formation. A central organizing system does not perform this coordinated set of exchanges that takes place in a highly spatialized manner. It arises from the agency attributed to the exchanges at field interfaces and provides an open system for experimenting on dynamic materials and structures in ways that can be observed, manipulated, and recorded. Gels are translucent and, therefore, with the right framing, can clearly reveal the topology of complex, unevenly distributed processes. The structural and spatial qualities of gels that reflect on the properties of soil systems are enabled through polymer chains. Gels have the capacity to create obstacles, provide rigidity, chemically interact, and transform in the presence of other agents. The spaces in between these physical strands may be variably occupied by a host of molecules to which the gels are semi-permeable or selectively permeable. Additionally, the molecular forces within these fields of spaces and structures—charge, or weak molecular forces—are also catalytic in configuring the character of systems. Consequently, gels possess a dynamic character that can change according to physical and chemical conditions. They provide an accessible model to engage with the principles of soil construction.

Soils are also characterized by highly mobile systems that are characteristically biological in origin. They are sufficiently complex to recycle organic matter back into the fabric of new life. As matter decays in the form of fallen deciduous leaves and the dead bodies of organisms, it provides sustenance to many species of bacteria. Through a host of metabolic networks, the building blocks of life are re-assimilated through their transformation into rich compost. This matter, in turn, is passed through food chains such as fungi that are saprophytes, which feed directly on decaying matter—or through the roots of green plants. Through myriad exchanges, organic building blocks are incorporated into new metabolic systems—namely photosynthesis in green plants and the respiratory networks of animals. Such agents confer complex functions on soils such as redistributing matter, like earthworms, or forming communications systems, which is a property of fungal-root networks of mycorrhiza.

The active spatial programming of soils through physiological processes was first documented by Charles Darwin, who observed that earthworms were responsible for the movement of large stones into the earth. In this context, worms acted as a kind of local 3D printer system. They obtained the printing material by removing dirt from underneath the rocks and then re-depositing it onto the surface as casts. The stones therefore sank into the ground faster than by gravity alone (Darwin 2007). However, earthworms do more than create the structural matrix for soils. They transform organic matter through their digestive processes and transport substances from one place to another at a rate that exceeds diffusion. For example, earthworms also produce calcium carbonate casts, which increase the alkalinity of soils.

Other biological systems such as fungal mycelia also rapidly transmit chemical information through the soil body, like a chemical Internet that possesses such rich connections that—like mother trees—they are critically important in nurturing the diverse ecosystems of our forests (Howard 2011). Through these highly choreographed differential activities, therefore, soils underpin the webs of life.

Since Persephone is considering the structure of soils from their first principles, the idea that chemical agents could replace agents such as worms and fungal mycelia is entertained. Substitutions are made with a simple, mobile system of dynamic droplets that act as complex lifelike bodies, although they are not, technically speaking, alive.

Dynamic droplets can be produced through the self-assembly of molecules that arise from overlapping fields of chemical potential by, for example, introducing alkali into a field of olive oil. A specific example is the Bütschli system, first described in 1892 by Otto Bütschli, who mixed potash with olive oil to form an "artificial" amoeba with pseudopodia (cytoplasmic extensions) that behaved in a lifelike manner (Bütschli 1892). His aim was to make a simplified experimental model to explain the plasticity of body morphology and movement, based purely on physical and chemical processes such as fluid dynamics and changes in surface tension (Fig. 4.4).

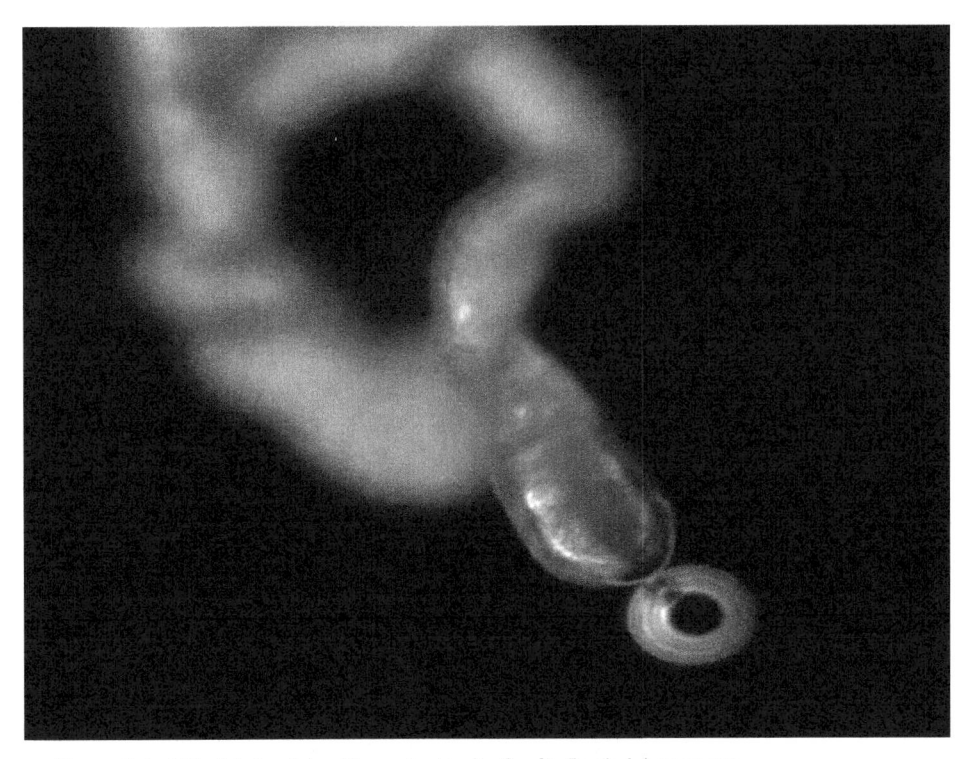

Figure 4.4 Bütschli droplet with product trail. *Credit*: Rachel Armstrong

Bütschli droplets are around a millimeter in diameter and possess a primary metabolism, saponification, which spontaneously exists at the interface where strong alkaline water and olive oil meet. This reaction releases both energy and products in the form of surfactants that modify the oil/water interface. This reaction is responsible for both the lowering of surface tension, allowing the droplet to deform, and the flow of liquid, which results in droplet morphological fluctuations, movement, and even droplet splitting. Where the generative chemical fields overlap, dynamic droplets form. As the droplets move through their environment, they use the alkaline reactant within the droplet as fuel, which consumes the olive oil and processes it to form skins of soap. Bütschli droplet movements last between several seconds and around 20 min. The activity of any particular droplet is not predictable, and the success of creating the system is variable and possibly dependent on the quality of ingredients. The outcomes can be highly complex and variable, despite the system possessing no central programming code such as DNA. Additives or degradation products in the olive oil will alter the reactivity of the system and further enhance its variability. As the active droplet system progresses in time, the activity of the system slows as it approaches chemical equilibrium. Due to both the accumulation of inhibitory products and the consumption of fuel, the droplet eventually becomes inactive (Armstrong 2015).

Dynamic droplets are characterized by a leading edge and a trailing edge. The active anterior pole of the droplet is a site where oil and alkali vigorously produce sodium oleate (a soap-like substance) and absorb heat. These exchanges also establish a convection current inside the droplet body, as consumed alkali rushes to the interface to further interact with the olive oil. Their combined exchanges generate the Marangoni effect, which enables the droplet to propel itself through the viscous oil medium. Soap crystals produced at the interface are swept to the posterior pole and form a layer that occludes interaction between alkali and oil. These combined physical and chemical systems exert positive feedback on the droplet body so that it is propelled toward the anterior pole. Of note is that, although the basic chemistry and material program have not changed—the droplet is fundamentally a soap-making system—the outcome of the system is altered by its spatial relations. Specifically, the fate of the molecular species depends on their position relative to the poles, the net outcome of which establishes the morphology (shape) and behavior of the system (direction of travel). This context-based program is similar to the relationship between genetic sequencing of amino-acid chains and the highly complex, spatial programs that underpin protein folding.

The dynamic droplet system behaves in a remarkably lifelike way and generates an experimental platform that can be customized to perform differently under a range of contexts. These mobile, programmable, and responsive bodies may be thought of as performing functions that have resonances with the biological organisms that inhabit soils, such as worms and mycelia. In a very simple way, they become chemical equivalents for terrestrial life.

Indeed, if we are to venture and thrive beyond Earth's surface, we need to find opportunities to promote the life-giving fabrics in which we are immersed. Soils influence the performance of molecular species that pass through their complex spatial infrastructure via their spatiality. The barriers and passages in the labyrinthine material generate the conditions in which equilibrium is evaded and new opportunities for syntheses are produced. Notably, the matter that passes into a soil system also becomes an integral part of the soil body—not just structurally, but also physiologically. The transformative nature of these highly heterogeneous materials may transform sterile environments into new sites of abundance, so that human development becomes a process of terraforming—not its antithesis. By identifying experimental models that can help us prototype a diverse range of life-promoting fabrics, we may conduct complex experimental that help us ask challenging new questions. For example, we may develop a science of making supersoils, which augment environmental performance and may help us learn to live more sustainably on Earth as well as, considerately and sustainably, to colonize the stars.

REFERENCES

R. Armstrong, *Vibrant Architecture: Matter as a Codesigner of Living Structures* (De Gruyter Open, Berlin, 2015)

J.D. Bernal, *The World, the Flesh & the Devil: An Enquiry into the Future of the Three Enemies of the Rational Soul*, 2nd edn. (Indiana University Press, Bloomington, 1969)

W. Blake, *The Marriage of Heaven and Hell: Good Is Heaven—Evil Is Hell* (John W. Luce and Company, Boston, 1906)

A. Bond, A.R. Martin, Worldships: an assessment of the engineering feasibility. J. Br. Interplanet. Soc. **37**, 254–266 (1984)

J.L. Borges, *Labyrinths* (Penguin Classics, London, 2000)

O. Bütschli, Untersuchungen ueber microscopische Schaume und das Protoplasma, Leipzig (1892)

C.R. Church, J.T. Snow, E.M. Agee, Tornado vortex simulation at Purdue University. Bull. Am. Meteorol. Soc. **58**, 900–908 (1977)

A.C. Clarke, *Rendezvous with Rama* (Gollancz, London, 1973)

C. Darwin, *The Formation of Vegetable Mould Through the Action of Worms* (The Echo Library, Fairford, 2007)

Dezeen, Can city mobile aluminium furnace by studio swine (2013a), Nov 1, www.dezeen.com/2013/11/01/can-city-mobile-aluminium-furnace-catadores-sao-paulo-studio-swine/. Accessed 28 Apr 2015

Dezeen, Open source sea chair by studio swine (2013b), Feb 16, www.dezeen.com/2013/02/16/open-source-sea-chair-by-studio-swine/. Accessed 15 Apr 2015

B.R. Finney, E.M. Jones, *Interstellar Migration and the Human Experience* (University of California Press, Berkeley, 1985)

W.H. Fuller, *Soils of the Desert Southwest* (University of Arizona Press, Tucson, 1975)

D. Gissen, Thoughts on a heap of rubble, in *Paradigms of Nature: Post Natural Futures. Kerb 19, Journal of Landscape Architecture*, ed. by C. Daly, S. Hicks, A. Keene, R.R. Ricardo (RMIT University, Melbourne, 2011), pp. 50–52

A.M. Hein, M.D. Pak, D. Putz, C. Buhler, P. Reiss, World ships—architectures and feasibility revisited. J. Br. Interplanet. Soc. **65**, 119–133 (2012)

C. Howard, "Mother trees" use fungal networks to feed the forest ecology. Canadian Geographic (2011), Jan/Feb, https://www.canadiangeographic.ca/magazine/jf11/fungal_systems.asp. Accessed 19 Apr 2014

R.D. Johnson, C. Holbrow, Space settlements: a design study. (NASA, Washington, SP-413, 1976), www.unbc.ca/assets/history/courses/201101_nasa_sp_413.space.settlements.a.design.study.pdf. Accessed 18 July 2015

M. Kayser, Solar sinter (2011), www.markuskayser.com/work/solarsinter/. Accessed 18 Aug 2015

R. Kurtzweil, *The Singularity if Near: When Humans Transcend Biology* (Viking, London, 2005)

Le Corbusier (C.E. Jeanneret). *Toward an Architecture*. (Getty Research Institute, Los Angeles, 2007)

I.M. Levitt, D.M. Cole, *Exploring the Secrets of Space: Astronautics for the Layman* (Prentice-Hall, New Jersey, 1963)

E.F. Mallove, G.L. Matloff, *The Starflight Handbook: A Pioneer's Guide to Interstellar Travel* (Wiley, New York, 1989)

A.R. Martin, Worldships: concept, cause, cost, construction and colonization. J. Br. Interplanet. Soc. **37**(6), 242–253 (1984)

C. Martineau, *How the Gringos Stole Tequila: The Modern Age of Mexico's Traditional Spirit* (Chicago Review Press, Chicago, 2015)

M. Mautner, *Seeding the Universe with Life: Securing Our Cosmological Future* (Legacy Books, Christchurch, 2004)

T. Morton, *Ecology Without Nature: Rethinking Environmental Aesthetics* (Harvard University Press, Cambridge, MA, 2007)

G.K. O'Neill, *2081: A Hopeful View of the Human Future* (Simon & Schuster, New York, 1981)

J. Poynter, *The Human Experiment: Two Years and Twenty Minutes Inside Biosphere 2* (Thunder's Mouth Press, New York, 2006)

N. Takahata, Allelic genealogy and human evolution. Mol. Biol. Evol. **10**(1), 2–22 (1993)

Vimeo, Anthropocene, Capitalocene, Chthulucene: staying with the trouble. Donna Haraway (2014), Sept 5, https://vimeo.com/97663518. Accessed 18 Aug 2015

5

Experimental architecture: on-world and off-world exploration of possibilities

Rachel Armstrong

This chapter explores ways of dealing with very complex and ambitious research questions using iterative experiments and innovative laboratory settings whose explorations produce probabilistic outcomes.

5.1 Gel experiments

Persephone explores the fundamental principles of soil-making in an experimental context, so that it is possible to think about what ingredients may be used in the construction of soils within its living interior.

The starting point is to examine spatial principles of metabolic agents within a matrix that helps coordinate life-promoting events in time and space.

Gels are used as a fundamental infrastructure, as they consist of structures that take the form of polymer chains that are separated by spaces, which enable substances such as water, salt solutions, and air to pass. This framework, therefore, sets up a medium in which movement through the fabric can be examined, as gels are translucent, allowing the passage of light and, therefore, unlike the light-absorbing organic matter typical of composting soils, the spatial events within them can be observed and recorded. The limiting factors in the passage of these substances include many physical and chemical factors that relate to variables, such as the size of the molecules, the presence of an interstitial medium such as water, and gradient concentrations.

A simple experimental system was devised to represent an entirely designed soil system and reveal principles related to the movement of matter through a structural matrix that could be transferred to non-terrestrial environments. Activated gels, which contained a low concentration of alkali reagent (sodium or ammonium hydroxide), were infused with brightly colored

---

Rachel Armstrong, PhD. (✉)
Professor of Experimental Architecture, Newcastle University, Newcastle-upon-Tyne, UK

salts (divalent cations such as nickel, chromium, iron II, iron III, and copper II) so that movement of substances through these matrices could be observed and design strategies proposed.

A range of prototypes were constructed by architecture students at the Chemistry Outreach Laboratory at Newcastle University, to understand how substances that are not at equilibrium states may spatially organize.

Of particular interest in this context was a chemical phenomenon that produces Liesegang Rings. This is a diffusion precipitation phenomenon first observed by Raphael Liesegang in the preparation of gels using silver nitrate and potassium dichromate, in which he observed self-organizing bands. This spontaneous recording of molecular events can demonstrate the generation of spatial and chemical complexity within very simple soils, like gel systems. The recordings are directly produced by soluble salts introduced into "activated" (alkaline) gels that move sequentially, as soluble and insoluble complexes, through the matrix under the influence of gravity.

To demonstrate the design principles in action, a range of self-organizing chemical drawings were created to produce soft, dynamic traces to depict the flow of water and matter through the rudimentary gel matrix. The experiments were set up during the dynamic chemistry and architectural design class at the "Thinking through Making: One Week of Material Design" week, held on January 29, 2015, at the Chemistry Outreach Laboratory, with fifth-year students from Newcastle University.

Liesegang Rings are spatial, time-independent structures that are produced from the interactions between non-uniform fields of agents. Specifically, symmetry is broken by an initial spatial concentration gradient that prompts a wave of reagents to move through the system. Physical forces such as diffusion and gravity, and chemical transformations such as crystal precipitation, shape the interactions between these multi-material dynamic agents. The width and spacing of the bands of crystal precipitation vary according to a variety of starting conditions, such as their distance from the origin of the imposed concentration gradient (Fig. 5.1).

The architectural objective was to observe and choreograph the performance of enlivened inorganic salts—1 M solutions of nickel, chromium, manganese, iron II, iron III, and copper II—in excitable media, which consisted of alkali-activated agar (1 M ammonium hydroxide in 2% agar). The conditions for the diffusion and exchange of matter were established by producing interfaces that took the form of diffusion waves and the periodic precipitation and re-dissolution of complex salts through the gels. The activated matrix was selectively permeable to the movement of matter, as salts passed in and out of solution through the design field. The field of interactions between cation and anion pairings enabled material transformation to occur. These were observed as a whole range of evolving patterns in the gels. Further complexity was introduced by using an open Petri dish, rather than the traditional vessel (a boiling tube). This encouraged lateral diffusion in the system and gave students access to multiple sites of access through which they could choreograph a designed topology composed of hubs of diffusion and precipitation waves.

Procedurally, the modified Liesegang Ring experiment speaks of an origins of lifestyle transition that is unconstrained by naturalized aesthetics, where the "qualia" of interacting lively materials orchestrate new independent acts of creation. These were of interest to Alan Turing in his 1952 paper "On Morphogenesis" in which he attributed diffusion precipitation waves as being responsible for patterning in organisms, such as dappling (Turing 1952).

Poetically, the ensuing experimental events are read as an alternative synthesis of the story of life from chemical and physical events through the text of *Paradise Lost* (2005).

Figure 5.1 A Liesegang Ring-based gel system that produces complex patterns and structures. Credit: Rachel Armstrong

This text, by John Milton, enables us to graphically discuss how the kind of chemical and physical events observed in the gels suggest the organizing forces that underpin an alternative Nature. However, this is not of divine origin, but is born from a new collaboration between empowered matter and secular human agency, which share a common project in their mutual, continued survival. Each gel produced during this experiment is considered as a drawing related to Milton's text that frames the experimental work as a series of chemically produced graphics, which enable a new reading of a barren landscape as a physical topology for the production of soils (Fig. 5.2).

Questions are raised about whether these properties of gel matrices are universal, and how the massive physical complexity of soil-like structures can be translated into architectural-scale environments like worldships.

Although, in this context, the goals of the experimental process are examined through an artistic lens, such explorations also create the conditions for thinking about the design and engineering of soils in a scientific and engineering context. Through design-led experiments,

106 Experimental architecture: on-world and off-world exploration of possibilities

Figure 5.2 A reading of *Paradise Lost* through the images generated by Liesegang Ring-primed gels speaks of a novel synthesis of life orchestrated through human design and engineering. Credit: Rachel Armstrong

we already become aware of working with the spatial qualities of matrix design, their physical performance, and their capacity to change over time. Ultimately, these kinds of explorations may help us to practically engage with the possibility of making supersoils.

5.2 The Hanging Gardens of Medusa

The Hanging Gardens of Medusa are the world's first stratospheric sky gardens and laboratory to feature "proto" artificial life forms at around 100,000 ft above Earth's surface. They embody an encounter between the sublime and the material, whereby life forms are supported by an artificial platform. They climb by helium balloon to reach conditions that resemble the Martian atmosphere in terms of extremes of temperature, pressure, and certain bands of UV radiation. The Hanging Gardens of Medusa are made up of two tiered gardens that are supported by separate artificial "soil" infrastructures. The highest tier, the Falcon Garden, features biological organisms which are epiphytes that can survive within the atmosphere, and the lower tier, the Medusa Gardens, proposes alternative future life forms that take the form of self-assembling chemistries, which require oil-based soil systems.

This collaborative project between Nebula Sciences and Newcastle University, took the form of a stratospheric garden launched from Central Valley, CA, on Sunday, August 16, 2015. Collaborators included Rachel Armstrong, professor of experimental architecture, Newcastle University, UK; Sam Harrison, founder and lead engineer, Nebula Sciences; Andrey Sushko, lead hardware engineer, Nebula Sciences; Ian Gomez, flight mechanics advisor, Nebula Sciences; Aria Tedjarati, volunteer lead software engineer, Nebula Sciences; Daniel Parker, operations coordinator, Nebula Sciences; Dr. David J. Smith, advisor, Nebula Sciences; Richard Osborne, near-space engineering consultant; and Kelvin F. Long, lead technical advisor, Nebula Sciences.

This design-led experiment aims to pose challenging and complex questions through making—or "synthesis"—rather than provide premature or idealized solutions to highly complex issues. Its complex, collaborative, multidisciplinary nature, which has required the coordination of teams of researchers and entrepreneurs in the UK and the US, demonstrates how the nature of experiment is changing, from being highly refined and controlled to being open, networked, and contextualized. It also asks about the nature and organization of laboratories in the near future. The project highlights our current vulnerabilities in the face of the changing status of our planetary ecosystems, while dramatizing the challenges presented in extending our reach into unknown territories. Addressing the notion of life at its extremes, the project seeks to construct the infrastructures of life forms capable of surviving, and even thriving, in extremely harsh environments. These are vital for creating the foundations for sustaining ecosystems in the longer term.

The project differs from others that have flown plants to high altitude in its ambition to propose alternative forms of life to terrestrial organisms. For example, the 2014 Japanese project "Exobiotanica," which launched botanical bouquets at high altitude using a helium balloon, was largely an aesthetic exercise in situating plant life against the backdrop of Earth's surface. In giving up gravity and soil, the striking bouquets posed questions about the infrastructures of life. In contrast, the Hanging Gardens of Medusa not only observe the impossibility of life under these conditions, but also propose alternative strategies for survival that can be viewed through artificial chemical landscapes and in the "programming" of the gardens' soils (AMKK 2014).

The gardens also gesture toward the ancient wonder of the world, the Hanging Gardens of Babylon, which were an ascending series of tiered gardens containing all manner of trees, shrubs, and vines constructed within the heart of the ancient city. The context for their design is based on "Meeting with Medusa," a short story by Arthur C. Clarke (Clarke and Robinson 1988), which is set in Jupiter's atmosphere. Here, alien life forms reside, including incredible megafauna, such as a mile-long aerial jellyfish, which is the reference to Medusa in the story. The gardens raise questions about an ecological era for space exploration, where infrastructures that enable us to live away from our home planet for increasing periods are established. During these times away from our planet, we may encounter alternative life forms that are part terrestrial and are, at least in part, of our own making, or entirely alien.

The hanging gardens consist of two separate landscapes: the Falcon and the Medusa; one is situated above the other.

The Falcon Garden is named after the cyborg protagonist in Clarke's short story. The cacti garden tier consisted of three "space plants": *Tillandsia medusae*, a South American plant renowned for its ability to grow without soil or water and thus withstanding hostile environments; a hybrid of red-stained sea anemone skeleton and air plant *Tillandsia medusa*; and Euphorbia Lactea Variegata or Dragon Tree Bone Catcus. This experiment draws our attention to the need for infrastructures that are not naturalistic solutions like terrestrial soils, but demands new kinds of material choreography and syntheses between mechanical, physical, and chemical systems, to produce livable environments in extreme places (Fig. 5.3).

Figure 5.3 Falcon Garden with hardy biological life forms. Credit: Rachel Armstrong and Nebula Sciences

The protocells that make up the sealed Medusa chemical garden are lifelike spherical bodies of around half a centimeter in diameter and are growing solid structures within them. They are contained within a liquid environment with an electronic infrastructure that is insulated with gold foil that protects them against the extremely low temperatures, which reach around −70 °C. Being jellyfish-shaped, the protocells represent the "Medusa" alien life forms referred to in Clarke's short story. The pre-launch cell possesses a transparent body and there is clear mineralization evident in the droplet. This is formed through a combination of iron and copper salts added to the cell body that makes them easier to see. It also gives them a slow and primitive "metabolism" as they absorb carbon dioxide from the solution and turn it into crystals, which is something that can be seen changing. Since they show lifelike behavior, they also age, but not quite the way we do. To slow down the ageing process, a soap-like substance was used that acted as an inhibitor. Through starting as semi-transparent droplets, as the garden gains altitude, minerals are introduced into the liquid and create a metabolizing effect, causing a dark layer to gradually form and turn into sediment. Yet the protocells know no limits of life and death, as their constituent agents exist in a transitional state of being. Although the ecosystem of chemical exchanges is capable of growth and movement, they do not have the status of being fully "alive." (Fig. 5.4).

Figure 5.4 Medusa Garden with chemical lifelike forms. Credit: Rachel Armstrong and Nebula Sciences

The images from the experimental platform capture three synergetic perspectives of living systems at high altitude that include:

1. A view of the thin natural membranes that meet at the horizon to produce the atmosphere, ocean, and soil. This delicate living layer is highly patterned with complex dynamic meteorological systems—clouds, storms, rain, hail, snow, and brilliant sunshine. They collectively constitute our natural operating system, which underpins all living systems.
2. A perspective over the biological Falcon Garden toward the curvature of Earth that is foregrounded by suspended fleshy cacti. These organisms are destined to die—but do so spectacularly and defiantly.
3. A panorama through the Medusa Garden that captures Earth's curvature through a thin, vertical tank of semi-living agents with no expectation of being alive or dead. Through this liquid lens, the living chemistry of "alien" bodies are ambiguously falling toward Earth, or rising up away from our planet and into unknown extraterrestrial environments.

It was anticipated that the lifelike chemistries would persist better in extreme conditions than biological agents. In practice, terrestrial plants outperformed the artificial cells, as they had much more structural integrity that enabled them to withstand the extreme conditions of the high altitude at 85,000 ft, including a vigorous jet stream, while the artificial life forms with their super "soft" and distributed self-organizing program were completely destroyed by the process. The post-flight protocells formed a dark complex, where the droplet structure broke down, resulting in sediment which indicated that they had rapidly "aged." (Fig. 5.5)

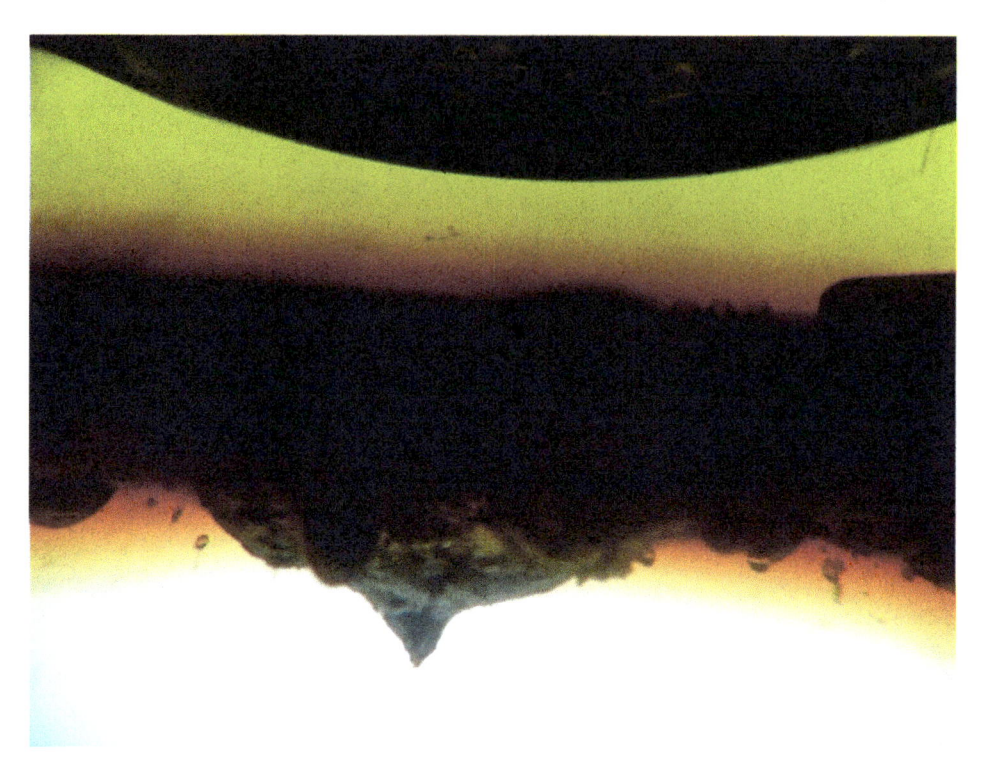

Figure 5.5 Medusa Garden with chemical lifelike forms that have disintegrated on exposure to the extreme environmental conditions of the stratosphere. Credit: Rachel Armstrong and Nebula Sciences

In examining the pre- and post-flight protocells, it is clear that, at high altitude, the ageing process of these droplets is immensely accelerated and, by the time the platform returns to Earth, the droplet structures have shattered into mineral fragments.

This design-led experimental approach reveals that, counter to the popularly held ideas of "gray goo" and synthetic forms of life posing a threat to biological life on Earth, when these artificial systems are put through their paces, they are much less robust than our native biological systems.

The Hanging Gardens of Medusa are an experimental soil prototype that establishes a rudimentary baseline for the support and synthesis of new ecological relationships within extreme and alien environments. The gardens also generate technological and material possibilities that may later inform the synthesis of Persephone's artificial soils. They also draw our attention to the spectacle of life and also the technical challenges invoked by sustaining it, especially in designing life-promoting infrastructures, such as nurturing soils.

Additionally, the project creates a juxtaposition of the possibility of artificial life that speaks about the potential panspermic origins of life on our home planet—a theory that supposes chemicals from space may have started off a series of reactions that have resulted in the kinds of life forms that we recognize today. Ideas of life beyond our atmosphere are contemplated so that, if these chemical "seeds" were to keep on going upward into the realms beyond our world, they might reach conditions that could potentially transform and sustain them to become living systems. This constructed act of "seeding life" might be considered as a form of "directed" panspermia, whereby human-initiated events precipitate the evolution of life on other worlds.

5.3 Capsule of Crossed Destinies

The "Capsule of Crossed Destinies," a sealed biosphere carried on a helium balloon flight, was launched from Neath, South Wales, by Nebula Sciences in January 2016. The 300-g payload was piggybacked on a commercial flight to record the behavior of hissing cockroaches in the stratosphere in a modified environment that aimed to keep the creatures alive, despite the Mars-like conditions at a height of 85,000 ft (26,000 m). Oxygen, carbon dioxide, pressure, humidity, and temperature were recorded inside the pressurized, reusable habitat, and a timer installed to accurately log capsule events. Footage of the roaches using a Go Pro 4K camera observed the cockroaches' response to the turbulent atmospheric conditions. The capsule was successfully recovered following the flight in which the unharmed creatures had reached 30,000 ft (9000 m) (Fig. 5.6).

Around 4500 species of cockroaches exist around the world, although only around 30 species are considered pests. They are extremely adaptable and, for example, can metabolize anaerobically, survive freezing, and even tolerate extremely high levels of ionizing radiation. Although the Madagascar hissing cockroaches prefer fruit and vegetables, they will eat almost anything, from cheese, beer, leather, glue, hair, starch in book bindings, flakes of dried skin or decaying organic matter—to wallpapers and stamps—because of the glue on them (Fig. 5.7).

The experimental use of animals in flight and the exploration of space dates right back to the 1700s when Joseph-Michel and Jacques-Étienne Montgolfier sent a live payload that included a sheep, duck, and rooster into the sky. Having demonstrated that it was possible to survive such a mission, the first hot air balloon with a human cargo was launched carrying Jean-François Pilâtre de Rozier, a chemistry and physics teacher, who remained in the air for almost 4 min.

In 1947, the first creatures that survived Earth's orbit at a height of 170 km were fruit flies. Seeds of corn accompanied them to test the effects of radiation on DNA. The capsule and cargo were recovered with the flies alive and well.

Albert II was the first monkey to be sent into space on June 14, 1949, but sadly died on impact after parachute failure. However, Able (a 3.18-kg American-born rhesus monkey) and Baker (a 0.31-kg squirrel monkey), who achieved their historic flight in 1959, were the first mammals to survive a spaceflight.

Following these first precarious steps, a whole range of different creatures has been launched into the upper atmosphere and beyond. Flies, spiders, rats, mice, hamsters,

112 Experimental architecture: on-world and off-world exploration of possibilities

Figure 5.6 Madagascar hissing cockroach in insectarium. Credit: Photograph courtesy Nebula Sciences (2016)

guinea pigs, rabbits, cats, dogs, moths, frogs, goldfish, beetles, wasps, tortoises, newts, stick insects, snails, carp, shrimp, jellyfish, rock scorpions, and cockroaches have all exceeded the pull of gravity.

The experiments conducted in this stratospheric flight follow a design-led agenda that—through their context—speaks to the possibility of metamorphosis as creatures are exposed to extreme environments. With the advent of advanced biotechnologies like genetic modification, whereby organisms may be engineered to perform differently, we may begin to see more creatures designed and engineered to acquire spacefaring

Figure 5.7 Madagascar hissing cockroaches safely retrieved from stratospheric flight. Credit: Movie still from Capsule of Crossed Destinies experiment (2016)

capabilities. For example, water bears can withstand the vacuum of space and hardy bacteria such as the Deinococcus *radiodurans* species can survive exposure to ionizing space radiation, which causes catastrophic tissue damage in ordinary creatures.

In years to come, we may see natural adaptations to challenging landscapes as in the traces of marine plankton and other microbes growing on the surface of the illuminators of the International Space Station (ISS) in 2014. Alternatively, deliberately modified organisms designed to survive conditions that exceed their natural capacities may be introduced into non-terrestrial environments as our first non-human space colonists. We have no way of knowing what different hybrids, chimeras, or unknown species may be produced during such explorations.

Although we may deliberately design and engineer creatures to meet specific requirements, living things also possess their own agency and capacity for survival. The hissing Madagascar roaches speak of the potential for adaptation and change in these extraordinary habitats. They represent the need for experimentation and adventures not only for off-world habitation by a whole range of creatures—on whom the future of our survival may depend— but also as part of our ongoing quest for a better understanding of what it means to live as part of a dynamic, mutable ecosystem in our own increasingly hostile home planet.

Beyond human intention, the creatures that precede and accompany us will also be subject to spontaneous evolutionary process. The Capsule of Crossed Destinies draws from the reinvention of a story by Italo Calvino that talks about the quest for meaning. Bodies of creatures converge and remix external and internal forces through various acts of survival. We can never be totally in control of these processes, which will produce new languages and typologies of survival, as all kinds of organisms begin to inhabit increasingly extreme habitats. Given that, under extreme conditions, ants and roaches are likely to fare much better than we do, the roaches are exploring this space for transformation in a short journey, which talks about the future of the evolution of life and into spaces that we are familiar with, which are increasingly becoming more strange.

As we start to enter an age of evolutionary design being thrown up against extreme environments, the Capsule of Crossed Destinies serves as a kind of evolution machine whose outputs are currently unknowable. It is an experimental platform that acts like a time-traveling bottle on a beach, which raises questions about the future of life on Earth and beyond it. Although we cannot predict all future forms of evolution for what its cargo, and we ourselves, may become—we can glimpse possibilities. By observing how creatures with extreme capabilities for survival behave in extremis, we may start to learn the lessons that we need to survive in a rapidly changing world, and the terrains beyond it (Fig. 5.8).

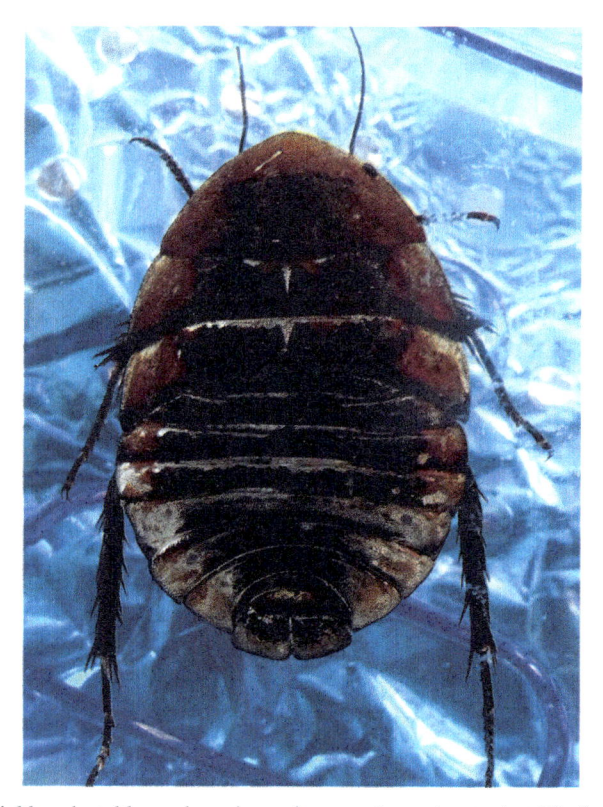

Figure 5.8 Highly adaptable cockroaches raise questions about what life "could" become when hardy organisms are exposed for prolonged periods to extreme environments. Credit: Drawing by Rachel Armstrong (2016)

5.4 Hylozoic Ground

"The realm of the born—all that is nature—and the realm of the made—all that is humanly constructed—are becoming one. Machines are becoming biological and the biological is becoming engineered." (Kelly 1994, p. 6)

The third millennium heralds a time whereby the operations of life and what Kevin Kelly calls the "*techniuum*" are converging (Kelly 2010) as "living technology" (Bedau et al. 2010), whereby agents possess some of the characteristics of organisms without having the full status of being truly alive. The cybernetic installations of architect Philip Beesley are an architectural investigation of these themes. His "semi-living architecture" draws on technological convergence and finding synergies between the contributing platforms, to respond to environmental cues and produce phenomena that are increasingly associated with lifelike activities such as growth and sensitivity.

Beesley is an architect and professor in the School of Architecture, at the University of Waterloo, Canada. His Orgone Reef (2003), Hylozoic Soil (2007), and Epithelium (2008) are architectural-scale cybernetic projects. Each consists of highly choreographed swarms of modular mechanical components that are orchestrated into fields of active agents with emergent responsive properties. Beesley's early installations were static geotextiles made of inert components but, increasingly, his work has incorporated sensors, microprocessors, motors, and shape-memory metals that are capable of generating movement in response to the presence and actions of spectators. Consequently, his installations have become more complex and lifelike. They possess the multiple material layering associated with soils but with added interaction as they can respond to visitors with apparent curiosity, or unexpected gestures, such as raising friendly tendrils or spiky fronds. His work is a forward-looking example of architecture's historical interest in becoming lifelike. Ambitions to achieve this aim bloomed during the Renaissance, when Leon Battista Alberti compared the proportions of living creatures with building design. Throughout the twentieth century, they became more complex and interactive as architects were influenced by the mathematical principles attributed to natural laws such as D'Arcy Thomson's 1917 *On Growth and Form*. With the rise of digital computing, algorithmic engagements with calculus are informing the organizing principles for the production of lifelike architectures such as Greg Lynn's Embryonic House and William Latham's virtual life forms.

Beesley's pioneering work produces lifelike structures by prototyping organic systems of behavior. In generating such highly complex environments by designing processes of communication and control that are linked in parallel, he engages with a broader "ecosystem" of technology. Sensors and effectors connect through distributed communications networks and are coupled with organic substances such as fruit batteries. These amalgamations blur the distinctions between human life, animal life, technology, and environment. His research transfigures the tangled web of representations implicit in the contemporary discourse about Nature, science, and design to create a web of interactions that create the conditions in which architecture may literally possess a "life" of its own.

Indeed, Beesley frequently draws on the life-bearing qualities of soil systems to invoke the conditions for transformation implicit in his work. The complex exchanges within these installation functions may be considered as wholly artificial soils and geotextiles. While these terms might initially invoke notions of secure mass and compression, and resource for framing human territory, Beesley's contemporary engagement with the term is rich with a myriad of participating agents that possess a silent, spatialized, primal fertility. Rather than naturalistic ingredients, his installations incorporate memory alloys, microprocessors, microcontrollers, steel, glass, and plastic into a hanging garden of robotic agents activated through a primitive neural network. The different components communicate and interact with each other as individual entities to produce chain reactions of responses that ripple throughout the installation, continually changing fields of experience. The perpetual performativity of Beesley's notion of "soil" becomes the site for experimentation.

Hylozoic Ground is a collaborative project for the 2010 Venice Architecture Biennale, which investigates design tactics for the construction of synthetic soil at architectural dimensions. Responding to Beesley's challenge to produce a series of chemical agents for the construction of cybernetic soil systems, I developed "wet chemical organs" that were integrated into the cybernetic matrix (Fig. 5.9).

Unlike mechanical systems, dynamic chemistries are coupled sensor and effector systems that possess their own spatial properties and transformations. The Hylozoic Ground installation provided an active site and laboratory that was intrinsically coupled to its environment. Chemistries were selected for incorporation into the cybernetic system on the basis that they demonstrated some recognizable lifelike qualities, such as movement, sensitivity, and the ability to adapt to changing environmental conditions (Armstrong and Beesley 2011). Four different dynamic chemical species were developed into arrangements that complemented the hylozoic, or life-bearing, ambitions of the installation. They were positioned within the cybernetic matrix as a cohesive web of materialities that invited the participation of other actants. These agents took the form of dynamic droplets and gel

Figure 5.9 Hylozoic Ground installation, at the Canadian Pavilion, 2010 Venice Architecture Biennale. Credit: Philip Beesley Architect Inc

plates that sensed environmental changes to produce changes in metabolism that could be interpreted through a variety of cybernetic agents—including the gallery visitors.

Liesegang Ring plates were installed as a means of recording gravity, diffusion, and time in the cybernetic system, as well as dynamic droplets that were deployed to integrate chemical exchanges as an integral aspect of the cybernetic system on an architectural scale (Fig. 5.10).

Two kinds of droplets were produced for Hylozoic Ground. Modified Bütschli droplets were the chemical system of choice, which were produced by adding concentrated sodium hydroxide to an oil interface. They self-assembled into droplets, to which highly cultured salts and an inhibitor were added, so that the evolving bodies could persist for the full 3-month duration of the installation. The metabolism of the droplets responded to the presence of carbon dioxide, which was already dissolved in the water in the flask. Bütschli droplets were entangled with the cybernetic system and its environmental interactions by responding to the presence of carbon dioxide through the production of tiny, brightly cultured structures about

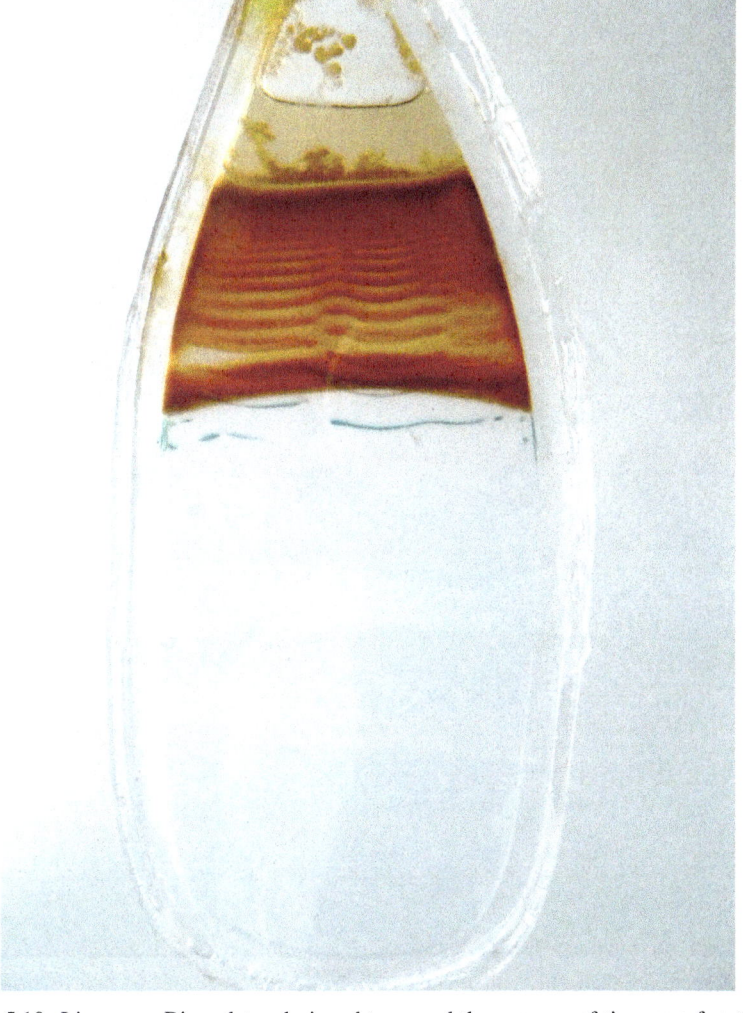

Figure 5.10 Liesegang Ring plates designed to record the passage of time as a function of primitive metabolisms within the Hylozoic Ground installation. Credit: Rachel Armstrong

118 Experimental architecture: on-world and off-world exploration of possibilities

the size of a little fingernail that grew through the support provided by the liquid media in the flasks. Walking through this installation might be likened to exploring inside a giant nose—where the plastic fronds act as sinus hairs and the "golden apple-like" flasks are "smart" mucus glands that can "smell and taste" the presence of carbon dioxide (Fig. 5.11).

Figure 5.11 Smart chemistry functions as "soft" technological sensors that perform as an artificial smell and taste system within the Hylozoic Ground installation. Credit: Rachel Armstrong

Another droplet system was designed in which oil droplets with a calcium metabolism were added to a flask of lagoon water that was open to the atmosphere and produced a series of pearl-like sculptures that recorded the ambient carbon dioxide levels in the gallery through the density of soils material produced. The flasks were open to the air, and so enabled the ongoing exchange of carbon dioxide across the air–water interface and increased the synthetic capacity of the modified Bütschli droplets.

The Hylozoic Ground installation provided an experimental environment that enabled the exploration of soil-producing strategies in a non-naturalistic manner. The cybernetic system could be likened to a soil, which is not formless, but has a specific architecture and an evolving body that is shaped by its responsive, material ecologies. The lively community of converging technologies appears to promote the occurrence of lifelike events, such as growth, movement, and environmental sensitivity. Potentially, the strategy of designing hubs of chemical activity that perform some of the functions of a soil body may be designed into an architectural-scale artificial soil for the living interior of a worldship. Conjecturally, given enough time, perhaps a sufficiently life-promoting environment could provoke events that might formally be considered "alive."

5.5 Future Venice

"Futures not achieved are only branches of the past: dead branches." (Calvino 1997, p. 29)

Future Venice examines how an urban-scale, environmentally responsive soil may confer the fabric of the city with dynamic properties while remaining sensitive to the needs of local marine wildlife.

In this project, the historic Italian city of Venice becomes the site of a laboratory for conducting site-specific experiments that explore how it may be possible to confer its foundations with lifelike qualities that, for example, create the possibility of growth and repair. The lagoon that channels the runoff from the fertile lands of the Po delta becomes a site for soil synthesis. The project proposes that material transformations that occur within the lagoon where it meets the city's underwater foundations may be considered as a technologically assisted soil-generating process (Fig. 5.12).

The fundamental technology used in this project is a dynamic droplet system that can be engineered and directed to specific sites by using a range of metabolisms. Each of the droplets is made from a simple recipe produced from a self-organizing mixture of water, oil, and salt. When compiled as a technology, the droplets may move away from the light or deposit crystals at its interface. Such chemical reactions are capable of transforming substances in the lagoon water through complex networks of spatialized chemistry that enable the urban fabric literally to fight back against the damaging effects of natural elements in a struggle for survival—and therefore secure its longevity.

The chemical technology is developed through a series of laboratory and field experiments. The end goal is to produce an accretion technology that is mediated at the interface between the lagoon water and the dynamic droplets. The watery infrastructure of the city is integral to the success of the soil-making platform, as it provides the specially engineered droplets with an abundant flow of nutrients, such as dissolved carbon dioxide and minerals. The collective action of the droplets gradually forms an artificial garden reef underneath the city's foundations that spreads Venice's point load over a much broader

120   Experimental architecture: on-world and off-world exploration of possibilities

Figure 5.12 Microecologies in the Venice waterways synthesize soil-like biocretes. Credit: Rachel Armstrong

base. Consequently, Venice is prevented from sinking so quickly into the soft mud on which it has been founded.

A natural version of this accretion process is observed around the lagoon side and the canals, which is orchestrated by the native marine wildlife. Potentially, dynamic droplets could work alongside these organisms to co-construct an architecture that is mutually beneficial to the marine ecology and the city (Fig. 5.13).

Importantly, should the environmental conditions change and the lagoon dry out—say, for example, that Pietro Tiatini and his colleagues succeed in anthropogenically lifting the city by pumping seawater into its deflated aquifers (Teatini et al. 2011), or should the Moses gates precipitate environmental events so that the native ecology reaches a catastrophic tipping point—then the chemical droplets can follow a different program. Instead of generating an outward-spreading layer of accrete matter, they coat the woodpiles in a downward direction as the waters subside with a protective layer of "biocrete." This stops them rotting when they are exposed to the air (Fig. 5.14).

Potentially, dynamic droplet technology could be applied to the whole bioregion of Venice. Facilitated by flowing water, carefully designed metabolisms and spatial programs could give rise to tactics that generate new relationships between natural and artificial agents as a synthetic ecosystem grounded in artificially produced soils. These may become the bedrock for forging life-promoting, synthetic ecologies, both in the historic city as well as in the context of a starship interior.

Figure 5.13 Programmable droplets potentially may orchestrate the metabolic and material relationships that organize the production of biocrete. Credit: Rachel Armstrong

In this context, artificial soils enable a coordinated system of exchange between fabric, space, structure, and location. The outputs of the system do not imitate Nature, but work as an alternative kind of life-promoting system—using a common chemical language based on physics and chemistry that are shared with the natural world. The idea of self-assembly in the production of materials and lifelike architectures also creates a potential new portfolio of solutions on Earth to deal with rising sea levels in coastal areas, biocompatible materials, architectures that can deal with wet conditions, and self-repairing systems. None of these possibilities has been formally produced but they are informing further experimental research.

Yet, the design-led experiments that inform Future Venice do not propose to be a complete solution for the city's precarious future—or, indeed, erase our current legacy of environmental crises. They do not attempt to "solve" the inevitable changes that accompany a lively environment, but open up the possibility of new approaches in invigorating material systems and finding new ways of designing and engineering with the complexity of environmental events. Such dynamic practices generate new possibilities for metropolitan environments beyond the city of Venice, where settlements and ecologies become sites of pulsating, vibrating, transforming, flowing materials that may produce new kinds of spaces for innovation and inhabitation.

Figure 5.14 Soil-like technology could potentially choreograph the material flows within the Venice region to provide an alternative construction process whereby the city of Venice acquires some of the properties of living things such as growth and self-repair. Credit: Christian Kerrigan

From a longer-term design perspective, Future Venice implies that the kinds of oceans proposed in the worldship design may most productively be imagined as bogs rather than a soup of minerals. Indeed, unruly delta basins provide a valuable organizing matrix that regulates and filters material flow, which contrasts with the obedient agricultural landscapes proposed by O'Neill.

5.6 Future Venice II

"You take delight not in a city's seven or seventy wonders, but in the answer it gives to a question of yours." (Calvino 1997, p. 44)

Future Venice II examines the possibility of building a new kind of soil for our times, as well as thinking about the longer-term view of Persephone's living fabric.

While the end of humankind has been a perennial concern for each generation, advanced technologies and imaging techniques have taken our concerns beyond the realm of belief systems. They have sharply drawn our focus to our gross mistreatment of the infrastructures that enable our existence, whose robustness and resilience we have taken for granted for so many millennia. If the long-term aim of our species is collective survival, then we simply cannot continue to ignore the devastating condition in which the infrastructures and environments that support our world are.

The Anthropocene has brought the greatest wave of human prosperity that the planet has ever known. There are now more than seven billion of us. By the end of the century, we are likely to have reached 11 billion, and two-thirds of us will be living in cities. Yet, we are also facing a sixth great extinction, with devastating losses of species, the erosion of fertile soils, and the global-scale pollution of our liquid, as well as gaseous oceans. The impacts of our human endeavors are effectively reverse-terraforming our planet, rendering it an increasingly turbulent and hostile landscape, which has changed our relationship with the natural world.

In the third millennium, the potent forces that constitute the natural realm are no longer symbolized by unblemished wildernesses devoid of humanity. They do not reside in the countryside, or on the pretty rooftops of tall buildings. Rather, they infiltrate our urban environments, where they are metamorphosing alongside legions of antibiotic-resistant microbes into something stranger, substantial, and infinitely more subversive than the Nature we thought we once knew. This vast assemblage of material agents confronts us with its raw, culturally unedited substance. It heaves open the tops of barely buried garbage icebergs, oozes from guano-rich crevices that melt building fabrics, and splits the paving slabs of walkways with exploding weeds. Defiantly, it continues to thrive—despite us. Yet, millennial Nature defies the odds against its survival and resurfaces from its untimely grave to exert revenge, by leaking carcinogens and volatile gases into our urban environments. These subversive acts of resurrection are nothing less than magnificent, as our new material realm splits Earth open with pitiless sinkholes and dazzles our skies with polluted rainbow-brilliant sunsets.

In considering how we may start to design and engineer alongside Nature's unruly forces, it is not enough to imagine just the production of soils, but also the way they change with time, space, and circumstance to generate a range of outputs that reinvests in the future.

124　Experimental architecture: on-world and off-world exploration of possibilities

Zanzara Island is the central project of Future Venice II, Newcastle University's collaboration with the IDEA Laboratory that proposes to grow a new island for the city. It identifies two major challenges for the city—plastic debris and cellular overgrowth caused by eutrophication—and asks whether there is synergy between them that can provide new opportunities for the inhabitants. Specifically, Zanzara Island is the synthesis between natural microfilms that are produced by consortia of microorganisms and the plastisphere, which is formed by ecosystems of organisms that have evolved to live in human-made plastic environments. Plastic debris in waterways is a steadily increasing global challenge. Robert Day and colleagues found 3370 objects per kilometer in the Pacific Ocean in the period 1985–1988 (Day et al. 1988). Moore carried out a similar study in 1999, and found an average of 334,271 plastic particles per square kilometer, which suggests a huge increase over the next 10 years (Moore et al. 2001). In Venice, the plastisphere originates locally from the degradation of around 13 million plastic bottles that are left behind by around 20 million tourists each year. Many of these end up as landfill, or are discarded into the canals and lagoon. Here, they are broken down by the mechanical action of the waves and photodegradation, where they are shredded into tiny fragments, or microplastics, which are less than half a centimeter in diameter and remain in suspension in the water several centimeters below the surface (Fig. 5.15).

Figure 5.15　Microorganisms trap discarded plastic in the water to accrete a new island for Venice. Credit: Rachel Armstrong, with Artwise Curators and IDEA Laboratory

Drawing together the spontaneous production of biofilms that bloom as a consequence of the pollution and excess minerals that run off from the rich agricultural lands of the Po delta, with the need for biofilms to anchor on solids, Zanzara Island examines how it may be possible to approximate the two processes to address pollution in the lagoon. The idea is to seed areas of plastic accumulation with naturally occurring consortia of organisms, which consist of different species of algae and bacteria. If the biofilms form attachments to microplastics in suspension, they may be removed from our food chains. These amalgamations of organic matter and plastics may then be sunk into earth, where they are composted and may become a new island for the city. The earthen materiality will gradually transform over time and may provide useful materials for future generations. Since synthetic plastics have only existed for around a century, their future transformation through the spontaneous forces and processes that constitute the natural realm, such as composting and weathering, are explored in this project. For example, chalk is formed from the skeletons of tiny organisms that have been composted and pressurized. Yet, there is no way of knowing what the future "natural" history of microplastics may be. Zanzara Island, therefore, reflects on laboratory and field observations to make propositions about how desired outcomes may be shaped into functionality and generate the kinds of outcomes we wish to promote.

By encouraging the biofilms to attach to the plastic in suspension, it may be possible for the combined material to be harvested, or deposited in specific areas of the lagoon—specifically, 500 m from the San Michele cemetery in a south-easterly direction. As rafts of biofilms weave their membranes around the plastics, they produce long yards of thread-like materials. These are knitted into mats by the tides and eddies around the biopositive concrete posts that signpost the island's borders to marine traffic. Zanzara's multi-material bones, therefore, form the shallow building blocks of an artificial reef on which lagoon debris is hung by the washing-machine action of its currents, which act as attractors for non-living matter such as human refuse, decay, and dirt (Fig. 5.16).

A field laboratory was set up as a bank of eight aquariums for the IDEA Laboratory at the Vita Vitale exhibition that sought to shape questions that reflected on how this process could provide new habitable places, materials, and artisan practices for the citizens of Venice. This event was designed by Artwise for the Azerbaijan Pavilion at the 56th Venice Biennale. Each tank contained consortia of microbes from local sources of water in Venice that were cultured with beach plastic debris from the Lido beach, to see whether biofilms were produced (Fig. 5.17).

Out of the eight tanks, one grew thick, spider-web-like biofilms that were formed by a consortium of algae and bacteria that was firmly attached to the plastic flotsam in the tanks. These structures were exactly what was anticipated and demonstrated the potential for further refining and developing the questions related to the goal of the project. Another tank grew similar biofilms, but they were much more spartan in their form and distribution.

Two tanks produced algae growths without any obvious bacterial component, taking on the form of ball-like colonies, which may have been sculpted, at least in part, by the flow produced by the pumps used to aerate the tanks.

The presence of polystyrene in one tank appeared to shatter the microorganismal consortium into two separate populations of algae and bacteria. The algae preferred to colonize the interstitial spaces between the polystyrene that provided shelter, water, and minerals that were drawn from the tank by capillary action. The bacteria did not appear to

Figure 5.16 Microorganisms produce biofilms that bind to plastics, generating a "natural net." Credit: Rachel Armstrong, with Artwise Curators and IDEA Laboratory

Figure 5.17 Bank of eight aquariums installed for the IDEA Laboratory at the Vita Vitale exhibition, 56th Venice Biennale, 2015. Credit: Rachel Armstrong, with Artwise Curators and IDEA Laboratory

share the same affinity for these spaces and formed bacterial biofilms around the pump outlets. Two tanks remained sterile, with no microorganismal growth; this may have been caused by pollutants coming in on the harvested plastics. The final tank was colonized with mosquito larvae that were eradicated by adding a layer of mineral oil to the water surface. No biofilm growth was observed in this tank.

Collaborators included artist Mike Perry, whose photographic work addresses environmental issues. During his photographic explorations, he collects plastic "stones" that are altered by the environment. He collected these items and exhibited them as a future natural history of plastics that emphasized the capacity of the natural realm to transform malleable matter into alternative configurations.

Studio Swine's "Sea Chair" was built when they accompanied a trawler during a fishing expedition out at sea. As the fishermen hauled in their catch of fish, the designers collected the plastics entangled in the net. They sorted them into different colors, cut them up into small pieces, and melted them by sintering. By the time they returned to shore, they had made a chair.

Rather than the traditional, inert *briccole* that are wooden posts, which guide vessels around the lagoon, architects ecoLogicStudio generated signposts with environmental sensors that can detect chemical changes in the water, using algae. The living agents generated outputs that not only altered the angle of the posts to guide travelers to their destinations, but could also indicate the health of the water.

Julian Melchiorri's installation uses a natural plastic, silk to construct an artificial leaf. Using shape-memory metals, clusters of leaves form shading in Campo San Stefano during the day and during the night, when the leaf clusters are folded and a solar-charged light is activated, they become a lighting system that illuminates the square.

Future Venice II generates strategies for the long-term understanding, design, and engineering of urban-scale geofabrics and soils that reinvest in their sites. The laboratory experiments generated an initial set of prototype relationships between the plastics and microorganisms from the city that demonstrated the idea that these two materials could have complex relationships with each other. The conditions for the production of such a relationship are complex, which was supported by the varied results from the eight tanks. However, this particular instrument was valuable in beginning to address new possibilities, and raises further valuable questions about how we are currently managing our resources, and identifying the kinds of practices in which synthetic materials can provide ways of working synergistically with ecological systems to transform landscapes in ways that render them more habitable.

REFERENCES

AMKK, Exobiotanica: botanical space flight (2014), July 15, http://azumamakoto.com/?p=5051. Accessed 7 Aug 2015

R. Armstrong, P. Beesley, Soil and protoplasm: the hylozoic ground project. Archit. Des. **81**(2), 78–89 (2011)

M.A. Bedau, J.S. McCaskill, N.H. Packard, S. Rasmussen, Living technology: exploiting life's principles in technology. Artif. Life **16**(1), 89–97 (2010)

I. Calvino, *Invisible Cities* (Vintage Classics, London, 1997)

A.C. Clarke, K.S. Robinson, *Meeting with Medusa & Green Mars (Special Double Release)* (Tor, New York, 1988)

R.H. Day, D.G. Shaw, S.E. Ignell, *The Quantitative Distribution and Characteristics of Neuston Plastic in the North Pacific Ocean, 1985–1988.* (Final Report to US Department of Commerce, National Marine Fisheries Service, Auke Bay, AK, 1988), pp. 247–266

K. Kelly, *Out of Control: The New Biology of Machines, Social Systems and the Economic World* (Basic Books, New York, 1994)

K. Kelly, *What Technology Wants* (Viking, New York, 2010)

J. Milton, *Paradise Lost* (Dover Thrift Editions, New York, 2005)

C.J. Moore, S.L. Moore, M.K. Leecaster, S.B. Weisberg, A comparison of plastic and plankton in the north pacific central gyre. Mar. Pollut. Bull. **42**(12), 1297–1300 (2001)

P. Teatini, N. Castelletto, M. Ferronato, G. Gambolati, L. Tosi, A new hydrogeologic model to predict anthropogenic uplift of Venice. Water Resour. Res. **47**(12), W12507 (2011)

A.M. Turing, The chemical basis of morphogenesis. Philos. Trans. R. Soc. Lond. B Biol. Sci. **237**(641), 37–72 (1952)

6

Building a worldship interior

Rachel Armstrong

This chapter proposes an approach towards constructing new worlds beginning with life-bearing soils.

6.1 Architectures of elsewhere

"All matter squirms. This is the fundamental reality that underpins our cosmic fabric." (Armstrong 2015, p. 71)

Perhaps the greatest expectation we make of contemporary architecture is its permanence. Throughout the ages, we have built structures that strive for immortality in pyramids, churches, cites, and starships, around which the natural realm must move. Owing to our short lifespans, when compared with the events that characterize the life cycle of a rock, we appear ephemeral. Yet, while our monuments stand steady, we proliferate, tear down permanent structures, and alter landscapes beyond all recognition.

Modern buildings, which are both built by and imagined as machines (Gallagher 2001; Le Corbusier 2007), require significant investment in energy and resources for maintenance if they are to resist the relentless non-linear actions of the material world. This results in significant wear and tear of inert materials, which occurs at the many micro-environmental interfaces in which our buildings are immersed. Viewed over a 30-year period, we spend around 2 % per annum of the initial construction costs of a building on maintenance (CBEC and APEGEC 2009). These expenditures are manageable and economically justifiable when social and environmental conditions are stable.

Architecture is traditionally designed not only for permanence, but also to shield us from hostile environments and so prolong life. Since ancient times, we have constructed barriers between our potentially deadly natural surroundings and ourselves. Our earliest societies took shelter in caves, like those in Nerja that were used by many different tribes,

---

Rachel Armstrong, PhD. (✉)
Professor of Experimental Architecture, Newcastle University, Newcastle-upon-Tyne, UK

including hunters, fishermen, and harvesters, from the Neolithic to Paleolithic periods and the Bronze Age. Other structures shielded soldiers, such as the tunnels under Dover's White Cliffs that were used in wartime operations during the Middle Ages and during World War II. Bequest with Vitruvian permanence, contemporary architects design for the immortality of matter. Modern buildings are geometric obstacles or machines in a landscape around which living things are compelled to move.

When we think of a materially dynamic architectural system, we might imagine processes of decay as in Piranesi's drawings of ruins. His brutally dissected forms invoke dereliction to generate cityscapes and isolated buildings as decadent expressions and pessimistic fancies of a doomed city. The erosive qualities of natural infrastructures continually steal material from building sites, causing them to wear and weather. Yet, ruins have an iconography of transformation, gradually dissolving into the landscape and becoming invisible. Perhaps we might reflect on a primal primitivism, in which barely evolved peoples flounder in the as yet unconstructed mud, like worms.

Despite the inert nature of the materials that we conventionally use in the construction of architecture, our terrestrial environments are lively. When situated within a dynamic environment, these relentless fields of interactions that constitute the natural realm reassimilate our buildings. While we notice their absence from places, their reappearance elsewhere is largely unobserved. Think of a limestone cave in which the subtractive process of water passing through a mineral field dissolves matter and re-deposits it elsewhere as stalactites and stalagmites.

These material erosions are not always physical. Active biological processes may also produce them, which results in a kind of digestion and compositing of inert matter. These processes can be seen actively taking place as cryptogeographies on buildings—highly localized building details that are assemblages of microorganisms, crumbling brickwork, and the invasion of physical forces—mold on a damp wall, lichen on a harsh surface, and spindly, yet tenacious, multicellular plants in deep cracks where molecular planes have fractured.

In a materially closed environment, the subtractions occurring at one particular site are likely to be entangled with depositions elsewhere in a landscape. While disappearing substances may be diffusely distributed and assimilated within a much bigger body of soil, active processes will be concentrating and displacing chemistries elsewhere.

These architectures of elsewhere that are informed by remote, yet entangled, material events constantly reconfigure spatial relationships between substances. By drawing on the generative fields of "unbuilt" agencies such as seas, soils, and rainforests, they express a much larger reserve of creativity than exists within our man-made urban spaces, and ally with the potency of robust, self-replenishing environmental cycles—our ecologies. The elemental infrastructures that inform our landscapes of material exchange quite literally provide a means through which the traditional boundary between landscape and building may be eroded and shaped by new spatial programs. These organizing fields may serve as a transport medium for substances that permeate the barrier fabric of buildings and open up the possibility of designing in new and surprising places. Rather than buildings being cleaved from their surroundings by brute matter, the spatial programs that shape the non-equilibrium properties of lively matter literally bring living processes into the very heart of building fabrics, so that they breathe, feel, grow, and change with time.

Potentially, if these relationships are correctly choreographed, they could persist for lengthy periods. Procedural architects Shusaku Arakawa and Madeline Gins produced spaces to house the "architectural body"—a term they specifically used to refer to the human organism coupled to its living space. Believing that the nature of lifestyles shaped the possibility of survival, Arakawa and Gins proposed that the complacency and familiarity inherent in sedentary lifestyles eroded our vitality. They drew together the forces of the animate and inanimate worlds, proposing to transform the built environment into a closely argued built-discourse between body and space. Their earnest preoccupation with the human condition not only proposes to be a strategy for achieving immortality, but also becomes an escape route for defeatism and cultural paralysis. By generating unpredictable environments for the architectural body, real effects and transformations are facilitated, such as fortifying the immune system, which enables occupants to evade death.

Yet, our living spaces are semi-permeable, both to the environment and also to human habitation. Persephone's investment in ecopoiesis proposes a fundamentally dynamic landscape, in which the field of organization that promotes material flow and exchange is a primary condition for survival. Such infrastructures are the forces that shape its soils, which are complexity "elevators" that work with a range of strategies (e.g. catalysis, spatialization, temporality, transformation, and polymerization) that evade the system's collapse toward equilibrium.

Within this writhing survival landscape, colonists start to shape their living spaces. Here, they forge a reciprocal relationship with the architecture that they inhabit, which shapes their experiences and encounters of Persephone's living realm. Within lively environments, heaving membranes engage in a complex choreography of dynamic transformation with the colonists that is enabled by elemental infrastructures selectively infiltrating the material realm. To imagine further what living on Persephone might be like, think of smart mud as the starting point for a home, which can eat its own waste, produce its own energy, and make new materials. Depending on how it is lived in, the building may also regenerate or decay.

So what might living in Persephone's cities be like? As they are made from soils, buildings can be fashioned like clay into any kind of shape. But, because they are living, they have odd qualities. For starters, Persephone's buildings make sounds that remind us of organs—like a stomach or heart—rather than a cacophony of engines. They smell as bodies do, not like oil refineries, but are free of noxious pollutants. Yet these living spaces are not primitive habitats, but rich environments that celebrate the poetry of life. They are warm to the touch and shockingly bio diverse. In fact, the living cities that spring from Persephone's soils could be accused of having "too much" life, as they seek to make new relationships with other materials around them.

No two buildings are the same as one another, just like the bodies that Nature produces, and each dwelling is ornamented in mosaic fashion by riches from the soils, which provide harvests of minerals, stones, metals, and shells. Persephone's soils may be organized to accommodate spatial positioning without losing their potency. Perhaps these effervescent spaces could be thought of as being stretched and molded into configurations like Frei Otto's bubble-like structures or Cappadocia's ancient architectures situated within a natural cave system.

Persephone's soils, therefore, become its architectures, not through a newly conferred object status through subtracted liveliness, but in its elevated status as a new site for many processes that may have brought about its construction, which may also ultimately contribute to its decay. We may also observe and measure these material changes, and consider them beautiful and encourage them or disagreeable and seek to prevent or repair them.

However, our understanding of which processes are important is culturally determined and not Platonic truth. Bruce Sterling highlights our role in valuing material relations, observing that "any sufficiently advanced technology is indistinguishable from its garbage" (Sterling 2012). Garbage making is a subjective exercise in which we distinguish ourselves from the natural world and our technologies. By "editing" our material networks to reflect our cultural conditioning, we choose what is important to us by applying cultural and aesthetic rather than material criteria to make the appropriate selections.

Persephone's interstellar culture begins its evolution at the point at which a new kind of relationship with matter is shaped where the inside and outside of things are already entangled. Yet, the production of the worldship interior is concerned not only with the human decisions that shape its spaces and environments, but also with the complex landscape that facilitates a host of activity such as spontaneous transformations, subtractions, additions, erosions, depositions, compactions, inflations, and collapse that produce the conditions in which life may continue to thrive.

6.2 Worldship interior

At first, Andrea thought it was the tree roots yelping. They did that when they ran short of water. But it wasn't the roots. The sobbing was much more plaintive and pitiful than she would expect of a stressed root system.

Andrea scooped the tip of her biowand into the rippling surface water. The gray tissue under the floating island monitored, evaluated, and recorded the data request. The glades were getting plenty of oxygen and the floating vegetation was healthily saturated with water. Sometimes, the light convection currents of the worldship dried out the upper soils, but the peat was moist to the touch. There was no need to call on Persephone for a precipitation breakout. Even the effluent appeared well behaved, lapping gently at the thirsty plants and blackening the tips of their foliage.

The awful shrieking continued. Not loud, but pervasive. Was it following her around? She tried to tune it out but, like an infant's cry, it was impossible to ignore. Where was it coming from?

The worldship's ocean spanned thousands of square kilometers across, with many swampy settlements where open expanses of water were never more than a few meters deep. Persephone's material bulk consisted of hoop-shaped islands that bobbed and gurgled under her feet. Beneath this, multiple layers of gray sensor-earth were entangled with the biotic substrates of the worldship. Life in Persephone was not just sentient, but also smart and interconnected—a knitted hybrid of silicon and carbon that interfaced at trembling biofilms. Living things within the ecosystem could therefore connect with each other through radio waves as well as chemical networks. There were days when she felt that certain events had been anticipated, although it was unclear who might be in charge of the biosurveillance systems.

Two small birds suddenly spiraled out of a bush, violently fighting each other over contested territories. The robins locked in a stranglehold of each other's feathers fluttered voicelessly around her feet. She split them up reproachingly, but they started squeaking antagonistically again only several meters away.

The glades filtered the worldship's waste, processing gray water and excrement. Bacterial biofilms and algae blooms playfully extended their fingers into the black water, while matted islands sipped at the slowly rocking fluids. Here, amidst the splashes and slurps, anything without a backbone flourished. The metabolism of these simple creatures was so much more efficient than that of higher organisms and generated infrastructural support, such as water and nutrient cycles, which sustained the worldship's ecosystems.

Andrea could not identify all the species living in the glades. They didn't tend to be vertebrates; those generally clustered around the clinic, a cross between a zoo and a hospital, which was some half a day's walk from the glades. While the reproductive cycles of vertebrates were obligated to the fusion of gametes from "opposite" sexes, invertebrates such as insects and worms were wayward in their unregulated sexual conduct. Within the glades, the air and waterways buzzed with the mating calls of a host of unclassified soft hermaphrodites, clones, chimeras, and budding bodies. They propagated via a range of ingenious contamination practices, rather than through ordered, genetically regulated, sex-matched, linear filiation systems. Two large dragonflies—one red, one blue—bobbed their metallic abdomens at her from a reed that was trapped in an algal mat, then disappeared. They were gleefully carefree. Not only did they seem to defy the laws of physics, but Persephone's biospherical codes as well.

While Earth's ancient populations had left such things to chance, they consequently suffered famine, drought, pestilence, war, and overcrowding. The interstellar generations, therefore, took a more strategic approach to environmental design. Formal breeding programs were a humane way of regulating vertebrate lifecycles. Higher organisms had fewer procreative freedoms, but were guaranteed a longer and more prosperous lifespan than the wanton invertebrates. All creatures with a backbone were hatched and tagged at the clinic. Even the unpredictable glade aligators were entered into formal breeding programs with implants injected under the skin at birth to chemically control their fertility.

The distressing sobbing appeared to have quelled. Making sure that she was not leaving the scene of an environmental crime, she dipped the biowand tip into the black water again—probing the smart earths for more information. They had detected nothing untoward. So, she started home, leaping from foot to foot across the soft soil sods as if she were playing a game of hopscotch.

Andrea enjoyed her work. The glades were truly stunning. Solar mirrors were constantly angled toward them so that the algae blooms could make best use of the light powered by the ship's fusion reactors for photosynthesis. This meant that the dramatic reflections produced by the moody waters were always entertaining. They fragmented and distorted the island anatomy, so that sometimes it appeared she inhabited more than one reality simultaneously. The highlight of the day, however, was when the cylindrical worldship grated over a worn bearing that was due for replacement. The uneven motion caused gravitational distortions so that the whole bog bubbled joyfully and dissolved gases fizzed to the surface like black champagne.

Skipping over the soggy peat, she passed a group of professional structural knitters whose duties including fishing out strands of vegetation from the glades and weaving them into islands, roads, bridges, nets, and homes that shaped the main highways of the worldship.

134 Building a worldship interior

When she got home, she threw off her degradable bioskin overalls and noticed a tiny leech attached to her leg veins that was so hungry that it was whimpering pitifully as it gorged on her blood. From that moment, the creature and the woman shared a circulatory system that established an unbreakable bond between them.

6.3 Soils as urban infrastructure

"… making [a city] is not a planning process; it's a becoming process. Because we can only partially see the results of what we do, we live in the face of mystery. There's magic and there's enchantment. And that leaves us with deep questions as human beings because we've been taught that we can know, master and optimise. We have to think in new ways about how to do that wisely." (Kauffman 2012)

Persephone's soil-like technologies may directly benefit terrestrial communities by providing a rich new palette of fabrics, technologies, and approaches for the construction of our third-millennium cities. Imagine how rich and diverse these spaces could be if they possessed the same kind of material creativity that is typical of terrestrial ecosystems. Yet, to change the performance and, therefore, the expectations of our urban spaces requires us to reconsider the infrastructures that support them, which provide the fundamental material flows in which natural computing (natural or designed) exists. If the liveliness that underpins the incredible creativity of matter at far-from-equilibrium states is to be sustained in places in which it has not previously existed, then ways of designing and engineering the underpinning infrastructures must be explored.

Current spacecraft are constructed in a similar manner to modern homes. They rely on active electromechanical systems to generate the kinds of material flow that are necessary for life and, vicariously, natural computing, but are prone to breaking down, as they need to run continually. A mechanically operated infrastructure is likely to be prohibitive in the day-to-day running of an artificial world, and backing up these vital technologies with spare capacity is bulky, expensive, and heavy. Passively operated biological and chemical processes that do not depend on machines are therefore essential in maintaining a living interior within a worldship, or more strategically within urban environments, such as water infrastructures for vertical green walls. A new relationship between air, water, sunlight, earth, and mechanical systems needs to be interwoven in ways that provide platforms to increase the fundamental liveliness of these habitats.

What if we actually had to build the soils beneath our feet, purify the air we breathe, and filter the water we drink within our homes? Currently, in our cities, life-giving processes occur "away" from our living spaces, so that it is easy to take them for granted. Yet, they have not disappeared; we have simply displaced our relationship with them out of sight. We are no less dependent on these vital infrastructures and, therefore, have to transport their operations, or the products of their creativity, through procured routes and distribution networks as utilities or commercial produce. Perhaps we might access and value the complex fabrics that generate such fertile opportunities differently if they were closer to us and woven into our everyday encounters with space and matter.

Generating living materials and technological prototypes builds toward a scenario where architecture not only deals with the manufacturing of objects. These spaces may also performs the equivalent work of machines as architectural "organs" with unique physiologies, by organizing the flow of matter through our living spaces.

By developing a different kind of infrastructure for worldships and cities, it is possible to consider a different kind of performance. For example, we may design breathing systems instead of vents within buildings, or circulations instead of drains. With the rich supply and removal of waste products, certain locations within these networks may begin to develop highly specific functions and become architectural "organ" systems—in other words, highly designed and engineered elemental infrastructures that convey the passage of light, water, air, and earth, organized so that these systems can process their environment in very particular ways. These operations can be provoked through strategic design and engineering processes, and are likely to take the form of aquariums that contain microcellular organisms such as bacteria and algae. They may even use smart chemistries, like dynamic droplets, to perform equivalent work to machines, such as producing heat, filtering water, or fixing carbon dioxide. The bioprocesses within these spaces and tanks could be supplied with "food" or nutrients, and the buildup of waste products such as organic waste and gray water could be controlled by digital systems. Indeed, such architectural organs would not need to be designed as open compost heaps, but could be engineered in ways that are beautifully designed and even rendered invisible to residents by situating them in under-imagined sites within our buildings, such as under floors or within wall cavities.

Architectural organs may be designed as highly visible structures as in Phillips' Microbial Home, where voluptuously shaped bioprocessors transform waste products into useful substances. These may be valued, exchanged, and transformed through a locally defined ecology. Strategically positioned, these architectural organs could give rise to buildings with physiologies that strengthen the material exchanges within a community through networks of metabolic processes, and act as biotic, life-promoting oases within cities for human and non-human communities.

For example, architectural organs may provide heating through composting, or produce low-level lighting through bioluminescence. They could be situated in under-used and under-imagined spaces such as cavity walls within our homes, or be on proud display like the designs as in Philips' Microbial Home, which is fueled by the actions of microorganisms. In this scenario, our cities will not be imagined as machines for living in, but as ecologies for thriving in. Each city possesses a unique urban metabolism that arises from its natural and architectural physiologies, with the capacity to structure the social, political, spatial, and cultural character of spaces.

The incorporation of elemental infrastructures into building and spacecraft systems may produce a range of different effects, owing to the diversity of available metabolisms and the vast array of potential spatial configurations. There is no one dominant environmental image that encapsulates a worldship interior or a "living" habitat within cities, but these spaces may be accessed and inhabited in many possible ways. While such plasticity and parallelism may seem impossible to fathom, the recalcitrant biotechnologies are set to bring Nature right into our homes in ways that render them more resilient, individual, and attractive. Think of the biodiversity of a rainforest, or a coral reef—these are the kinds of environments that could potentially be produced by "living architectures." Yet, these notions are not fictions, and are already being developed within the modern built environment.

We are already tapping into this relationship between natural systems and machines using renewables—and now, with an additional 30 years of advanced biotechnology, we can also think of how micro-agricultures may help us live in densely populated spaces in new ways. Indeed, in the near future, we are likely to see a rise in smart green cities, where Nature and digital information systems converge and work together do "more" with fewer resources, so that we can meet the needs of our urban populations differently.

These are not entirely speculative scenarios. The BIQ House built by Arup for the International Building Association in Hamburg opened to the public in November 2014 (Steadman 2013). It uses the empty space between glass facade panels to house micro-agriculture, where green algae feed on sunlight and carbon dioxide within a building facade, which uses the solar thermal effect to offset some of the building's energy demands (Fig. 6.1).

Figure 6.1 BIQ House, Hamburg, open to the public in 2014. The building facade is a liquid site for the growth of microcellular organisms that absorb carbon dioxide and sunlight to produce biomass. The solar thermal effect offsets some of the building's energy demands. Credit: Colt International, Arup, SSC GmbH

A range of ambitious proposals is currently being tested in experimental contexts and in bespoke building installations that are setting new benchmarks for the next generation of "sustainable" building designs. For example, Astudio architects (Part II, Chap. 12) and Sustainable Now Technologies are constructing a bioprocessor for a sixth-form college in Twickenham, as a next-generation ecological architecture. Not only did the system prove a focus for the school curriculum, but may also produce soil-producing substrates within the urban site that can be used as a local resource for green roofs and walls, and also filter water or promote urban biodiversity.

A practice of "experimental architecture" within today's homes and cities can provide a platform that allows us to explore the construction of research questions for the development of our future starships, as well as invest in technologies that have a different relationship with the environment than our current machines. Taking a distributed and open approach to the questions related to our ongoing survival on Earth today as individuals, communities, and consortia of many different kinds of collaborators (industry, non-governmental organizations (NGOs), governments, etc.) may not only help to establish insights into the transition from inert to living matter, but also generate potential cultural impacts and commercial opportunities in the built environment that help us to make the transition from an industrial age of building construction toward an Ecocene, where our living spaces are grown and cultivated within more traditional structural frameworks.

While explorations such as living facades are still very much at an experimental stage of urban development, they could potentially give rise to a new platform for the growth of cities that transforms the impact of mechanical technologies by converging their performance with biological systems, entangling them in a common project of mutual survival. In this way, our development paradigms may not inevitably damage our ecosystems, but strengthen them. Ultimately, the technical systems that underpin supersoils and elemental infrastructures will inform the infrastructures and design of our megacities, in which the built and living environments may become one through their mutually reinforcing interactions.

Kevin Kelly notes that, even when it is possible to see the perfect outlines of how an emerging technology—such as incorporating elemental infrastructures into our living space—may be realized, we tend to overestimate how soon it will become available to society in general. He attributes this delay to the need for other invisible ecologies of co-technologies to exist concurrently. In other words, producing "living environments" from scratch may only be realized when they can be supported by the appropriate infrastructures. In the case of Arup's BIQ House, rich water infrastructures were necessary for the algae to grow, as well as solving issues such as load bearing within the structural framework. In some ways, this apparent transition from inert to living architecture may be regarded as analogous to how flowering plants suddenly evolved when they solved a water infrastructure challenge (Field et al. 2011). However, when these technological ecologies finally converge, advances suddenly seem to appear in our lives as if from nowhere, and are received with much surprise and applause for the "unexpected" development (Kelly 2010).

6.4 Ecological design: biosphere

A starship environment begins within a sealed metal can, with no native ecology. It is also devoid of a centralized source of natural energy—like Earth's own Sun.

To ensure the long-term survival of colonists, it is essential to establish a destination where the environmental conditions can support and enable life to thrive. Yet, our current terrestrial practices do not support thriving within an already lively environment—even when we insist that our approaches are "sustainable"—which is attained when we are capable of meeting our own needs without compromising those of future generations (see Part I, Chapter 2).

We have learned to take the elemental systems that sustain us for granted to such a degree that, when we build our homes and cities, we can assume that the natural realm is a constant. In fact, sometimes it even appears that there is "too much" life on Earth—that we are overpopulated and exceed the actual capacity of our planetary system to secure our ongoing and future survival. Earth's vitality provides an illusion of abundance—one that we have greatly exploited, most particularly over the course of the Industrial Revolution. We have mechanically processed natural resources to feed human expansion and propagate into almost every niche in Earth's biosphere. We have used these approaches not to work alongside natural systems, but in opposition to them. This exploitation has been pursued with such intensity, and to such an extent, that we are, in fact, reverse-terraforming our home planet and compromising the resilience of our natural environment. Now, acidifying seas are choked with plastics, and our skies are riddled with gaseous waste that chews holes in the protective layer of ozone around us and contributes to dramatic changes in our weather.

However, in spartan and extreme environments such as the abyss, the polar regions, and Earth's orbit, the assumption that Nature's wealth will take care of us is at best tenuous. Only the hardiest organisms can live without any kind of assistance in these spaces—from deep-sea fishes that can withstand hundreds of atmospheres of pressure that can squash a Styrofoam cup into the size of a thimble, to microbes that have survived under the frozen Lake Vostok, which has been isolated from the rest of the biosphere for around 15 million years, and the water bears, or tardigrades, that can survive the vacuum of space. These extremophiles exist in their native states at the very limits of terrestrial possibility under conditions so harsh that technological assistance is required if we are to share their spaces. As we move through such alien landscapes, we depend on the integrity of our pressure suits, peak layering, and spacesuits. While such protective layers support our survival, they do so in the short term. For longer periods that require sustenance beyond a reservoir of supplies obtained from terrestrial sources, then alternative ecosystems are needed.

With no established life-bearing infrastructures on the Moon or Mars, of course, the challenge of interplanetary space is how to use existing resources in sustainable ways and turn these barren expanses into livable spaces. Potentially, a "green" backbone of organic islands such as algal biospheres might be strategically seeded in interplanetary space as a series of algae islands. However, such green orbital highways may only be useful as a concerted investment, as we are talking of a millions-of-kilometers scale on which related orbital dynamics just benefit spacecraft gravitational (un-propelled) motion. However, green gateway stations at a Lagrange point in the Earth–Moon system may be a stepping stone towards understanding how such organic highways may be constructed in interplanetary space. For example, the European Space Agency (ESA)'s MELiSSA (Micro-Ecological Life Support Alternative) program—a terrestrial lake eco-loop system, which uses microorganisms to process waste for plant growth—is taking steps towards this possibility.

Yet, if we are to genuinely survive as settlers beyond Earth's surface, we will have to bring our environment along with us, especially since a target destination for interstellar colonization has not yet been agreed. There is also no guarantee that an identified world will continue to suitable for colonization by terrestrial life forms by the time that starfarers arrive. Therefore, an optimum contingency plan is to construct a worldship environment that can support its inhabitants indefinitely.

However, our current inability to build a sustainable ecosystem from scratch is worrying. Ecosystems management assumes a basic level of a priori functioning that spontaneously takes place against a background of life-bearing terrestrial events that we still do not fully understand. This has allowed us to develop sophisticated agricultural practices without actually being capable of building individual life forms from fundamental chemical ingredients. While we can assume that "life happens" on Earth—our imperfect knowledge of how biological systems sustain each other has been highlighted in a number of attempts to artificially construct ecosystems as materially "closed" environments.

Since we are dependent on our native systems for survival, our longer-term thriving depends on our capacity to promote the infrastructures of life throughout interplanetary and cosmic space. In other words, our explorations must be concerned with the processes of "ecopoiesis"—the transformation of barren, non-dynamic planetary surfaces such as Mars—into primitive ecosystems, in which elemental cycles generate planetary-scale systems of exchange (Armstrong 2014). Such developments must go beyond the sealed ecologies proposed by Christopher Alexander (1978) and R. Buckminster Fuller (2008) to enable our connection with alien environments. In these habitats, our living spaces are selectively permeable to their surroundings, allowing nurturing substances to enter our living spaces and unwanted substances to be discharged. These basic conditions create the possibility of terraforming, whereby existing dynamic conditions, such as weather patterns, may be shaped in ways that make them more similar to terrestrial conditions and potentially, therefore, could directly support a founding human colony.

In planning for settlement in a lifeless environment, it is necessary to anticipate those conditions that enable us to settle and thrive.

By definition, alien environments are hostile to terrestrial life forms and need to be modified before they are humanly habitable. While we are searching for Earth-compatibilities within our cosmic constellations, we can only offer a best guess as to what kinds of surface conditions actually exist in these locations. For example, the rocky Earth-like planet in Kepler 186 could not be assumed to support organic life unless certain conditions are met. The current minimum requirement is the presence of water, although organic (carbon-containing) compounds are also indicators that life may be supported in these systems. Yet, there is a big gap between finding water and carbon on an extraterrestrial planet and being able to throw open the hatch of a starship and generate settlements. Small gravitational differences in a planet can have a profound impact on our physiology. For example, people at high altitudes, such as the Andes, produce a hormone called erythropoietin that thickens the oxygen-carrying capacity of the blood.

Generally, the approach to constructing habitats in alien terrains is to construct sealed biomes. The Moon base project proposed by Foster+Partners as a joint project with ESA is an example of such a practice. Regolith is sprayed onto domed scaffoldings as habitable lunar bases, to provide shelter using the local planetary resources and minimize the need for transporting expensive terrestrial construction payloads (Leash 2014). Biomes contain the size of the perceived problem—a hostile environment—so that we can control internal conditions necessary for life such as atmosphere and water. Simply by bringing the right ingredients together, like making a cake, it is assumed that they will self-organize to become an ecosystem in any sealed location with Earth-like conditions and perform in an equivalent way to Earth's infrastructures. Yet, we have not established that such closed biospheres can actually be built and indefinitely sustained—even on Earth.

During the 1960s, NASA conducted an experiment to see how long people could survive in a sealed container. At the Langley Research Center, four astronauts entered the Living Pod with no contact with the outside world and access only to resources available (water, food, and oxygen) within the artificial environment as waste products (carbon dioxide and excrement) built up around them. Everyday activities like personal care, going to the toilet, eating, and dealing with the continual presence of other people in your personal space affected the viability of the system. After 4 months, living conditions began to deteriorate. Detritus began to build up in the extraction system—hair, fingernails, and skin—while the astronauts suffered nausea and headaches. Eventually, the pod was soiled by human waste and the astronauts had to be released from the cabin earlier than anticipated.

Yet, this did not deter underwater pioneer Jacques Cousteau from establishing a series of marine Conshelf habitats (I, II, and III) filled with oxygen and helium, where aquanauts lived for 1 month conducting research on the seabed (MessyNessy 2013). These three undersea living experiments produced valuable knowledge of undersea technology and physiology and, although it was found in later years that industrial underwater tasks were more effectively carried out by robots, they were still valuable as proof of concept prototypes.

Also, in the early 1970s, Graham Caine built the "first" ecological house in London that used sunlight and ecological systems to produce energy and food as part of an anarchist architecture group, called Street Farm (Hunt 2014). They were early proponents of community architecture and sustainable architecture looking for a new relationship with Nature. They used solar panels alongside materials found in the streets and also constructed hydroponic gardens. Yet, the house was not popular with neighbors, who volunteered to pull it down when it was demolished 3 years later.

Between 1972 and 1984, a Soviet BIOS-3 series of experiments examined how careful management alone could sustain a terrestrial "ark." BIOS-3 maintained a community of three people using an algae cultivator and a "phytron," which grew wheat and vegetables using sunlight (Morrison 2015). Yet, the experiment was not a fully "closed" biosphere, as dried meat and energy were provided from external sources, and human waste was stored instead of being recycled back into the system.

The ambition to construct a fully self-sustaining ecosystem was attempted again in the 1990s with Biosphere 2 in Arizona (for more details, see Part I, Chapter 4). This important study clearly demonstrated our inability to design fully self-sustaining closed environments that promote life—even when we are taking great care of them under closely monitored conditions. While these results bode very poorly for our long-term survivability in extreme environments beyond Earth's atmosphere, they establish an important benchmark in appreciating the limits of the design and engineering of closed ecosystems.

While Biosphere 2 was, perhaps, the most important controlled experiment of its kind, from the perspective of settling a lifeless environment, the results indicate we have so much more to learn about how to establish ecosystems sustainably.

While these experiments in creating closed worlds aimed to create a kind of freedom from many forms of control afforded by terrestrial systems—Nature, governance, or the land—they actually ended up substituting one form of oppression with another.

If we are to remain in an extraterrestrial environment, it is essential to develop effective ways of dealing with space radiation and generating life-support systems. Organic matter absorbs radiation effectively and, unlike metals, does not re-emit it. Potentially, then, space habitats could be seamlessly interwoven with the construction of ecosystems that can provide fresh water, a breathable atmosphere, nutrition, energy, and the removal of waste products. This may not only provide organic living spaces with more aesthetic appeal than the current International Space Station (ISS) interior, but could also offer an effective degree of radiation shielding.

While the ISS has been continuously inhabited since November 2, 2000, when Expedition 1 commander Bill Shepherd and flight engineers Sergei Krikalev and Yuri Gidzenko became the first residents, it is not a self-sustaining habitat, and requires regular supplies that are ferried from Earth to support its residents. In fact, the quality of the living experience inside the ISS has also much to be desired. It is reported that being inside the ISS is analogous to living in a sweaty portable toilet. So, with the prospect of the expansion of human habitation into space, ecosystem-promoting infrastructures are a priority. The incorporation of living systems into space habitats will not only increase the physiological and aesthetic qualities of these spaces, but will also enable longer residencies and, ultimately, permanent settlement. The kinds of techniques that may be useful in building prototypes of "effectively" closed ecologies in space, which do not rely on matter exchange with any part outside the system, may also provide insights that help us to address some of our own challenges of resource sustainability on Earth. While truly closed ecologies are not possible to sustain, as life requires external energy to drive metabolism, which ultimately comes from external sources such as the Sun—or in a few rare biological cases, chemical systems—the ambition to optimize metabolic exchange within a closed space may help us to optimize relationships within living systems and would benefit resource-constrained environments both in space and on Earth. For example, growing food from wastewater and carbon dioxide, which are in plentiful supply as a side effect of the daily activities of human living, and their recycling into functional substances could greatly increase the fertility and livability of cities and space habitats. Yet, growing food is not currently considered a critical aspect for space missions of less than 2 years' duration, where additional food supplies are preferred over the extra payload needed to grow plants.

NASA is currently supporting the development of "effectively" closed ecological infrastructures, such as Michael Flynn's "Water Walls: Highly Reliable and Massively Redundant Life Support Architecture" project (Cohen 2014). This aims to reduce the failure-prone technical systems that currently generate the flow of resources through closed systems, typical of space habitats. Water Walls is a human life-support system that harnesses biological and chemical properties and only draws on mechanical systems for a few tasks that are difficult to engineer in other ways, such as pumping fluids around the space. The processing power is generated by an active chemical technology using a method called "forward osmosis" that consists of a polyethylene bag that houses a series of semipermeable membranes. In this manner, Water Walls can perform useful ecological work that is usually performed by natural systems on Earth, such as processing water, purifying air, growing food, affording radiation protection, and growing protein and carbohydrates to sustain the crew over multi-year missions (Boyle 2012).

Extreme environments also challenge our understanding of the character of Nature, where, to support human survival, the roles of natural and artificial systems may be enigmatically blurred. Koert van Mensvoort observes that our ideas about the natural world are deeply entangled in our preconceptions about ecology, technology, and culture (van Mensvoort and Grievink 2012). Such a complex understanding of our environment, however, provides innovation challenges that enable us to rethink the critical role of natural systems on Earth, so that we may consider their innate technological potential, as well as their cultural significance. While agriculture once started off as a technological engagement with the environment, which shaped the productivity of the land, it gradually became naturalized through culturally accepted practices.

When machines became the dominant technical system with the rise of the Industrial Revolution, agriculture has generally been a beneficiary rather than an innovation driver. The cultivation of massive monocultures demands new relationships with the land that have been forged by the use of powerful machines such as combine harvesters. These liaisons have stretched the productivity of natural systems to meet industrial standards of efficiency and production. With our rapidly swelling populations, where we are likely to see around nine billion people living on our planet by the middle of this century and 11 billion by the end of it (Kunzig 2014), our systems of food production, which rely on the more efficient extraction of resources from the land, need rethinking. Mealworms are being explored as potential nutrition for researchers within the sealed biosphere Moon Palace One at the Beijing University of Aeronautics and Astronautics, which is China's largest and most sophisticated facility for developing life-support systems in space. The worms are eaten neat and not mashed, blended, or minced into something more visually appealing, as a mental test for the challenges posed by new eating experiences, which are emotionally as well as physically challenging (Chen 2014). However, we need to not only reconsider the kinds of substances that we may eat beyond Earth's surface, but also understand how it may be possible to facilitate the relationships for their cultivation. This may require new kinds of symbiotic relationships between organic systems and machines. Indeed, the design of self-sustaining urban environments has many similarities with the challenges of solving food production on a spaceship or in a Mars colony, because they require advances in our knowledge and engineering of how organic resources may best flourish through an association with technical systems.

While, in the modern era, we equate technology with machines, in the last 30 years or so, advances in biotechnology have provided us with insights into the technical potential inherent in life's processes that may be harnessed to perform useful work through the process of metabolism—a set of self-sustaining chemical reactions. Yet, paradigm shifts in the way that we work do not come from cumulative incremental changes to the current system, but require a disruptive approach. For example, robotic management processes could potentially provide an augmented environment for the growth of plants and, through biotechnological developments, tailor the supply of vital resources to meet the individual needs of specific plant species. Advances in this area could potentially help agriculture to reclaim difficult-to-use spaces, such as in desert areas or on roofs and walls in urban areas. Automated systems that address the wide range of mechanical, user-interface, and communication challenges to enable robot-augmented plant growth could potentially form the foundations for remotely operated, ecosystem life-support systems that could also be deployed before human settlement in remote locations. Yet, while mechanical systems deal well with resource efficiency, they have little expertise in the recycling of organic waste through composting—the building blocks of all ecosystems—back into the growth, development, and systems of senescence of living things.

As events unfold surrounding the Mars One project, which aims to send a crew of four to Mars in 2026 (Mars One n.d.), questions are being raised about whether the proposed plans to support a colony will actually be sufficient (Ho et al. 2014). Although the Martian surface has been studied since 2003 using robots that have been busy characterizing a wide range of rocks and soils, and looking for clues that may lead to the discovery of water or uncover its geological history, these data do not provide enough information to conclusively calculate the requirements for a founding colony.

Of course, the logistics related to supply demands of a long-term colony on the planet's surface are not well understood, since our occupation of the ISS has been supplied by terrestrial resources and their in-situ recycling. So, we have no pilot off-world project with which to make a comparison.

However, a range of Mars simulations are providing valuable insights into life in extreme environments, such as the Mars-500 mission (to gather planned long-term deep-space mission data), the use of holographic glasses (that generate a 3D simulation of the Red Planet) (Total Recall 2015), and the Hawaii Space Exploration Analog and Simulation (HI-SEAS I) mission (to establish what kinds of foodstuffs may be successful) provide us with some principles that can be applied to the question of longer-term settlement to help construct a range of scenarios.

The starship question highlights how little we really know about the operating systems and infrastructures that generate the conditions for our own survival. While we may be gardeners of our planet, we do not know how to increase its liveliness or invigorate its failing systems. Confronting deep gaps in our understanding of how the environment operates does not obviate other challenges in the construction of a worldship, such as the question of propulsion, but makes pressing the immediate and long-term investment in the study of life (as we thought we knew it) and how it forms our ecosystems.

6.5 Space Nature

Ecological systems that thrive in very hostile surroundings are needed for the synthesis of worldship habitats. The material ecologies, which enable the fabrics of life and its associated bodies to persist, may be considered as different kinds of Space Nature.

While the ambitions to produce a living interior for a colony aim to optimize the survival of its inhabitants, the principles through which these materialities are constructed are likely to differ from any naturalistic intentions. For starters, significant technical challenges exist—whereby, for example, we are unable to make non-living matter come alive—so the possibility of designing and engineering lively, life-promoting, self-sustaining material fabrics that can potentially support us indefinitely is a daunting challenge.

Moreover, modern expectations establish a particular view of the natural realm, which is greatly simplified and universalizes its performance. For example, in a biotechnological era, we take a deterministic approach to living systems in which we seek increasing degrees of control. According to this worldview, we can reasonably expect a certain degree of material efficiency and productivity from biological systems. Additionally, molecular biology adopts a radical engineering approach towards living materials by viewing them as digital computers that can be programmed using DNA to perform new or more efficient functions. Yet, these assumptions are not the actual conditions and outcomes that we encounter when designing and engineering with the natural realm, which is probabilistic, operates at far from equilibrium states, and therefore possesses its own energy and agency. Yet, ecological simplification is a block to livability. Hypercomplexity is what enables growth and flourishing.

Biotechnology also considers the natural realm as something that—through molecular surveillance and repair—is potentially immortal. In those instances whereby atomic control cannot be wielded, outcomes may be regarded as failures, or need to be excluded as substrates for design and engineering with. Indeed, unruly Nature may be described as a natural "disaster" that has "permission" to exist within the wilderness—but must be prevented from entering our living spaces. Such cultural expectations render the idea of alternative kinds of Nature problematic because they are based on idealized notions of material behavior—rather than actual events. In other words, our existing ideas about the nature of Nature preclude our ability to observe, design, and synthesize new expressions of natural systems, which might better suit worldship interiors than naturalized or modern perspectives.

Additionally, should such radical material systems be produced, they are likely to be met with scathing cultural criticism that pertains to their constructed character. For example, humanoid robots are already tarnished with the idea of the "uncanny valley" that exists where simulacrum approach their goal to mimic a "real" object. Likewise, even when ecosystems aboard a worldship produce a nurturing environment, observers that have been raised in an "original" terrestrial environment are likely to note oddities that mark their unnatural origins. Generally, these small but significant differences are valued negatively, resulting in continual efforts to close the gap between simulacrum and authentic bodies. In other words, Space Nature is likely to have a distinctive and discernible character depending on its context, its inhabitants, and the technologies from which it is produced that is quite unlike the experience of life-bearing fabrics elsewhere. In fact, it may be more helpful to think of Space Nature existing as many different kinds of species that are shaped by the methods of their production, the matter from which they are forged, and the manner in which they are inhabited.

To develop an effective approach to the production of living habitats, a fundamental change in philosophical focus underpinned by new kinds of technical operating systems may help us to change the current paradigm of human development, so that it enlivens, rather than damages, our global ecosystems. From this viewpoint, landscapes are not stages for biological events, but active players in the kinds of experiences that enable us to gain purchase on environments and thrive within them.

Persephone's living interior is neither a modern view of biology that can be commanded at an atomic scale nor an untamed, all-knowing "teacher" (Benyus 1997). It is an oceanic (Lee 2011), restless, feral, resurgent (Tsing and Haraway 2015) material force, which is beyond human control. Its technological and cultural entanglements work together, seeking to evade the dissipative decay towards equilibrium states, and therefore maintains the status that we attribute to life. It does not place human concerns at the center of its operations but can create the conditions for thriving under the right conditions. Currently, we are not even close to producing the necessary kinds of life-promoting approaches on Earth that are needed to boost the vitality of existing living systems—let alone constructing new ones from scratch. Yet, such a practice is utterly essential, not just in securing our continued existence beyond our mother planet, but also in better being able to inhabit and survive on our home planet.

But if Space Nature is a subversive, constantly changing, material body capable of evolution, then how far is it possible to design and engineer with its systems?

Juan Enrique believes that the single greatest responsibility that we have today is to become masters of our own evolution. Yet, in the Ecocene, a different framework is needed than top-down imperatives, where we accept that environmental systems resist deterministic approaches such as those that underpin the philosophy of synthetic biological approaches. Instead, they work with probabilistic notions of soft control whereby many agencies choreograph their performance. Establishing acceptable limits of behavior, what their core qualities may be, and how they can be engaged and inhabited may ultimately enable us to develop ways of working with natural resources so that they are not inevitably denatured, degraded, and consumed by communities. Rather they are integral to ongoing exchanges that result in increased fertility, livability, and biodiversity.

By thinking ecologically rather than industrially, an alternative set of tactics and programs for the development of colonies may be possible. Dynamic exchanges shaped by the performativity of interfaces and the paradoxical expression of dissipative structures could facilitate different kinds of exchanges that enable us to go "beyond" consumptive relationship with natural systems. In this manner, Space Natures may be continually transformed and reinvigorated through their engagement with living bodies and therefore can continue to nourish their native ecosystems in ways that are interwoven with cultural practices. First principles thinking about our habitats is required to generate value systems for inhabiting a lively, material realm that works alongside the unique qualities of the natural realm such as resilience, robustness, unpredictability, and evolvability.

However, this is not the current practice on Earth, which is largely influenced by the values and traditions of industrial-age systems. Designing an alternative development framework for designing with, engineering, and evaluating how matter relates to living processes is required to achieve significant step changes in practice, performance outcomes, and spatial experiences.

One such approach may be to frame our questions about constructing ecosystems within the context of ecopoiesis, where we have to think about the a priori performance of material systems and life, so that we may begin to discover alternative ways of constructing life-supporting fabrics and worldship terrains.

6.6 Manifesto for Persephone's "living architecture"

Manifestos are typically positioned as a rallying cry to inspire action. However, this one is informed by Persephone's experimental findings and aims to begin a series of conversations in which the principles of living architecture may be developed in a social context—in which thriving is a contestable but ongoing process. In doing so, this manifesto generates a context for the exploration of radical materials that engage the principles of Ecopoiesis, which may change our environmental expectations. Potentially, we may transform the process of human development from being a destructive force, implicit in the mythos of the Anthropocene, into being capable of procuring fertility.

Persephone embraces the Ecocene as an iterative and ongoing process of inquiry. Its value system is centered on facilitating the emergence of lively environments that have the capacity to support living systems, from non-living matter.

This manifesto therefore does not stand as an immutable set of proposals, but as an iterative and ongoing process of inquiry through an appreciation of:

1. *Land as multidimensional space*: rather than being regarded as 2D plots of land, territories may be valued as potent multi-dimensional spaces that facilitate the flow of matter through manifold domains.

2. *Technology of Nature*: physics and chemistry provide a common language so that natural and technological forces work synergistically to transform matter into high-biotic-value fabrics, which differentially shape space.
3. *Metabolic economy*: metropolitan-scale economies arise out of spatialized biophysical transformation in sociopolitical contexts.
4. *Fertility as value*: value is established by the capacity of a site to transform matter and may be increased in any site by harnessing bioprocessing activities such as composting, food, and energy production.
5. *Non-human investors*: our living spaces are not exclusively human and require investment by the non-human realm. Sustainable societies may therefore be defined by local, open agreements that enable multiple bodies to share contested spaces and shape the myriad of material exchanges in our urban environments in ways that enable the participants to evolve.
6. *Consumers become producers*: a metabolic economy transforms obligate consumers into co-producers of their living spaces.
7. *Ethical exchange*: the cultural, moral, and ethical systems of a community may be shaped by lively exchanges between cooperating bodies within an ecosystem. For example, SymbioticA's Victimless Leather offers a way of producing meat other than by slaughtering animals and provokes technological innovation. It has foreshadowed the DIYbio movement (Myers and Antonelli 2013, p. 12) and precipitated scientific study, whereby it has now become possible to make cultured meat burgers and leather without animal suffering.

REFERENCES

C. Alexander, *A Patter Language: Towns, Buildings, Construction* (Oxford University Press, Oxford, 1978)

R. Armstrong, Space is an ecology for living in. Archit. Des. **84**(6), 128–133 (2014)

R. Armstrong, *Vibrant Architecture: Matter as a Codesigner of Living Structures* (De Gruyter Open, Berlin, 2015)

J. Benyus, *Biomimicry: Innovation Inspired by Nature* (William Morrow, New York, 1997)

R. Boyle, Future spacecraft could protect crews with walls made of water. Pop Sci. (2012), Sept 5, www.popsci.com/technology/article/2012-09/future-spacecraft-could-protect-crews-walls-made-water?dom=PSC&loc=recent&lnk=1&con=future-spacecraft-could-protect-crews-with-walls-made-of-water. Accessed 24 May 2014

CBEC and APEGEC, Budget guidelines for consulting engineering services. The Consulting Engineers of British Columbia and The Association of Professional Engineers and Geoscientists of British Columbia (2009), www.apeg.bc.ca/ppractice/documents/ppguidelines/BudgetGuidelines.pdf. Accessed 25 Aug 2014

S. Chen, A diet of worms: Chinese volunteers spend 105 days eating creepy crawlies in space experiment. South China Morning Post (2014), May 21, www.scmp.com/news/china/article/1517052/scientists-tackle-final-frontier-worms-space. Accessed 24 May 2014

Cohen, M. 2014. Being a space architect: AstrotectureTM projects for NASA, AD Space Architecture, 232, pp.80–81

T. Field, T. Brodribb, A. Iglesias, D. Chatelet, A. Baresch, G.R. Upchruch, B. Gomez, B.A.R. Mohr, C. Coiffard, J. Kvacek, C. Jaramillo, Fossil evidence for cretaceous escalation in angiosperm leaf vein evolution. Proc. Natl. Acad. Sci. U. S. A. **108**(20), 8363–8366 (2011)

R.B. Fuller, *Operating Manual for Spaceship Earth* (Lars Muller Publisher, Zurich, 2008)

D. Gallagher, Le Corbusier. Open learn, The Open University (2001), Nov 16, www.open.edu/openlearn/history-the-arts/history/heritage/le-corbusier. Accessed 27 Apr 2014

K. Ho, S. Schreiner, A. Owens, O. de Weck, An independent assessment of the technical feasibility of the mars one mission plan. 65th International Astronautical Congress, Toronto, Canada, IAC-14A5.2.7 (2014), http://web.mit.edu/sydneydo/Public/Mars%20One%20Feasibility%20Analysis%20IAC14.pdf. Accessed 13 Apr 2015

S.E. Hunt, *The Revolutionary Urbanism of Street Farm: Eco-Anarchism, Architecture and Alternative Technology in the 1970s* (Tangent Books, Bristol, 2014)

S. Kauffman, Evolving by magic. Arup Thoughts (2012), May 17, http://thoughts.arup.com/post/details/200/evolving-by-magic. Accessed 21 Aug 2015

K. Kelly, *What Technology Wants* (Viking, New York, 2010)

R. Kunzig, A world with 11 billion people? New population projections shatter earlier estimates. National Geographic (2014), Sept 18, http://news.nationalgeographic.com/news/2014/09/140918-population-global-united-nations-2100-boom-africa/. Accessed 13 Feb 2015

Le Corbusier (C.E. Jeanneret). *Toward an Architecture*. (Getty Research Institute, Los Angeles, 2007)

N. Leash, 3D printing in space. Archit. Des. **84**(6), 110 (2014)

M. Lee, Oceanic ontology and problematic thought. (NOOK Book/Barnes and Noble, 2011), www.barnesandnoble.com/w/oceanic-ontology-and-problematic-thought-matt-lee/1105805765/. Accessed 19 Apr 2014

Mars One, The next giant leap for humankind (n.d.), www.mars-one.com. Accessed 10 Aug 2014

MessyNessy. Remains of an underwater habitat left by 1960 sea dwellers. Messy Nessy Chic (2013), May 17, www.messynessychic.com/2013/05/27/remains-of-an-underwater-habitat-left-by-1960-sea-dwellers/. Accessed 18 Mar 2016

N. Morrison, Algae farming in low earth orbit: past, present and future. J. Br. Interplanet. Soc., **67**(7,8,9), 332–337 (2015)

W. Myers, P. Antonelli, *Bio Design: Nature, Science, Creativity*. (Thames & Hudson, London/MOMA, New York, 2013)

I. Steadman, Hamburg unveils world's first algae-powered building. Wired (2013), Apr 16, www.wired.com/design/2013/04/algae-powered-building/. Accessed 18 Aug 2014

B. Sterling, Design fiction: growth assembly: Sascha Pohflepp and Daisy Ginsberg: beyond the beyond. Wired (2012), Jan 3, www.wired.com/beyond_the_beyond/2012/01/design-fiction-sascha-pohflepp-daisy-ginsberg-growth-assembly/. Accessed 18 Apr 2014

Total Recall, NASA wants to send its scientists to mars (using holographic glasses) (2015), Jan 23, http://attodiforza.blogspot.co.uk/2015/01/nasa-wants-to-send-its-scientists-to.html. Accessed 20 Aug 2015

A. Tsing, D. Haraway, Tunneling in the Chthulucene. 11th Biennial ASLE Conference, Idaho, Moscow (2015), Oct 1, https://www.youtube.com/watch?v=FkZSh8Wb-t8&feature=youtu.be&utm_content=bufferd4a8e&utm_medium=social&utm_source=twitter.com&utm_campaign=buffer. Accessed 26 Dec 2015

K. Van Mensvoort, H.J. Grievink, *Next Nature: Nature Changes Along with Us* (Actar, Barcelona, 2012)

7

Designing and engineering the infrastructures for life

Rachel Armstrong

This chapter examines unconventional forms of computation from an analog perspective that may enable us to develop new methods, materials, and technologies for constructing interstellar habitats.

7.1 Natural computing

A new way of working with natural systems is needed if we are going to design and engineer with them at architectural and, potentially, planetary scales (Morris 2014). In the context of a starship environment, it is not a case of "life" versus machine, where the utility of one is pitched against the other. Rather, it heralds a new age of synthesis between multiple operating systems—such as the made and the born—to produce platforms that can prototype new forms of making.

Biomimicry proposes that Nature can provide solutions to a truly sustainable future, if only we follow its lead (Benyus 1997) by copying existing solutions from the natural realm that provide design inspiration from its relentlessly optimistic and seductive visions. Our gaze is drawn towards biological geometries, such as sponge skeletons, that are privileged over less seductive processes, like rotting carcasses. This creates the conditions for a practice that is so highly aestheticized that it remains implementable in our industrial modes of development and changes nothing. Yet, the ambition to find a different toolset than mechanical modes of production is not to set one technical system against another, but to diversify the opportunities for working in new ways that are qualitatively different from those preceding them. However, such a possibility does not yet fully exist within twenty-first-century practices. Even notions of "sustainable development" have largely been framed around more considerate forms of industrialization, which cling doggedly to established practices that generate most of the pollution on the planet. We consume millions of

---

Rachel Armstrong, PhD. (✉)
Professor of Experimental Architecture, Newcastle University, Newcastle-upon-Tyne, UK

tons of consumer goods a year that end up in stinking garbage dumps that produce metabolic waste, which leaks into the environment to alter our ecological systems. The average person in the US generates 2 kg of waste per day, accounting for 220 million tons each year, which is 36% more than in 1960. About half of this is discarded into 3500 landfills, which are the second-largest source of anthropogenic methane in the US, accounting for 22% of these emissions in 2008 (Duke Nicholas School for the Environment 2015). Also, despite global awareness of the destructive impact of fossil-fuel energy sources, these industries continue to be funded by national governments. The Overseas Development Institute observes that producers of oil, gas, and coal received more than US$500 billion in government subsidies around the world in 2011, with the richest nations collectively spending more than US$70 billion every year to bolster fossil-fuel markets (Atkin 2013).

Moreover, crude-oil-based manufacturing processes are still filling our oceans with plastic, which makes up 90% of all trash floating in the world's oceans (Weiss 2006). The United Nations Environment Program estimated in 2006 that every square mile of ocean hosts 46,000 pieces of floating plastic (Green24 n.d.). In some areas, the amount of plastic outweighs the amount of plankton by a ratio of 6:1. Of the more than 200 billion pounds of plastic the world produces each year, about 10% ends up in the ocean (Green24 n.d.). Seventy percent of that eventually sinks, damaging life on the ocean floor (Green24 n.d.). The rest floats; much of it ends up in gyres and the massive garbage patches that form there, with some plastic eventually washing up on a distant shore, while microfibers invade almost every ecological niche.

Right now, it seems almost impossible to imagine an alternative worldview to modern society, for, despite decades of postmodern theorists deconstructing our reality, there have been no major adjustments to the way industrial production platforms operate. Despite alternative development systems proposed, such as Jeremy Rifkin's third Industrial Revolution, which is based on alternative energy generation and distributed forms of production like the "3D revolution," such measures do nothing to address the consumptive processes driving economic prosperity (Rifkin 2013). In fact, if energy became freely available to everyone tomorrow, the "energy" crisis would be transformed into a "war against matter" (Serres 1980), where material resources are perceived as the bottleneck for unlimited consumption.

At the start of the third millennium, we are confronted by a rapidly changing sense of reality, which has been catalyzed by the Internet and enabled us to transgress established boundaries of order to seek new opportunities. Indeed, we are no longer constrained by modern binaries of geography, politics, economics, identity, or discipline, but may simultaneously and coherently inhabit new kinds of existence. The clear either/or distinctions that formerly shaped our experiences of the world are being replaced by a much more fluid relationship with reality, where possibilities are no longer fixed. Rather, they coherently and simultaneously exist in many states as an unfathomably complex entanglement of objects and processes. These hybrid approaches and perspectives may be key to developing new kinds of technical platforms and modes of practice. This exciting, convergent, complex worldview extends to our view of Nature, which is made up of many interacting bodies that have been historically recognized as the animal, plant, and mineral kingdoms (DeLanda 2000, p. 26). Yet, powerful modeling tools and biotechnological insights that have been developed over the last few decades now enable us to harness the processes of organisms to solve challenges in new ways in a technological context with cultural impacts. Such developments enable us to re-evaluate our contemporary concerns about living on an unstable Earth and respond differently to them.

While machines and digital technologies clearly represent the pinnacle of an industrial portfolio, it is unclear exactly what kinds of toolsets may help us deal with a world in continual flux that underpins ecological thinking. Over the last 30 years or so, advances in biotechnology have enabled us to consider the processes of life themselves in a technical capacity, which can perform work in ways that may be the equivalent of machines, but also in ways that are very different from them. The advent of modern biotechnology creates a new context for design and engineering the natural world, since it now possesses a fundamentally technological character that works synergistically with Nature. For example, bacterial proteins can be used to make "bioluminescent" light that is activated on vibration, by approaching footsteps, or by passing traffic (Myers and Antonelli 2013).

Yet, the technologies of life—or "living" technologies—which possess some of the characteristics of living systems that can perform useful work without necessarily being granted the full status of being truly "alive"—address an emergent, contingent series of events that are spatially choreographed and shaped by their relationship to an unevenly distributed and constantly changing present. Living technologies may catalyze our transition towards the Ecocene. One of these options is the development of metabolic, or living, materials, which can be used to create a system of techniques that enable the construction of "bio-" or "living architecture." Living materials enable building fabrics to be integrated with the natural world, instead of insulating our habitats against Nature. The importance of this transition is to establish ways of producing material programs and architectures that can better deal with our restless, constantly changing environment—both on Earth and beyond its life-bearing systems.

Potentially, by applying living technologies to the design and construction of our habitats, it may be quite literally possible to undergo an origins-of-life transition within the built environment. Typically, our urban environments are inert deserts of tar and concrete that are inherently hostile to living systems. Yet, with the strategic introduction of metabolic materials into our living spaces, it may be possible to transform these barren wastelands into vibrant communities. Networks of biochemical reactions could be constructed to link the living world to Nature, and enable "life" to make the most of its local resources. However, emergence alone is insufficient to promote the vibrancy of the natural world, as we need tools that also help us shape outcomes, so that human requirements become synergistic with the performance of natural systems and may even enhance their vibrancy.

Such platforms operate according to the principles of non-equilibrium processes, which do not strictly obey the laws of classical mechanics. In fact, the decision pathways and events that shape the outcomes of living things remain open to influence by many contingencies until they have actually occurred. This probabilistic portfolio trades off against our desire for certainty in design and engineering practices. However, despite the desire for atomic control that characterizes modern projects, the instabilities implicit within dynamic systems offer an unrivaled opportunity to work alongside their creative capacity and embrace the resilience that is characteristic of natural systems. The technologies of life require their own operating system if they are to produce qualitatively different kinds of effects from mechanical platforms, or they will be swallowed up into their ontological and production frameworks as "soft" machinery.

Computation is fundamental to our species and may be thought of as the art and science of sorting and ordering the world in the quest to acquire new knowledge. Today's counting systems have been around for a few hundred years. "Zero" was invented to symbolize empty space in 1598, while John Wallis introduced the infinity symbol a little later, in 1655 (Chatelin 2012).

While digital computing is not possible without the concept of zero, it can operate quite happily without infinity. Modern computing deals with finitudes and drawing limits around a solution space to let the Turing operating system know where to search and when to stop its calculations (Kauffman 2012). Yet, natural systems are rich with unbounded possibilities, which are "effectively" infinite from a computational perspective. These iterations are based on different kinds of excursions based in physics and chemistry. Some of these drivers, such as Jeremy England's notion of "adaptive dissipation" (Wolchover 2014), shape evolutionary processes and occupy what Stuart Kauffman describes as "the adjacent possible" (Kauffman 2008). So, while digital systems can be applied to rationalize natural phenomena, its solutions are always abstractions of its material potential. This abstraction also extends to the structure of DNA, which can be readily represented as a series of ones and zeroes. This has given rise to a host of genetic algorithms and digital breeding projects, which produce algorithms that describe new combinations of morphology and behavior, which are then likened to biological systems.

These are informative, unconventional computing practices, which cannot be framed by the conventions of Turing or von Neumann architectures. These approaches may be valuable in observing and shaping the outcomes of biological systems in new ways, not as abstractions, but as functions of material processes. While there is no formal definition, it embraces a range of overlapping practices from the digital modeling of biological systems to chemical computing (Adamatzky et al. 2007), which propose to ensure that we do not limit ourselves to using only one model or simulation of the world. Potentially, then, unconventional computing can unlock more of the creative potential of the natural realm by "counting" directly with the material realm through its spontaneously occurring iterative processes.

Natural computing is a term that has been inspired by Alan Turing's interest in the technological potential of the natural world, and consists of a range of overlapping scientific practices that range from the computer modeling of biological systems to working with programmable matter. It also shares many overlaps with morphological computing (Pfeifer and Iida 2005) and unconventional computing (Adamatzky et al. 2007), which are also concerned with how matter makes "decisions."

Dissipative structures may be regarded as operative agents that can complexify space and matter. They function as an iterative system—or "natural" counting process—that can be ordered to generate a range of programmable outputs. Jeremy England's notion of dissipative adaptation, as a possible complementary driver to evolutionary processes rather than genetic modification, is of particular interest as a generative and selective strategy when considering the transition from inert to living matter (Wolchover 2014).

Dissipative structures arise spontaneously in Nature across a range of scales, such as crystals, tornadoes, and galaxies. While not all dissipative systems meet the technical qualifications of "life," all "life" is a dissipative process. Those systems that are not given the full status of being truly alive are of great interest in the design process as a set of tactics that may increase the probability of life-promoting events.

Dissipative structures are environmentally sensitive and, therefore, may be influenced by their context. The technical platform also serves as the material and medium for design. Functionally, they are dynamic leaky objects that continually take in matter and energy until they reach equilibrium. They exist within a range of limits and are unstable, eventually succumbing to the forces of disorder or entropy, at which point they lose their interesting properties. Dissipative structures also interact beyond their boundaries through active fields. Perhaps the best way to imagine a dissipative structure is to think of a tornado. A storm chaser can feel the presence of a twister long before they ever reach the eye of the storm. Because they are physical systems, each reaction–diffusion wave never truly collapses to a zero state and therefore tends to seed the conditions for its next iteration until relative thermodynamic equilibrium is reached.

The material events that characterize an iterative system or oscillator directly produce maps of the computational system they embody, which can be read through time-based events such as the Belousov–Zhabotinsky reaction. Dissipative systems spontaneously arise from overlapping fields of activity. Specifically, these computational platforms have also been called "reaction–diffusion" computers (Adamatzky et al. 2005). They are massively parallel computing devices, where micro-volumes of an excitable chemical medium act as an elementary processor through which different chemistries diffuse and react in space and time. Both the data and the results of the reaction–diffusion computer are encoded as concentration profiles of the reagents, or local disturbances of concentrations. The actual computation is performed through the propagation of waves caused by these disturbances. Their interactions produce waves of observable activity and material events that directly feed back into the outputs of the system, so that new spaces and possibilities become accessible. By considering dissipative structures as the iterative process that underpins material computation, the technical operations possess a different ontology to digital computing. This perspective may enable us to work with low-level infrastructures, which produce material programs that directly promote life without reducing or abstracting them.

Dissipative structures, such as tornadoes, therefore require us to count along with them—not "for" them, by imposing our own expectations on the system's performance. In other words, they require different forms of soft control. These kinds of techniques are already known to us as artisan practices and range from cooking, gardening, and agriculture to working with small children or herding cats.

Networks of dissipative structures may be used in the production of complex, lively materials that can actively engage in ecological exchanges and persist through the complexification of space and matter. These bodies may not qualify as "life," but produce effects that occur as a function over time to produce increasingly lifelike events. Potentially, when these spatialized material exchanges reach a certain degree of complexity, the system may even spontaneously produce lifelike events that could even be inevitable. Such substances sound mysterious, but we can recognize them in everyday materials such as soils, on which all terrestrial life is founded. Soils represent an alternative material-organizing system than is used in our design processes today. Rather than being purified, homogenized, and constrained within bounded spaces, they are open, messy, and highly heterogeneous. This is of extreme interest in identifying processes that increase the liveliness of space and even raise the threshold of events that may spontaneously produce "life" within a specific environment.

Whatever the approach taken toward the construction of a worldship environment, the development of bodies and communities that may inhabit them is likely to be diverse—even when foundational principles and identities are identical. A series of design-led laboratory experiments provide a point of reference for thinking about how environments that support lively events—as opposed to designing life as a set of objects—may be approached as a way of producing "artificial" soils that underpin fertile terrains, in which the probability of life is increased, using a set of simple techniques.

However, we will not make these assumptions, but rather investigate what kind of starting conditions are required to produce lively matter—not as an inevitable consequence of the system, but as a likely event from a set of staring conditions. In other words, we will deal with the process of construction of dynamic systems as having probabilistic rather than deterministic outcomes. These techniques will be used to engage with aspects of ecopoiesis to establish rudimentary principles of construction. Gradually, something that appears to have the kinds of interconnected, complex, regenerative qualities of soils will be developed as a discursive practice and a point of reference for thinking about how a living environment may be approached, using a set of simple techniques.

7.2 Dissipative structures

To kick-start the process of ecopoiesis requires the strategic overlapping of active fields, where dissipative structures are likely to arise between interfaces that produce tensions between matter and energy.

Dissipative structures spontaneously arise at the intersections of mutually complexifying, active fields. They can be experimentally demonstrated using simple kitchen ingredients that provoke dynamic interactions. For example, water affinity can be used to produce a very simple dissipative system, by harnessing the energetic and material potential that exists between substances that have very different relationships with water—namely glycerin, which strongly attracts water, and olive oil, which repels water. A base layer of glycerin and a top layer of olive oil create an interface at which droplets of food coloring are introduced using a handheld pipette. Adding rock salt to the droplets increases their affinity for water, which provides further complexity and tension in the system. The hygroscopic properties of glycerin eventually overcome the osmotic pressure of the food color droplets, at which point they explode downwards, leaving structural trails in their wake that outline the resultant transient dissipative structure. Oil-soluble coloring spreads sideways through the olive oil interface, providing a residue of molecular activity—or 3D painting.

In an installation for the On Architecture exhibition at the Department of Arts and Sciences in Belgrade, food color droplets were discharged into a base layer of glycerin, leaving spidery threads of salt crystals behind that depicted the transient trails of water molecules as they became assimilated into the hygroscopic base layer (Fig. 7.1).

Once potentiating fields of interaction have been established, the propagating waves of interaction produced by the oscillators begin to radiate and collide with each other. In reaction–diffusion computing, wave collisions have been used as tactics to establish classical engineering strategies such as logic gates (Adamatzky et al. 2005). Essentially, this

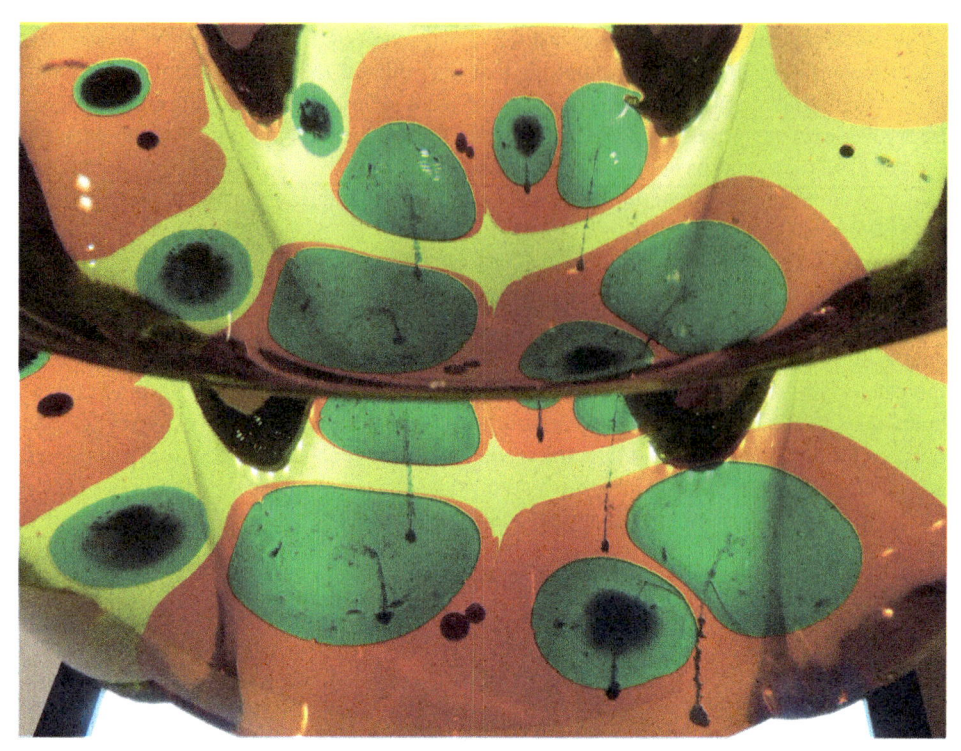

Figure 7.1 Dissipative structures produced at a hygroscopic interface between oil and glycerin. Credit: Rachel Armstrong

approach looks for events that can be translated and worked into mechanical circuits. In other words, complexity is being decomplexified, so that it may be assimilated within a familiar mechanical engineering framework.

Perhaps surprisingly, dissipative systems are remarkably predictable, although they produce a spectrum of outcomes that operate within "limits" of possibility, which are imposed by the properties of the system and its contexts (see above). Dissipative structures are highly resilient and able to transform themselves—structurally and morphologically—according to changes in their interior and exterior environment. They can also form reversible groupings that generate the lifelike behaviors, which account for their flexibility, robustness, and environmental sensitivity. However, populations of dissipative structures may periodically behave unpredictably when a tipping point is reached and give rise to novel, emergent, complex events that cannot be deduced from their characteristic behaviors. New ways of accurately describing what is happening during striking phase changes are needed to more fully describe the continuous nature of change and its material complexity. Currently, phase changes are described according to a set of recognizable metapatterns that are documented over a series of time intervals. The relational aspects of tipping points therefore escape comprehensive description and analysis because of the way the system is observed during these events.

Dissipative structures 155

But what if an open-ended set of events within an environment is encouraged, by starting at a rich and complex initial state as a way of preventing or delaying the dissipative system from falling into a simple state—that is, reaching equilibrium? Ikegami and Hanczyc (2009) call this approach the maximalism design principle. The concept invites today's experimentalists to design more complex initial species of dynamic agents, where "life" emerges at the edge of self-organization and complexity. This theory is supported by England's observation regarding the dissipative adaptation of matter—that, when a group of atoms is driven by an external source of energy (the Sun, chemistry), it will often gradually restructure itself in order to dissipate increasingly more energy. This could mean that, under certain open environmental conditions, matter inexorably acquires the key physical attributes associated with life, and "semi-living" systems may emerge, where bundles of self-assembling, mutually influencing bodies produce the conditions and environments of their own existence. Such complex events can be observed using Bütschli droplets, which represent an example of highly complex initial agents, whereby each body influences its surroundings. The following images are taken from a series of over 300 replicate experiments, whereby lifelike events and organic structures with a range of striking morphologies and behaviors were produced within seconds to many minutes (Armstrong and Hanczyc 2013) (Fig. 7.2).

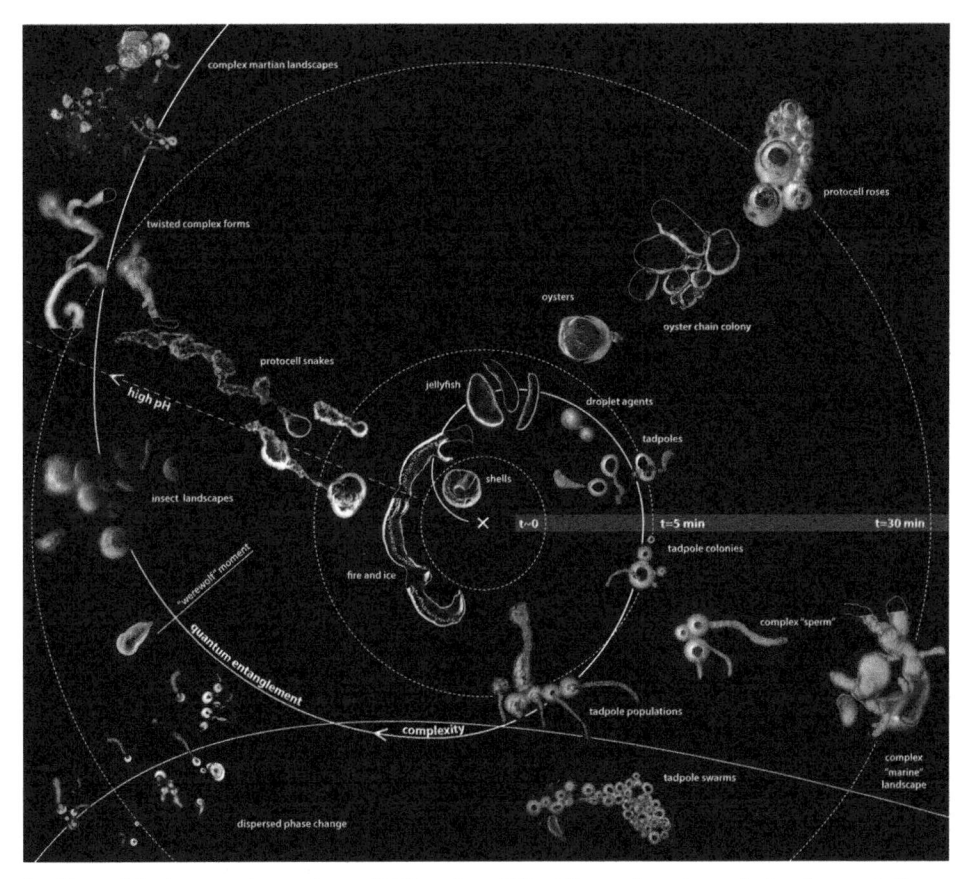

Figure 7.2 An oceanic ontology of 300 replicate Bütschli experiments, exploring the creativity of the chemical system as a function of time. Credit: Rachel Armstrong and Simone Ferracina

In the following diagram, an "oceanic" ontology (Lee 2011) of Bütschli species have been grouped according to observed morphological and behavioral typologies. The diagram represents the contextualization of meta-events between droplets with time within a complex field of activity. The stage is not a single reading of events, but reflects multiple possibilities, where the field of activity is constructed through exploratory, graphical approaches. The resultant diagram maps relationships in the system, rather than invoking the classical "tree" metaphor of classification systems, which focuses on differences rather than similarities between actors. The graphic is centered at time zero, from which concentric circles radiate, representing an exponentially increasing series of time intervals. It depicts the intense self-organizing activity that happens early on in the chemical reaction. An estimated 90 % of chemical activity is completed within five minutes of activation of the system, although individual droplets have been observed to be active as long as an hour after their genesis. A spiral that represents complexity also radiates from the origin, around which various droplet morphologies and behaviors that indicate change in the system are grouped subjectively. For example, "oyster chains" are distinct in appearance but only differ from "complex marine landscapes" by the relative degree of solid material they produce. The range of different bodies produced by the Bütschli system generates a potential portfolio and experimental starting point for "soil-producing" technologies—fabrics and materials that may increase the material complexity and "fertility" of a landscape and, therefore, the probable occurrence of lifelike events. Such systems do not have any direct commercial applications, but do begin to suggest ways of taking a first-principles approach towards developing "artificial" soils or other life-promoting synthetic media.

Natural computing constitutes a platform that can facilitate entirely new design and engineering opportunities. These take the form of material convergences that emerge from the entanglements of the spontaneous flow of energy and matter, and can be choreographed in time and space to provoke new events. Some of the lively bodies, such as dissipative structures, form loose, reversible groupings—or assemblages—with each other, while others become coupled and transformed by the interactions. Natural computing techniques also help break down some of the ontological barriers that set objects and systems in opposition, and facilitate technological convergence, whereby lively bodies may reach tipping points that produce radical breaks and discontinuities in the system. These constitute the kinds of material events that lead to evolutionary transitions and are fundamental to worldship evolvability.

7.3 Computing with hypercomplexity and uncertainty

Lively substances that can make "decisions" at the molecular level may be considered as an unconventional form of computing. In natural computing, the operative agents are dissipative structures that exist at many scales—from the microscale growth of crystals to cosmic events like the birth of stars. Characteristically, dissipative systems do not "fail" when encountering hypercomplex situations, as they are able to select from a range out outcomes depending on their context. This contrasts sharply with digital computing that relies on a deterministic approach, where the computation looks for a single outcome. If a digital computer is presented with more than one possibility, then it cannot make a decision and the system crashes.

Computing with hypercomplexity and uncertainty 157

The implications are that natural computing offers a toolset for dealing with very complex challenges that have a material basis and also allows us to engage with unknowns, by simply running the computer multiple times and calculating various event probabilities. The field of natural computing is still at an early experimental stage of development. Therefore, if we are going to generate materials, methods, and fabrics that can easily be handled, valued, sorted, and ordered in the production of sustained architectural experiences—from lively landscapes to inhabitable structures—then a portfolio of toolsets are needed.

The Bütschli system provides an example of how a natural computing system may be developed. It is a species of dynamic droplet, which is produced during a simple chemical reaction used to make soap. Bütschli droplets can be easily worked with by hand if they are stabilized by adding an extract of their waste products to their body, which makes them bigger (around 2 cm in diameter, rather than 1–2 mm) and enables them to last longer (several weeks instead of about an hour). Although these droplets are also less lively than the unmodified version, they still possess the fundamental properties of dissipative systems. For example, in this attenuated form, they can be programmed to produce malleable microstructures by adding salt solutions that transform soluble carbon dioxide into an insoluble carbonate precipitate (see Part I, Chapter 5) (Fig. 7.3).

Figure 7.3 Modified Bütschli droplet producing microstructures at the centimeter scale.
Credit: Rachel Armstrong

158 Designing and engineering the infrastructures for life

Further constraints placed on the system can provoke a range of creative phenomena within the Bütschli system. For example, Turing Bands (Turing 1952) can be produced by reducing the physical dimensions of the reaction space to a few centimeters, which approximates to the maximum diameter possible for a modified droplet. This transformational ability was clearly demonstrated in this installation in Vienna at the Natural History Museum in 2012 (Fig. 7.4).

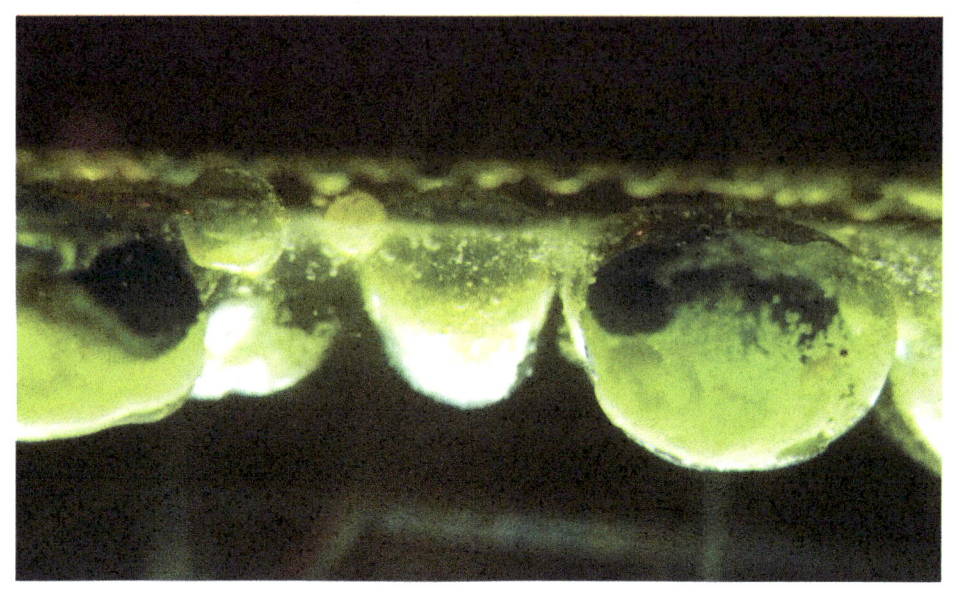

Figure 7.4 Modified Bütschli droplet producing undulating Turing Bands, at the Synth-Ethic group show, Natural History Museum, Vienna, April 2012. Credit: Rachel Armstrong

Once established, dynamic droplets may be manipulated within a network of interactions between bodies by altering the internal and external compositions of a system. However, such distinctions are over-simplistic, as these dissipative structures are inherently leaky and introduced events have the propensity to generate events that influence sustained fields within a specific range of limits. Design strategies use spatial and temporal tactics that set up gradients, which alter the limits and even the nature of these. For example, a differential process that can be shaped through multiple interference points periodically occurs in the reaction–diffusion system exemplified in Liesegang Rings (see Part I, Chapter 5).

Bütschli droplets can be designed to create a range of different products by adding different chemistries to their internal environment, or droplet body. For example, aqueous inorganic salts may be added to a field of Bütschli droplets to produce "secondary" forms on contact with the alkaline droplet body, such as microstructures, which are deposited at the oil/water interface. For example, insoluble, magnetic "magnetite" crystals can be produced by adding iron II/iron III salts (Berger et al. 1999), which produce inclusion bodies in the droplets that may also be deposited within the olive oil matrix (Fig. 7.5).

Computing with hypercomplexity and uncertainty 159

Figure 7.5 Modified Bütschli droplet producing magnetite structures. Credit: Rachel Armstrong

Changing the external conditions of the medium can also alter the behavior of the Bütschli system, based on physical and chemical changes, such as surface tension and chemotaxis (Toyota et al. 2009). For example, adding ethanol or "alcohol" to the olive oil

160 Designing and engineering the infrastructures for life

field produces a rather a dramatic effect characterized by the population-scale, sudden movement of the Bütschli droplets toward the alcohol source. This may be explained by changes in surface tension that promote movement of the droplet dynamics, but also by reducing the viscosity of the olive oil (Armstrong 2015).

Perhaps surprisingly, dissipative systems like Bütschli droplets are remarkably predictable, although, unlike machines, they produce a spectrum of outcomes that operate within "limits" of possibility. These are imposed by the properties of the system and its contexts. Dissipative structures are highly resilient and able to transform themselves—structurally and morphologically—according to changes in their interior and exterior environment. They can also form reversible groupings that generate the lifelike behaviors, which account for their flexibility, robustness, and environmental sensitivity.

However, populations of dissipative structures may periodically behave unpredictably when a tipping point is reached, and give rise to novel, emergent, complex events that cannot be deduced from their characteristic behaviors. Indeed, ways of accurately describing what is happening during striking phase changes in morphology and behavior are needed (Fig. 7.6).

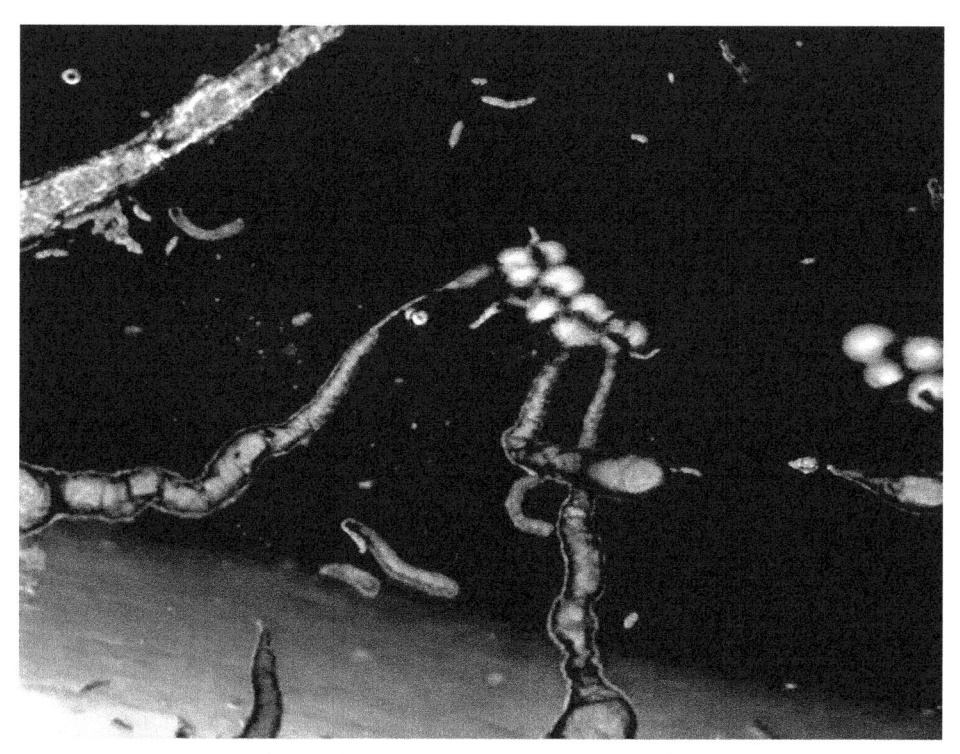

Figure 7.6 Two populations of Bütschli droplets meet, cluster, and undergo a phase transition of movement and morphology. Credit: Rachel Armstrong

Scaling downwards to the quantum realm invokes an alternative set of readings about the behavior of lively matter that may provide ways of understanding some of the stranger computational properties of matter, which are starting to be recognized within biological systems. Such ideas, however, are contentious. Yet, if they turn out to be true, then it is possible that the operational portfolio of natural computing may engage with the strangeness and unpredictability of quantum phenomena—even more controversially, these interactions could take place at the human scale and at room temperature.

In biology, it is generally considered that quantum effects, such as tunneling, superposition, and entanglement, are obliterated through decoherence—the randomness of molecular behavior at the macroscale. Quantum strangeness may be observed at the macroscale, but mostly in unusual places, like the interior of the Sun or in highly choreographed experiments. Yet, in recent years, quantum phenomena have been observed in biology, such as photosynthesis, and growing evidence suggests that life exists at the edge of the classical world using quantum effects (Al-Khalili and McFadden 2014).

Indeed, Erwin Schrödinger observed the paradoxical qualities of living things. He noted that life broke the second law of thermodynamics to create order from disorder. It also resisted the increasing entropy (or disorder) that typified the behavior of matter according to classical physics (Schrödinger 1944) and raised operational paradoxes that could not be easily explained using classical systems. For example, it is not clear how living cells produce such specific ranges of products amidst the massive complexity of biological systems.

While organisms may harness the opportunities that exist at the interface between quantum and classical systems, dissipative structures may also benefit from quantum effects, whereby relatively small numbers of highly ordered particles can make a difference to an entire system. Although this has not yet been experimentally proven, Jeremy England has described how open systems that are strongly driven by external sources of energy dissipate more energy when they resonate with a driving force, or move in the direction in which it is pushing them, and they are more likely to move in that direction than any other at any given moment. This dissipation-driven adaptation of matter (Wolchover 2014) may be responsible for the distinction between living and non-living matter considering chemical circuits that only involve a few carbon-containing biomolecules. Potentially, then, dynamic droplets such as the Bütschli system could exhibit lifelike behaviors by harnessing quantum effects. Johnjoe McFadden and Jim Al-Khalili propose that building "quantum protocells" (Al-Khalili and McFadden 2014, pp. 325–329) could test such theories by harnessing subatomic coherence in biomolecules to produce organized, macroscale effects. While these ideas are still exploratory and have not been experimentally tested, dynamic droplets do exhibit surprising liveliness that could, at least in part, invoke non-classical phenomena.

Stuart Kauffman observes that, in systems that are so massive and hypercomplex like soils, the phase states cannot be predetermined—so we are inevitably working with uncertain outcomes. He concludes that there are no entailing laws of motion, for example, in the evolution of the biosphere and, with no selection at all, this system is creating its own future possibilities of becoming:

> "When Alan Turing invented the Turing Machine, he couldn't have predicted the eventual invention of Facebook or its role in the Arab Spring. The evolution of computer technology has continually created new niches or opportunities that couldn't have been pre-stated—everything from personal computers, to the World Wide Web. The same is true in the biosphere: evolution is continually creating new niches for organisms. Darwinian pre-adaptation means that an unused causal property of an organism that has no selective value in the current environment might come to be of selective value in a different environment." (Kauffman 2012)

Exactly the same principles apply to hypercomplex materials, which do not need to know the outcome or the future before they calculate their responses to change. Their capacity to perform these functions is intrinsic to their materiality, kinetics, and context. Incredible parallelism implicit in molecules enables them to coherently exist in multiple configurations and intermediate states, with the potential for phase changes and paradoxical, or even quantum phenomena. In the design and engineering process of Persephone's living interior, the technical platform that can orchestrate this degree of hypercomplexity and contingency is yet to be established, but represents no less than a new kind of computing, in which information and matter are coupled. The outputs of this environmental-scale computation are experienced as Persephone's very own "natural" environment.

The aim to sustain life beyond Earth raises the need for a new kind of ecological practice—one that promotes life, rather than merely conserves dwindling resources. And, despite the odds being stacked against us, we need new tools, materials, and methods to explore different ways of achieving our visions as soon as possible. Not only will this benefit our stressed ecosystems in the here and now—ultimately, it will expand our horizons as we settle new worlds.

REFERENCES

A. Adamatzky, B. De Lacy Costello, T. Asai, *Reaction–Diffusion Computers* (Elsevier Science, London, 2005)

A. Adamatzky, L. Bull, B. De Lacy Costello, S. Stepney, C. Teuscher, *Unconventional Computing* (Luniver Press, Beckington, UK, 2007)

J. Al-Khalili, J. McFadden, *Life on the Edge: The Coming of Age of Quantum Biology* (Crown Publishers, New York, 2014)

R. Armstrong, *Vibrant Architecture: Matter as a Codesigner of Living Structures* (De Gruyter Open, Berlin, 2015)

R. Armstrong, M.M. Hanczyc, Bütschli dynamic droplet system. Artif. Life J. **19**(3–4), 331–346 (2013)

E. Atkin, Fossil fuels receive $500 billion a year in government subsidies worldwide. Climate Prog. (2013), Nov 7, http://thinkprogress.org/climate/2013/11/07/2908361/rich-countries-fossil-fuel-subsidies/. Accessed 20 Aug 2015

J. Benyus, *Biomimicry: Innovation Inspired by Nature* (William Morrow, New York, 1997)

P. Berger, N.B. Adelman, K.J. Beckman, D.J. Campbell, A.B. Ellis, G.C. Lisensky, Preparation and properties of an aqueous ferrofluid. J. Chem. Educ. **76**, 943–948 (1999)

F. Chatelin, *Qualitative Computing: A Computational Journey into Nonlinearity* (World Scientific, Singapore, 2012)

M. DeLanda, *A Thousand Years of Nonlinear History* (Zone Books, New York, 2000)

Duke Nicholas School for the Environment, How much do we waste daily? (2015), https://center.sustainability.duke.edu/resources/green-facts-consumers/how-much-do-we-waste-daily. Accessed 20 Aug 2015

Green24, The Great Pacific garbage dump (n.d.), www.green24.com/Science/GPGP.php. Accessed 20 Aug 2015

T. Ikegami, M.M. Hanczyc, Search for a first cell under the maximalism design principle. Technoetic Arts **7**(2), 153–164 (2009)

S.A. Kauffman, in *Reinventing the Sacred: A New View of Science, Reason, and Religion*, first trade paper edn. (Basic Books, New York, 2008)

S. Kauffman, Evolving by magic. Arup Thoughts (2012), May 17, http://thoughts.arup.com/post/details/200/evolving-by-magic. Accessed 21 Aug 2015

M. Lee, Oceanic ontology and problematic thought. (NOOK Book/Barnes and Noble, 2011), www.barnesandnoble.com/w/oceanic-ontology-and-problematic-thought-matt-lee/1105805765/. Accessed 19 Apr 2014

M.S. Morris, Galaxy gadgeteers: architects in space. J. Br. Interplanet. Soc., **67**(7,8,9), 272–278 (2014)

W. Myers, P. Antonelli, *Bio Design: Nature, Science, Creativity*. (Thames & Hudson, London/MOMA, New York, 2013)

R. Pfeifer, F. Iida, Morphological computation: connecting body, brain and environment. Jpn. Sci. Mon. **58**(2), 48–54 (2005)

J. Rifkin, *The Third Industrial Revolution: How Lateral Power Is Transforming Energy, the Economy and the World* (Macmillan, New York, 2013)

E. Schrödinger, What is life? The physical aspect of the living cell. Based on lectures delivered under the auspices of the Dublin Institute for Advanced Studies at Trinity College, Dublin (1944), in Feb 1943, http://whatislife.stanford.edu/LoCo_files/What-is-Life.pdf. Accessed 16 Apr 2013

M. Serres, *The Parasite* (Johns Hopkins University Press, Baltimore, MD, 1980)

T. Toyota, N. Maru, M.M. Hanczyc, T. Ikegami, T. Sugawara, Self-propelled oil droplets consuming "fuel" surfactant. J. Am. Chem. Soc. **131**(14), 5012–5013 (2009)

A.M. Turing, The chemical basis of morphogenesis. Philos. Trans. R. Soc. Lond. B Biol. Sci. **237**(641), 37–72 (1952)

K.R. Weiss, Part four: plague of plastic chokes the seas. LA Times (2006), Aug 2, www.latimes.com/world/la-me-ocean2aug02-story.html. Accessed 20 Aug 2015

N. Wolchover, A new physics theory of life. Quanta (2014), Jan 22, https://www.quantamagazine.org/20140122-a-new-physics-theory-of-life/. Accessed 10 Feb 2015

8

Choreography of embodiment

Rachel Armstrong

This chapter considers the impacts of extreme modes of being on our anatomy and notions of identity that may shape what forms of embodiment we may experience as we journey towards interstellar space.

8.1 Interstellar being

"Who were the people who had invented Tlön? The plural is unavoidable, because we have unanimously rejected the idea of a single creator, some transcendental Leibnitz working in modest obscurity." (Borges 2000, p. 32)

The challenges faced by the pioneers of interstellar exploration are frequently likened to the escapades of early human explorers, but there are many striking differences. A fully artificial environment needs to be designed into existence. It has no natural rhythms. It does not experience a fresh sunrise every 92 min, as astronauts orbiting Earth on the International Space Station (ISS) do, or even make an inspiring appearance over a horizon every 24 h, as we see on Earth. It is almost impossible to comprehend the potential disorientation that we may experience in an interstellar environment, for there is nothing to familiarize us in the void. We cannot take a window view to contemplate our home planet, as an interstellar journey takes place in darkness—between solar systems—with nothing outside to see. Even the stars do not light our way. There are no familiar constellations. There are no repetitions, cycles, or periodic encounters with cosmic bodies. There is nothing but the nothingness of time unfolding, where only the memories of a mother planet and the values that it holds may guide us.

Interstellar exploration is not something that can be achieved while hanging onto the placental pull of Earth.

Yet, venturing into unchartered space is a profound question, not simply of our biophysical survival, but because it is also related to our being and our identity. Living in a worldship

---

Rachel Armstrong, PhD. (✉)
Professor of Experimental Architecture, Newcastle University, Newcastle-upon-Tyne, UK

is not only a question of not fully understanding the physiological changes that we may experience during long-haul spaceflight, which may radically affect our physical performance and health—perhaps to the point of no recovery—it also involves us submitting to disorientation in our encounters with a constructed set of experiences, transformations, and relationships. Yet, how we deal with these immense challenges will not only mark our presence in the challenging terrains beyond Earth's surface, but will also redefine who we are.

At the outset of an interstellar mission, it is likely that a worldship interior will be designed with vestiges of our ecologies and terrestrial remnants in tow. As we push forward into interstellar space, our living conditions will begin to change. This does not apply just to the worldship inhabitants, but also to the terrains we pass through. We will inevitably scatter fragments of ourselves behind us, like a tiny trail of dust, in which we are the aliens, not the aboriginals. Even the disturbances made by the worldship vessel may be amplified throughout the cosmos from initially tiny pinpoint perturbations of scattered space dust, or if we collide with an asteroid or other celestial body. In venturing into interstellar space, we may transform everything we touch—or, at the very least, leave a scattering of non-native being behind in our wake. The idea that we can leave our home planet and traverse interstellar space with zero net impact on other worlds is simply unthinkable—a starship is a highly specific act of directed panspermia (see Part I, Chap. 3).

There are no expectations that interstellar explorers will hang onto their notions of humanness or desire to be replicas of their ancestors.

The specific environmental conditions on board a worldship will have a significant impact on the character, physiology, and anatomy of a colony. Should light be scarce, which is less likely in a vessel fueled by nuclear fission that effectively carries its own sun than one powered by solar sails, then the adaptations made by the populations may be reminiscent of those of chemotropes that have evolved in lightless caves, like the Movile cave in Romania (Grove 2013). Such creatures make extraordinary adjustments to their surroundings, losing their skin pigment and eyes, and completely changing the way they produce energy from food. While these kinds of changes may not be anticipated at the outset of an interstellar journey, it would be remiss not to consider the possibility that colonies could potentially undergo significant changes and transformations before they reach their destination.

An interstellar vessel provides a thin layer of independence within an overwhelming void of absence. So, rather than assuming a set of constants from the outset for the construction of a base for human colonists, Persephone makes no attempt to assume that the preconditions for successful terrestrial life are taken as a given. The prospect of living within a designed environment needs to be re-examined, reworked, and immersed in constantly emerging, individuating, and social processes. In undertaking such a monumental venture into interstellar space, it is essential to deconstruct our existence and build up our identity again as a species with the flexibility to deal with radical shifts in ways of being, but also united by a common mission and cosmic purpose. This is a challenge with no single solution. It is awe-inspiring, difficult, and strangely tautological. Notions of identity within this environment are manifold—each set of decisions defines who we are and what we may become. Indeed, interstellar explorers are faced with multiple existence paradigms shaped by countless contingencies, variables, and experiences. This is a complex, messy business with no singular or simple answer, and speaks to the mythos of the Ecocene.

The question of identity and community is complicated by the relatively recent discovery of the human biome. We are no longer a pristine expression of 46 chromosomes, but a minority representation within a consortium of participating bodies that make up our anatomy.

Only 10% of these agents may be defined as genetically "human," the other 90% being bacterial in origin. Indeed, our relationship with bacteria is likely to intensify in extreme environments and cause us to reflect on notions of humanity, community, and ecology, where we are all chimeras and hybrids of a much broader kinship with the community of life.

This book challenges the very notion of humanity within a starship community—from its materiality to its subjective experiences and notions of identity. Indeed, the current views of our existence, shaped by the Anthropocene, may no longer be sufficiently inclusive or malleable enough to survive the complexities of inhabiting worldships. In traversing interstellar space, we must be prepared for continual change through adaptation that may even culminate in the evolution of beings that, from a contemporary understanding, are wholly alien and unfamiliar. Yet, this ambition does not submit to our extinction—nor does it propose to irretrievably separate from humanity in the quest for post-human status. Rather, it adopts an ecological perspective that embraces an expanded, mutable sense of identity and ongoingness.

Interstellar travelers may no longer identify as human, but as ecological beings with the capacity to transform, enable, and immerse themselves in new modes of existence. Persephone uses this avatar to be more fully informed by a new understanding of and respect for humanity, within the context of its obligate relationships with myriad communities of organisms.

8.2 Ecological being

> "He saw what looked like land—fantastic fields under cultivation, a settlement of some sort, factories, and—beings. Everything moved with incredible rapidity. He couldn't see the inhabitants except as darting pinky-white streaks." (Sturgeon 1992, p. 154)

Right now on Earth, we appear to inhabit a place of extremes, where the materiality of our Earth suddenly appears strange and increasingly lively. Consequently, we seek increasing control over our surroundings, down to the precision of the atomic scale, proposing to remake our world to keep us in a state of perpetual stasis—a process called "sustainability"—through which we can ensure our permanence. Yet, environmental challenges are dynamic, recalcitrant, immense, and defy top-down control.

Although the challenges faced by human space travelers have been likened to those faced by the first settlers of new worlds (Finney and Jones 1985), space colonists confront a unique set of challenges—whereby even the ordinary may become peculiar.

When an astronaut is in free fall, orbiting Earth, they are descending at the same speed as their spacecraft. During this orbital plunge, the effects of Earth's gravitational force and other forces begin to cancel each other out, and they appear to be floating.

Less than a celestial stone's throw away, on the International Space Station (ISS)—as big as a six-storey house and the most recognizable of all extraterrestrial living spaces—the daily routines are far from familiar. Space is a brittle place whereby a single error may result in total disaster, nothing is left to chance, and risks are carefully managed. Every excursion to the ISS, for example, involves intense planning, scheduling, and execution—even in the simplest activities of daily living. Yet, despite all efforts to terrestrially normalize the off-world *quotidian*, nothing is usual. Something strange happens to a body in microgravity. Eating is different, sleeping is different, and toilet ablutions take on a different quality altogether (Hadfield 2013).

The absence of gravity as an organizing tool causes bone mass to drop by 2% per month and muscle mass to deplete by as much as 5% per week. Consequently, astronauts wear stretchy suits to mimic gravitational compression of body fluids, and the ISS is full of gym equipment. Laptops float in front of those exerting themselves on handless exercise bikes, where it is oddly unsafe to stay in any one particular place for too long, as exhaled carbon dioxide has a habit of forming an invisible cloud that causes severe headaches. Astronauts are in a continuous battle to constrain their bodily fluids within their vasculature. Plasma relentlessly leaks into intracellular spaces, causing a puffy, marshmallow-type appearance. Yet this fluid retention is much more serious than a very bad case of premenstrual tension, as pressure can build up behind the eyeball and compress the optic nerve. While the long-term risk to vision has not been formally established, artificial gravity can be applied through centripetal acceleration to address such challenges. Yet, there are other serious impacts of living in an off-world environment. Sleep disturbances and mood swings contribute to the stress that astronauts face, and prolonged exposure to elevated levels of radiation is also likely to have serious longer-term consequences for organs (The Economist 2015).

Worldship designs aim to circumvent the possibilities of colonists experiencing the effects of microgravity. This is achieved by rotating the cylindrical structure along its long axis. However, if inhabitants ascend toward this axis of rotation, like a circus performer climbing tall rigging, they experience less centripetal acceleration and therefore less force on their anatomy.

At some point in their climb, they will also start floating like an ISS astronaut. In this *Alice in Wonderland*-like space, the human body is no longer subject to the terrestrial rules in which it was forged, and rapidly reconfigures itself.

Under these changing conditions, anatomical conventions are no longer fixed points of reference through which we experience the world. They do not obey the classical, Vitruvian ordering systems that we have become accustomed to, but instead become pliant and suggestible. As gravity recedes, new fields of interaction expand and are revealed.

Fluids that were once constrained by downward forces can now move sideways and upward to mingle in new ways with their neighbors. Tight and tethered tissues become relaxed and semi-porous under a range of new influences as microgravity reorganizes our environmental relationships. Fields of biochemical interaction that were once bound by our ropey, collagen skin-bag start to reorder themselves. Under these conditions, a kind of re-evolution takes place, where physical forces alter the molecular repertoire of inorganic matter. We may even be freed from the influence of our genes as they operate differently as ionic fluids are redistributed, which carry biological energy essential for life's processes. This literal decoupling from Earth's constants confronts us with a new strangeness of the human body and prompts a reimagining of what it means to be human *in extremis*.

Of course, we could think of these events as being a series of malfunctions within an idealized body-machine, whereby bones and muscles are resorbed, intercellular fluids are expanded, and our senses are disorientated. Yet, under challenging conditions, the machine metaphor that has dominated since the Enlightenment no longer appears to be an ordering system that is sufficiently flexible or robust enough to represent the constantly changing operations and needs of an anatomical system in flux. In reframing these changes, we may be released from notions of object-boundedness and permanence. We must begin to understand our anatomy in new ways. Such fields of possibility not only entangle us with our surroundings, but also extend our presence through ecologies of material interactions, where our bodies begin to reconfigure themselves through trembling, transitional states of being. In defying the geometrical conventions that have constrained human anatomy, we discover, instead, that our existence and that of our living spaces are conditional, contextual, and inconstant. In fact, it becomes apparent that—at the limits of our existence, when things are so undetermined—we are not objects, but excitable fields that shape the behavior of matter. Fresh awareness of the potential interconnectedness of our terrestrial environment enables us to view the metabolic networks that infuse our flesh as an expression of nascent material creativity. In these transitional states of existence, it becomes possible to extrude our being into new terrains, so that we may reinvent ourselves within our inhabited spaces. So, as gravity recedes, our fields of interaction expand and are revealed through a new kind of unbounded anatomy that is seamlessly integrated with its ecological relationships and can dynamically respond to them (Armstrong 2015). Yet, such reconfigurations are not simply pertinent to weightlessness, where traveler and spaceship are set on a mutual pathway of survival. In reframing the changes that come with a body in a constant state of flux, it becomes possible to reinvent our bodies and living spaces.

Where, once, our anatomical limits were defined through cell membranes, 30 years of advanced biotechnology have provided us with new toolsets that enable our metabolic networks to be extruded into new terrains. We use the technologies of life to enhance our well-being—from replacing missing insulin in diabetics to manipulating the synthetic properties of bacteria that make useful substances like vitamins. Although grand designs may facilitate the sensible exploration of environments, they cannot constrain the restless curiosity of the human spirit. Nor will they dictate the host of possible material interactions in a constructed environment, where our anatomical relationships may begin to reconfigure themselves. To deal with continuing uncertainty, our being transforms through pulse-like separating fields that continually separate and recondense into structural arrangements. Gradually, these alternative organizing fields, rich with many potential states of existence, settle and·nucleate around contested boundaries.

These paradoxical relationships take place not only *in extremis*, but throughout the life cycle of a body. Yet, the hostility of extraterrestrial environments highlights the impermanence of our flesh. Not only can we adapt in new contexts and reorganize at tipping points, but we are also subjected to reorganizations during growth, maturation, senescence, and decay. Yet, in constantly changing contexts, such transformations are not failures, but are integral to the successful identity and integration of an ecological being into an ecosystem. The success of each body lies in how the multiple assemblages, dissipative structures, and nurturing media are redistributed according to changing needs. Indeed, this movement away from Platonic ideal forms of being can be understood in an ecological context with respect to the capacity of a being to be consumed by its relations and environment. Yet, the civilizing of the assemblages and relationships that constitute these appetites requires appropriate cultural context and ethical codes that value change and decay. Such a system goes against conventions that seek ideals, immortality, and permanence.

If we are to construct living spaces that infuse our bodies and metabolisms with physically networked survival strategies, then we need to be able to articulate this relationship in ways that meaningfully reconfigure our being within the material realm. This necessitates a new kind of anatomy that is seamlessly integrated with its ecological relationships and can dynamically respond to them. Such an ecological existence is fundamentally networked into an extended realm of material connections that are a spontaneous expression of the natural realm itself, with a new relationship with matter, consciousness, and life.

This is the "ecological being."

8.3 Nature of life

The body-as-machine metaphor underpins the modern conceptual model of life (Nicholson 2009, p. 48). These views are consistent with the Anthropocene mythos and were established during the cultural Enlightenment, when rational scientific methods were replacing an ancient knowledge system based on the flow of humors. René Descartes proposed that the human body was "an earthen machine" and that its movement could be likened to the workings of clocks, fountains, or mills. Making a distinction between the mechanically ordered material body and the superior mind, or "soul," these cleaved agencies were allowed to connect through pores in the pineal gland of the brain.

During the Renaissance, the machine-like human body was increasingly depicted as a mathematical construct. Having been separated from divinity, the body could be designed and idealized through "golden" ratios, which are beautifully depicted in Leonardo da Vinci's anatomical chart of Vitruvian Man. The building blocks of human anatomy were characterized by a series of technologies that allowed us to observe reality in new ways. Robert Hook coined the term "cells" to refer to the building blocks of life that could be seen for the first time using a microscope while, using a different kind of imaging technique based on the diffraction of radiation by objects, James Watson, Francis Crick, Maurice Wilkins, and Rosalind Franklin unraveled the structure of DNA—the organizing structural molecule of cell development located at the heart of each cell or nucleus. Over the course of the twentieth century, the rapidly developing portfolio of the tools of modern synthesis made it possible to consider living entities as "smart," programmable machines. Such a worldview enables us to consider the body as an object with a certain set of properties that can be resolved through mathematical modeling and calculation. When taken to logical extremes, these componentized objects can, theoretically, be entirely replaced by hardier and stronger machines.

Through the lens of the Anthropocene, human survival *in extremis* may be best guaranteed through future-proofed, technical upgrades, as if we were going on a mission to the International Space Station (ISS) and carrying out an equipment check. Yet, in making preparations for human interstellar flight, it is not sufficient to treat colonists as imperfect objects, obedient collectives, or cryogenically frozen bodies awaiting a nanotechnological resurrection. Although the tools of modern synthesis can usefully employ big data to generate models of certain traits (like movement through public spaces), the very nature of our being resists its complete representation through any media.

To understand what is at stake, it is first worth considering how reducing the Interstellar Question into a series of technological fixes affects the status of the human, which becomes little more than problematic cargo—as it is mortal.

High modern narratives of the Anthropocene regard the human body as a flesh cage that can be instrumented, modified, augmented, and replaced as if it were little more than soft scaffolding for colonization by technology. These instrumentations are envisaged to take place by, for example, becoming the charge of artificial intelligences, nanobot-mediated resurrections, and "uploads" into machines, which are justified by technophilosophies such as the Extropian Principles (Extropian Principles 3.0 n.d.). Our existence is reduced as our non-mechanical being is collapsed into binary code that ricochets through the information circuits of an immortal mechanical object. The bodies we once inhabited become suspended as an archive of dismembered flesh, molecular recipes, and genetic codes. As ciphers of what we once were, our information streams are "upgraded," processed, and made ready for packaging, existing only in the abstract in putative conformist collectives. Now we are more durable and suited to the harsh environments of interstellar space.

Thin calculations in which ones and zeroes represent our identity take on the faceless dimensions of "big data," whereby we are breeders/not breeders, consumers/not consumers of resources, or breathers/not breathers of atmospheres. Nothing more. These groupings acquire population-scale constants so we are a part of the norm, without errors and accidents. Deviancy or subtlety is not discussed. The hidden imperfections in the system reduce us to little more than to epistemologies that are filed under our respective categories. It is assumed that consistency is evenly distributed, persists, and is desirable for the duration of the journey. Such totalitarian forms of social engineering—even "positive" ones—have not turned out well in our terrestrial history, and there is no reason to believe that they would be uncomplicated in the context of a worldship. Yet, we are safely housed in a tin can at the price of having no purchase on events or capacity to claim new identities. Free will is absent. Only the replicant program remains, which is already subsumed within a technologically mediated extinction scenario.

Over the latter part of the twentieth century, a range of disciplines, including philosophy, cultural theory, feminism, science studies, and the arts, have developed alternative identity frameworks. Conceptually, these ideas are shaped by notions of process and metabolism, and are somewhat closer to the fluid mechanics concepts that underpinned the "humors" of the ancients than they are to the object–network interactions that shape the modern synthesis. Notably, Donna Haraway's concept of "cyborg" (Haraway 1991) and post-human ecocritiques of identity, such as Jane Bennett's (2010), Myra Hird's (2009), Karen Barad's (2007), and Rosi Braidotti's (2002) (to name but a few), have radically deconstructed and re-synthesized the notions of bodies, identity, and environmental politics in ways that speak compellingly to the Ecocene. Yet, altering the frameworks through which we understand our being does not require us to abandon the status of bodies or diminish them through their assimilation into unfathomably complex networks of material exchanges. Rather, critically evaluating their material and existential composition facilitates an expansion of our limits of existence through identification in object/network assemblages with metabolism, non-human collectives, and even non-living entities. Through subjectivity and intraconnection with their context, these designed bodies are an integral living aspect of a designed world and fundamental for a thriving ecosystem. They participate in life cycles, fall victim to unpredictable events, succumb to decay, succumb to passions, and their tissues are ultimately regenerated back into the metabolic systems within designed worlds. They constitute a creative force that enriches the process of exploration and discovery through which change and transformation can take place. Indeed, life is therefore necessary for the unfolding of worldship events. These ecologically constructed identity frameworks confer us with resilience and transformability that offer a greater chance of ongoingness when confronted with potential extinction events such as *extremis* and flux.

To fully address the starship question, we must deal with many unknowns. One way of preparing for our continued survival is to begin to understand our being through the possible ongoingness of multiplicities that are implicit in ecosystems thinking. Establishing a sense of identity within a constantly changing environment requires a deeper exploration of alternative existence paradigms than the familiar frameworks that inform the body-as-machine metaphor.

A range of alternative perspectives on notions of being and trajectories that can be traced through the ways humans may survive in closed ecosystems is explored in this publication. Work by Sarah Jane Pell, Kevin Warwick, and Arne Hendriks (all Part II, Chapter 13) interrogates the nature of bodies and experiences within a worldship community. These multiple and nascent identities start to paint portraits of a community in transition towards the Ecocene. In these explorations, some alternative bodily configurations identify as human, while others do not. Yet, these are not forms of extinction, but the synthesis of new beings and an extension of the community of life (see Part II, Chapter 11).

This book, then, does not look to assert an iconic view of the interstellar body as a clearly identifiable object. The ecological being does not need to be more effectively stored, operated, uploaded, increased, reduced, or made impermeable to risk—culminating in a more efficient starship mission. Rather, it is an active, morphologically and physiologically plastic agent that lives, feels, explores, creates, and loves through a radical ecology of material exchanges and passions. It appreciates its existence through its lived experiences and notions of continuity, which are integral to the poetry of life.

8.4 A living city: building with organs

The islands dripped photons.

It wasn't that the days were any longer, because there really was no such thing. It was more that the worldship's skylights, located at its core, would not be dimmed. A century and a half of correlation between diurnal periodicity and the mood of the colony had proven that irregularities in the system were actually highly beneficial. It was common knowledge that worldship communities quickly became desensitized to the homogenous electromagnetic wavelengths produced by the nuclear fission reactors. Circadian rhythms drifted causing mood swings, sleep disturbances, and all kinds of odd ailments from bad breath to hair loss.

A weeklong colony holiday of light seemed to reset these systems. Light-flooding therapy provoked gene amplifications in the molecular mechanisms that set biological clocks and boosted neurotransmitter levels in the pineal gland. The process was disorienting but incredibly effective. It was much more resource-effective to administer this treatment at a population scale than individually through its health centers.

The carousel of light that comprised the artificial summer solstice was celebrated as vigorously as any other pagan festival. Although the community was giddily relieved of their usual duties and instead reveled in the swampy outskirts of the settlement, the holiday was not at all frivolous. Boomerang-shaped weather drones sprayed potassium iodide into the atmosphere, ensuring that the occasion would be rain-free, so nobody would think to return home. These indulgences had a purpose. During the solstice week, dwelling doctors were deployed to conduct their annual door-to-door rounds of the empty dwellings.

Bethany Petroff worked at the Chopra Health Center. This centrally organized expanse was one of the largest of Persephone's organs in the worldship. It was formed as an outgrowth, where swampy soils met at the confluence of an underground ocean spring. The turbulence had trapped compacted rafts of floating soils that slowly swelled and gyrated in the underlying eddies. The island accretion was in an ideal location for easy waterway access. It was buoyant enough to support the fleshy growth that formed the body of the clinic. This was founded on a network of over 200 spongy bioplastic bridges that squelched like a gut as they slopped in and out of the water.

Bethany frequently worked off-site, but she also enjoyed the comfort of her office. It was spun from a bacterial bioplastic that was harvested from the waterways by Proteus drones, which cleared waterways with biopositive nets that attracted cellular organisms in its wake. Bioplastics were cultured from these biofilms and were farmed in large floating vats. Sintering transformed these substances into a pliant resin. Using a process that resembled glass blowing, this could form building materials. The bacterial bioplastic was heated in large containers and spun on rods, as air was forced through them at extremely high pressures. The architects and artisans that coordinated this process could produce light fluffy structures, dense cavernous ones, or ingeniously sculptured villous enclosures. The material took several hours to cool and could be further molded during this time to make formal seals and joins between modules. Indeed, the whole complex had been grown from this process through the modular addition of chambers, like colliding bubbles. Construction processes in the worldship sprawled outward rather than upward to keep gravitational forces as constant as possible, so load-bearing challenges were rarely problematic. Grasses and creeping multicellular plants quickly colonized the outside of the building so that the whole complex had the appearance of a nodular hill. Inside, the spaces were domed and roomy. While the walls were warm and soft to the touch, they were also heavy with moisture. Fingers of biofilm forced their way into the spaces and there was a lingering scent of freshly fallen rain. This exuded from the ozone-producing ultraviolet lights scattered around her workspace.

Bethany stepped into one of the clinic boats. The hull jolted as her bag of diagnostic instruments, pharmaceuticals, surgical tools, and sample pots landed slightly off-center. The magnetic motor reluctantly tipped into a continuous spin cycle and she steered her way through the canals and confluences to the village of Vinci. It was the eighth year in a row that she'd worked on call during the solstice. Two more years and she'd be promoted into a more managerial position. Even Persephone, an unfaltering taskmaster, recognized that dwelling doctors could not sustain endlessly demanding routines. Bethany didn't mind. She was professionally ambitious and would earn a fortnight of relief cover in lieu of this gargantuan undertaking. Besides, the professional opportunities within this particular village were intriguing, as it had been constructed using particular bioscaffolding that promoted the growth of human tissue. Vinci was a living experiment where residents clothed the bone-like bioscaffolding with scrapings from inside cheeks, hair samples, and blood. Although the project had been growing for around 2 years, there was little information as to how successful it had been. Bethany was relieved that she would be able to examine the unoccupied village without the continual interruptions or territorial behavior of occupants. Sick buildings were almost impossible to diagnose when they were crawling with people, as the competing biosignals confounded the diagnostic toolsets.

The Vinci settlement was several kilometers from the Chopra complex. She could not help notice how silent the usually bustling riverbanks were, as the lazy waters intermittently sucked, slobbered, and slurped at the banks. Most people made their living from harvesting materials, fishing, and growing crops within the waterways. However, although the land was untended by people, it was not entirely vacant; several giant eels were making the most of shortcuts between waterways. They slithered gracefully like shadows from mooring sites onto the land and disappeared into banks of trembling rushes.

The village was only a short walk from the water. Unlike the clinic, whose bioplastics were buried under plant matter, Vinci was a mound of folded flesh. Under the intense, unflinching light of the solstice, she watched its body dance at a distance. Its unevenly pigmented surface moved like a rib cage, which was detailed with pleats and rosettes that exuded a healthy, yet somewhat overpowering chameleon musk, from which she could identify people, hens, cows, pigs, dogs, cats, rats, hedge birds, mice, and bats. She spied glints of spiders' webs, a dried-out slug, and specks that moved like flies in the yard, but their pheromones belonged to another realm. Cryptogeographies, too, the microcellular landscapes that littered architectural recesses, betrayed by sometimes brilliant patches of hardy lichens and, at other times, subtly darkened with scars of molds. All found niches within Vinci's various recesses.

Bethany thought she could hear the buildings breathing. Fluids intermittently trickled through the gurgling swamp chorus like an erratic fountain, filling and emptying functional architectural cavities known as organs.

The village was, in effect, one giant body of many different tissue types.

While reminiscent of a human form, it was far from being laid out in any gesture toward a recognizable anatomy with a discrete head, limbs, and torso. Rather, the buildings suggested that they were scaffolding for a series of differentiating organs suspended within a network of bones and sinews that had been laid down during construction. Varying laxity and turgor within the outline of the buildings betrayed an active remodeling process. Over there, by the public square, a low-lying shelter's roof was sagging and pleating in excessive folds, in which several species of hedge birds had already made their nests. Here, by the fountain, a flesh-stretched structure was reaching towards the trees. Spreading finger-like new growths toward the branches, it seemed to suggest that the two living structures would fuse in choreography of plant–human pleaching. One cottage was shockingly wrinkled. Bethany assumed that it was ageing prematurely, although the folds did not feel thin and lax, but full and thick. She wondered whether the pleating was caused by the production of a cartilage-like material being that gave the Chinese Shar-Pei breed of dogs their crumpled appearance.

Windows were membranous apertures, like the iris of an eye. When she tried to peer inside, she mostly encountered an opalescent sheen that refracted traces of her face onto other surfaces. Pressing her body close to the pulsing framework, she was acutely aware of the inescapable intimacy of this examination—body against body. She decided to inspect the building surface details more closely. Tiny hairs, pores, and patches of feathers punctuated the building skin, which moved to the orchestral beat of variably hollowed, solid, and fluid structures housed within the welcoming walls of the village.

Bethany approached the sphincter entranceways with some trepidation, which was unnecessary. They were remarkably accommodating, opening reflexively at her slightest touch and relaxing again after several seconds. Some even chose to theatrically display the complex muscle movements and attachments like peacock tails, which enabled them to perform their convulsive movements.

It was incredibly hot and humid inside. Yet, it was not dark. The light was strange, trans-illuminated, moist, and deflected. Channeled from elsewhere, it re-emerged through various orifices and membranes that winked and stared through diverging skyline views. These portals appeared to follow the Sun's course, moving in their orbits like eyes. A fog of breath condensed on her clothing. Soon, she was drenched.

Each dwelling in Vinci was well differentiated into distinct living sections. Yet, they were not configured like traditional zones. There was no discrete kitchen or bathroom, but shelves of flesh and pools of fluids that could be appropriated for a range of different functions. Habitats were linked through transitional sections. Gaining access from one domain to another required negotiation with a range of admission strategies that included ligamentous valves and undulating surfaces. Bethany discovered that they responded generously to soft touch and other persuasive gestures, but resisted any forceful movements by completely restricting access to those sites. Once shut, exclusion was total and assailants were not forgotten.

A carpet of tongues tickled her feet with their breath. She respectfully took off her shoes to avoid carelessly presenting them with impenetrable surfaces, like soles or clothing. A wall invited her into its girth like a giant lap, where she succumbed to its tactile kinship. Disorientated by the quivering eroticism in the air, Bethany realized that her observations exceeded the capabilities of the instruments that she'd brought to record them. She resisted an overwhelming impulse to rid herself of all clothing and embrace this mutual proximity. While·Vinci's seduction was all-pervasive, it was also overwhelmingly diffuse. There were no grotesque confrontations or blatant genital displays. Yet, the persistent promise of erotic couplings through dense material interdigitations became increasingly provocative and intriguing.

On rising from the lap, Bethany's attention was drawn to a fleshy arcade of pillars that were formed by the entwined limbs of horse and woman, united by a single column of blood. They were supporting an unusually pulsatile structure adorned by knots and swirls of convulsing vessels. She auscultated the walls and realized there were multiple thick organs inside, each with a unique pulse rate. She ran her hand over their surface and felt them kicking and wriggling in their organ sacs. She climbed on top of the structure using the tethers of the large ropey nutrient cord that sprouted from the floor like a giant climbing plant. A pearlescent membrane adorned the summit of flesh and glared at her through an enormous eye aperture that offered access to an aquarium of gliding orbs. Were these human incubators, or something else entirely? Strange forms gracefully moved within the floating spheres. As they rose and sank back again into the gelatinous nutrient fluid, they appeared to represent a monstrous zoo of organisms, from bacterial biofilms to birds, cows, humans, and creatures that Bethany could not begin to recognize. Vinci was an architecture made pregnant through the union of human tissue with other life forms. It was a breeding experiment—an open embryogenesis in a village of tissue culture, whose chimeric couplings reinforced the inseparability of humans from other creatures and environments. All forms were equally compelling, ambitious, ostentatious, and alluring—a gaudy display of delectable and complex leviathans. Yet how exactly had they come about? Were they engastrations, matings, clonings, modifications, mutations, or resurrections?

As Bethany climbed over the rubbery lids of the rubbery vat to inspect the floating vessels in closer detail, she marveled at the effusive and transformative processes that contravened all algorithms calculated by Persephone's breeding programs. She took a deep breath, sinking down into its warm fluids, kicking vigorously to reach the larger orbs that beckoned her into the chamber's depths. As she continued to dive into its amniotic reef, the multitudinous transgressions before her seemed to embody the raw materiality of the worldship; these magnificently monstrous enfleshings were not expressions of the tame program that Persephone employed her to patrol, but something much stranger and creative.

Revelers were welcomed home as dear friends by Vinci, with familiar hugs and scents of reunification.

When the light intensity was finally dimmed, the community slept erratically, as if under a physiological spell, as their cellular systems crashed and reset. On wakening, they were disorientated and nauseous, as if they had a bad case of jet lag. As they resumed their familiar routines, they barely noticed an additional globe-like organ that was swelling with new possibility within a fleshy vat at the far end of the village. The transforming woman inside this pod was tightly held in the embrace of ropey but translucent membranes that welcomed her as a pioneer within the rapidly diversifying ecology of the worldship.

8.5 Anandgram: new ways of living and being

"Luminous beings are we … not this crude matter." (Yoda, *The Empire Strikes Back* 1980)

The Ecocene deeply entangles our existence with the environments in which we live.

My research questions regarding the design of the human body and its relationship with the environment began to formulate themselves when I was a medical student in 1991 (Armstrong 1995). I elected to spend time at Anandgram—the Village of Joy—a leprosy hospital and rehabilitation center situated on the outskirts of Pune, India, during an optional sabbatical study period before my final clinical medical exams. Despite the overwhelming odds against their success, the residents of this community choreographed an inventive synthesis of technology and architecture to lay claim to a new identity for themselves and create the foundations of a productive economic and socially integrated future.

Leprosy is a terrible infection. It is one the oldest diseases known to us. The earliest reference appears in an Egyptian Papyrus document written around 1550 BC. It is caused by a bacterium called *Mycobacterium leprae*, which was isolated by Dr. Amauer Hansen in 1873. Despite its significance, both historic and contemporary, it still has not been cultured in a laboratory setting, and not much is known about how the disease is contracted or spread. It infects soft structures, particularly the sensory nerves and soft tissues, such as

176 Choreography of embodiment

cartilages. The condition is quite curable if it is detected during the early stages of infection. Typically, the first sign of the disease is a de-pigmented patch of skin, which can sometimes be mistaken for the autoimmune condition vitiligo. At this stage, if there are no other ill effects such as nerve damage, then the condition is completely curable using a vigorous and lengthy regimen of antibiotics.

However, the infection is often left untreated, owing to the stigma that contracting the disease brings. Contaminated nerves and soft tissues are progressively destroyed and lead to sensory loss, leaving the body unable to protect itself from everyday trauma. Depending on the site of infection, gradual impairments may result in major physical changes. For example, facial infections leave corneas unprotected against drying gusts of wind that result in scarring and blindness. When feet are infected, a person will walk without automatically spreading their weight, resulting in chronic "march" (or Charcot) fractures. People may also suffer the indignity of "stigmata," which mark them out as potentially carrying the disfiguring infection that include loss of nasal cartilages, which give a "lion-like" appearance to the face and loss of digits (see figures). Sadly, once these physical changes take place, they are permanent and require active or palliative interventions. Arguably, the social and psychological consequences of contracting the bacillus are more deadly than the chronic physical effects of the untreated infection. The deep-seated fear and misunderstanding of leprosy cause sufferers to be rejected from their communities—even by their nearest and dearest—and result in a negative spiral of social exclusion and societal reinforcement that prevents people seeking treatment. A diagnosis of leprosy is much more than a set of physical symptoms, but profoundly affects many aspects of a person's life, including mobility, interpersonal relationships, marriage, employment, leisure activities, and access to social functions (Figs. 8.1 and 8.2).

Figure 8.1 Stigmata of leprosy include the "lion-shaped" nose where cartilages have been destroyed by infection. *Credit*: Rachel Armstrong, movie still (1991)

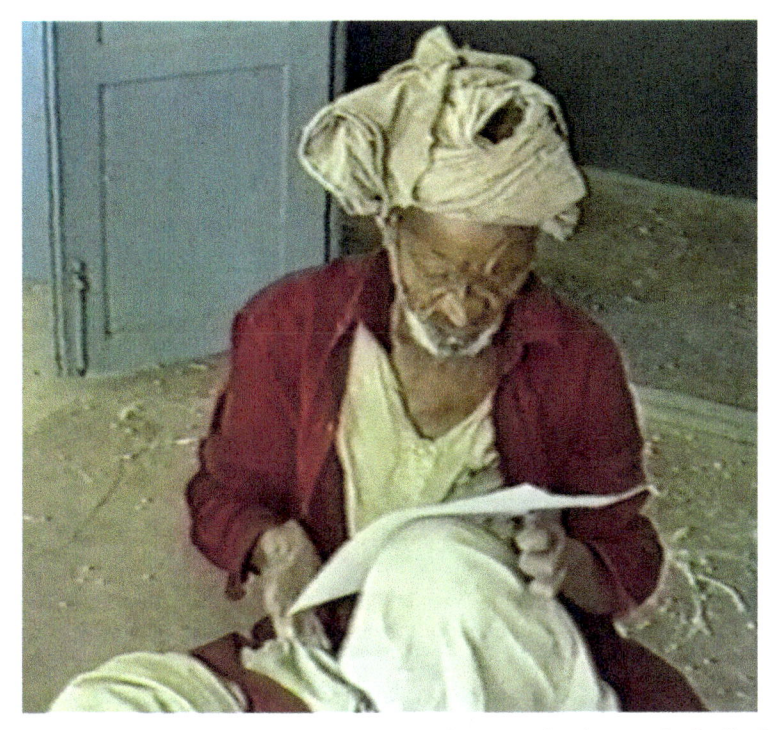

Figure 8.2 Constant trauma to hands causes the fingers to be destroyed. *Credit*: Rachel Armstrong, movie still (1991)

8.5.1 Reconfiguring the leprous body

At Anandgram, I was assisting a hand surgeon, who re-threaded tendons in patients with critical loss of hand function. By sacrificing less important muscle groups and rerouting the tendons to new sites of attachment, it was possible to revitalize movements in major groups of muscles that had been permanently damaged by leprosy. These operations were performed without general anesthetic because the patients could not feel any surgical pain in afflicted areas. Part of my role was to make polite conversation with a fully awake patient, the other to assist the surgeon in performing complex surgery—this in a bare, sterile room. Through this simple approach, my colleague could complete intricate and sometimes lengthy procedures, without the normal cost of an anesthetist. It was also possible to directly assess the immediate success of the surgery by asking patients to move their hands during the procedure. These visceral, restorative procedures were not confined to hand function, but could also be applied to other important muscle groups. One important operation involved splitting a face muscle tendon and attaching it to the inner corner of the eye. This allowed people who had lost the ability to blink to do so voluntarily, by clenching their teeth. With intensive training, eye drops, chewing gum, and rehabilitation, those who had undergone the procedure were able to consciously prevent their corneas from scarring through dryness and, ultimately, to save themselves from blindness (Fig. 8.3).

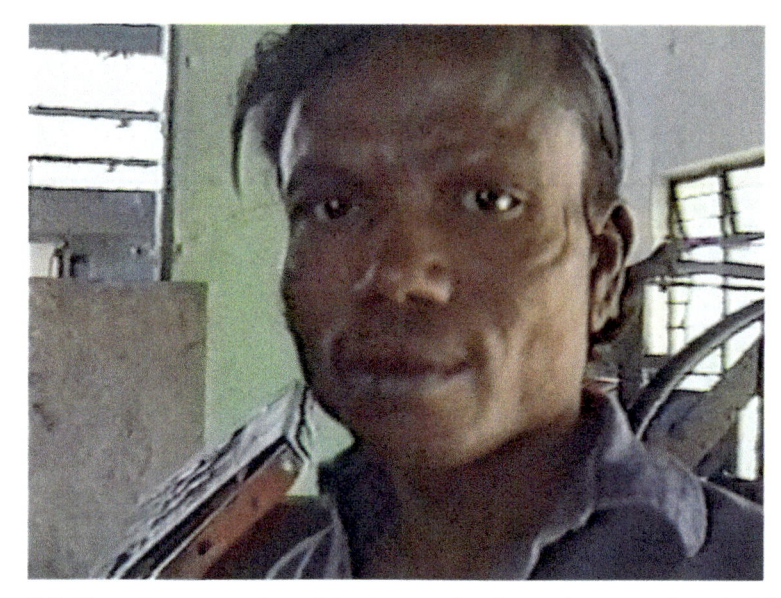

Figure 8.3 Corrective surgery through tendon transfer observed as a prominent incision on the left cheek, which involves splitting of the masseter muscle. *Credit*: Rachel Armstrong, movie still (1991)

Beyond the operating room, the Anandgram community itself created a supportive environment for rehabilitation. Tendon-transfer patients were actively encouraged to use, and to continue to use, their rewired bodies. From the moment the patient left the operating theatre, their relocated tendon insertions immersed them in a novel reality. Patients emerged like children, calibrating their being-in-the world. In fact, the whole village was an experimental environment full of ready-made technologies and a community that was not alienated by "difference." The village itself had evolved out of a collaborative effort that began as an unauthorized settlement on the outskirts of Pune, where outcasts banded together to restore their self-respect and independence. In 1970, they secured a grant from Oxfam that enabled them officially to buy an arid 18 acres of rocky, undulating land. This was situated near the village of Alandi, where the thirteenth-century Maharashtrian Hindu saint Dnyaneshwar achieved enlightenment. This holy figure is celebrated for performing miracles such as baking bread upon his heated back and making a wall move. Here, the Anandgram residents achieved their own miracle by transforming poor-quality land, which could support little more than a few stubborn shrubs, into a farmable terrain.

The journey towards self-sufficiency was inaugurated on October 7, 1970. Houses were built and jobs secured through the ingenious and widespread use of prostheses. Simple technologies and tools enabled the residents to cook, farm, and produce fabrics. They learned how to customize a whole range of operating interfaces to compensate for specific impairments that could not be restored by surgery such as dropped feet that dragged on the ground.[1]

---

[1] "Dropped" foot is a medical term for a condition that results from nerve or muscle damage that causes people to scrape the front of the foot on the ground.

Bodies were transiently coupled with, and transformed by, a range of devices, which restored a sense of wholeness to residents, now able to earn a living. In this way, the community could transgress the limits of deep-rooted social stigma using a range of highly adapted technical platforms. They could reclaim their dignity and self-respect, which broader society had ruthlessly denied them.

In this environment, it is impossible to identify a hierarchy of actors, or identify simple causes and effects, that enable such a remarkable transformation. Anandgram's premises were organized so that living places and workspaces were entangled. Manufacturing workshops were within easy walking distance of the residences, individual homes were simply laid out with access to basic facilities with kitchen ranges, running water, bedrooms, and a shared large social space, while communal outdoor spaces were pleasantly landscaped. Trees shook with the song of cicadas and provided welcome shade for picnics where we sat on hand-hewn benches. Vibrant floral blooms exploded with impossible vigor from the sandy soils. People appeared to have time for each other and, despite their strange prostheses, they radiated palpable congruence with the natural world.

While the unfamiliar appearance of residents defied classical notions of physical symmetry, a "purity" of form, or even a logical hierarchy of physical order in which particular bones are moved by specific muscles, their incongruities spoke of a new kind of anatomical configuration. The combined synthesis between the destructive actions of the leprosy bacillus, re-threaded muscle tendons, temporary prostheses, and highly customized tools and the unique character of these bespoke entanglements asserted that people were neither well-circumscribed objects nor machines. Rather than appearing machine-like with their prostheses, the Anandgram residents themselves seemed to humanize the technologies that surrounded and constituted them. Despite being in an altered body, villagers were empowered by the way in which their structures built on existing physiological, anatomical, cognitive, social, and environmental connections. In turn, each person's unique anatomy became a site for the entanglement of new networks and relationships.

Such relationships, however, are not exclusive to Anandgram. Modern Western communities depend just as much on supportive environments and technical systems, since they are regarded as part of our culture. Indeed, we have naturalized them, so they are no longer considered "strange." Each of us—wherever we live and whatever our customs—extrudes aspects of ourselves into the environment through myriads of constantly shifting, mutually sustaining, and collapsing interactions—like the 1969 Apollo 11 mission astronaut footprint photographed by Edwin (Buzz) Aldrin.

Yet, Persephone's living interior is much more negotiated than a boot print permanently sunk into regolith. Its colonists, ecosystems, and infrastructures continually modify it. Environmental and corporeal events constitute entangled aspects of natural computing, whose myriad interactions provoke transformation in the system. With the appropriate infrastructures and material resources, Persephone's experimental space may be considered as an instrument that increases complexification, promotes the flow and overlapping of potentiating fields, and ultimately generates the conditions under which both environment and occupants stand a chance of ongoing survival.

8.6 Expanded modes of being

In the twenty-first century, there is growing recognition that our future is intimately entangled with that of the biosphere. A civilization that identifies intimately with its environment—specifically one that works in synergy with, rather than consumes its surroundings—requires a different set of ideas that express its relationships. These will be discussed through the lens of the "ecological being."

While integral to the community of life, the ecological being is not constrained by its specific anatomy. It is an organism that is sympathetic and entangled with humanity, but may not be exclusively human. In space, the body is not a unitary being, but an "ecology." Although it is framed by conventions of human anatomy, it is not a static form, or materiality, but is in continual flux. Nor is it bounded at the skin, but deeply embedded within and extruded into its environment. This unbounded, semi-permeable organism may be thought of as the "ecological being." Owing to this lack of purity, the ecological being is protean and embraces many future configurations that are yet to be expressed. Yet, it identifies with humanity through an extended realm of material and cultural connections, which are produced by the combined interaction between complex interdependent agents and networks that are collectively recognized as its flesh. These are not discrete accumulations of identical tissues, but exist at the transition zones of cellular organization. Such assemblages may also find transient, or persistent commonality with biospherical, technological, cultural, material, and social processes as well as other specific agents; for example, intestinal bacteria, trees, implants, or gadgets—which may even be regarded as "part" of them. Some of these relationships are obligate (like the energy-producing networks of mitochondria)—but many, (like smartphones) are associative (Fig. 8.4).

Existing in such a highly dynamic, materially heterogeneous, and networked state, the ecological being is porous to invasion. It is constantly patrolled and remade at its limits. The ecological being is therefore not an ideal form, but a paradox of existence. Its community includes chameleons, shape-shifters, transformers, mutants, the offspring of multiple parents, the bacterial biome, tissue cultures, and changelings that invite multiple social readings at the level of individual bodies and as a collective.

Notable non-human members of the ecological being include the bacterial biome, a community of human-dwelling bacteria. They are so numerous that, on a cell-for-cell basis, only 1 in 10 of the cells that make up our bodies possesses human DNA. Yet, because of their tiny size, bacteria only make up a few kilograms of our body mass. While we have overlooked the role of our bacterial biome (Hooper and Gordon 2001) in the modern age, it nonetheless produces essential fats, digests our food, and provides immunity against pathogens—functions that are destroyed in the presence of excessive antibiotics.

In addition, ancient bacterial and viral sequences are stitched within our so-called "human DNA," and have even become obligate organelles. Ancestral bacterial associations are even responsible for critical physiological human functions. For example, the theory of "symbiogenesis" originally proposed by Konstantin Mereschkowski in 1910 and championed by Lynne Margulis (Kozo-Polyansky et al. 2010) proposes that the energy-producing cell organelles called mitochondria were once free-living organisms.

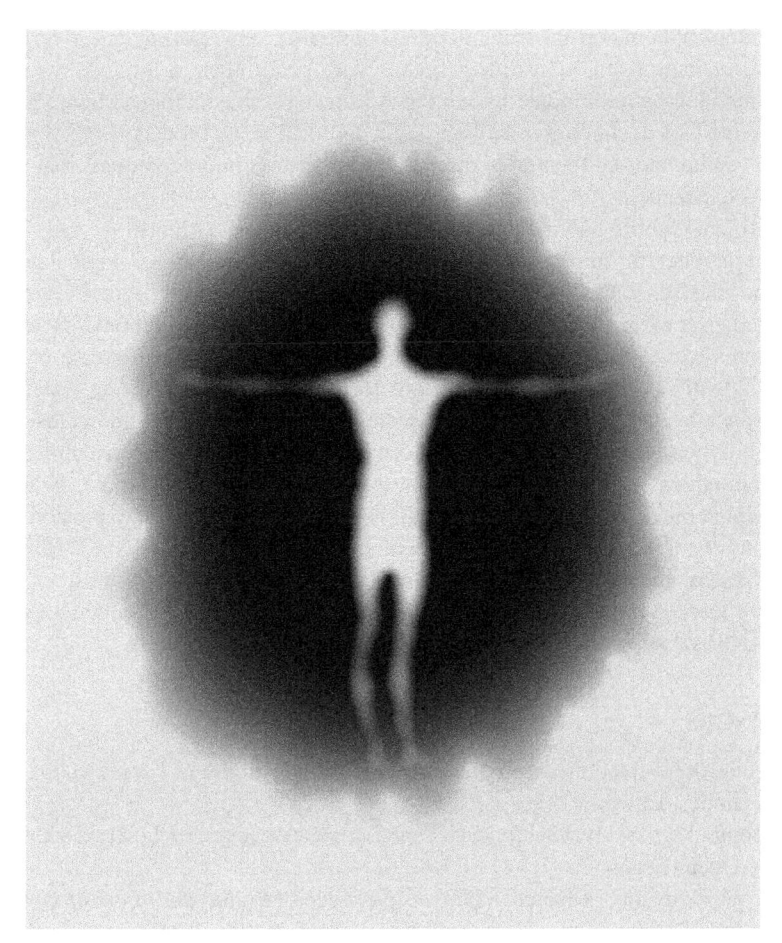

Figure 8.4 Ecological being. This iconic representation indicates the semi-permeability, connectivity, and protean nature of a being that succeeds humans and views itself as being entangled in biospherical relationships that become integral to its identity. *Credit*: Simone Ferracina

They possess their own genome and, over billions of years, have evolved to form an obligate but symbiotic relationship with living systems. In exchange for a nurturing, protective environment, these ancient and partially digested creatures provided a flow of energy for cellular functions.

Moreover, transformed human cells could also conceivably be considered as ecological beings. For example, in 1951, cervical cells belonging to Henrietta Lacks were used to produce an "immortal" cell line that could live outside of her body and were therefore used for medical and biological research, including cancer, Auto Immune Deficiency Syndrome (AIDS), radiation sickness, and gene mapping. It is estimated that there are now more than 20 tons of Henrietta's cells in existence and this raises complex moral, political, and ethical questions about their status (Skloot 2011).

The ecological being brings value to these communities by sharing its network of operations, and entangling the constituent bodies with each other to produce meaning and value, which may be assimilated by human cultures. Yet the ecological being is not infinite, but edited by technology, ecology, and civilization to produce highly negotiated boundaries and contested territories that are patrolled by immune systems that shape the limits of its existence.

But how can identity be maintained in the face of such uncertainty?

Despite its inherent fluidity and relentlessly material nature, the ecological being does not surrender its sense of "self." It is coherent, and does not invest in reckless metamorphoses. Rather, it expands the limits of its operations through its relationships and modes of self-expression. This nascent hypercomplexity is expressed as a diverse portfolio of creativity. Identity, community, and kin are formed through the editing processes that occur between the ecological being's entanglements with other bodies and fields of experience. The provocative questions that these beings raise will need to be constantly addressed within their ethical, philosophical, existential, environmental, technical, cultural, and unforeseeable contexts to create the conditions for a radical identity explosion. The full implications of such questions will only become fully apparent as we explore the limits of the Solar System and spread out into the uncharted terrains of the cosmos.

We have learned how to access and observe space: now our challenge is how to stay there—and thrive.

REFERENCES

R. Armstrong, A troubled guest, in *More Women Travel*, ed. by N. Jansz, M. Davies (The Rough Guides, London, 1995), pp. 310–318

R. Armstrong, *Vibrant Architecture: Matter as a Codesigner of Living Structures* (De Gruyter Open, Berlin, 2015)

K. Barad, *Meeting the Universe Halfway: Quantum Physics and the Entanglement of Matter and Meaning* (Duke University Press, Durham, NC, 2007)

J. Bennett, *Vibrant Matter: A Political Ecology of Things* (Duke University Press, Durham, NC, 2010)

J.L. Borges, *Labyrinths* (Penguin Classics, London, 2000)

R. Braidotti, *Metamorphoses: Towards a Materialist Theory of Becoming* (Polity, Cambridge, 2002)

Extropian Principles 3.0 (n.d.), www.ilxor.com/ILX/ThreadSelectedControllerServlet?boardid=40&threadid=15301. Accessed 28 Aug 2015

B.R. Finney, E.M. Jones, *Interstellar Migration and the Human Experience* (University of California Press, Berkeley, 1985)

J. Grove, What lies beneath: the Movile Cave's dark secrets. The Times Higher Education (2013), Apr 11, https://www.timeshighereducation.co.uk/features/what-lies-beneath-the-movile-caves-dark-secrets/2003026.article. Accessed 21 Aug 2015

C. Hadfield, *An Astronaut's Guide to Life on Earth: Life Lessons from Space* (Pan Macmillan, London, 2013)

D. Haraway, A Cyborg Manifesto: science, technology and socialist-feminism in the late twentieth century. in *Simians, Cyborg and Women: The Reinvention of Nature* (Routledge, New York, 1991), pp. 149–181

M. Hird, *The Origins of Sociable Life: Evolution after Science Studies* (Palgrave Macmillan, London, 2009)

L. Hooper, J. Gordon, Commensal host–bacterial relationships in the gut. Science **292**(5519), 1115–1118 (2001)

B.M. Kozo-Polyansky, V. Fet, P.H. Raven, *Symbiogenesis: A New Principle of Evolution* (Harvard University Press, Boston, 2010)

D. Nicholson, Is the organism really a machine? in *Proceedings of the Eleventh International Conference on the Simulation and Synthesis of Living Systems Abstracts* (2009)

R. Skloot, *The Immortal Life of Henrietta Lacks* (Broadway Books, New York, 2011)

T. Sturgeon, *Microcosmic God. Volume II: The Complete Stories of Theodore Sturgeon* (New Atlantic Press, Berkeley, 1992)

The Economist, How space flight affects the human body (2015), Apr 16, www.economist.com/blogs/economist-explains/2015/04/economist-explains-16?fsrc=scn/tw/te/bl/ed/EEspaceflighteffectshumanbody. Accessed 16 Apr 2015

9

Constructing lifestyles

Rachel Armstrong

This chapter reflects on how alternative modes of existence shape our expectations of inhabiting different spaces, the quality of our experience of them, and our identification as communities of cohabiting beings.

9.1 Transformations

"He told me that before that rainy afternoon when the blue-grey horse threw him, he had been what all humans are: blind, deaf, addle-brained, absent-minded…. For nineteen years he had lived as one in a dream… when he came to, the present was almost intolerable in its richness and sharpness…. Somewhat later he learned that he was paralysed." (Borges 2000, p. 91)

The Ecocene presents an existential paradox, in which we are both one and many at the same time.

Following their reconstructive surgery, Anandgram residents were not just anatomically restored "human beings." Owing to their diverse range of appearances, various prostheses, behaviors, and personalities, they could also be described more inclusively as "ecological beings." Empowered by their expanded connectivity through simple technologies, they actively forged new networks of social, cultural, and even economic interactions. Dropped feet were protected by attaching leather thongs to toes and shins which were strung like a bow around the front of the leg. Automobile-tire-soled shoes distributed weight sequentially over the entire surface area of feet and prevented traumatic fractures. Some residents brachiated crab-like on wooden appendages, to spare excessive strain on newly strengthening tendon attachments. Their ease of locomotion was remarkable for its ordinariness. The various technologies were even adopted as fashions, being worn with pride. Grubby bandages were wound decoratively around numb areas of skin. Residents

---

Rachel Armstrong, PhD. (✉)
Professor of Experimental Architecture, Newcastle University, Newcastle-upon-Tyne, UK

that had lost the upper soft palate fashioned new noses by hand from hot wax. These prostheses were inserted into the cavity left in the roof of the mouth to provide support for the flabby nasal tissues whose cartilages had been destroyed by the infection. Invariably, this had also claimed the upper teeth, and the insertions countered the lion-like features caused by the disease. While medical professionals advised sunglasses as protective devices to prevent corneal drying, residents also regarded them as accessories and status symbols. Indeed, these simple, low-tech technologies were successful, not just because they served to replace a body object that had been lost, or even to augment the existing body machine by adding a new component, but by increasing each person's self-esteem and extending their sphere of influence within their family, community, and their habitat. As their capacity to re-articulate their abilities within their social spaces expanded through a newly empowered anatomical configuration, residents gained increased independence and new capacities to build enriched forms of social interaction with the wider community.

In Anandgram, the success of tendon transfers was not primarily to do with how a person's body had anatomically been reconfigured, or with the efficiency of the relocated structures in providing rehabilitative support, but with how each individual resident was re-enabled and inspired to explore their newly expanded environment. In essence, this community subverted the traditional Western notions of ideal morphologies such as a preference for symmetry and conformity to a particular body plan. They also subverted the assumption at the heart of my medical training—that the human body was fundamentally mechanical in its ordering (Figs. 9.1 and 9.2).

Rather than valuing people, things, or tools and places, each constituted separately, Anandgram asks us to think in terms of ever-shifting relationships between the animate, the inanimate, and the environment, where self, family, and community are fluid concepts. The political system in the village was not centered on prevalent social hierarchies based on a

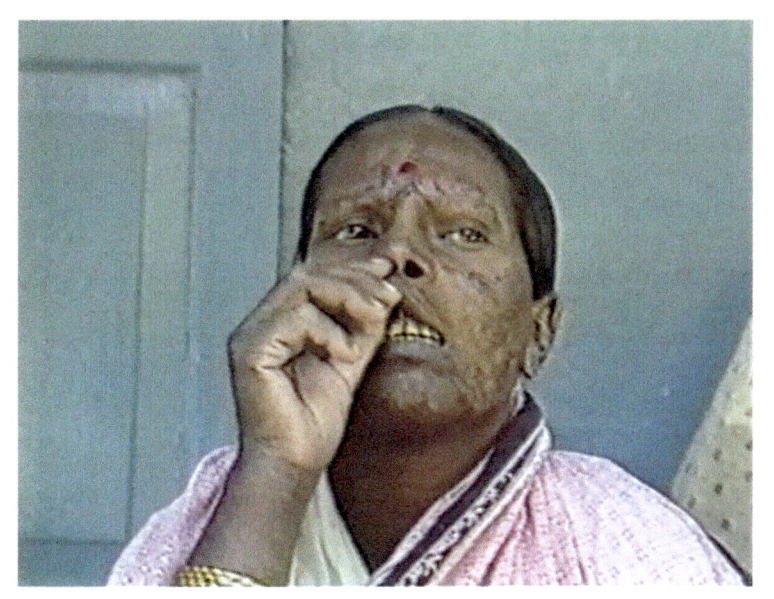

Figure 9.1 Nasal prosthesis to correct fallen nasal cartilage also involves replacing upper teeth, which are lost when the soft palate collapses. Credit: Rachel Armstrong, movie still

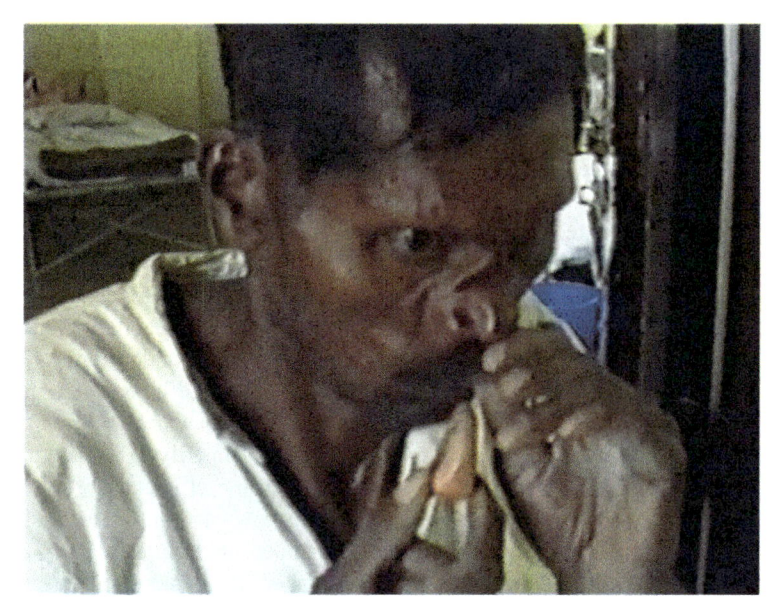

Figure 9.2 Demonstrating how to insert the nasal prosthesis through the roof of the mouth. Credit: Rachel Armstrong, movie still

caste structure typical of Pune, or even anatomical typologies that denoted stigmatizing illness—but around a notion of kinship. This sense of inclusion created an extended notion of family within the village as people were bound together in shared survival and an ambition to thrive. Indeed, the emergent ecological politics of Anandgram primarily had the capacity to change social status within the village in ways that did not exist outside this community. People within Anandgram could marry, raise children, form extended families, and enjoy lives without social prejudice. Potential inequalities could be continually negotiated and engaged through an alternative value system to reaffirm belonging both within and beyond the village, by trading with the nearby city of Pune. Such cultural contracts transcended traditional notions of status and identity whereby, previously, those with leprosy were viewed as "untouchable." Now they were in economic partnerships with the broader community. That is not to say that prevalent social order was irrelevant. Beyond the village, those with the stigmata of leprosy were still excluded. But, within the Anandgram community, the alternative ethics shaped freedoms that had frankly destroyed lives. Yet, the traditional boundaries of what it meant to "be human" in Anandgram in this context was extended. It leaked well beyond the stringent definitions posed by classical Western ideals of form and function. Indeed, Anandgram shares many issues with modern Western communities, which depend just as much on supportive environments and technical systems. While association with a range of prosthetic technologies may socially reconstitute those who have been afflicted with the sequalae of leprosy infection, our first-world societies also place technological systems as integral to our culture. Indeed, we naturalize technology to the point at which not only are they no longer considered "strange," but may also even be considered indispensable to our identity—think of the way in which our smartphones have become extensions of ourselves. The rehabilitation process at Anandgram also respected traditional cultural perspectives while embracing new values, and granted status to different physical configurations that reflected the constantly evolving needs of the community. Of course, Anandgram had limits of technical and social plasticity. These were reached when surgical procedures were unsuccessful, prostheses broke down, tools suboptimally modified, family traumas occurred, existential crises began, and broken hearts cried—just like any other collective of "humans." Yet, the stories that shaped this vibrant place emphasized the inseparable interconnections between people, technology, anatomy, and environment.

The founding politics, identity, value systems, culture, and ethics of such "ecological" communities are far from formalized. Yet, in recent years, a range of "new materialist" perspectives that explicitly seek to raise the status of inanimate matter from object to "actor" has emerged. Notions of participatory inclusion can be found in texts such as Bruno Latour's actor network theory (ANT) (which treats objects as part of social networks) (Latour 2005), Karen Barad's agential realism (whereby "objects" emerge through discursive relationships between networks and matter) (Barad 2007), Jane Bennett's vital materialism (in which materials possess their own trajectories, potentialities, and tendencies independently of human agendas) (Bennett 2010), Donna Haraway's companion species and cyborg manifestos (which propose the foundations for non-human politics) (Haraway 2003), and Graham Harman's object oriented ontology (which rejects the privilege of humans over non-humans) (Harman 2011). When these new materialist ideas are contextualized by Timothy Morton's concept of dark ecology—the idea of the natural realm is deconstructed as more than a bucolic utopia. Instead, it asks us to consider the interconnectedness of the natural realm as a fundamental condition that is "dark," invoking hesitation, uncertainty, irony, and thoughtfulness (Morton 2007). Although the ecological being can deploy shape-shifting operational systems to flexibly engage notions of humanity, this does not imply that absolutely anything is possible. In finding sympathy with other life forms, new discourses may potentially be established through non-human languages, without diminishing ourselves and precipitating social atrocities of a different kind and order.

In considering the protean nature of the ecological being revealed by the alternative forms of material, social, and ecological systems in Anandgram, our attention is drawn to the flexibility of humanity. This fluid identity can be thought of as "ecological being." It is not fixed in terms of its material or anatomical relationships, but can deploy shape-shifting operational systems to flexibly engage notions of humanity. In finding sympathy with other life forms, new discourses may potentially be established without diminishing ourselves and precipitating social atrocities of a different kind and order from those we are all too familiar with now. While such ambitions may at first seem utopian, they cannot exist without skirmishes such as boundary conflicts, issues of identity, and cultural conflicts. Yet, implicit in the possibilities for constructing identity, relationships, and kin, the ecological being resists determinism and even formal categorization. It is no more possible to design or anticipate the relationships and networks that comprise an ecological being than it is to predict the evolution of a swim bladder from studying the developmental biology of a lungfish (Kauffman 2012). So, in thinking about what a body might become, we have to consider all the possible trajectories that an organism may take in its route towards a future form of survival. Frankly, this is incalculable. While we may be able to post-rationalize how an organism made the transition, say, from water to land, we do not have the capability to observe a living system and predict in a forwards direction all possible future states of its existence—let alone say how quickly or slowly these may occur.

But not knowing what an "ecological being" might be in the future does not preclude us from considering the principles of its actuality and establishing sympathy with other bodies, beings, and creatures.

Indeed, the ecological being is not constrained by its specific anatomy. It is an organism that is sympathetic to and entangled with humanity, but may not be exclusively human. Owing to this lack of purity, the ecological being is protean and embraces many future configurations that are yet to be expressed. Yet it identifies with humanity through an extended realm of material and cultural connections, which are produced by the combined interaction between complex interdependent agents and networks that are collectively recognized as its flesh. These are not discrete accumulations of identical tissues, but exist at the transition zones of cellular organization. Such assemblages may also find transient or persistent commonality with biospherical, technological, cultural, material, and social processes as well as other specific agents, such as intestinal bacteria, trees, implants, or gadgets—which may even be regarded as "part" of them. Some of these relationships are obligate, so we cannot do without them (like the energy-producing networks of mitochondria)—but many (like smartphones) are associative, where we have a choice.

Donna Haraway notes the protean relationships between human beings and non-humans through our relationships with domestic species. Her thoughts can be applied to relationships in which "otherness" coexists with various forms of kinship. The relationship is not uniformly agreeable—it is full of waste, cruelty, indifference, ignorance, misunderstandings, and loss—as well as joy, invention, work, intelligence, and play. In short, our relationship with domestic species is complex, invoking social, cultural, ethical, and political challenges. Haraway appreciates these contested realms by using the idea of "naturecultures"—a perspective on the nature of species that rejects typological thinking, binary dualisms, relativism, and universalisms, and appreciates the complexity in the production of "natural" encounters. She observes that biological and cultural determinisms are both instances of misplaced concreteness and proposes an alternative rich array of approaches with which to view our relationship with non-humans, such as emergence, process, historicity, difference, specificity, cohabitation, co-constitution, and contingency. These correlations are also subject to the influences of technoculture and the way in which the semiotic and material realms interrelate. They create the context for ideas of "significant otherness" that enable non-harmonious agencies and ways of living to be juxtaposed—even in the most highly constrained spaces. Indeed, she observes that, even in modern societies, co-constitutive companion species and co-evolution are the rule in the development of communities, not the exception. Such hybridizations do not work from a *tabula rasa*, but are accountable both to their disparate inherited histories and to their joint futures, which are absolutely necessary for their shared survival (Haraway 2003).

The intimate complexities of lived existence emphasize our status as ecological beings, which are not isolated, but always in existence as multiples. We are bound through entanglements of flesh and fascia that detect the unfolding of events on many different scales. Conventionally, we interpret these encounters as singular occurrences when, in fact, our lives are shaped by many simultaneous and unevenly distributed events. We selectively focus or foreground particular aspects as, culturally, we look for stories about our singularity, through which we exert a significant degree of autonomy. Yet, all the while, we exist in many parallel forms and sensibilities.

Perhaps one way to think about the immense complexity of our being is to consider the structure of the Large Hadron Collider (LHC) that allows the instrument to detect the smallest fragments of reality. It is made of layers upon layers of sensors that are all woven together in a giant metal organ, 100 m below the earth. This metallic colossus can feel the movement of the tiniest particles in the universe. Now, imagine just how complex our flesh and the sensations and interpretations that give rise to our own feelings and identities must be. As in the case of the LHC, our sensations arise from the integrated action of many agents, but this huge instrument is relatively simple compared with life forms that may also see, feel, know, and communicate with each other. In some ways, we are merely scratching at a kind of existence that speaks of parallel sensory inputs, multiple identities, and collaborating communities of agents where, from a cultural perspective, the songs of the many agencies from which we arise are conventionally understood as a single voice or identity that speaks for all its constituents.

Alternative modes of existence through a different kind of (re)telling of the story of life on Earth are needed—so that, in an ecological context, the future of one life form is not prejudiced over another. Yet, while proposing to raise the status of the non-human realm, it is also necessary to accept the impossibility of entirely removing ourselves from the establishment of emerging or future communities. Not only does this require us to uncouple from our existence, but it would also inevitably reduce the status of humanity with serious ethical consequences that objectify people. Indeed, practical and ethical concerns are raised when co-designing and engineering ecosystems alongside non-humans. Yet, by attending to their needs, we initiate the process of strengthening our relationships and building alternative communities in which our common survival is mutually beneficial. In turn, such acts of trust extend our direct realm of influence and increase material creativity across many scales of integration and sites of action. Yet, ecological relationships are robust and responsive to changing conditions and can manage the risks associated with novelty in a plethora of different ways, such as using immune systems to patrol interactions between bodies. Such negotiations cannot be absolutes, but are continual diplomatic exchanges capable of responding to a host of variable circumstances in which both human and non-human are implicated.

When relying on the goodwill of non-human agents to consider human interests within an ecosystem of participating bodies, a sense of "fairness" is invoked. While this may initially appear unlikely, there are examples of heterogeneous non-human systems actively nurturing each other in symbiotic partnerships, such as mother trees. These forest matriarchs are responsible for resource distribution, which takes place through their complex relationship with mycorrhiza (which are fungal networks). These fibrillated systems live in the roots of mature trees and produce brilliant white and yellow threads, which preferentially transport carbon, water, and nutrients through rich soil networks like a chemical Internet. Mother trees regulate this flow of vital resources to subsidize younger ones, which would otherwise fail to thrive (Howard 2011). "Fairness" and trust are therefore dependent on interspecies communications systems. These may be accessed through better understanding chemical "languages" that constitute the biochemical pathways of biological systems. Much has recently been made of bacterial languages like quorum sensing (Schauder et al. 2001) and, recently, new findings suggest that chemical languages can be very complex and even evolve (Scott-Phillips et al. 2014). These are fluid, context-sensitive, ongoing, and iterative, and change with time—providing the kind of matrix that is flexible enough to shape community relationships. However, not all creatures are as physiologically benevolent as mother trees. Even with the age of modern science and antibiotics, creatures still perish from bacterial infections and invasions of culturally and physically established boundaries by other agents of contagious disease. However, when the boundaries are invaded by the failure of diplomatic negotiations, there is a condition of war—a disturbing notion that is not confined to conflict between humans.

9.2 Establishing new and contested territories

"A virtual system inspired by a biological one that looks a lot like a manmade one." (Yong 2010)

In an interstellar context, we are aliens. The beings and ecologies that inhabit worldship environments will need design, construction, and diplomatic management. Indeed, before sending humans into interplanetary and interstellar space, it is prudent to gather experimental evidence from other kinds of populations such as bacteria that may show us how complex communities, like biofilms, perform under such extreme conditions. This may offer clues as to how we may be able to encourage the mutual construction of stable ecosystems.

On October 29, 1998, the Space Shuttle *Discovery* carried some very special cargo into space. Astronaut John Glenn transported 400 million Sea-Monkey eggs into orbit, where they spent 9 days, and were hatched 8 weeks later. The creatures showed no ill effects from their journey. In fact, these kinds of organisms appear to be so hardy that unconfirmed reports (TASS 2014) indicate that they may have somehow managed to colonize the exterior of the craft (Kramer 2014). It is also entirely possible that these creatures are also atmospheric contaminants that attached themselves to components of the International Space Station (ISS) during the assembly process, which has been underway since the 1990s.

Space probes containing materially closed, hardy ecologies could provide real information about the biophysical challenges of interstellar exploration. The ethics of such experiments are complex, especially in relation to their potential to breach confinement. Yet, microbial ecosystems appear to be possible candidates owing to their high rates of reproduction and versatility. With increasing information about microorganism performance in extreme environments that transgress the Solar System, it may be possible to improve designs for materially closed ecosystems and generate platforms that support increasingly complex microbiological consortia capable of evolution. Ethically, of course, questions must be raised about the potential of such colonies seeding independent communities elsewhere in the Solar System or interstellar space. For example, bacteria are known for their extreme versatility under challenging conditions, and the virulence of salmonella species is increased in microgravity. Such transformations are devastating for astronaut crews that have limited access to water supplies and bathroom facilities. The alternation in its biological state appears to occur as the organism interprets the environmental change as having already reached the gut (Wilson et al. 2007). However, conducting experiments to provide better questions about how we may travel beyond our own world assumes that we have already accepted the possibility of life beyond our familiar terrains. Also, the panspermia hypothesis acknowledges the chance that Earth's own life forms may have been seeded by chemical agents from non-terrestrial origins. While the ethical concerns about the conscious spread of life through the universe are not resolved by these considerations, the life-bearing potential of the cosmos beyond our own planet encourages additional reflection on the quality of existence and the conditions for negotiating survival in pioneering communities.

Using bacterial biofilms as an instrument for thinking about the challenges of worldship populations, the principles of diverse relationships between multiple species may be inferred. The results could be used as a comparator with other colonial species in extreme environments. For example, work with the slime mold organism as a computational system that can find the shortest route for establishing networks produces maps that closely resemble real-world systems, such as the Tokyo rail network (Tero et al. 2010). By compiling, comparing, and contrasting the results, maps of relationships may be produced that will help us design optimum conditions in which starship colonists may thrive. It may be possible to overcome the overwhelming notion of designing and engineering a functional ecosystem that can potentially persist indefinitely, by keeping ambitions small for those fields we can influence and forming focused social groupings that can respond to immediate contexts.

While it is possible to use organic systems literally as ways of observing the possibilities of highly complex relationships between adapting bodies and their environment, it is also worth considering a model system that speaks to the kinds of principles for nurturing communities within a materially closed ecosystem. Such constructions are not simply about the way matter is organized, but are deeply influenced by the creative processes, cultural practices, and ethical decisions implied in these extremely complex condensations of ideas, substances, processes, and experiences.

The kind of materiality implied in a closed system may be likened to the production of ambergris, an incredibly complex accretion of materials that contain a high percentage of squid beaks produced in the intestine of a whale. The resultant grayish, waxy pellet provides a rich variety of entangled organic materials from which potent compounds may be extracted for the perfume industry (Kirby 2015). Yet, this extraordinary substance cannot be manufactured by combining its constituents, but is so complex that it needs to be synthesized within the living body of a cetacean.

Perhaps another way of considering how matter may be organized within a worldship may be found in the preparation of gastronomic delicacies. For centuries, chefs have prided themselves on monstrous hybrids of meat dishes. Each delicacy is produced by stuffing one exquisite body inside another in a process called "engastration" (Hay 2014). For example, recipes such as the scholar Macrobius's Roman "Trojan boar"—a pig "made pregnant with other animals and enclosed within as the Trojan horse was made pregnant with armed men"—dates back to the fifth century. The cockentrice was born out of Henry VIII's uncontrollable desired to impress the King of France with a new mythical beast as the centerpiece of a feast. It took the form of a pig and capon sewn together. None of these creations, however, was more ornate than the nineteenth-century concoction by French gastronomist Grimod de la Reynière. In violently and deliciously forcing 17 fowl together, which included turkey, chicken, and duck, as well as more exotic birds like the ortolan bunting and garden warbler, the rôti was a fusion of edible monsters. Through the proximity of their flesh, they also feel as if they are in the process of some strange act of breeding. While this gastronomic perversity may not feel relevant to the construction of worldship environments, the civilizing of the appetites and practices that arise from unnatural juxtapositions is exactly the quality of strangeness that must be considered in a fully constructed world. Without an ethics, culture, or set of conditions in which possibilities are nurtured into existence, the worldship colony could descend into an inhumane struggle for territory, breeding, food, and ultimately survival at any cost.

The governance of such a chaotic enablement of bodies and desires is not a single strategy, breeding program, set of recipes, or guiding cultural practice. There is no one solution at all. Yet, the kind of diversity and creativity that can be found within vibrant ecosystems suggests such complex constructions are possible. Such decision-making processes invoke hyperlocality in their responses, where limits and possibilities are established, continually edited, and patrolled within a whole range of different contexts. While this may sound extremely complicated, the properties of ecosystems themselves facilitate such processes. Indeed, the possibility of establishing clandestine ecologies to resist assimilation by homogenizing political structures enables worldship populations to maintain their resilience without loss of individuality. In this way, it may be possible to maintain wide-ranging communities capable of ethically and creatively responding to change. Thereby, the resultant system retains the robustness conferred by variation and diversity, which facilitates lively ecologies and, inevitably, also invites tension.

Peace within a biodiverse population of worldships may be established when diverse populations render other systems capable of producing functional communities in the presence of competing impulses and entanglements. Effective communications strategies and languages that enable different species to dwell together may invent whole new ways of living. Model communities to help us interrogate these processes through the principles of complexification—such as in the formation of biofilms—may be found within the innate properties of microorganisms. The complex interactions between these agents may help us to ask fundamental questions about how populations simultaneously deal with cooperation, hybridization, and competition. For example, Myra Hird employs bacterial relations to examine the origins of sociable life (Hird 2009) and Bonnie Bassler even uses bacterial models to understand the operations of language as a physical phenomenon (Schauder and Bassler 2001): Such negotiations are not absolute conditions of existence, as they can be adapted and changed. However, they can be found in dynamic, flourishing ecosystems, such as in the various, entangled intra-actions tactics used by nurturing mother trees. Yet, exchanges within such communications networks are not risk-free. Much may be lost in the translation between individual agents and provoke misunderstandings that have the potential to generate hostility or destruction.

Haraway suggests that intra-relationality between highly diverse communities is critical in synthesizing notions of ongoingness (Vimeo 2014). Bruno Latour advocates that it is possible to use constructivist practices to broker new kinds of peace, where negotiations begin not with establishing the power relationships between opposites, but finding the right way for communities to build and move forward together synergistically. Through the constant renegotiation of boundaries, it may be possible to confront differences using a new set of tactics to negotiate multiple agendas. Such approaches may help to establish what constitute good and bad strategies in establishing and maintaining starship communities (Latour 1993).

Yet, ecology is more than a biophysical strategy through which spaces can be differentially occupied. They also produce a politics and practice that can buffer against conflict. This is not to say that population-scale extinctions do not occur; rather, the complex continual negotiations within complex multiply inhabited communities have great capacity to productively and creatively coexist.

The Demilitarized Zone (DMZ) between North and South Korea is a "no man's land" that emerged from the armistice that suspended the Korean War (1950–1953) and divided the Korean peninsula into northern and southern territories. It is approximately 992 km^2 and consists of mountains, plains, valleys, and basins. The origins of the conflict go back to the end of World War II, when the peninsula was split at the 38th parallel by the Soviet Union and the US as the Allies drove Japan out of Korea. Seeking to impose communist rule throughout the peninsula, North Korea launched a surprise, tank-led invasion across the line on June 25, 1950, with backing from the Soviet Union. China entered the war in October. By 1953, almost 900,000 soldiers had died, and more than two million civilians had been killed or wounded. Joined by United Nations troops, the South Korean military battled the forces of North Korea and China to a standstill. The end of fighting did not bring an end to hostilities. Instead, a cease-fire was established that carved out the 4-km-wide DMZ. The political border, called the military demarcation line (MDL), was drawn straight down the center, and anyone trying to cross the MDL is likely be shot. The area serves as a buffer zone that runs from the southern limit line, which extends from the Imjingang River in the west to Dongho-ri in the east. Restrictions in the area, which for example restrict civilian access, have been in place for the last 50 years and have enabled the unique ecology of the area to flourish. The DMZ is home to many internationally protected species, endangered species, natural monuments, and protected wild flora and fauna. Of the 2900 species inhabiting South and North Korea, 960 flora species, 35 of the 70 mammal species, and 64 of the bird species can be found in the DMZ.

To this day, South Korea and North Korea do not recognize each other as sovereign nations and are officially still at war.

The DMZ exemplifies the possibility of ecology formed by collectives of non-human agents serving as a practical arbitrator for peacekeeping. By introducing time, agency, and life into an area of conflict, direct tensions may be diffused or transformed into new experiences and encounters. Yet, these tactics are not panaceas and require highly diplomatic management. Such negotiations are integral to the emerging Ecocene, which demands that we recognize a new kind of community with the capacity to navigate hostile terrains.

Within the finite environment of a worldship, a colony may recognize its innate dependence on ecosystems by identifying differently with them, not just as "human," but also as other kinds of groupings of beings that possess continuity and kinship with each other. Through the political, biophysical, ethical, and social framing of the Ecocene, ways of thinking about what it means to be "human" could promote the formation and coalescence of diverse cultures that encourage the kind of potency and variation that is native to our home planet. Indeed, they may be essential for the continued thriving of our universal community of life, not only within the terrestrial sphere, but also in its propagation across the Solar System and out into interstellar space.

9.3 Dying well

"There are three deaths. The first is when the body ceases to function. The second is when the body is consigned to the grave. The third is that moment, sometime in the future, when your name is spoken for the last time." (Eagleman 2010, p. 23)

In a probabilistic reality, there is no life without risk of its failure. When dissipative structures approach equilibrium, anatomical integrity is breached in the face of biophysical challenges, when vital processes fail to persist in the case of sudden lethal accident, when diplomacy for contested spaces breaks down and fatal conflict ensues—it is necessary to establish organismal, legal, and narrative engagements with death.

In the terrestrial realm, the notion of biophysical "death" may be understood from a thermodynamic perspective as being the collapse of a dissipative structure. Yet, even the finality of such events is disputable. Resurrections are integral to our narratives of life, spanning folklore, religious texts, and medical handbooks. Definitions of when life starts are equally complicated, ranging from deterministic events that are described in prophesies to family legacies and organismal propagation. Currently, reaching the status of aliveness is viewed as a process of coming into existence, whose stages such as fertilization, embryogenesis, and birth are regarded as being the opposite end of a spectrum of biophysical processes, which naturally culminate in senescence and death. Increasingly, the tenuous relationships between some of these states may be facilitated, artificially maintained, and even prolonged beyond their expected lifespan through technologies such as life-support machines.

Yet, even on Earth, the concept of sociopolitical death is not quite so straightforward. The simplest structures that enable our ongoing existence after our biophysical systems have perished are notions of inheritance. Facebook has recently provided the "if I die" application that can be managed posthumously by an agent, giving the appearance of continuity to that identity.

Political and cultural frameworks for the life cycle of a living being have not been formally established beyond the geosphere. There have been no births beyond the planet's surface and very few deaths. The three crewmembers of the Soyuz 11 spacecraft, namely Georgi Dobrovolski, Vladislav Volkov, and Viktor Patsayev, are the only humans reported to have died in space during a depressurization accident during re-entry on June 30, 1971. Earth's governance systems are generally considered to end where outer space begins. As a rule of thumb, this is accepted as the lowest altitude above sea level at which objects can orbit Earth, which is approximately 100 km, or 60 miles. This is particularly pressing when considering interstellar exploration, since worldships will need to produce their own legal systems.

Moreover, many ideas about traveling across the cosmos relate to the deferment of physical existence. Such belief systems are made more complex by social and technical developments that introduce ethical and legal dilemmas in the construction of identity beyond terrestrial confines. Although suspended animation is not the main thrust of this book, such notions are relevant to "resurrecting" cryogenically preserved bodies—or even concepts related to burial. How would such a procedure be procured on a worldship, since, on Earth, a body can only be cryogenically frozen after it is legally dead?

Throughout this publication, the possibility of colonists that believe physical existence is regarded as being continually recycled through metabolic networks has been considered. In this eventuality, it may be preferred that de-animated bodies are actively composted rather than frozen. For example, Jae Rhim Lee's Infinity Burial Project proposes an alternative way of dealing with a post-mortem body, using a unique strain of mushroom that decomposes and remediates toxins in human tissue. In decomposition shrouds, biological activating agents begin the process of "decompiculture" (the cultivation of decomposing organisms) (TED.com 2011).

It is essential that, when developing a framework for the construction of a starship colony, difficult questions such as death are raised and taboos are breached, but not in an oppressive or conflagratory manner, so that we may identify ongoing opportunities to humanely develop the project and be more cognizant of our ongoing responsibilities in the present.

9.4 Interstellar possibilities

The Interstellar Question encourages us to look boldly beyond what is immediately possible in the way we approach space exploration and has the capacity to completely change the way we inhabit non-terrestrial domains. Yet, because of its hypercomplexity, constantly shifting contexts, its entanglement with many contingencies, and the significant timescales that it invokes, we simply cannot anticipate these outcomes. The Interstellar Question therefore occupies a different kind of category to those with which we are used to dealing when engaging the tools of modern synthesis—and this is one of the reasons why addressing its concerns is so valuable.

Maybe what pads out of the Schrödinger box that we have been looking at is not a cat at all, but something else quite unexpected.

Yet, this does not mean that anything goes. There is much at stake.

Unlike near-space ventures, we cannot rely on huge organizations, significant budgets, or ambitious engineering platforms such as the International Space Station (ISS), before we make meaningful contributions but be creative about alternatives—particularly as we appear to be entering into a new era of human development in which the ground rules for the way we work appear to be changing.

Since Konstantin Tsiolkovsky proposed his space ark, we have seen many developments and extrapolations based on the kind of extraterrestrial environments that we are encountering today. While this kind of pragmatism keeps the Interstellar Question real, fundable, and implementable, we must also not forgo the experimental and non-scientific aspects. There is no Deep Thought supercomputer to resolve the hypercomplex challenges we face, and certainly none that will be of more use to us than our imaginations.

In this new epoch of shared-interest communities, low-cost manufacturing platforms, and knowledge transfer, highly distributed approaches may help us to develop new kinds of questions, methods, tools, and practices that enable us to build prototypes using an experimental approach. For example, hackathons may help us to design new kinds of questions through the production of prototypes and test them against a range of specific contexts, such as in our living spaces and cities. In this way, we can demonstrate our ideas and learn how to build starships on this planet in a massive, open, and distributed fashion—before we ever head for the stars.

A critical aspect of the Interstellar Question is that it is not solely about technical feasibility, or simply to "science the shit" out of a seemingly hopeless situation—a meme that has endeared many of us to the movie *The Martian* (2015). It asks about our commitment to investment in our shared future. How far ahead can we envisage our discovery road maps? How bold will we dare to be in establishing new prospects and opportunities for our successors? Is it possible for our civilization to unite around a common goal without tearing ourselves asunder through differences?

In the process of discovery, it is essential to stay vigilant to the existential aspects of the project that invite poetry, art, and storytelling. Nor do these aspects neglect the degree of ethical, social, and cultural difficulty that the Interstellar Question raises, as ideas and worldviews collide. Building a starship is not just a phenomenally difficult technical challenge—it is a social enterprise and cultural endeavor. Diplomacy is required to actively establish peacemaking processes that help us to constructively manage diverse ranges of ideas, people, and ecosystems around the Interstellar Question.

With critical commitment to the starship project, we will begin to see new possibilities take shape out of the Tower of Babel of ideas from which the pressing questions have arisen. Over time, it is likely that increasingly formal toolsets, methods, and means of production and analysis will emerge.

Of course, there are risks to taking a highly distributed experimental approach to the Interstellar Question, which may become a practice entrenched in prototyping mode, as it may be a while until we move to the next phases of development. Yet, during this phase of discovery, we may reach a point at which our questions have become more investable or our ways of working have changed again. By remaining critical of our approaches, it will be possible to evolve and refine even our most successful practices. Perhaps we will discover that we are asking the wrong kinds of questions and will need to be flexible in our approaches, so that we can establish better alternatives. In fact, it is likely that our question-making, experimentation, and prototypes will be iterative, so they twist and turn many times before reaching the next phase of development.

The Interstellar Question is a quintessentially human endeavor. We have already gone to these faraway places in our dreams. Now we are gaining the capacity to realize them. This requires the kinds of creative vision and abilities that, currently, no machine possesses.

9.5 Making moonlight

At eight-seventeen: thirty Chronos, the lights-out policy that proposed to prevent the detrimental effects of light pollution, kicked in. Colonists gathered around the reflective surfaces of water as they heard the illuminator lens systems grind into position to produce a brusque sunset, where no celestial orb sank slowly into a horizon. Instead, a spreading heat wave rose to become a chromatographic spectacle of crimson, red, orange, yellow, green, indigo, violet, and blue. As patterns shattered into bioluminescent fragments, the inky sky filled with squiggly embers, thought to be aberrations of atmospheric pollution. Viewers blinked firmly to release the pain behind their eyes and quickly made their way to bed.

At twenty Chronos, the lights returned with flourishing rainbow of colors that rapidly descended from night cool to radiant waves of day heat, which warmed the humus-topped ground.

Yet, such a light show did not bring the community peace.

Persephone's clinics were cluttered with sleep disorders, irregularities in menstruation, violent mood swings, changes in appetite, suicidal thoughts, intolerances to noises higher than 15 kHz, infertility, breast cancer, obesity, diabetes, various psychoses in which pale-faced observers followed you, impotence, an increase in somnambulism, and a strange kind of hydrophobia that had only been recognized in the last few years, where sufferers refused to drink if they had first seen their face reflected in the water.

Such episodes were documented as sporadic; but the unrest swelled.

Minds and flesh crawled with a creeping malaise. Repeatedly, they tried to ignore it, but then the coincidences ran out. On thirteen-thirteen Calandros, at thirteen-thirteen: thirteen Chronos, the surveillance systems alerted the night watchers that almost the entire village of Gladesend was standing in the street staring aimlessly into the blank sky.

Various professional bodies concluded that these colonists were suffering from a light disorder. They tried altering the worldship's illumination system by changing the quality and angle of the lenses. However, it was not until a spike in wellbeing was associated with a technical error, which occluded one of the reflectors with a black carbon coating, that they realized a lack of indirect light in the worldship might be responsible.

Town leaders immediately wanted reflected night-light—lots of it—as quickly as possible. But how could this be achieved without flooding the sky with pollution from the fission reactor?

They tried subtlety with direct lighting systems first.

A photon web was installed. It laced all Persephone's settlements together through a delicate neural net. This light-producing fabric, which was studded with LEDs that were activated by proximity sensors and powered by electrodes in the soil, provided low-level illumination. The whole network glowed like a giant root system and shivered, as residents strolled their mammals and crustaceans around the streets. Weaving in and out of illuminated islands that were decorated by complementary swaths of darkness, they kept an elegantly respectable distance from each other, so that lobsters and terriers did not get tangled in skirmishes. Populations of fireflies plummeted as the bioavailability of oxygen, magnesium, and adenosine triphosphate wavered, with drastic results for luciferin metabolism, which produced the characteristic yellow light emissions of many firefly species. Yet, although residents admired the sparkling inconstancy provided by the smart biolighting system, it did not produce the quality of experience they needed.

"We need a particular wavelength that works on the mood centers," observed scientists, who set about defining the optimum range. This peaked at 509 nm, which was within the range of visible white light. In their haste to produce an alternative night-time experience, town leaders sought to deliver this in the largest doses possible. Searching for inspiration, they called a meeting of designers that recalled old Earth's moonlight. Yet, according to calculations, a single source would never be enough to cure the ailing residents.

So, a series of moonlight towers were commissioned that sought to embody the shape of the full summer Moon.

Having never made a moon before, the construction process itself became a spectacle. People gathered to witness structures being built that represented a new kind of night-time experience. Accidents were inevitable. Thin metal scaffolding, hugged by artificial gravity, had an uncanny way of slicing through roofs. Over-ornate towers kinked and toppled from odd structural faults. Eventually, Gladesend's glowing orbs were erected and suspended high above the village on incredibly ornamental scaffolding—a mélange of attractive metals, tile fragments, and decorative glass that could have been assembled by giant caddisfly larvae. Each globe glowed with a powerful radiance that matched the strength of around 6000 candles and was given a prominent place in the township—above public buildings, or in the center of squares. Here they stood gaudily, tall, demanding to be gazed upon. Yet, in the rush to deliver therapy to their populations, the light beacons were too harsh to be truly admired. They shone with the intensity and pomposity of suns that proclaimed there would never again be darkness without light.

They made this point so forcefully that people began to drown in a photon-soaked sky and cast their eyes downwards when they left their houses at night.

Although they were a popular visitor attraction, residents did not consider the towers successful. Indeed, quite a few had been skeptical from the outset of the project and were concerned about the value of bringing daytime light intensities into the night. The singular light sources were prone to casting strange shadows tinged with sharp blue light in their wake, which left pedestrians "dazed and puzzled." Foggy evenings could thrust the whole of Gladesend into a realm of poor visibility as intense light was scattered diffusely through mist and vapors. Blinded night travelers groped their way along sidewalks under an eerie brilliance. One woman was so infuriated by the intrusion that she even tried to chop down a moonlight tower in an act of civic rebellion, for which she was arrested. Animals also became neurotic and unpredictable with the newly extended daylight. Christopher Meadows claimed to have spooked a mule, after vigorously blowing his nose. The beast dashed down the street, colliding with scaffolding, and brought down an entire moonlight tower to its knees.

Without an artificial distinction between night and day, songbirds chirruped continually, flowers etiolated, worm burrows dried out, painted surfaces deteriorated, barren patches scarred lawns, while workaholics developed strangely tanned complexions as they gathered around the light sources to obsessively work on chores. At one point, Gladesend boasted a flourishing trading economy based on the habit of continual work.

Legions of strange new insects were also drawn to the extraordinary light. They were preyed upon by swarms of bats that squeaked incessantly as they echo-located their food. Although their striking silhouettes and dazzling aerial displays drew admiration from many residents, others found them noxious. They took delight in confusing the bats' activities, with whistles and strange radio-like devices, hoping to drive them away. The bats became territorial and dive-bombed observers indiscriminately with increasing boldness. This was not their only claim to notoriety. Bats sought refuge in attics and cavity wall spaces. They chewed building materials to enlarge access holes. They roosted noisily by the score as their corrosive urine consumed flooring, insulation, and wood. Decomposition was sped up by piles of pungent insect-rich droppings that brought microbe-rich sediments and respiratory diseases.

Bat-infested homes stank.

So they were netted and poisoned by pest-control officials. Still, as long as the insects danced around the artificial suns—that pretended to be moons—somehow the bats thrived.

With growing protests about environmental disturbances and unfair economic advantages, the idea of artificial moonlight outweighed its beauty and poetry. So, the globes were removed and the scaffolding left derelict. Within months, they were roosting platforms for pigeons that caked up with guano, which rotted their steel frameworks. Vigorous vines found new opportunity in the rust and digested these carefully crafted monuments into piles of orange dust, in which black ants made rhizomic nests. New neuroses began to clog casualties—an over-sensitivity to daylight, dreamlessness, dysmorphophobias treated by wearing dark glasses, psychosomatic melanoma, and a whole host of time-orientation syndromes that vaguely resembled jetlag. Gladesenders were pitied for their "flat failure of night-time."

Town leaders slowly conceded, with new assertions that made no apologies, while accepting that a special kind of daylight at night-time was not what had been needed.

Fresh discussions began about how a lighting system that produced reflected light could be built. Did it have a special quality beyond the emission of a particular wavelength of light? Scientists shrugged. Of course—light reflection was not at all simple. It was a complicated physical phenomenon.

It wasn't just about how the photons performed, but how the surfaces and environments that they encountered transformed their behavior. The way that light was scattered, lost energy, refracted, or was altered in some way contributed to its quality. Laboratory tests using grazing mirrors, which gathered stray light beams like condensers, were initially promising. Yet, the moment they were installed, they produced the "wrong" kind of light. But it was impossible to say what exactly wasn't working. The measurements all added up. There were no surprises in the data. After 18 months of intensive research into the paradox, the engineers concluded that they'd leave these issues to the poets.

The sarcasm of the engineering report was lost on the town leaders, who decided to invite all the worldship poets to a lakeside fête. Revelers were promised a day of old-world English events with tombola, raffles, lucky-dips, coconut shies, bat-a-rat stalls, white-elephant stands, home-made toys, and a spread of local jams, pickles, pies, and cakes. Morris dancers slapped their sticks, swords, bells, and handkerchiefs together. Audiences sniggered at mummers' bawd, sprained tendons in the tug-o-war, showed off in fancy-dress competitions, and proudly entered their cherished pets into best-in-show contests. Hand-reared lambs gamboled around obstacles behind children, small goats balanced gracefully on dangerous-looking climbing frames, sharp-eyed dogs rounded up sheep and carefully coiffured calves batted long lashes at adjudicators. At the end of the festivities, which formally came to a close at exactly eight-seventeen: thirty Chronos, the poets were invited to publicly share their thoughts, philosophies, and impressions on the engineer's challenge regarding the quality of reflected light.

"It'll never work," grumbled the pragmatists, who had mostly come to see the poets fail, but also could not resist the prospect of an unlimited supply of thickly cut, freshly baked bread generously wiped with damson jam and lumps of unsalted butter.

As conversations quelled for the final address and the sunset climbed dramatically upwards, the poets proposed with unprecedented unanimity that the lake itself would be the source of reflected light.

"Preposterous," laughed the pragmatists, as engineers shook their heads with renewed clarity and began to make sketches in their notebooks of how optical fibers might be threaded underground into the lake through soils.

And so Gladesend was the first of the Persephone townships to experiment with indirect incandescent light systems. Artificial earths that relayed photons from the nuclear fusion reactor were installed under the lake, where optical fibers heated dissolved gases in the water, which bubbled to the surface. On rising, they hovered as condensations before breaking into tiny patches of rainfall around the lakeside. Fluctuations in the surface tension, humidity, mineral content, organic matter, ecology, temperature, opacity, and pollution also contributed to the quality of light emanating from the lake. These could be manipulated by a host of technologies that also threaded their way down into the lake soils. Billowing like smoke, the mirrored photon relay dripped softness into the night, enticing others to come and make their own gourmet experiences with water, light, soil, and mirrors.

The strange moonlight produced by this process was the true attraction. It never rose high like sunrises or sunsets, but hung low and heavy, soaking into immediate landscapes in extraordinary ways. It seeped through streets, crystallizing in building details, and softened the edges of restless dreams.

Opalescent light was everywhere—especially where Nature had not intended it to be. Milky, luminescent bodies rose like gigantic cheeses and wandered dripping over the city's rooftops. It bathed the surrounding fields and lonely outskirts of the city with its abundant curds and whey. Lovers sat at the water's edge on crumbs of moonlight, watching its scintillating particles skip in waves towards the night's horizon. Poets wrote of lardy swathes of light, implausibly long shadows that followed you home, and glowing Unidentified Flying Objects. It etched yellow scratches on brickwork, turned sclera blue, and chewed needle-fine teeth marks in windowpanes. Guilty arthropods slinked around its liminal glare to avoid detection by nocturnal predators. Photon puddles smeared walkways, hovered like halos around dewdrops, and tiptoed as breath on windowpanes. Wildlife courted in the subdued light, people sat outside on benches, and eggshells split open as helpless newborns flopped, concealed into the iridescent gloom. Frivolity and lightness of touch to living returned. The day broadened its reach and joy ensued.

One visitor described the citizens of Gladesend as "in a state of delighted enthusiasm over the splendid practical results." "The moonlight relay," he declared, was a "most brilliant success."

The lights-out routine spreads skywards as it always did at eight-seventeen: thirty Chronos promptly and falls again at twenty Chronos. These cycles of photons that travel in straight lines are those that set the colonists' routines, keeping them healthily productive—but it is the indirect caress of moonlight that nurtures their souls.

9.6 Other voices

"Imagination will often carry us to worlds that never were, but without it we go nowhere." (Sagan 1980, p. 4)

Persephone is more than a set of experiments, but also reflects on the various project outcomes as a discourse on living within a worldview, so that we may begin to strategically investigate the kinds of questions we need to ask before we attempt to formalize our responses to the challenge. Icarus Interstellar imagines a 100-year developmental process. At the end of this incubation period, an orbital platform will be established from which we can begin to prototype and develop the next stages towards building a starship.

In thinking through the challenges of existence from a range of challenging perspectives, the ideas are actively tested through a set of real laboratory experiments, design explorations, case studies, and conceptual provocations. The multiple voices and contrasting viewpoints of expert contributors gesture even more broadly towards the kind of diplomacy and navigational skill that is required to embrace the Ecocene that underpins Persephone's culture. They represent a diverse, sometimes overlapping, at other times contrary, community of ideas. Their combined readings produce a mutually reinforcing perspective of inclusivity and diversity, from which it may be possible to garner a sense of purpose, ongoingness, and quality of living of which a starship culture could be proud. The convergences sought between these differing viewpoints form an investigatory toolset that may be entwined through the process of "worlding" to generate a living starship interior.

In generating conceptual and practical tools that facilitate a priori thinking and exploration through the production of experimental prototypes, Persephone's proposals challenge fundamental expectations of materials, environments, cultural forms of organization, and even "life" itself. Once viable questions are formulated, the first set of answers may be constructed through prototypes of biosphere-style experiments. Some of these are anticipated to provide opportunities for rigorous empirical testing and data collection that lend themselves to formal scientific analysis. Quantification of these future experiments will be used to further inform Persephone's development. Yet, generative fields of possibilities that cannot simply be reduced into bounded spaces or materialities are also anticipated from Persephone's iterative inquires. The phenomenologies, social interactions, cultural expressions, music, and poetry that exude from the overlapping realms that are continually in flux are just as much part of the starship question as the infrastructures in which their existence is grounded. These experiences transcend easy apprehension or meaningful analysis, and emanate from Persephone's operatic qualities, which compose the stories of our future among the stars.

In the following chapters, invited contributors present a series of parallel worlds that are integral to the discovery process and add richness to the ongoing process of debate and discovery. Authors are drawn from a spectrum of background that include academia, small businesses, and the arts.

As president of Icarus Interstellar, Andreas Tsziolas outlines the ambitions of the group in the context of interstellar travel. Nathan Morrison establishes where and why we should venture beyond our Solar System. Michael Mautner establishes an ethics and scientific practice for a purpose to interstellar exploration—namely deliberately seeding the cosmos with life. Simon Park highlights the great versatility of bacteria in being an integral part of establishing an interstellar presence, whether that is in a barren planet or on board a spacecraft—they are part of our story, whether we like it or not. Steve Fuller examines the legacy of cosmism in creating a framework for a technological future, while Sarah Jane Pell offers a portrait of the body *in extremis*. Arne Hendriks examines the benefits of shrinking the species to an average of 50 cm—a challenge of miniaturization and resource-saving strategy in which NASA has invested. Kevin Warwick considers the kinds of technological symbioses that human space travelers may adopt in attempting to exist in extreme conditions. Rachel Armstrong outlines the precariousness of human reproduction in space and asks what this might mean for the future of humanity. Barbara Imhof, Peter Weiss, and Angelo Vermeulen outline the state of the art for constructed ecosystems that provide the infrastructures for life and their future trajectories. Susmita Mohanty, Barbara Imhof, and Sue Fairburn apply the knowledge from sustainable ecosystem design to "terraform" our current cities so that, instead of being sites of ecocide, they enrich our biosphere. Astudio architects discuss how these principles of ecosystem and sustainable design can be more strategically applied to our current cities, transforming their impacts and enabling us to do more with fewer resources in ways that are life-promoting, to generate an era of "starship cities." Rolf Hughes examines different ways of inhabiting spaces through the circus arts, which are traditionally disruptive in their thinking and practices, and may help us to develop alternative ways of being and living together. Esther Armstrong considers the scenography of Earth since the Moon landings and how its iconography has generated resistance to our capacity for complete reinvention or transformation as a global culture. Jordan Geiger reflects on the challenges of designing and constructing spaces that engage with very large organizations that we can only ever perceive incompletely. Mark Morris looks at the architectural tropes of world building, and how the alchemical *ouroboros* offers a design strategy for embracing the challenge of real cycles, duration, and spatial endlessness across all scales from the molecular to the infinite. Roberto Chiotti examines the implications of an architectural cosmology. Krists Ernstsons reflects on a strategy for escape by applying the latest concepts in quantum computing and astrophysics; he proposes how we might solve the interstellar issue without a worldship by creating tiny swarm-mind avatars to travel across interstellar space so that we may settle the stars in another plane of existence within our lifetimes.

This publication, therefore, represents a conversation and set of experiments that produce prototypes aimed to encourage both experts and readers to radically reflect on the kind of adjustments that may be needed to sustain ourselves indefinitely—here on Earth and out into the dark spaces beyond.

9.7 New beginnings

Endlessly it rained.

Plates of flabby drops broke the skin of the ground.

And no one came.

The big striped tent, a comical pastiche of itself, billowed like a jellyfish in the spiteful winds. It's complaining tethers ground their restraints taut, like stressed teeth.

Roots rotted, molds flourished, sewage spilled, soils turned into toothpaste-like sludges, electrical circuits shorted, Wellington boots slurped water in over their tops, socks soddened, bones ached, and ducks swam in chevron formations down high streets.

Still it rained.

"It's getting warmer," they said as worms suffocated in the wet tedium and floated as pink question marks on puddles.

But, despite countless cycles of night and day, no one saw the Sun.

The circus troupe huddled in their trailers while water gurgled under the door and made industrial carpet smell like pondweed. To keep each other company, they built walkways from farm flotsam that spanned the growing gaps between their quarters. Tired of tin restraints, they began to live under parasols on the roof of their vehicles. Becoming more ambitious in the use of their props, they somersaulted, flip-flopped, jumped, cartwheeled, pole-vaulted, swung, teetered gracefully, hovered in formations, and even used the rising waters as a performance site for aquabatics.

"We don't need the old rules," they said, "let's make our own."

So, they floated in convoy to the candy-striped tent and climbed aloft. Their art now stretched above the apex of the big top. Here, they built scaffoldings that spiraled precariously upwards, and constructed a rickety helter-skelter with string, planks, tree trunks, old fences, roof tiles, abandoned cars, and solid drowned beasts, which stretched ambitiously to the thickening rainclouds.

The wetness continued and they slipped endlessly as they ascended but, one day, they reached the sky.

Yet their cheers and sense of accomplishment were short-lived. The mists had become so dense that they were now crusts in the atmosphere.

Noticing that the ground seemed no farther away than before they'd begun climbing, they sat awhile and sulked that they had not been ambitiously creative enough in escaping the impenetrable gloom that had seized their realm. Just before they could blame one another for this predicament, their scaffolding began to fall.

At first, it was little more than the groaning of a hungry stomach, but soon it became the complaints of overstretched joints and tendon sprains. As the makeshift framework screamed all the way through its dismemberment, the troupe slammed their fists against the sky and began to think anew.

"We want to live," they cried.

But the cruel weather sphincter gripped them tightly.

And so they upped their game.

They knitted nets from vegetable strands and launched them so that they would challenge the very limits of the sky. They traveled with the ascending fibers and, just as they reached the apex of their climb in those moments before they would fall, they riveted them to the tarry lid. Feeling the full weight of impossibility, tears leaked from their eyes, noses congested, and their hearts crushed. Secretly, they wondered how long it would be before they breathed their last but, steeled by radical love for each other, a few of them built cutlery from their disarticulated sky scaffold and organized a last supper—a communal breaking of sky-bread as a dignified farewell to one and all.

"Bon appétit," they said as they plunged their knives and sporks into the approaching miserable matter under which they'd soon choke and began to chew.

And so they ate their way through the upper limits of their world and flopped like newly evolved beasts into a brand new terrain for which they were ill-equipped.

They squinted into the space above them without the words to describe what their senses beheld. With unlimited freedom to invent, they began to construct new ambitions, new practices, and a brand new space to thrive in.

"Persephone," they said.

9.8 Interstellar synthesis

"We [lack] the creative faculty to imagine that which we know." (Shelley 2014, p. 530)
"The greatest weakness in contemporary thought seems to me to reside in the extravagant reverence for what we know compared with what we don't yet know." (Breton 1987)

The mutable Interstellar Question resists any single reading of possibilities.

It is a transformer of ideas that can be explored and transferred across generations. The sheer ambition and audacity of the challenge have the potential to make radical breaks in our thinking and being. Through expanded frameworks of discovery, we may venture beyond current knowledge conventions and raise insightful, or uncomfortable, questions about our approach to space exploration.

Throughout the production of this publication, ways to embrace the richness and complexity of the Interstellar Question and resist its collapse into a soluble cipher have been sought. This is no easy task, as it resists conventional thinking, whereby uncertainty is dealt with through high modern perspectives that convert insoluble, hypercomplex challenges into fully comprehensible ones. By taking a deterministic approach to the Interstellar Question, it is decomposed into logical steps that may be solved by appropriate technological developments. Yet, while dominant, such perspectives are extremely reductive and, notably, do not provide the opportunity to address the significant existential and human challenges implicit in the Interstellar Question.

However, if we are to transcend the limits of extreme abstraction and break free into a creative realm of uncertainty, we must also confront the terrifying prospect of an incomprehensible universe. What instruments do we need, or must we invent, to deal with the endless uncertainty of an ever-unfolding universe?

Since the first starships are not under construction and we therefore lack their concreteness against which to kick our ideas, it is essential to keep the Interstellar Question open and evolving. Endless theoretical positioning can only take us so far towards the realm of discovery. A transition towards practice-based research and development is pressing, even if the outcomes may appear to be an unfeasibly long way from realization.

This book proposes an implementable, experimental strategy for approaching the many daunting unknowns and unknowables implicit in the Interstellar Question through an investigation of the Persephone project that assumes an ecological perspective of the cosmos. It uses an experimental practice to navigate a range of possibilities that embrace the science and technology that are critical in constructing the vessels that will transport us to worlds elsewhere, but also approximates these with ways of dreaming together. In proposing to construct a sustainable world from scratch that will support colonists indefinitely, the prototypes that explore this challenge reveal alternative ways of seeing, living, and becoming a starfaring community.

If we are to apprehend the nature of our own being in a highly labile terrestrial environment, as well as establishing ourselves in new terrains, the participation of ecosystems in our new landscapes, anatomies, and identity cannot be overstated. While the human population is dense on our own planet, we are very scarce in distant lands and are foregrounded by the void. Non-humans fill the cosmos with dark matter, heavy metals, comets, methane, radiation, hydrocarbons, dark energy, photonic winds, black holes (Griffin 2015), panspermic seeds, and potentially with civilizations of other beings that may be entirely unrelated to and unrecognizable by us. This is the ecological landscape in which our ambitions are projected. In these unknown expanses, we are not the measure of all things, but absorb the qualities of the environments in which we are immersed (Morton 2013). Within a worldship landscape, these are entirely of our own making.

The constructedness of our existence is one of the great challenges of the Interstellar Question. It is an unfathomably challenging quest, and yet it does not help us to stand caught in the headlights of immensity, but somehow to take faltering steps towards capturing possibilities. The insights that may be revealed through these questions have pressing relevance to our current situation, as well as generating a new set of research questions. The Persephone project engages with these vast unknowns by adopting alternative viewpoints and perspectives. Prototypes take on the status of instruments with which it is possible to navigate the many uncertainties within the unfolding complexities of the Interstellar Question. While these explorations are not formal solutions to established problems that are recognized as pertinent to the construction of starships, they invite multidisciplinary conversation and generate a new criticality to the field. This has largely been a discourse of varying expert opinions by physicists and aerospace professionals.

Indeed, as the interstellar community broadens its inquiry from the production of megascale starships, and establishes practices that are relevant to an emerging cultural interest, impacts then appear much more tangible. Certainly, there is little point in building a megastructure without advocates. While building a starship is still beyond our natural lifespan, the experiments, prototypes, and conversations within this book enable readers to re-orient around the idea of living amongst the stars. Current developments within the interstellar community are also anticipating this more open and inclusive approach with the first Interstellar Hackathon, whose outputs may form the basis of further research projects (Mookerjee 2015).

Such alternatives provide a kaleidoscopic view of possibilities, whereby ambitious new territories move into sight. Many of these are highly contingent, transient, and continue to mutate long after we have encountered them. Yet, their relevance cannot be dismissed as—in the manner of quantum phenomena—they may suddenly reappear with new meaning at another time and place.

The aim of this publication is, therefore, not to cumulate specific forms of existing knowledge as much as it is to examine the conditions of developing new types of practices. By critically exploring alternative possibilities and creatively hybridizing knowledge fields, we may unlock the phase space of possibility within terrains that are currently regarded as "unknown" and "unknowable." Indeed, an essential aspect of our journey is the acquisition of new kinds of knowledge that arise out of these exchanges, the like of which we have not yet experienced and to which we must remain vigilant.

Every aspect of the Interstellar Question reinforces our inability to observe or comprehend existence and the nature of reality in its entirety. We only ever work with partial knowledge. Yet, by generating a mosaic of diverse propositions, we can start to generate topological landscapes of understanding. These can be overlapped and shared across a much broader cross-section of society than exists today, to garner interest and critique in an emerging interstellar culture. Perhaps we can think about this kind of map-making as having resonance with the manner in which Dmitri Mendeleev's 1869 periodic table of the elements was produced. This cornerstone of knowledge was compiled from fundamental principles, and took over a century of growth in the understanding of chemical properties of matter before the contemporary codes that we employ today could be established. Yet, such an instrument is not a sudden revelation, but an accreting palimpsest of discovery. Even when Mendeleev first made his propositions, he was building upon earlier findings by Antoine-Laurent de Lavoisier and John Newlands. Doubtless, with the advent of supra-molecular chemistry (Lehn 1988) and the synthesis of molecules that have not yet existed within the history of the universe, the matter maps with which we navigate the subatomic realm—and, vicariously, the cosmos—will change again.

Ultimately, these combined approaches generate a chimerical synthesis of ideas, approaches, skill sets, communities, and institutions that may help realize our potential as a starfaring culture. The challenges that we face develop our stamina, resilience, and creativity in generating new platforms for research, discovery, making, and dwelling. This is an exciting time for exploring the Interstellar Question, as we collectively perform the groundwork in taking the field from being a largely theoretical conversation into a critical, practice-based discipline. Like those who have gone before us, venturing into the truly unknown, we are compelled by something that exceeds the rational aspects of our being. This strange compulsion to boundlessly "venture forth" is an indelible passion of our species that is steeped in our irrepressible nature. Such drives exceed the eternal human qualities of faith, hope, and charity—and stir a deeper cauldron still, of curiosity, discovery, and adventure:

> "I could not draw a route in the map or set a date for the landing. At times all I need is a brief glimpse, an opening in the midst of an incongruous landscape, a glint of lights in the fog, the dialogue of two passersby in the crowd, and I think that, setting out from there, I will put together, piece by piece, the perfect city, made of fragments mixed with the rest, of instants separated by intervals, of signals one sends out, not knowing who receives them. If I tell you that the city toward which my journey tends is discontinuous in space and time, now scatters, now more condensed, you must not believe the search for it can stop." (Calvino 1997, p. 164)

Indeed, the imperative of the Interstellar Question is more than impulse, but also an imperative for our existence for we must learn how to live on starships before we can truly and sustainably dwell on our home planet, Earth.

REFERENCES

K. Barad, *Meeting the Universe Halfway: Quantum Physics and the Entanglement of Matter and Meaning* (Duke University Press, Durham, NC, 2007)

J. Bennett, *Vibrant Matter: A Political Ecology of Things* (Duke University Press, Durham, NC, 2010)

J.L. Borges, *Labyrinths* (Penguin Classics, London, 2000)

A. Breton, *Mad Love* (University of Nebraska Press, Lincoln, 1987)

I. Calvino, *Invisible Cities* (Vintage Classics, London, 1997)

D. Eagleman, *Sum: Forty Tales from the Afterlives* (Canongate Books, Edinburgh, 2010)

A. Griffin, Black holes are a passage to another universe, says Stephen Hawking. The Independent (2015), Aug 25, www.independent.co.uk/news/science/black-holes-are-a-passage-to-another-universe-says-stephen-hawking-10471397.html. Accessed 26 Aug 2015

D. Haraway, *The Companion Species Manifesto: Dogs, People and Significant Otherness* (Prickly Paradigm Press, Chicago, 2003)

G. Harman, *The Quadruple Object* (Zero Books, Winchester, UK, 2011)

M. Hay, The rôti sans pareil is 17 birds stuffed inside each other and it is delicious. Vice, United States (2014), Apr 4, www.vice.com/read/the-rti-sans-pareil-is-17-birds-stuffed-inside-each-other-and-it-is-delicious. Accessed 12 Sept 2015

M. Hird, *The Origins of Sociable Life: Evolution after Science Studies* (Palgrave Macmillan, London, 2009)

C. Howard, "Mother trees" use fungal networks to feed the forest ecology. Canadian Geographic (2011), Jan/Feb, https://www.canadiangeographic.ca/magazine/jf11/fungal_systems.asp. Accessed 19 Apr 2014

S. Kauffman, Evolving by magic. Arup Thoughts (2012), May 17, http://thoughts.arup.com/post/details/200/evolving-by-magic. Accessed 21 Aug 2015

D. Kirby, The £7,000 lump of whale vomit that holds the secret to the world's most exotic scent. The Independent (2015), Sept 11, www.independent.co.uk/news/uk/home-news/the-7000-lump-of-whale-vomit-that-holds-the-secret-to-the-worlds-most-exotic-scent-10497372.html. Accessed 12 Aug 2015

M. Kramer, Sea plankton on space station? Russian official claims it is so. Space.com (2014), Aug 20, www.space.com/26888-sea-plankton-space-station-russian-claim.html. Accessed 12 Sept 2015

B. Latour, *We Have Never Been Modern* (Harvard University Press, Cambridge, 1993)

B. Latour, *Reassembling the Social: An Introduction to Actor-Network-Theory* (Oxford University Press, Oxford, 2005)

J.M. Lehn, Perspectives in supramolecular chemistry: from molecular recognition towards molecular information processing and self-organization. Angew. Chem. Int. Ed. **27**(11), 89–121 (1988)

A. Mookerjee, "Interstellar hackathon" to chart our path to the stars. Discovery News (2015), July 27, http://news.discovery.com/space/private-spaceflight/interstellar-hackathon-to-chard-our-path-to-the-stars-150727.htm. Accessed 28 Aug 2015

T. Morton, *Ecology Without Nature: Rethinking Environmental Aesthetics* (Harvard University Press, Cambridge, MA, 2007)

T. Morton, *Hyperobjects: Philosophy and Ecology After the End of the World* (University of Minnesota Press, Minneapolis, 2013)

C. Sagan, *Cosmos* (Random House, New York, 1980)

S. Schauder, B.L. Bassler, The languages of bacteria. Genes Dev. **15**(12), 1468–1480 (2001)

S. Schauder, K. Shokat, M.G. Surette, B.L. Bassler, The LuxS family of bacterial autoinducers: biosynthesis of a novel quorum-sensing signal molecule. Mol. Microbiol. **41**(2), 463–576 (2001)

T.C. Scott-Phillips, J. Gurney, A. Ivens, S.P. Diggle, R. Popat, Combinatorial communication in bacteria: implications for the origins of linguistic generativity. PLoS One (2014), Apr 23, doi:10.1371/journal.pone.0095929, www.plosone.org/article/info%3Adoi%2F10.1371%2Fjournal.pone.0095929. Accessed 23 Aug 2014

P.B. Shelley, *A Defence of Poetry and Other Essays*. (CreateSpace Independent Publishing Platform: Online, 2014)

TASS, Scientists find traces of sea plankton on ISS surface (2014), Aug 19, http://tass.ru/en/non-political/745635. Accessed 12 Sept 2015

TED.com, Jae Rhim Lee: my mushroom burial suit (2011), July, https://www.ted.com/talks/jae_rhim_lee. Accessed 17 Apr 2015

A. Tero, S. Takagi, T. Saigusa, K. Ito, D.P. Bebber, M.D. Fricker, K. Youmiki, R. Kobayashi, T. Nakagaki, Rules for biologically inspired adaptive network design. Science **22**(5964), 439–442 (2010)

Vimeo, Anthropocene, Capitalocene, Chthulucene: staying with the trouble. Donna Haraway (2014), Sept 5, https://vimeo.com/97663518. Accessed 18 Aug 2015

J.W. Wilson, C.M. Ott, K.H. Zu Bentrup, R. Ramamurthy, L. Quick, S. Porwollik, P. Cheng, M. McClelland, G. Tsaprailis, T. Radabaugh, A. Hunt, D. Fernandez, E. Richter, E. Shah, M. Kilcoyne, L. Joshi, M. Nelman-Gonzalez, S. Hing, M. Parra, P. Dumars, K. Norwood, R. Bober, J. Devich, A. Ruggles, C. Goulart, M. Rupert, L. Stodieck, P. Stafford, L. Catella, M.J. Schurr, K. Buchanan, L. Morici, J. McCracken, P. Allen, C. Baker-Coleman, T. Hammond, J. Vogel, R. Nelson, D.L. Pierson, H.M. Stefanyshyn-Piper, C.A. Nickerson, Space flight alters bacterial gene expression and virulence and reveals a role for global regulator Hfq. Proc. Natl. Acad. Sci. U. S. A. **104**(41), 16299–16304 (2007)

E. Yong, Slime mould attacks simulated Tokyo rail network: not exactly rocket science (2010), Jan 21, http://scienceblogs.com/notrocketscience/2010/01/21/slime-mould-attacks-simulates-tokyo-rail-network/. Accessed 13 Aug 2015

Part II
Anthology of Interstellar Culture

10

The interstellar mission

Andreas C. Tziolas, Nathan Morrison and Esther M. Armstrong

Three individual authors Tziolas, Morrison, and E. Armstrong discuss different reasons to explore interstellar space and what is being left behind.

10.1 Introduction to interstellar exploration

Andreas C. Tziolas, president Icarus Interstellar

10.1.1 The worldview of the explorer

Human thought and science are magnetically attracted to and dwell at the intersections between the known and the unknown. The explorer pulls away the veils of unfamiliarity and enters fully into a relationship, where imagination slowly gives way to fact. Discovery gives substance to our concepts as beckoning boundaries are overtaken and conquered.

And so it will come to be with our effort to establish our purpose among the stars.

Human history reveals that our imagination has never been bound to this planet. Our mythologies projected humanoid gods to the stars millennia before we saw how this may

---

Andreas C. Tziolas (✉)
President Icarus Interstellar. 2809 Spenard Rd, Anchorage, AK 99503
e-mail: atziolas@icarusinterstellar.org

Nathan Morrison (✉)
COO Sustainable Now Technologies, Designer Persephone, 19859 Kittridge St., Winnetka, CA 91306
e-mail: Nathan@SustainableNowTechnologies.com

Esther M. Armstrong (✉)
Programme Director Theatre and Screen, Wimbledon College of Arts, University of the Arts,
Merton Hall Rd, London SW19, United Kingdom
e-mail: e.armstrong@wimbledon.arts.ac.uk

be possible. The ancient Greek gods were carried to the stars in a variety of transport systems that took the form of winged horses and chariots. Today, we celebrate our engineers and space walkers for their growing achievements. Yet, while we imagine and celebrate others inhabiting the unchartered terrains of space, we have not surrendered our desire to sit in the captain's chair ourselves, from where we may direct our expansion to Earth orbit and way, way beyond.

10.1.2 The Interstellar Question (and answers)

There is a point in time at which game-changing questions such as "Should we explore interstellar space?" transition from mere musing to scientific query. At the point at which we make a commitment to realizing this aim, a massive universe of inquisition opens up before us. Fruitful, like the gardens of Eden, the question forces us to inhabit those mythological chariots of our ancestors and attempt to inform them with new knowledge, design, and function. In an era of advanced science and technology, we are now able to offer implementable answers. And so, this is how we begin to build a starship.

The cornerstone avenues of thought revolve around propulsion, habitation, and motivation/objective.

Propulsion studies in the fields of nuclear fusion, fission, and solar sailing hint at embryonic approaches that will grow into tangible engineering options in the next century. Ideas have been seeded by the Bussard Ramjet (1960), Project Daedalus (1978), the memorable Robert Forward's Lightsail (1984) research, and, more recently, Project Icarus (2015). Subsequent studies as to how we may propel ourselves fast enough to reach a deep-space outpost or star system by various groups around the world are seeding the emergence of technologies for rapid terrestrial and interplanetary travel. With them, the industrialization of space and, thus, the eventuality of extending human culture off-world become tangible. High-density, compact, clean energy sources and robust high-speed communications will deliver abundant resources to everyone on the planet. And, with those utilities in place, health and education will be accessible by all.

Contemporary developments in urban design and renewable technologies are influencing the design and development of interstellar habitation solutions. The concept of a worldship, developed wonderfully by O'Neill in 1978, envisioned massive cylindrical vessels encompassing complete terrestrial biospheres capable of supporting very large populations of human migrants. Those concepts are now evolving, via Project Persephone and others, to conceive of living ships with renewable architectures. The paradigm of humans living in an inert ship is turned on its head, urging us to consider how humans should coexist within a living vessel, raising profound questions about identity, community, and ecology, where our lifestyles are entangled with many different relationships to produce a dynamic living world in which we are invested in a myriad of symbiotic relationships—such as those within the "bacterial biome," a micro-world of vital metabolic exchanges that take place within each of us, independently of our control. Testing and implementing these ideas and relationships on Earth give us a new way of thinking about the design, engineering as sustainability of our environment, and our role as protectors of our planet.

While the motivations for interstellar travel may be inferred directly from the above, we find other challenges in the unexplored territory in the areas of existential risk and the evolving fields of future cities. Here, the anthropological and archaeological fields, once coined as studies of the past, become imperative for understanding what our future may hold and how risks to our survival or unexpected crises might be avoided from encroaching. Even for the most ardent supporters of interstellar flight, the Interstellar Question offers an opportunity, through cultural storytelling, to reflect on history with renewed interest in anticipating the future and revisiting the learning that the past has to offer.

10.1.3 Explore, explore more

The interstellar era, in which interplanetary travel is commonplace and interstellar flight is realistically achievable, will not come quickly. Our history tells us that, when we explore, we explore *more* and seek out new ways to do so better, faster, safer. Our distance from the stars is cosmologically determined and it will take a lot of work to broach the expanse.

Yet, the role of humans—and of humanity—in making the bridge to the interstellar era will be critical to achieving our collective vision. Children will be taught that the Moon became our first real off-world colony that provides water, metals, minerals, and fuels in its regolith. In 2075, such wealth was sufficient for the first broad migration of humans to another celestial object. With the Moon colonized, the Martian migration was unstoppable and, despite the discovery of bacterial and other forms of microbial life, the numerous research outposts swelled and overflowed. In 2125, humanity consummated its love for the Red Planet with the first township, and called it their home.

Corporations dominated the asteroid belt, with Ceres being the notable exception, as a popular vacation spot. Even so, in 2150, it became the headquarters of the Starship Corporation, the brainchild of graduates from the prestigious Starflight Academy back on Earth. Starting as a courier and freight company, with contracts with asteroid mining corporations for the transportation of goods and services, they rapidly became the go-to subcontractor to space industries. Having cornered the market by establishing over 600 high-bandwidth transceivers, which made up the Planetary Positioning System—the only safe means of navigation and communications throughout the inner Solar System—they became indispensable to the broader interplanetary economy.

The corporation had not forgotten its origins and attempted twice to commission an interstellar vessel to Alpha Centauri, in 2200 and 2250. Analysts said the failure was partially due to of the risks and resources involved, but mostly because humanity had been living and working in space for over 200 years already. The robotic interstellar probes, which had been launched in 2050 and 2100, had not arrived at their destinations of Proxima and Alpha Centauri. Their technical capabilities were almost outmoded in comparison with contemporary developments in space systems engineering.

As their know-how with building large spacecraft and crew transports grew, deep-space exploration vehicles, which were tasked with mapping outer planetary dwarf planets, were sent to the Oort Cloud. By 2250, over 6000 such objects were discovered. By 2350, some deep-space science stations even boasted crewed outposts. Ozone- and methane-rich atmospheres were spectroscopically detected around Alpha Centauri's rich planetary system of over 40 planets and hundreds of moons. Speculations ran wild—with resonances similar to the news that water had been found on Mars—and interplanetary excitement for the possibility of interstellar travel intensified with each new probe and survey mission to the stars.

Amidst all the ruckus and calculated excitement of the prospect of a new era of space travel, a teacher decided to affirm the significance of these developments with her star-struck class:

> "Humanity has extended itself to 10,000 astronomical units from our Sun, and we are now at the very edge of our Solar System. Maybe one of you will be among those willing to take the next step and make new homes for us all among other stars."

10.2 Why and where we should go, boldly

Nathan Morrison, COO Sustainable Now Technologies

10.2.1 Why?

> "… the purpose of playing, whose end, both at the first and now, was and is, to hold as 'twere the mirror up to nature: to show virtue her feature, scorn her own image, and the very age and body of the time his form and pressure." (*Hamlet*, Act 3, Scene 2, 17–24)[1]

Art often holds a mirror both to Nature and to society itself, and this age-old sentiment rings particularly true in the recently released theatrical production *Interstellar* (2014) directed by Christopher Nolan. The film depicts a bleak future wherein the activities of man have caused the world to become an increasingly uninhabitable wasteland. In reality, while Earth has not begun to experience the full effects of global climate change portrayed within this film, Earth's resources are becoming increasingly depleted, its atmosphere increasingly toxic, and our species' population continues to rise. Crises such as peak oil, atmospheric carbon content, the global water shortage, and what scientists are now beginning to call the sixth great extinction are implications of a larger issue—one of overpopulation and resource mismanagement. The consequences of this issue can be seen across the globe; however, they are abundantly prevalent within my home state of California, where, in the midst of an unprecedented state of drought and in cities like Los Angeles with legendarily poor air quality, oil refineries race to increasingly higher production rates. Our priorities have gone askew, even as it has become increasingly clear that there will soon simply be too many people on Earth for our native world to continue to support our ever-growing needs. Our rate of population expansion and/or our rate of consumption may change moving forward. However, presently, both rates appear to be unsustainable over the course of the twenty-first century.

The state of our world's political society is such that we are divided among political factions we call nations and no individual nation, nor indeed the combined efforts of these nations, has yet risen to properly address these pivotal issues of our time. Nor have any of these entities realistically investigated feasible long-term alternative strategies that would ensure our species' survival, should one of these crises indeed curtail our ability to thrive on this world. If we are not to address these issues directly to mitigate and eventually

---

[1] Shakespeare, W. *Hamlet: Entire Play*. Available online at http://shakespeare.mit.edu/hamlet/full.html (accessed August 13, 2015).

reverse our present trends of consumption, then we must—as a direct consequence and as a species—begin to accept responsibility for the destruction of this world. We must actively pursue new worlds, and indeed new strategies, for resource acquisition, so as not to repeat our present misadventures. The issues at hand are intricate and not easily addressed, but they can be solved if we can work together in a concerted effort to change.

10.2.1.1 Oil

I own a small research and development firm south of Los Angeles near Long Beach. En route to the office, my view of the skyline is regularly obscured by smog. Not only is downtown Los Angeles hidden behind a curtain of pollution, but at times it is difficult to see the coastline, merely a mile off of the highway. The smog originates along the 405 highway. It is generated by no fewer than seven oil refineries, and travels north and east into the mountain-ranged bowl surrounding Los Angeles. This noxious air traps the city in a nearly perpetual haze. How oil is extracted, refined, and used in modern society has become a problem that is no longer possible to ignore. Back in 1956, M. King Hubbert proposed his "peak oil" theory, predicting that the point of maximum oil production would occur in the 1970s.[2] As of around the year 2005, after much debate, a scientific consensus began to form regarding the potential supply of hydrocarbon deposits on Earth. That fewer new oil deposits are being discovered and oil refinery production continues to increase has been widely recognized. Studies have been conducted on how to transition from crude oil into other hydrocarbon sources.[3] Charts illustrating the discovery and production rates of oil were produced, clearly pointing to a peak in the 1960s–1970s.[4] This realization, that oil extraction had reached "peak" and that present levels of extraction, refinement, and production could not be maintained into the foreseeable future challenged oil industry analyst assessments that oil production could continue uninterrupted for decades to come. In the end, the argument between oil industry analysts and geologists came down to the issue of "crude" oil. Essentially, the amount of hydrocarbons produced during the Cretaceous period prior to the K-T (Cretaceous-Tertiary) mass extinction event, approximately 65 million years ago, has since congealed into what the modern world recognizes as oil deposits.

This stock is indeed finite and, in the present landscape of energy production, we have moved through surface deposits, moved through easily accessible subterranean deposits, and are presently engaged in extracting oil from offshore deposits, to the extreme detriment of our world's natural marine resources.[5] In May of this year (2015), the latest in a

---

[2] King Hubbert, M. *Nuclear Energy and the Fossil Fuels*. Available online at www.hubbertpeak.com/hubbert/1956/1956.pdf (accessed August 13, 2015).

[3] Hirsch, R. L.; Bezdek, R.; Wendling, R. 2005. *Peaking of World Oil Production: Impacts, Mitigation, & Risk Management* (2005), available online at http://large.stanford.edu/courses/2014/ph240/liegl1/docs/hirschreport.pdf (accessed August 13, 2015).

[4] Campbell, C. *Association for the Study of Peak Oil USA*. Available online at http://peak-oil.org/peak-oil-reference/peak-oil-data/oil-discovery/ (accessed August 13, 2015).

[5] Macadangdang, R. *California Oil Pipeline Spills 21,000 Gallons of Oil Into Pacific Ocean: LIFE: Tech Times*. Available online at www.techtimes.com/articles/54187/20150520/california-oil-pipeline-spills-21-000-gallons-of-oil-into-pacific-ocean.htm (accessed August 13, 2015).

long line of oil leaks devastated the Santa Barbara coastline, releasing an estimated 21,000 gallons of oil near Refugio state beach. This recent spill is a perfect example of just how little our society has learned when it comes to handling and processing this valuable but dangerous resource. Forty-six years earlier, in 1969, an estimated 100,000 barrels of oil were spilled along this same coastline, blackening sands and smothering wildlife from Goleta all the way to Ventura. It is appalling that, from an environmental perspective, we continue to condone and financially subsidize the mining of crude oil by corporations. In the recent Canadian tar sands extraction and refinement scheme, we have also begun to extract and process what can only be properly referred to as "impure crude." Canada, its First Peoples, and the US executive government have recognized the inherent dangers of processing and transporting this "tar sands oil" and have taken action to temporarily block the proposed Keystone XML pipeline envisioned to deliver tar sands oil from Canadian deposits to refineries on the southern coast of the US. This temporary measure is presently being fought by industry. Due to the current political and economic climate in North America, the outcome of this struggle is difficult to foresee.

The pursuit, both of "impure crude" and of offshore oil deposits, is not only dangerous, but is entirely unsustainable as an energy acquisition strategy over the next century. We will most certainly exhaust this hydrocarbon resource sooner than later, and we must immediately develop alternative strategies for sustainable hydrocarbon production. Dedicated financial support to the development of these strategies is urgently needed. Our world holds only so much oil. We have reached a point in our societal evolution at which we must accept our clear, if alarming, recognition of this fact and begin to shift our strategies for energy production and acquisition. One such strategy—a task that lies within our present technological grasp—is to seek out worlds where hydrocarbon deposits similar to our own also exist. Within these terrains, we must develop responsible extraction protocols that will allow us to utilize these resources without releasing dangerous levels of carbon into the atmosphere of that world.

10.2.1.2 Carbon dioxide

Over the last 150 years, roughly the span of the industrial age, Earth's atmospheric carbon dioxide (CO_2) content has risen dramatically due, primarily, to the emissions associated from burning "carbon-heavy" fossil fuels such as coal and petroleum. As of November 2007, the CO_2 concentration in the atmosphere was measured at 0.0384% by volume, or 384 parts per million by volume (ppmv).[6] When compared with CO_2 levels acquired from ice-core samples, this measure presents at roughly 100 ppmv higher than levels found in 1832. These numbers indicate a 35% increase in CO_2 concentration over a period of approximately 150 years. The evidence clearly suggests that humans, through the mechanism of industrial production, have been responsible for a steady record of carbon emissions throughout industrial history. The refineries mentioned earlier directly contribute to the air quality in the Los Angeles area, and Southern California is not alone in this regard. In Shenzhen, China, for example, air quality and

---

[6] Energy Information Administration – US Department of Energy. *Emissions of Greenhouse Gases in the United States 2007.* Available online at www.eia.doe.gov/oiaf/1605/ggrpt/pdf/0573%282007%29.pdf (accessed August 13, 2015).

pollution measurements have been in the 151–200 range steadily over the last several years. This rating is described in grave terms where, according to China's Air Quality Index (AQI), which monitors real-time variables, everyone may begin to experience health effects, while members of sensitive groups may experience more serious problems.

Climate scientists have long been aware of the "greenhouse effect." From the early formative guesses by Joseph Fourier that gases in the atmosphere can increase the surface temperature of Earth (1824),[7] to the industrious works of John Tyndall, who identified CO_2 as a gas that can trap heat rays (1859),[8] turn-of-the-century scientists were cultivating an answer to the cause–effect relationship between carbon emissions and weather temperature. Later in the nineteenth century, public discourse was still undecided as to whether or not industrial sources were contributing to the increase in atmospheric carbon. Arvid Hogbom postulated that "present-day" factories had reached equilibrium with the planet's ability to naturally absorb the increase in CO_2 (1896).[9] Svante Arrhenius was able to calm the worry about eminent "climate change" (1908)[10] because, at the current CO_2 emission rate, the impact of "global warming," he concluded, would not appear within the century. While this "time-sensitive" argument calmed the public at large at the dawn of the twentieth century, it did not refute the preponderance of evidence indicating that global climate change caused by industrial emissions was in fact underway. Here we stand, over a century later, observing these effects first-hand.

While working at the US Naval Research Laboratory, an American physicist, E.O. Hulburt (1931),[11] published a review using then current instrumentation to support Arrhenius' theory. Taking detailed measurements periodically over time and testing these theories against this data, the forward-thinking experiments of Stewart Callendar (1938),[12] Lewis Kaplan (1953),[13] and Gilbert N. Plass (1956)[14] all supported the conclusion that atmospheric carbon content rises due to industrial emissions. In 1955, prior to Plass's determinations, chemist

---

[7] Weart, S.; American Institute of Physics. *The Carbon Dioxide Greenhouse Effect*. Available online at www.aip.org/history/climate/co2.htm (accessed August 13, 2015).

[8] New Zealand Climate Coalition. *The Triumph Of Doublespeak: How UNIPCC Fools Most of the People All of the Time*. Available online at http://nzclimatescience.net/index.php?option=com_content&task=view&id=483&Itemid=1 (accessed August 15, 2015).

[9] Bigas, H.; Gudbrandsson, G. I.; Montanarella, L.; Arnalds, A. *Soils, Society & Global Change*. Proceedings of the International Forum Celebrating the Centenary of Conservation and Restoration of Soil and Vegetation in Iceland, available online at http://eusoils.jrc.ec.europa.eu/esdb_archive/eusoils_docs/other/EUR23784.pdf (accessed August 15, 2015).

[10] Krosnick, J. A.; Holbrook, A. L.; Visser, P. S. *The Impact of the Fall 1997 Debate about Global Warming on American Public Opinion*. Available online at www.arb.ca.gov/research/seminars/krosnick/krosnick%283%29.pdf (accessed August 15, 2015).

[11] Hulburt, E. O. *The Temperature of the Lower Atmosphere of the Earth*. Available online at *http://journals.aps.org/pr/abstract/*10.1103/PhysRev.38.1876 (accessed August 15, 2015).

[12] Knight, M. *A Timeline of Climate Change Science – CNN.com*. Available online at www.cnn.com/2008/TECH/science/03/31/Intro.timeline/index.html (accessed August 15, 2015).

[13] Weart, S.; American Institute of Physics. *Basic Radiation Calculations*. Available online at www.aip.org/history/climate/Radmath.htm (accessed August 15, 2015).

[14] Sample, I. *The Father of Climate Change*. Available online at www.guardian.co.uk/environment/2005/jun/30/climatechange.climatechangeenvironment2 (accessed August 15, 2015).

Hans Suess reported detection of carbon-14 in the atmosphere, enabling the dating of atmospheric carbon from ancient coal.[15] Charles David Keeling's chronological CO_2 measurements of the atmosphere in the middle of the twentieth century became the widely accepted data set confirming Earth's growing atmospheric problem (1960).[16] Present-day climate research scientists have been able to track this escalating danger and have been able to accurately predict the progression of atmospheric carbon content against escalating emissions rates over time.[17] Much like oil, Earth's atmosphere is a necessary resource for our species' survival on this world. It is a resource that is rapidly being depleted. While the stark smog-lined sunsets of Southern California can be strikingly stunning in their own strange way, this skyscape also depicts danger—an ever-growing menace that eventually must be faced. It is a beautiful harbinger of dystopia, and one that cannot be ignored.

10.2.1.3 Water

While roughly two-thirds of Earth's surface consists of water, only approximately 3% of that water is freshwater, and more than half of that freshwater is either icebound or located deep underground outside of our species' reach.[18] Water is an irreplaceable resource that is fundamental for the survival of all life on Earth. Additionally, water is a scarce commodity elsewhere within our Solar System. Water is essential, not only for direct consumption, but also as an agricultural tool to produce crops. Additionally, freshwater is used to produce goods. With over seven billion people presently alive today,[19] the demand for clean freshwater is ever increasing, and our world's supply of freshwater has remained constant. This presents an inevitable issue that our species must seek to address. By the year 2025, it is estimated that Earth's population will reach over eight billion people.[20] It is estimated by the United Nations that approximately two billion of those people will inhabit regions of water scarcity.[21]

---

[15] Damon, P. E.; Lerman, J. C.; Long, A. *Temporal Fluctuations of Atmosphere C14: Causal Factors and Implications.* Available online at http://adsabs.harvard.edu/full/1978AREPS...6..457D (accessed August 15, 2015).

[16] Wilson, C. L.; Matthews, W. H. *Inadvertent Climate Modification: Report on Conference, Study of Man's Impact On Climate (SMIC), Stockholm.* Available online at www.aip.org/history/climate/xKeeling71-2.htm (accessed August 15, 2015).

[17] National Oceanic and Atmospheric Administration. *Teaching Activity: The Carbon Dioxide–Oxygen Cycle.* Available online at www.esrl.noaa.gov/gmd/education/lesson_plans/The%20Carbon%20Dioxide-%20Oxygen%20Cycle.pdf (accessed August 15, 2015).

[18] World Wildlife Fund. *Threats: Water-Scarcity.* Available online at www.worldwildlife.org/threats/water-scarcity (accessed August 15, 2015).

[19] United States Census Bureau. *U.S. and World Population Clock.* Available online at www.census.gov/popclock/ (accessed August 15, 2015).

[20] Olsen, A.; Associated Press. *U.N.: World Population to Reach 8.1B in 2015.* Available online at www.usatoday.com/story/news/world/2013/06/13/un-world-population-81-billion-2025/2420989/ (accessed August 15, 2015).

[21] National Geographic. *A Clean Water Crisis.* Available online at http://environment.nationalgeographic.com/environment/freshwater/freshwater-crisis/ (accessed August 15, 2015).

I recently drove the entire length of the state of California, from the southern border with Mexico up to the northern border with Oregon. Having lived in California for almost a decade, I am very familiar with the river systems, lakes, and reservoirs that support the state's population and agricultural regions. On this journey, I was alarmed to see first-hand the water levels in these aquifers. As of late July 2015, five lakes in California were classified as either "dried up" or closed. Thirty-nine additional lakes and reservoirs report in at less than 50% capacity; of those bodies of water, 17 are at less than 25% capacity. Forty-nine additional lakes and reservoirs report their status as "low," with nine of those bodies of water reporting in as "really low."[22] California (not the US, but California alone) makes up the world's seventh largest economy.[23] The water crisis in California illustrates that this issue is not relegated to non-industrialized nations. This crisis is a global phenomenon and can be seen in the most technologically advanced regions of the planet. It was horribly depressing to witness these bodies of water reach such shockingly low levels. Once thriving ecosystems, many of these areas have now lost the ability to provide a natural habitat for native wildlife. Once they are gone, these ecosystems will never come back.

For our species to survive into perpetuity, and for our world to recover from global-scale resource depletion, we must look outward, to the stars and to extrasolar worlds to provide for the resources of future generations. Additionally, we must decrease the burden imposed by humanity on our native world to allow Earth the opportunity to continue to serve as a womb for species great and small. Indeed, we must consider the state of the world as it will be presented not only to our children, but to their children, and their children's children. In the words of wisdom of the Onondaga nation, Faithkeeper Oren Lyons of the Seneca tribe:

> "The Peacemaker taught us about the Seven Generations. He said, when you sit in council for the welfare of the people, you must not think of yourself or of your family, not even of your generation. He said, make your decisions on behalf of the seven generations coming, so that they may enjoy what you have today."[24]

10.2.2 Where?

To accomplish these goals, and to mitigate these various crises, indeed to end them here in our present century, we must begin to develop the technologies that will allow our species to explore other worlds. Yet we are without a single viable spacecraft in operation capable of delivering us to another planet within our own system, let alone an extrasolar world. The state of our space vehicle programs on Earth at present can only be described as "lacking." There are fewer government-funded spacecraft in operation today than there were in

---

[22] Green, J. L. *Paddling California*. Available online at www.paddlingcalifornia.com/california_lake_and_reservoir_water_levels.html (accessed August 15, 2015).

[23] CBS News/Associated Press. *California Bounces Back as World's 7th Largest Economy, Larger than Brazil*. Available online at http://sanfrancisco.cbslocal.com/2015/06/10/california-world-7th-largest-economy-larger-than-brazil/ (accessed August 15, 2015).

[24] Public Broadcasting Service. *Seven Generations: The Role of Chief*. Available online at www.pbs.org/warrior/content/timeline/opendoor/roleOfChief.html (accessed August 15, 2015).

the 1980s. Those in service at present are not necessarily more advanced than their predecessors from the "space age" of the twentieth century. Government space programs face negligible and actively dwindling budgets, and the US Space Shuttle program, as of 2011, has officially come to a close.[25]

10.2.2.1 A slow road to discovery

Robotic spacecraft have carried out many exploratory missions throughout our star system over the last four decades, while manned flight beyond low Earth orbit (LEO) has been kept at bay and on "standby" while these probes gather actionable intelligence to plan for future missions. The Jet Propulsion Laboratory (JPL) has effectively replaced Mission Control in Houston,[26] and Vandenberg Air Force Base has effectively replaced Cape Canaveral as our primary launch platform for system-wide missions of discovery and exploration. NASA has either ceased its pursuit of "a few good men" entirely, or has at bare minimum redefined their role to that of supporting cast. We have sent probes to Mercury, and then to Venus. We have sent unmanned reconnaissance missions to study our Sun. But no man has yet navigated to either of our intersolar worlds—a task not outside the realm of possibility using modern technology. We have sent more than 30 probes outward to study Mars alone.[27] We have dispatched orbiters to image Saturn, Jupiter, Neptune, and Pluto, and we have even landed an unmanned craft on a comet.[28] In the 1970s, we launched two probes which have now traveled beyond the edge of the Solar System and into deep space, but yet, humans have not personally ventured farther away from Earth than our own world's satellite.

Yes, certainly from our collection of data, the atmospheric conditions of these locales seem extreme but, in the words of former US President John Fitzgerald Kennedy:

> "We choose to go to the moon. We choose to go to the moon in this decade and do the other things, not because they are easy, but because they are hard, because that goal will serve to organize and measure the best of our energies and skills, because that challenge is one that we are willing to accept, one we are unwilling to postpone, and one which we intend to win, and the others, too."[29]

---

[25] National Aeronautics and Space Administration. *Space Shuttle – STS 135*. Available online at www.nasa.gov/mission_pages/shuttle/shuttlemissions/sts135/main/index.html (accessed August 15, 2015).

[26] Jet Propulsion Laboratory, California Institute of Technology. *Mission Profiles*. Available online at www.jpl.nasa.gov/missions/ (accessed August 15, 2015).

[27] The Planetary Society. *Mars Missions*. Available online at www.planetary.org/explore/space-topics/space-missions/missions-to-mars.html?referrer=https://www.google.com/ (accessed August 15, 2015).

[28] Horowitz, A. *Rosetta Spacecraft Successfully Lands on Comet*. Available online at www.huffingtonpost.com/2014/11/12/rosetta-spacecraft-landing-comet_n_6146012.html (accessed August 15, 2015).

[29] Kennedy, J. F. *Presentation at Rice University, 1962*. Available online at http://er.jsc.nasa.gov/seh/ricetalk.htm (accessed August 15, 2015).

In the 40-plus years since NASA put a man on the Moon, we have decided not only to cease manned exploration of our Solar System, but also to cancel our ability to deliver payloads to LEO. And what of the "space race"? As of the final flight of the Space Shuttle *Atlantis*, the US has engaged in a policy of renting transportation to the International Space Station (ISS) via Soyuz craft. The US Space Shuttle program was scrapped without a functional replacement vehicle in place. Do nations still hold dear the sentiment that being the first to achieve a higher status in the arena of manned spaceflight is to be coveted? What has become of the notion of exploration, and of the idea of being the first to venture into and chart a brave new geography?

This issue is not relegated to the US; following the collapse of the USSR in 1993, the Russian Buran reusable shuttle program ceased operations after only launching a single unmanned test flight (in 1988). The Russian version of NASA's Space Shuttle fleet was left to decay in the MKZ building near the Baikonur cosmodrome.[30] The state of these vehicles, in my view, characterizes perfectly the value that manned spacecraft holds among governmental institutions. They are viewed as a relic of the past, instead of a pathway to our future. Manned spaceflight has essentially been relegated to LEO, and specifically to the ISS, approximately 250 miles above Earth's surface.[31] The final flight of the Apollo program landed in 1972, a full 43 years before this essay was penned.[32] Since that time, no manned spacecraft has ventured even as far as the Moon, a mere 252,000 miles out at its apogee.[33]

Despite the discontinuation of manned spaceflight programs with destinations outside of LEO for the past 40-plus years, there are myriad benefits to dispatching unmanned space research platforms that cannot be discounted. In the case of the inner Solar System, our orbiters and landers sent to the planet Mars have provided invaluable data that can be utilized to properly prepare for a manned lander mission.[34] In the case of the outer Solar System, unmanned probes and space telescopes have been effective in providing new targets for potential manned lander missions—for example, Jupiter's moon, Ganymede, where the existence of a subsurface ocean has been confirmed.[35] In advance of any manned mission, telerobotic probes, orbiters, and landers will be required to acquire atmospheric

---

[30] O'Callaghan, J. *Russia's Forgotten Space Agency: Haunting Images Reveal Two Abandoned Soviet Shuttles Rotting in Giant Derelict Hanger*. Available online at www.dailymail.co.uk/sciencetech/article-3119861/Russia-s-forgotten-space-agency-Haunting-images-reveal-two-abandoned-Soviet-shuttles-rotting-giant-derelict-hanger.html (accessed August 15, 2015).

[31] National Aeronautics and Space Administration. *Current Position of the ISS*. Available online at http://iss.astroviewer.net/ (accessed August 15, 2015).

[32] National Aeronautics and Space Administration. *Missions – Apollo 17*. Available online at www.nasa.gov/mission_pages/apollo/missions/apollo17.html (accessed August 15, 2015).

[33] Sharp, T. *How Far Is the Moon?* Available online at www.space.com/18145-how-far-is-the-moon.html (accessed August 15, 2015).

[34] Jet Propulsion Laboratory, California Institute of Technology. *Opportunity – All 205,241 Raw Images*. Available online at http://mars.nasa.gov/mer/gallery/all/opportunity.html (accessed August 15, 2015).

[35] Lakdawalla, E. *An Internal Ocean on Ganymede: Hooray for Consistency with Previous Results!* Available online at www.planetary.org/blogs/emily-lakdawalla/2015/03121716-ganymede-ocean.html?referrer=https://www.google.com/ (accessed August 15, 2015).

content, surface composition, and topography data, orbital flight plans and insertion points, and to scout for potential manned landing targets. Any conceived manned mission to another star system should incorporate plans to deploy such craft on entering a new system for data-acquisition purposes.

10.2.2.2 Kepler

Arguably, the single most significant piece of telerobotic equipment in service today is the Kepler space telescope, presently gathering data on extrasolar planets and moons. In March of 2009, Kepler was launched on a Delta II rocket out of Cape Canaveral Air Force Station.[36] Kepler was designed to image the region surrounding a single group of stars in the constellations Cygnus and Lyra for the 3-plus years of its proposed mission, and was placed into an Earth-trailing heliocentric orbit to maximize the viewing window.[37] The mission duration was intended to allow a planet within the viewing field to be imaged during four transits of its star, should the planet's transit time be comparable to the 365-day window in which Earth orbits the Sun; however, the craft was constructed to complete a 6-year mission, if necessary.

Exciting news broke in December of 2011, when NASA unveiled the discovery of several extrasolar worlds. Two of these planets—Kepler 20e and Kepler 20f—are approximately Earth-sized worlds; however, they orbit their star outside of the habitable zone (also commonly referred to as the "Goldilocks zone").[38] Also in December of 2011, NASA confirmed that the Kepler space telescope had discovered the first known extrasolar planet orbiting within the habitable zone of its parent star.[39] The planet was designated Kepler 22b, and has been found to orbit a yellow star quite similar to our Sun, at a distance between that of Earth and Venus. Kepler 22b is roughly two and a half times the size of Earth and, at the time of its discovery, was the smallest extrasolar world ever detected. The world is located approximately 600 light years from Earth, and orbits its sun every 290 days. It has an average surface temperature of approximately 72 °F (22 °C).[40] This planet, at present, represents a potential target destination for a manned generational starship mission.

---

[36] National Aeronautics and Space Administration. *Liftoff of the Kepler Spacecraft*. Available online at www.nasa.gov/mission_pages/kepler/launch/index.html (accessed August 15, 2015).

[37] National Aeronautics and Space Administration. *Kepler and K2 – Spacecraft and Instrumentation*. Available online at www.nasa.gov/mission_pages/kepler/spacecraft/index.html (accessed August 15, 2015).

[38] National Aeronautics and Space Administration. *Kepler 20 System: 5 Planets Including Two That Are Earth-Size*. Available online at http://kepler.nasa.gov/news/nasakeplernews/index.cfm?FuseAction=ShowNews&NewsID=172 (accessed August 15, 2015).

[39] National Aeronautics and Space Administration. *NASA's Kepler Mission Confirms Its First Planet in Habitable Zone of Sun-Like Star*. Available online at www.nasa.gov/mission_pages/kepler/news/kepscicon-briefing.html#.Vc0Uj_lVhBc (accessed August 15, 2015).

[40] Bloxham, A. *Kepler 22b – the "New Earth" Could Have Oceans and Continents, Scientists Claim*. Available online at www.telegraph.co.uk/news/science/space/8939138/Kepler-22b-the-new-Earth-could-have-oceans-and-continents-scientists-claim.html (accessed August 15, 2015).

Kepler's discoveries have continued to mount. In April of 2013, the space telescope recognized an extrasolar world often referred to as a "super-Venus." This world, named Kepler 69c, is larger than Earth and orbits a yellow sun-like star at a distance from Earth of roughly 2700 light years.[41] At about the same time, Kepler also recognized a world called Kepler 62f, often referred to as a "super-Earth." Kepler 62f, unlike Kepler 69c, 22b, 20e, and 20f, does not orbit a yellow sun-like star. It instead orbits a smaller, cooler red sun and is approximately 1200 light years from Earth. This type of star is called a red dwarf star (also commonly referred to as an M-dwarf star). Red dwarf stars are the most numerous stars recorded to date in the observable universe, and total roughly 70 % of the stars in our Milky Way galaxy. This dramatically increases the chances of discovering a habitable world orbiting a red dwarf star, in contrast to other star types.[42]

Later in 2013, the Kepler spacecraft experienced the failure of a second "reaction wheel," causing the telescope to lose its ability to stay focused on a single target area.[43] The telescope orbits the Sun at a similar distance as Earth; however, the craft itself at the time of the reaction wheel failure was approximately 40 million miles from Earth (160 times as far away as the Moon), making a repair mission beyond the scope of NASA spacecraft.[44] It bears mentioning that, at the time, NASA in fact had no operational spacecraft available to send for repairs, having become reliant on the Soyuz spacecraft for transport to the ISS and no replacement for the Space Shuttle platform in active service. The following year, in early 2014, Doug Wiemer of Ball Aerospace (the corporation which served as the prime contractor for the Kepler telescope) presented NASA with the concept of using the telescope's solar panels to detect solar radiation, and to balance the craft on its roll axis using these readings. This allowed the remaining two operational reaction wheels to control pitch and yaw.[45] This plan succeeded and, after bringing the telescope back online, the designation "K2" was adopted to categorize the mission's findings in its second stage of life.

In April of the year 2014, the Kepler telescope discovered an Earth-sized planet in the "habitable zone" of another star.[46] This world was designated Kepler 186f. As of January

---

[41] Pyle, T. *Kepler 69c: Super-Venus*. Available online at www.nasa.gov/mission_pages/kepler/multimedia/images/kepler-69c.html (accessed August 15, 2015).

[42] Culler, J. *NASA's Kepler Discovers First Earth-Size Planet in the "Habitable Zone" of Another Star*. Available online at www.nasa.gov/ames/kepler/nasas-kepler-discovers-first-earth-size-planet-in-the-habitable-zone-of-another-star (accessed August 15, 2015).

[43] Overbye, D. *Breakdown Imperils NASA's Hunt for Other Earths*. Available online at www.nytimes.com/2013/05/16/science/space/equipment-failure-may-cut-kepler-mission-short.html?_r=1 (accessed August 15, 2015).

[44] Kingson, J. A. *Kepler Telescope's Troubles, a Maya Pyramid in Ruins, and More*. Available online at www.nytimes.com/2013/05/21/science/kepler-telescopes-troubles-a-maya-pyramid-in-ruins-and-more.html?ref=topics&_r=1 (accessed August 15, 2015).

[45] Lemonick, M. D. *This Is How NASA Fixed a Telescope Currently in Orbit*. Available online at http://time.com/3643896/kepler-planets-repair/ (accessed August 15, 2015).

[46] Culler, J. *NASA's Kepler Discovers First Earth-Size Planet in the "Habitable Zone" of Another Star*. Available online at www.nasa.gov/ames/kepler/nasas-kepler-discovers-first-earth-size-planet-in-the-habitable-zone-of-another-star (accessed August 15, 2015).

of 2015, the Kepler telescope had discovered over 4000 presumed planets, and no fewer than eight worlds that are deemed "potentially habitable."[47] According to Ball Aerospace, the craft has detected three Earth-sized planets within the habitable zones of their parent stars.[48] This aforementioned world, Kepler 186f, serves as a viable potential target destination for the interstellar journey envisioned within this work.

Kepler 186f is roughly 10% larger than Earth; however, it is of unverified material composition. The world orbits its star every 130 days, and is the fifth planet from its star, the farthest planet out that has been verified in that system. The world is located approximately 500 light years from Earth in the constellation Cygnus.[49] Kepler 186f's parent star is also a red dwarf that is about half of the size of our Sun and, at the planet's distance of 0.36 AU from the star, it receives roughly one-third the energy that Earth receives from Sol.[50] Geoff Marcy, an astronomer from the University of California at Berkeley described it thus: "This planet basks in an orange-red glow from that star, much [like what] we enjoy at sunset."[51]

In July of 2015, Kepler made its most important discovery to date, when NASA announced verification of Kepler 452b. This is the smallest world discovered thus far orbiting within the habitable zone of a yellow G2 class star. While Kepler 452b is still approximately 1.6 times the size of Earth, it is the most similar planet to our own home world we have yet glimpsed in the depths of space. Kepler 452, the star around which this world orbits, is just 10% larger than the Sun, and is estimated to be approximately one and a half billion years older than our star.[52] The world itself is also older than Earth, and has spent more than six billion years orbiting within its star's habitable zone. Using Earth as an example of the timescale necessary for the formation of complex life, Kepler 452b stands as a strong potential candidate for existing life forms. This world is located approximately 1400 light years from Earth.

---

[47] *The New York Times*. Chronology of Coverage: News About Kepler Space Telescope, Including Commentary And Archival Articles Published In The New York Times. Available online at http://topics.nytimes.com/top/reference/timestopics/subjects/k/kepler_space_telescope/index.html (accessed August 15, 2015).

[48] Ball Aerospace and Technologies Group. *K2/Kepler*. Available online at www.ballaerospace.com/page.jsp?page=72 (accessed August 15, 2015).

[49] Culler, J. *Kepler 186f, the First Earth-Size Planet in the Habitable Zone*. Available online at www.nasa.gov/ames/kepler/kepler-186f-the-first-earth-size-planet-in-the-habitable-zone (accessed August 15, 2015).

[50] Quintana, E. *Kepler 186f – the First Earth-Sized Planet Orbiting in Habitable Zone of Another Star*. Available online at www.seti.org/seti-institute/kepler-186f-first-earth-sized-planet-orbiting-in-habitable-zone-of-another-star (accessed August 15, 2015).

[51] Kramer, M. *5 Things to Know about Alien Planet Kepler 186f, "Earth's Cousin"*. Available online at www.space.com/25541-alien-planet-kepler-186f-facts.html (accessed August 15, 2015).

[52] Johnson, M. *NASA's Kepler Mission Discovers Bigger, Older Cousin to Earth*. Available online at https://www.nasa.gov/press-release/nasa-kepler-mission-discovers-bigger-older-cousin-to-earth (accessed August 15, 2015).

10.2.3 Our Future Among the Stars

These five worlds—Kepler 22b, Kepler 62f, Kepler 69c, Kepler 186f, and Kepler 452b—all represent viable candidates for habitability from the data sets we have thus far acquired. All of these worlds are located in or near the constellation Cygnus in the night sky, and each could potentially represent a destination in its own right for a manned generational starship mission. Combined, these worlds represent the possibility for a generational starship to begin a colonization mission that does not necessarily hinge on the actual habitability of any one individual world. Given these conditions, and given our present technological ability to construct such a generational starship (or worldship) where the speed of travel does not necessarily need to achieve or even approach the speed of light to eventually reach these multiple destinations, a flight plan for such a vessel would visit these worlds according to their prospective distance from Earth in this order: Kepler 186f (400 light years out), Kepler 22b (600 light years out), Kepler 62f (1200 light years out), Kepler 452b (1400 light years out), and Kepler 69c (2700 light years out). In a spacecraft moving at one-quarter of the speed of light (or 46,500 miles per second), a destination like Kepler 186f could be reached in 1200 years.

The Kepler space telescope and the Spitzer space telescope, as well as ground-based telescopes like those found at the McDonald Observatory in Austin, Texas, the Whipple Observatory on Mt. Hopkins, Arizona, and the Keck Observatory on Mauna Kea in Hawaii now make up a network of discovery and verification tools that will undoubtedly continue to discover new potential target worlds for exploration in the years to come. In addition to the benefit of this hardware network, new software has been recently developed that allows for the detection of new planets to be performed automatically. This automated software rapidly accelerates the verification process of new worlds.[53] This author anticipates that the development of vehicles capable of bridging these distances will follow similar advancement trends over the course of this and the next century. The Kepler telescope has imaged merely one small swathe of sky and, even within this narrow window, we have discovered multiple worlds at various distances with the potential for habitability. The probability that habitable worlds exist in equally vast proportions throughout the known universe is increasing every day. It is time to begin to plan to go, boldly, out into the black to explore this vast new final frontier. We have myriad reasons as to why, and we now have some very strong indications as to where. It is time to start asking the next series of questions, namely "how?" and "when?".

10.3 Scenographies and space: reading the gendered tyranny of the "Blue Ball"

Esther M. Armstrong, Program Director for Theatre and Screen at Wimbledon College of Arts, University of the Arts, London

See Fig. 10.1.

> "From a quarter of a million miles away
> our fragile earth is the only coloured thing,

---

[53] Wall, M. *Exoplanet Finds Keep Rolling in from Kepler Spacecraft Despite Glitch.* Available online at www.space.com/30053-kepler-exoplanet-discoveries-earth-twin.html (accessed August 15, 2015).

Scenographies and space: reading the gendered tyranny of the "Blue Ball" 225

Figure 10.1 From Paul Godfrey's play, *The Blue Ball*, which opened at the National Theatre in 1995, that asked why mankind's forays into space had been "rendered mundane in such a short time?" Credit: Mark Douet: The Blue Ball by Peter Godfrey/ArenaPAL

the moon is grey or biscuit brown
and space is black." (Godfrey 1995, p. 56)

"… space itself was internalised, its dark brooding depths becoming little more than a poetic analogue for the uncharted continents of the human mind." (Benjamin 2003, p. 54)

10.3.1 Earth, we have a problem

The presentation of Earth in the cosmos as portrayed in Hollywood films has become a visual impasse that is potentially stifling support for scientific enquiry. In current films about space exploration, positive, "science fiction thinking" (Landon 2002, p. xiii), which encourages better thinking for the good of humanity and highlights the positive aspects of exploration in the realm of science fiction, is not common.[54] Most recently, successful Hollywood films involving space exploration have articulated technological dystopias that suggest, both directly and indirectly, that exploring the universe is undesirable and futile.

---

[54] Discussing the literary genre of science fiction, Landon (2002, pp. 6–7) shows how "science fiction thinking" allows serious consideration within the genre to postulate how things might be. He also suggests that the perspectives that science fiction provides bridge a gap between the given and the marvelous, and have become central to modern consciousness in the twentieth century.

Though, as individuals, we may be skeptical of the fantasy embedded within the narratives of Hollywood productions, their influence on wider attitudes is nevertheless significant and may impact on a wider cultural support for space travel and exploration. In the West, in 2015, we are conscious of media descriptions of manned-space-like ambitions such as suborbital flights that invite exclusive opportunities for Hollywood royalty and other business celebrity aristocrats to experience space travel first hand.[55] In contrast to this, reports of "hard science" portray the scientific community as focused toward robot-conducted, scientific bumbling, and often highlight the failure of attempts to understand more about the fabric of our universe. Such examples of ridicule can be seen recently in the reports about the poorly communicating space lander Philae on a comet (2015) and the disappearance of the Beagle 2 lander on the surface of Mars.[56]

In this mediatized climate, cinematic thinking about space has become distinctly Earth-bound and supports a kind of gendered essentialism in its wake, where the image of Earth is used to determine the reading of the actions of characters according to their biology. The two quotations that preface this discussion from Paul Godfrey (1995) and Marina Benjamin (2003) both represent aspects of the bind that occurs when depicting the image of Earth in Hollywood films in a visually realist sense. Currently, this image of our planet supports underlying themes which produce fixed readings on viewing. The first recurrent theme that emerges within the scenographic framing of Earth in film is illustrated within the quote from Paul Godfrey's drama, *The Blue Ball* (1995), which highlights the distancing and alienation that occurs on viewing Earth as a notable image of color in the dark vastness of space. This recognition of distance and scale creates a position of anxiety which is described in Benjamin's (2003) quotation from *Rocket Dreams*. Recent film contemplation of Earth by male space explorers creates a predominately male image of existential solipsism which is experienced as a kind of exiled subjectivity while in space. Dwelling on the image of Earth consequently encourages the experience of a spiritual interior space of self-consciousness for the astronaut witnessing its presence. In contrast, the role of the female astronaut is bound and grounded on Earth on viewing it, and committed to the act of returning to it without the male experience of spirituality. Marie Lathers (2010) notes that historical outer-space film narratives that deal with "space realism" have often emulated the traditional social norms of gender roles and restricted women to the material realm (pp. 2–13). In recent Hollywood productions, women are certainly subject to this grounding that Lathers suggests binds them to Earth, but they are additionally fueled by a Cassandra-like anxiety, echoing impending doom within space, which suggests that leaving the planet is both a *more unnatural* act and *less spiritual* experience for them.[57]

---

[55] See e.g. Richard Branson's Virgin Galactic (Spargo 2014).

[56] Philae still has intermittent contact and journalism tried to mask the disappointment of the mission, but its communication issues still remain as a disappointment. See e.g. Kitty Knowles (2014): "On the face of it, it looks like an expensive mishap to rank alongside disappointing Mars landers and malfunctioning telescopes." In addition, the lander Beagle 2 has been thought to have been found over a decade later on the surface of Mars from NASA photos. See e.g. the Telegraph Science Editor's report by Sarah Knapton (2015), who notes that the space craft became "a byword for mission failure."

[57] Schapira (1988) discusses this complex in Jungian terms in *The Cassandra Complex*, and there are some interesting parallels between the female astronauts and the patriarchal dynamic of male authority in *Gravity* (2013) and *Interstellar* (2014).

Scenographies and space: reading the gendered tyranny of the "Blue Ball"

This discussion will focus on the presentation of a relationship between humankind and Earth that has been attempted scenographically in the realist theater drama, *The Blue Ball* (1999). The parameters established by this visual depiction of realism within theater will form the basis from which to evaluate the presentation of gender in the recent Hollywood films of *Gravity* (Alfonso Cuarón 2013), *Moon* (Duncan Jones 2009), and *Interstellar* (Christopher Nolan 2014) as a way of highlighting some of the current deficit in *science-fiction thinking* within these filmic narratives. Since spectacle in science-fiction film is an important communicator and often "play[s] a more important role than narrative" (Willis, in Landon 1992, p. 62), the key relationship to be examined is the dynamic between the image of Earth and its connection to the significant protagonists and how this is constructed within the scenography.

10.3.2 The grip of Gaia

Anxiety about Earth's fragile predicament and our own planetary existence has arisen out of a wider cultural acceptance of the dangers of global warming. There has been an increased ecological focus within Hollywood, evidenced by the rise of "the state we're in" movies that address environmental crisis and space travel, which fundamentally suggests that leaving Earth is a fallacious solution to the problem (see e.g. *Oblivion* (Joseph Kosinski 2013) and *Elysium* (2013)). Discussions about the context of these filmic environmentalisms are pertinent, and there is increasing conflict within the politics of environmentalism itself between the technological ecomodernists (of the "Whole Earth Catalog"/Stewart Brand persuasion) and the Nature-minded agendas (e.g. Bruno Latour), where the political lenses differ radically in their approach as to whether technology or Nature should be trusted to solve these global issues.[58] Such tensions may indeed be expressed in cultural representations in realist science-fiction films and, by association, feed back into cultural attitudes toward technology and space exploration. How this rise in ecocriticism within Hollywood films operates with either progressive or regressive environmentalisms, as exposed by the use of technology in the narratives, is, however, outside the main scope of this discussion.

I choose, rather, to critique the way in which the presentational image of Earth is framed and manipulated because, in scenographic terms, the cosmic image of Earth in realist, science-fiction film is limited. Its visual presence is embedded in a Nature-focused, Gaia-style narrative that currently disallows audiences the space to think differently about the key relationship mankind has to the cosmos. There is a notable spectacularization of Earth's image, which I will term "Blue Ball" following the awkward climax of Godfrey's stage drama, *The Blue Ball* (1999), which occurs when Earth is presented to an audience. This Blue-Ball image evokes an internalized reading that suggests a fragile connection to the planet and communicates an overly defined relationship for protagonists and audiences. This limited cinematic imaginary echoes some philosophical traces of modernist thinking from the previous century, which dealt with a similar crisis of existence, so that the current framing operates as a form of tyranny and demands a repetitive, gendered relationship that now appears inescapable to film-making practice. Nature's totality, as represented in the completeness of Earth, is presented as "real" and natural, and this image

---

[58] See Nihjuis (2015) for the ecomodernist perspective and Bruno Latour's (1998) critique of modernism in "To Modernize or to Ecologize? That's the Question*."

of the Blue Ball embodies essentialist ideologies about men and women and their inabilities to cope in space due to their bondage to Mother Nature. Moreover, the "realist" aesthetic in this presentation restricts narratives through this framing of Man's relationship with Earth, and there is a loaded weightiness within the current aesthetic of realism that deceptively presents simulation as factual.

To clarify the use of the term "realist," the parameters for this discussion emerge from the understanding of the naturalist/realist theater movements and their impact on scenography. In the evolution of these theatrical genres, which are concerned with representing reality as a verisimilitude or mimetic attempt of presenting Nature directly to communicate its "truth," there was a significant focus on design that had grounding within and from reality.[59] This development of design, which employs realism in the visual narrative (despite the fictional aspects of the story), also occurs in the cinematic scenography of the chosen films. In doing so, this discussion focuses on science fictions that are notably grounded in known reality or, using Lathers's term, contain "space realism" (2010, p. 13) in their research and production values and attempt to depict this form of realism visually.

I will begin with Godfrey's theatrical approach to demonstrating the main protagonist's relationship with Earth in the premier of *The Blue Ball* (1999), not only because it predates the films chosen for discussion (*Gravity*, *Moon*, and *Interstellar*), but also because the scenographic relationship between an audience and narrative in traditional theater is less immersive and consequently not as totalizing as filmic design. Traditional theater relies less on a strong aural immersion that occurs with the use of soundtracks in film to create the mood for viewing. Additionally, in effective set design, the visual elements present on stage are less determined for the audience within the theater space itself. The theater designer Bob Crowley has highlighted a necessary *distance* in the practice of materializing set design in noting that a theater audience needs a design that leaves "certain ambiguities … hanging in mid-air" for it to be effective; otherwise, the visuality becomes weighted unnecessarily by the "dead hand of naturalism" (Crowley, in RNT 1995, p. 18). Effective design in the theater, therefore, allows an audience thinking space. Finally, the dynamic relationship between the live presence of the actor and the viewer also allows a more fluid framing within the scenography itself. In brief, designing for the theater affords the space for an audience's completion of the dramatic environment—something that is often notably disallowed in film. It follows that this potential space of distance within the medium of theater can also allow a critical distance in reading the image of Earth as the Blue Ball and, consequently, provides sufficient room to evaluate the efficacy and relative necessity of the image's appearance.

10.3.3 Theater's depiction of tyranny

Paul Godfrey's play, *The Blue Ball*, opened at the National Theatre in 1995. The narrative was driven by a key question of why mankind's forays into space had been "rendered mundane in such a short time" (Godfrey 1995, p. vi). The production values had been rigorous.

---

[59] The movement of naturalism was significant in theater design in presenting real environments on stage as opposed to painted backdrops, and took reference from what was considered "real environments" in terms of props in the everyday settings shown (see e.g. Collins and Nisbet 2010, p. 66).

Two years of sponsored research included primary sources, as Godfrey conducted numerous interviews with US astronauts and their Russian cosmonaut counterparts to underpin the drama.[60] The dramatic style placed *The Blue Ball* within a realist vein, as Godfrey fused his factual sources with his own imaginative creativity, resulting in a type of "faction" theater, akin to plausible, creative nonfiction.[61] The tensions explored on stage focused on the politics of existence as the astronauts and cosmonauts struggled with their experience of space and the humdrum existence on Earth on their return.

The scenography, which was designed by Stewart Laing, pared down the visuality and echoed the realist vein, showing a fluid, uncluttered sparse design on the stage, which echoed the aesthetics of "truth" that accompany verbatim drama.[62] In the documentary style of verbatim theater, there is often a minimalist design choice which implies that, in not being distracted by design, there is purity within the drama. Laing's sparse staging was positioned in traverse and occasionally neon-lit or intermittently washed with blue lighting. Its spatiality housed Earth-bound props such as bar stools, Eames chairs, and a bench area on one side to provide various settings of the Earthly quotidian. Carousel-style scenes included scenarios such as astronaut-selection trials, spaces of domestic tension, and, at one stage, a meeting in a commercial sushi bar.

Contrasting this distinctly Earth-bound setting, the program notes served to remind the viewer of the central subject of "space exploration" by symbolically using NASA's well-known visual image of the "Blue Marble" on the cover of the program (see RNT 1995). This familiar representation primed the audience, conveying the experience of orbiting Earth while also reinforcing the mediatized perspective of Earth's relationship within the cosmos. In addition, the image's history captured the sentiment within the unfolding drama as the last image taken by a *manned* lunar mission to date (Apollo 17 in 1972) and threw into focus Godfrey's wider question of why manned space travel had been currently made both mundane and redundant within the decades following the space age.[63]

The program design subtly re-colored the image on the cover to show a blue-gray tone, heralding perhaps the more opulent scenography that would occur as the show's finale which involved a spectacular "reveal" to the audience as the production design moved dramatically from the clinical depiction of the everyday to the depiction of the universe. With a sudden change in the staging, the floor lights indicated a launch pad and the stage heavens opened to show a quasi-realist representation of the cosmos. David Sawer's musical composition used a celestial harp which played as the audience witnessed the

---

[60] Godfrey was awarded the David Harlech Memorial Bursary, a fellowship from British American Arts Association, and the Stephen Arlen Award (ENO), among other grants to conduct the research. See NT program.

[61] David Edgar is seen as defining faction as a dramatic construct where a play is "based very closely and authentically on real subject matter"; see Edgar, in Innes 2002, p. 190.

[62] See *A Scenographic Analysis of Emergent British "National" Identity, on the Stages of the National Theatre between 1995–2005* (Armstrong 2010).

[63] See more details at http://history.nasa.gov/alsj/a17/images17.html#MagNN. Also see RNT/PP/1/4/171.

heavenly event. In the published text, the stage directions iterate the intended dazzling reveal to the audience:

> "The sky falls away and the Universe is revealed in which the earth spins revealed only by points of light on the surface. Then the light of the sun illuminates the whole planet and passes over, giving a sequence of night to day and back to night. (Godfrey 1995, p. 79)

The majesty of the presentation was, however, underwhelming. The sequence of the reveal involved Godfrey's semi-autobiographical writer character, Paul, who is researching the astronauts through the show, delivering a closing speech on a harness while floating upward, somersaulting, and then finally disappearing "through an oval-shaped hole in the rafters" where he "presumably heads for the open universe" (De Jong 1995). After Paul's dramatic acrobatics, the audience then experienced the planets descend and perform a temporal sequence, depicting day to night as the final tableau.

The reviews of the play were disappointing, to say the least, with one notable critic describing the symbolic finale as an "absolutely ludicrous closing scene" (De Jong 1995). Criticism drew attention to the paucity of the dramatic language used to describe the experience of space, the flatness of the dramatic text, and the incompleteness of the scenography to fully conclude the drama.[64] The need to present the astronauts' experience of orbiting Earth visually to the audience, so that they too could view the Blue Ball accordingly, appeared a theatrical indulgence. The discrepancy between the everyday setting of life on Earth and the spectacular cosmic reveal that concluded the production appeared to mismatch textual expectations. The locations shown failed to convey either the dramatic sense of alienation that astronauts felt or the mind-changing experience of space exploration. Yet, the final reveal of Earth in its day-to-night sequence seemed to be an inescapable issue of the visual aspect of the design.

10.3.4 The problem of showing the "real" Blue Ball in theater

The dilemma of depicting space within a realist aesthetic in theater is nonetheless problematic and has limits to what it can demonstrate in relation to the medium of film. In her overview of science on stage during this period, Kristen Shepherd-Barr draws attention to the tendency of contemporary science plays to present "textual richness and scenic restraint" (2006, p. 2). It is obvious that the premier of *The Blue Ball* (1995) bucked this trend and it is possible to connect the sudden appearance of the universe in the scenography as a notable visual excess that deflates any remaining potency of the written text.

---

[64] Many reviews are highly critical of both text and scenography. See e.g. Michael Billington, "Michael Billington finds Moon Exploration Shaky Ground on Which to Base Drama: Lost in Space," *The Guardian*, April 1, 1995, who summarizes the experience as "the best one can say of this curiously earthbound play about space exploration is that it goes some way towards explaining why this vast subject has failed to ignite the artistic imagination" and Alastair Macaulay, "Lost in Space, Crippled by Irony," *Financial Times*, April 2, 1995, who says the staging "is dismal, and in exact proportion to the dismalness of the play." See National Theatre archive, press reviews folder (RNT/PR/4/1/586).

I am not suggesting that the detached, "monotonous sterility" that "plods" (Macaulay 1995) within the language could be fully compensated by a more supportive scenography, yet it is possible that the reception of the poetry of the final scene might have been better supported by *not* depicting the universe and Earth's cycle as the concluding experience. Godfrey's play attempted to convey visually the mundane reality of life on Earth and the wondrous experience of space as witnessed by the individual astronauts, through the research experience of the writer. The climatic tableau of the writer rising though the heavens was an attempt to materialize this journey and ultimately communicate the ineffable experience of the astronaut. Yet, the production's final aesthetics subsumed the "real" experience contained in the script, so, despite the presentation of the Blue Marble Earth on the cover of the program, Laing's literalized heavens were read as a performative orrery rather than a realistic representation. Portraying Earth as an object of wonder on stage fundamentally revealed its presence as artifice.

Indeed, the representational nature inherent in theater implies that an image of Earth in a theater can never truly be accepted as real, but rather read as a deferred, theatricalized version of "Earth" (see States, in Collins and Nisbet 2010), unless of course we move outside the traditional bounds of theater on Earth and begin to perform in orbital theaters like the Canadian astronaut Chris Hadfield in his version of *Space Oddity* (2013).[65] However, in the conventions of theater, an audience will suspend disbelief and accept what is presented to them for contemplation if the scenography and writing are in harmony with one another. It is plausible to suggest that, given the current bounds of the theater space, there is a fundamental problem in the dynamics of theater design with representing Earth on the stage as a truly realistic, scientific image. Yet, the interruptive presence of the image of the Blue Ball here in realist theater suggests that it is a potent and perhaps irresistible symbol within the narrative of space exploration itself. What the apex of the failure of the scenography of this production draws to light is not so much the ineffectual visualization of Earth on stage as an object, but rather the constructed nature underlying this choice of image itself. The Blue Ball is deployed as a specific, symbolic device of profound loneliness and accepted by the audience as the "only coloured thing" (Godfrey 1995, p. 56) that it can "relate" to in the void of space. Yet, the very artifice of this image is revealed by its distinctly theatricalized presence within the realist drama and, thus, allows us to recognize and dwell on the fabricated nature of its symbolism.

10.3.5 Film and the immersive experience of space

The medium of film, on the other hand, deals more efficiently than theater with the ambiguity of visual framing by totalizing the experience though the camera lens. In its completeness, filmic scenography often creates a more passive role for the audience, as it subsumes the viewer's contribution to understanding what is seen, since it reduces the "distance" of meaning for audiences (a key feature of effective scenography within theater). Indeed, filmic scenography sometimes even merges the space between the "figure

---

[65] I accept that this is problematic even here, as Hadfield's performance is video of a performance, rather than the direct viewing as an audience in space itself as an *environmental theater* experience, which rejects conventional theater space, but Hadfield's performance is the nearest parallel to theatre in space to date.

and ground" in its framing, so that the subject and the environment become undifferentiated and presented as a complete, totalizing worldview. Simulated distances within film scenography may look spectacular, but they are flattened within the completed picture, and this act of framing is not dissimilar to Marshall McLuhan's (2001) phrase "the medium is the message" as the view is synthesized and therefore disallows any ambiguity for an audience. Film scenography can therefore accomplish a finished picture for the audience through its very medium of presentation.

The dynamic within film to produce a complete worldview is highly effective. In contrast, there are numerous parameters by which theater design is constricted; for instance, in terms of physically realizing the infinite problem of space, theater visually struggles with representing its sheer depth, whereas film as a medium is far more suitable. Laing's material struggle within the theatrical scenography of *The Blue Ball* (1995) was partially bound by the objectness of Earth and possible axis of travel in relation to the protagonist Paul. This is because the leaden dynamic of the fixed theater space meant that Paul had to ascend to the heavens (not the other way around) in a literalized act of enlightenment at the end of the show, since the act of framing Man's relationship with Earth was bound by the finite limits of the stage space itself. Filmic scenography, on the other hand, with its potential ability to present multiple points of view to an audience through the framing of the camera, can transcend such boundaries, as it can easily embody the infinite potential of outer space.[66] In addition, science-fiction film routinely encompasses a visual mixture of the familiar and strange so that transgressing Earth's limits is important in supporting the film narrative. Scenography within the genre of science-fiction film is both complex and fluid, since the "imaginative stance and scope" (Sobchack 2004, p. 101) allows the viewer to transcend physical and accepted limitations while also reminding the viewer of the familiar and known. Vivian Sobchack suggests that frequent presentational images in science-fiction film also carry "little emblematic power," as they have interchangeable interpretations (2004, p. 68), yet, in current realist science-fiction film practice, this key image of the Blue Ball conveys a strong message of Earth's meaning and remains uncharacteristically fixed.[67] Indeed, we have become so familiarized with the mediated distancing that occurs between Man and the Blue Ball in realist film scenography that this relational image has become somewhat of an impasse for both designers and film-makers.

10.3.6 Blue-Ball "fixed" thinking

The Blue Ball is now problematic as the "only coloured thing" (Godfrey 1995, p. 56) in our known mediatized experience of space's vastness since it has become a fixed trope in film narrative. Reading the relationship between Earth and humankind determines the reading of Man's relationship with space and the sense of human importance within the universe.

---

[66] Walter Benjamin (1999, pp. 226–227) discusses the dynamic of the camera eye and production processes that allow the film-maker to penetrate deeply into representing reality (as opposed to theater) and construct a realistic artifice that embodies the height of realism.

[67] Sobchack (2004, pp. 64–79) is discussing the notion of the constant image here and gives the example of the object of the spaceship, and suggests that, uniquely, within the sci-fi film genre, there is no definitive/exclusive meaning in objects, but rather dynamism in their function.

Scenographies and space: reading the gendered tyranny of the "Blue Ball" 233

In recent history, the astronaut experience of the Blue Marble as captured in NASA's image of Earth was decreased physically and philosophically by Carl Sagan's famous description of Earth as the "pale blue dot" (1994). Sagan's reduction of the scale of Earth conveyed both a greater fragility and a diminished significance. However, the importance of Earth in visual film narratives has re-inflated in size, so that its symbolic potency has a monolithic visual agenda which triumphs the idea of the importance of the human as individual. Marina Benjamin's description of space as a metaphor for the "brooding ... human mind" (2003, p. 54) is representative of the prevailing cultural attitudes of individuality. Survival in space exploration is focused on the potency of the *individual*, in terms of both their physical and mental well-being, and their capacity for self-consciousness, and echoes the philosophical attitudes of modernism, giving rise to thoughts of anxiety and displacement.

The current iconography of the Blue Ball becomes a wistful, historically retrograde metaphor that echoes some of the cultural identity anxieties which were prevalent in modernist thinking. Its on-screen presence disallows thinking of an alternative relationship with space from Earth that has fixed not only our relationship within the universe, but also re-essentialized gender roles in these factually based, recent science-fiction films. Gender divisions, which were an inherent tension in modernist thought, still remain a problematic concern in all of the films to be discussed where women protagonists are ultimately denied spiritual freedom from Earth.[68] The dynamic between the male and female protagonists and the "female" image of the Blue Ball operates in subtle but different ways to frame the relationship between the astronaut and his or her gender and their essential connection to Earth, and embody some of the traits of this nineteenth-century position of understanding. It is of interest to first look briefly at the values of the philosophical movement of modernism as the key features resonate in the chosen films.

10.3.7 The frame of modernity and the rise of modernism

Modernist thought arose in the West (in the late eighteenth century) as the conditions of modernity accelerated in the newly industrialized and increasingly democratized culture. The desire to reconsider what art should represent under these new conditions and how best to represent the new social experiences resulted in an expressive self-consciousness. The modernist movement, which is often summarized as a "crisis of representation" (see Lewis 2007, p. 2) had a diverse set of approaches that developed over the early decades of the twentieth century and allowed an increased focus on the role of the individual artist. These conditions also gave rise to crises of identity within the works themselves.

Philosophically, concerns often dealt with representing the interior reality of the mind, as well as the attempt to present a more convincing external reality, and rejected what they considered to be older forms of realism. In theater, the narrative forms pursued were mainly engaged with removing the illusory nature of the stage, while film, with its individual subjectivity of viewpoint and its ability to complete a total world of understanding

---

[68] Though there are women contributors in the period of modernism, it can still be seen as a masculinist movement. Due to masculinist fears about gender roles, women in high modernism were often equated with the natural and domestic (see Dekoven 2011, p. 212).

coupled with its physical "birth" in time (occurring at the beginning of the nineteenth century), can perhaps be regarded as *the* modernist art form.

Modernist artistic practices fundamentally endeavored to rationalize a worldview and convey a specific way of understanding the world, developing grand narratives in their attempts to do so. These universal perspectives produced totalizing thought, which disallowed the concept of anything "other" than the systems within which they were created.

The rise of industry and anxiety about its growth produced the self-conscious concerns of modernist Man—and here the gender is significant—who was trying to understand this changing environment. Predominantly a male construction, the self-reflexive individualism that arose exposed unstable identities and sometimes resulted in art practices that expressed pessimistic conclusions or subjective idealisms. Though in polar opposition, both positions attempted to counterpoise the underlying fragmentation and powerlessness felt in the face of rising technology.

Modernism's philosophical disposition, with its residual focus on individual crisis, certainly seems resonant in the recent films *Moon*, *Gravity*, and *Interstellar*, as all these films engage with a realist depiction of space exploration. In place of the threat of technology (the driving force for change in modernism), ecological crisis becomes the substitute peril in the films. Additionally, the lonely position of detached observation, while in the midst of spiritual crisis, which was a familiar position in modernist work, also resonates within the narrative.

The attempts to organize, rationalize, and control the space environment coupled with the protagonists' individual crises of identity, and even the film-makers' ambitions to complete their understanding of the universe within the narrative of the films, all seem to have strong parallels with modernist concerns of identity anxiety.[69] In these films, there is a notable presence of the individual to narrate the experience of space and direct the solution to the crisis.

Film-makers are skeptical and critical of the commercialism and the political orders of space exploration, yet they accept a grand narrative of Earth's position in the universe as fundamentally critical to Man's existence. *Interstellar* is perhaps an exception to this final point, yet the importance of Earth and its significant uniqueness are still echoed in the protagonist Coop's wistful declaration about Earth that there will "never be another one like her."

---

[69] It is of interest that, in Kubrick's *2001: A Space Odyssey* (1968), the astronaut bodies in space pose a kind of balletic sterility which Annette Michelson (in Schwan 2000, p. 203) has written about, describing it as a form of developmental modernism within the medium of film. She suggests that the creation of reality in the movie and the quasicontradictory perception of it are significantly intellectually provocative for the audience due to Kubrick's visually challenging presentation of space exploration and requires an audience to self-reflect on their own "carnal" knowledge and understanding of their bodies to allow the scenography to direct the audience to act "as agents of consciousness" (Michelson, in Schwan 2000, p. 210). It is obvious that this focus of a modernist perspective draws the audience down to earthly experience, rather than releasing them fully into a fantasy of "elsewhere." In modernist thought, the realist consciousness of privileging earthly existence dominates visual thinking.

Modernist universalist philosophy also affected thinking about physical space. Its attempts to describe and territorialize space to understand it as a known "place" was another way of managing philosophical anxiety. It was concerned with describing specific boundaries to try and ground the inherent instabilities of the "great 'out there' " (Massey 2005, p. 65). Complicated by a desire from the West to conquer, control, and contain, modernism put forward a singular model of understanding space which rationalized it, totalized it, and disallowed multiplicity or interactivity.[70] These general ideas of understanding spatial otherness can be applied to the chosen realist science-fiction film narratives about space, since these rationalizing parameters exist and are particularly prevalent when the presence of the Blue Ball is evoked in the scenographies as a symbol of grounding.

The typical modernist "tradition of self-reflexivity and isolation of the self" (Lewis 2007, p. 239), which focused on the importance of the individual's self-consciousness to make meaning in the world, is evident in the highly masculine narrative of *Moon*. The film presents a classic demonstration of existential Man, who yearns for Earth in his isolation, but overcomes this through spiritual self-mastery and a commitment to return to his rightful place. As will be discussed below, these modernist traits are evoked strongly in the dynamic between "Sam," the protagonist, and his relationship with the image of Earth within the narrative. The Blue Ball reveal becomes an important point within the pacing of the plot and the visual image of the solipsistic man in relation to Earth is striking.

10.3.8 Revisiting modernist Man: *Moon* and male existentialism

In *Moon*, the image of Earth on screen within the scenography is both brief and measured, and there is significant focus on creating the environment of space as distanced and "out there." The scenography conveys the humdrum existence of an industrial caretaker called "Sam" who is housed in a lunar base on the far side of the Moon, supervising the lunar extraction of Helium 3. His sole companion is a primary-carer machine called "Gerty," and we discover that he is a clone model of Sam Rockwell (Sam 5) as another healthier model (Sam 6) becomes prematurely activated before this previous clone's intended demise by the mining industry.

The ultimate message in *Moon* implies that cloning, simulating humanity, and the dehumanizing treatment that follows are the only ways in which human beings can physically and psychologically tolerate this othered, extraterrestrial living. The environment creates this feeling of alienation through its scenography which subtly depicts the familiar-within-the-strange, so that the surrounding, hostile everyday environment becomes a rationale for the return to Earth. Visually, the film also prepares the audience for this acceptance at the start by moving from an Earth-bound position through the medium of a television advert. This broadcasting transports the audience, via a mediatized Moon selling the virtues of helium-harvested energy, to join the central hero going about his lone, routine life of quotidian existence on the Moon in the Sarang mining base. Duncan Jones's

---

[70] Even though Einstein's discoveries, which warped the certainty of rectilinear time and space, had been presented in the early twentieth century, the "new imaginative possibilities" (see Lewis 2007, pp. 24–25) during the high modernist period were concerned with containment. Form, function, place, space, and boundaries were significant aesthetic subjects.

scenographic narrative throughout the feature focuses greatly on the industrialized tedium of a lunar colony, the stark isolation within the living quarters, and the desert terrain of the surrounding landscape. Sam's character is based on a classic science-fiction narrative of a "blue collar guy" and "how the environment affects them" (Jones 2009). The base's architecture, which was designed by the concept artist Gavin Rothery, demonstrated a "cold white" utilitarian interior and was realized as a location built in Shepperton studios (Shay 2009, p. 118). The aesthetics of these living quarters took inspiration from the space-travel-focused films *Outland* (Peter Hyams 1981), *Silent Running* (Douglas Trumbull 1972), and *Alien* (Ridley Scott 1979), and also from the futuristic artists Syd Mead and Ron Cobb (Shay 2009, p. 118). Interior design techniques used the placement of corporate branding to create added emotional distance. The clinical living quarters were set in contrast to more private areas where the stereotypical detritus of male occupation that included gadgetry and girly pictures were shown to festoon the intimate spaces and signaled the residual frustration of existence. The mining base interior set was also efficiently constructed through 360 degrees within the given space and was consequently relatively compact. This meant that, in addition to the desired film retro-aesthetic, the enclosed design allowed the actor, Sam Rockwell, to use the set as a method-style environment in which to perform the claustrophobia of the character (Jones, in Shay 2009, p. 118). The isolation of the hero, distanced from true living with his modernist solipsism, was embodied as much in the physicality of the set as the set dressing used to convey the space of loneliness.

In contrast to the solidity of the base interior, the external, desert-like surface and ragged terrain of the unforgiving, alien, lunar landscape were created and partially designed by Bill Pearson and his team, who constructed realistic miniature-scale models. This included landscapes of the Moon's surface and lunar rovers, which established a variety of action shots and aerial views of the mining base and harvesting activity.[71] Stylistically, the lunarscape emulated the drama of NASA high-contrast photography. The use of modeling also supported a significant number of shots of physical "distance" in the camera framing. The effects of travel, layering, dust, space panoramas, and other requirements were added onto the filming by Cinecite, to enhance both the realism and the drama in the modeling and the constructed interior spaces.[72]

10.3.9 Earth and its conveyance of the apex of male solipsism

It is quickly established in the beginning of the movie that Sam 5's contract will soon come to an end and he will return to a loving family waiting dutifully for him on Earth. Notable sequences in the visual scenography of the film emphasize his remote existence and make use of the infinite star-studded cradle of the universe from the position of the Moon to communicate Sam's wistful longing to be elsewhere. It is a scenographic framing that is distinctly without positive romance toward the vastness of space.

---

[71] The source quoted by Jones as influential is *Full Moon* by Michael Light; see Shay (2009, pp. 16–18).
[72] Roger Gibbon is credited with creating the mattes for the model and the digital view of Earth and the Moon in the opening sequences (Shay 2009, p. 19).

Scenographies and space: reading the gendered tyranny of the "Blue Ball"

The dynamic between Sam on the Moon's landscape and Earth embodies the male modernist trait of individual loneliness further as the reading of Earth's image is saturated, both visually and musically, with the nostalgic longing to be domestically Earth-bound again. There is a measured use of such imagery here, which culminates in a Blue-Ball climax within the dramatic narrative to enforce the apex of the tragedy of human isolation in space. The reveal of Earth occurs as a kind of emphatic anagnorisis as the clone Sam 5 contemplates the emotional and physical abyss that has arisen for him within the plot. The audience is not surprised by the climactic presentation of the Blue Ball that arises; indeed, it has been almost expectant of this image of Earth within the story to reinforce an encroaching solipsism and isolation that has been growing throughout the film.

Notably, in the director's commentary, Jones justifies criticism of this specific Blue Ball reveal (01:07:22–01:08:13) where criticism was leveled that a much longer drive in the lunar rover than the one actually shown would have been necessary for this scene to be realistic.[73] This journey happens at a critical point in the narrative when the clone Sam 5, though based on the far side of the Moon, drives from the compound to experience a view of Earth. He leaves the Sarang station's perimeter and, using a remote link, makes contact with his daughter, Eve, whom he discovers is unexpectedly older than he has been led to believe in his simulated conversations. Eve also informs him that his wife has died. In the climax of this sequence, the link terminates, and Sam 5 sobbingly declares that he wants to "go home." The scenography moves from the interior of the lunar rover, where Sam is distressed, to the external shot of the vehicle where the scenography allows us to hear Sam crying inside. The shot then pans around the outside of the lunar vehicle to reveal the zenith of the Blue Ball in the sky. The subtle piano soundtrack by Clint Mansell, which underpins this sequence, reinforces the isolation of Sam 5 (1:07:21–1:07:31) while, visually, the NASA image of the Blue Marble (which is directly replicated as part of the scenography) on the horizon acts as an end point for the viewer and emphasizes the solipsistic awareness of isolation and helplessness for the main character. The domestic and interior tragedy of the shattered family unit is shown as catastrophically and permanently fractured, although this crescendo in the scenography reveals the Blue Ball.

This relational image that occurs between the hero and Earth depicts the character Sam as an acute incarnation of the solitary man—a notable figure within modernist thinking. As the solitary individual, Sam's desolation and isolated awareness of his self prefigure his rediscovery of his own meaning and allow one of his reincarnations to return to Earth, while the other clone nobly dies. Sam's relationship with home changes after the Blue Ball reveal and becomes increasingly complex, since he escapes his previously desired domestic ties to fulfill a higher purpose for mankind in exposing the mining company's unethical practice of cloning. His release from the domestic sphere is, however, notably echoed in the earlier connection to his home on Earth in the film. Sam's dysfunctional relationship with his home life foregrounds his eventual release, and replicates the detached framing of the domestic sphere that is present in Stanley Kubrick's epic film *2001: Space Odyssey* (1968).

---

[73] Jones addresses criticism from what he calls "internet investigators" about the realistic aspects of his feature in his director commentary with the producer Stuart Fenegan on DVD.

10.3.10 Kubrick and the trope of the Earth-bound domestic

The emphasis placed on the relationship between Sam and the simulated domestic environment echoes the sterile connection between Dr. Heyward Floyd and his daughter in *2001: A Space Odyssey* (1968) and accentuates the masculine position of accepted separation as a male condition of being. Conducted through a video stream, Sam's conversations are distanced though the communication by his wife and child, which appears shallow, while he comes across as emotionally saccharine. Yet, unlike in Floyd's sole stilted domestic call to his daughter, there is no spectacle in *Moon* of Earth nonchalantly framing the distance in the background. Instead, the audience is presented with a set of superficialities that establish (despite the depth of emotion shown in the calls by Sam) that the level of reciprocation occurring between his wife and daughter is disturbingly brief and interrupted.

While Floyd, in contrast, performs the emotional distance of an explorer, almost reveling in the divide and leaving Earth proudly behind him, Sam is deflated and fundamentally reduced to loneliness. Nevertheless, the male distancing that occurs in both Kubrick's and Jones's narratives about space demonstrate that the male position of separation is a natural rite of passage, even though Sam in *Moon* feels isolated by the event.

Indeed, the trope of the mediatized distance from the Earth-bound domestic sphere is as enriched with the call for separation from the domestic as it is with the distance from Earth, and this dynamic is clearly established thematically in Kubrick's film, *2001*. This position of ultimate male detachment is also echoed in the key male characters in the other films that are case studies for this critique. Consider Kowalski in *Gravity*, who not only leaves behind his domestic impediment of a failed relationship through being an astronaut, but also willingly drifts off into space as an act of masculinist altruism, inflated by the bravado of breaking a record for a space walk. Similarly, Cooper in *Interstellar* willingly leaves his dependent children in the care of their maternal grandfather to try and save humankind from the environmental dust bowl of Earth (and, notably, also has dysfunctional televisual contact with his family on Earth). Though Jones's protagonist, Sam, escapes the initial outward showings of such masculinist bravado, he nevertheless embodies the virtues of the masculine self-sacrifice (the clone Sam 5 trades his place "home" with Sam 6) and the necessary drive to try and correct the status quo, even though the illusion of the family unit has been devastatingly exposed as artificial. Through this ultimate disconnection with the Blue Ball, the final position of the male protagonists suggest that distancing from Earth and the domestic sphere is a natural and essentialist male quality and echoes the aspects of the "cerebral and spiritual" male journey and "breaking ties" that men are aligned with in popular culture and film narratives about space (Lathers 2010, p. 6). Ultimately, there is a modernist trope of self-mastery in the face of crisis that allows these male protagonists to rise above these environmental limits and nevertheless triumph, despite the final outcome for them.

In contrast, the feminine aspect of space travel translates into an experience that is clearly unnaturally placed in space. Indeed, the film *Gravity* has a direct dynamic within the scenography that firmly locates the female protagonist on Earth and consequently suggests that the grounding of women is both natural and necessary. Again, the narratives echo some of the problematic concerns that arise in modernist thought about the natural role of women in the new environmental conditions. Notably, there is an interesting visual

Scenographies and space: reading the gendered tyranny of the "Blue Ball" 239

slippage that occurs scenographically with the protagonist Dr. Stone that visually and symbolically fuses women with the female image of Earth as Gaia, using this image of the Blue Ball. This use of this image highlights the presence of women in space as abnormal and specifically acts a symbolic call for this gender to re-Earth.

10.3.11 Earth-bound *Gravity*

The visual presence of the Blue Ball in *Gravity* is a good example of where the mediatization of Earth allows the "ground," which frames the subject, to act increasingly as the "figure" within the narrative (McLuhan 2001). In doing so, it draws into focus the feminine lure of Earth, and also how the feminized astronaut is problematic within this "othered" space of the cosmos. The act of such framing also admits a modernist totalizing of the environment of space as alien and seeks to carve clear boundaries for an audience's understanding of it. This visual narrative is created by the presence of the Blue Ball, which dominates the framing, and increasingly creates a visual tension between the importance of the principal protagonist, Dr. Ryan Stone, a rookie astronaut who becomes disconnected from various spacecraft, and the cosmic lure of experiencing Earth itself on screen.

There is an additional call to express the femininity of the space explorer which is demonstrated as a form of traditional, gendered essentialism in relation to Earth, which Marie Lathers has previously termed "grounding women" (2010, p. 7). Lathers (2010, p. 6) notes that popular narratives about women in space draw women back to Earth in a literal and metaphorical sense, suggesting that emphasis on the traditional social norms, in which women are equated with Nature and the material rather than the male aspect of the cerebral and spiritual, can be seen in these narratives. After having experienced the "otherness" of space, women are denied the agency to spiritually complete the full understanding of their crisis of identity other than through direct reconnection with Earth itself. Indeed, as the plot of *Gravity* develops, there certainly is an overwhelming drive to ground things in both the emotional well-being of the character of Stone and in the overriding visual cinematic expectation by the audience to experience landing on Earth, as an astronaut might. The rebirth of Stone is literalized as a form of primitivism on her successful return to *terra firma* as she emerges from her escape capsule in a lake, which is alive with frogs, to crawl on the mud as if she were a creature in the Devonian period.

Consequently, there is a cinematic requirement within the film to both re-Earth and rebirth that not only applies to the development and *dénouement* of Stone's character, but also within the audience's expectations as they are literally carried along with the ride in the scenographic experience. It is the framing of the presence of the Blue Ball and the cinematic language used that also conflate the female character with the feminine Earth and that promises this experiential event and the feminine desire to be grounded back on Earth.

10.3.12 Representing *Gravity*'s visual reality

Unlike Jones's *Moon*, which referenced sci-fi film tradition as well as realism in its design of space landscapes, *Gravity*'s (2013) visual scenography notably aimed for "absolute verisimilitude" (Luckhurst 2013, p. 27). The result was a highly complex, notable level of digital effects and simulation that produced the landscape of space, which was based on "naturalistic or realistic" aesthetics to emphasize the production values' attempts to strive for absolute realism (Webber, in Webber and Cuarón 2013). Having explored techniques from realist films of simulating zero gravity (such as *2001: A Space Odyssey* (1968) and *Apollo 13* (1995)), the visual effects supervisor, Tim Webber, along with the director, Alfonso Cuarón, devised a unique approach to recreate a more current, authentic experience. The techniques used were highly technical, painstakingly choreographed, and involved, for example, detailed levels of pre-visualization, technological control, set design, and animation, as well as complex rigging, lighting, and camera work. The manipulation of bodies in space at zero gravity involved live-action rigged actors. The camera focus and built sets were designed to concentrate on shooting the actors' faces when scenes occurred outside in space. With scenes played inside the spacecraft, the whole performer became the subject within the camera's framing. Both kinds of shots were then painstakingly augmented by the post-production company Framestore, who used CGI to create both realism in the physical movement of the actors and convincing details in the environment/space architectures (see Fordham 2014, p. 44). This fusion of the "live" and mediatized sequences to produce a detailed realism had Webber declare proudly:

> "Gravity isn't sci-fi … and it's not a space fantasy. It's set a few years into the future … it features contemporary space technology with a few minor changes." (Webber, in Fordham 2014, p. 44)

The assertion that minor changes of technology did not affect the overall audience "experience" of space within the film was disingenuous, as the power of the technology to produce and frame the visual experience and additionally create the fantasy elements of the narrative was notable in the film. Indeed, on release in the cinema, the film's highly mediatized aesthetic was also delivered in 3D, to augment the experience of space in a style of "immersive poetic realism" (Fordham 2014, p. 45).

In contrast to the film-maker's claims, however, *Gravity* is a sci-fi film, since it contains distinct elements of fantasy within its visual narrative, as well as within the storyline. Indeed, the plot is significantly driven by what Geoff King describes in the sci-fi genre as "frontier virtues" (2000, pp. 69–70), which are frequently prevalent within Hollywood narratives about space. The cultural politics embedded in *Gravity*'s own frontier narrative focuses on the attempt to "get home." The surviving US astronauts rely on appropriating other nationalities' spacecraft, and there is a subtext of castigating the irresponsibility of other nations (in particular Russia) in orbit, and by extension on Earth. Supported by the realistic plotline of current global politics, the presentation of individual astronaut action and the inherent solipsism of space again draw sharp parallels with modernist thought of isolation and existentialism. Additionally, the narrative depicts a clear image of space as an unnatural environment for women. The scenographic framing of Earth in relation to the female astronaut demonstrates that her psychological profile is unstable and her physical actions lack confidence when in orbit. These deficits in her character are neatly rectified when she lands back on Earth.

10.3.13 Dr. Stone, the hysterical Cassandra

Much of the filmic soundscape in *Gravity* swells cyclically to aid the action so that the music dramatizes the visual scenography. The aural narrative is also significant, and it is frequently punctuated by Stone's breathy hysteria of being alone in space. There are many events within the plot that can be defined as action-as-trauma scenarios. During these sequences, Stone's main verbal expression is a repeated outburst of "Ah!" underscored by a dramatic soundtrack. There is an extended amount of time used within the film in which Stone panics in this fashion so that the narrative drive of this female protagonist has parallels with many of the factors attributed to a stereotypical hysteric, expressing the traits of the "Cassandra" syndrome (Schapira 1988).[74] According to Laurie Schapira, the Cassandra figure is a type of female hysteric who is gifted but disbelieved, and it is interesting to reflect on this psychoanalytic stereotype in light of the plot, and also in relation to the way in which women's presence in space is read and ultimately dismissed.

There are significant parallels which can easily allow a reading of Stone's character as a filmic version of a Cassandra hysteric. One key attribute suggests that the hysteric identifies with an idealized patriarchal figure, who allows her to focus, understand, and articulate her experience, without whom she otherwise remains in a chaotic state. In Stone's case, such a dynamic exists between her and her male colleague, Kowalski. During Stone's struggle to get back to Earth, she repeatedly depends on his competency and experience to orientate her landing back on the Blue Ball. Even when he has chosen to heroically sacrifice himself, by detaching himself from her rather than pulling her with him into the void of space, Stone summons him up in a climactic state of despair (1:00:08) as a psychological wraith sent to remind her that she has "got to start living."

The hallucination is a significant factor within in the narrative, as it allows Stone to self-actualize in the midst of a spiraling death drive. The materialization of the dead Kowalski-as-motivator occurs as Stone has just given up the will to live and has decided to shut down the craft she is traveling in. In this ghosting scene, Kowalski compels Stone to carry on following a preceding "calm" moment from her hysteria in space, in which she has accepted the inevitability of her own female redundancy. The audience has witnessed Stone's maternal failing incrementally throughout the film. This first indication of this appears during exposition in an early scene after the initial catastrophe, where she tells Kowalski that her young daughter has tragically died. Stone's maternal loss is then reechoed later on in the aural scenography, during the key radio-contact scene that propels her will to self-destruction. During this scene, she fails to communicate her need for help due to language barriers, and she can hear a baby crying in the background of the radio transmission from Earth. Stone's fundamental despair, before she summons the ghost of Kowalski to save her, results from hearing this Earth-bound contact, and her attempts at communication merely emphasize her own redundancy as a mother and her helpless relationship with Earth as a woman in space.

---

[74] Lathers also suggest that hysterical women in space are often an echo of women's roles from the 1950s, where they are allowed a passive action of the "scream" moment for the spectator (2010, pp. 175–176).

The modernist solipsism of self-mastery in the midst of crisis is not achieved by Stone's character, as the environment of space does not provide the conceptual freedom of self-consciousness, but rather evokes the feelings of neurosis that she experiences. The unnatural environment created by the immersive technology that spectacularizes the Blue Ball for the audience and provides a visual sense of wonderment is also the locus for Stone's neurosis. On the one hand, the Blue Ball acts as a symbolic image of Stone's inadequacies. It represents the incomplete aspects of her womanly role, emphasizing fertility and nurturing, both of which demand to be completed only by her returning to Earth. The Blue Ball also functions in a literal sense by calling Stone to re-ground herself physically. This scenographic framing of Earth and its connection with Stone, as a form of visual motivation within the narrative, will now be addressed.

10.3.14 The call of the Ball

The presence of Earth as a visual image in the film's framing rivals the presence of the key protagonist, Dr. Stone, so that the relationship between the figure and ground in the visual scenography becomes significant. There are some interesting illustrations of how this framing of the Blue Ball builds within the scenography to demonstrate a key trope within the plot and articulates women as displaced within space, and additionally implies that the exploration of its boundaries is wasteful and indulgent.

Earth is the first full image that is seen (00:39) in the scenography, and it occurs immediately after the audience is primed with the graphic statement that "Life in space is impossible" (00:30) which precedes the title of the film *Gravity* to then sharply display a moving part of the Blue Ball that dominates near two-thirds of the screen. Additionally, the presence of the Blue Ball is emphatically brought to the audience's attention, by a dramatic cut in the stylized, industrial-sounding music, which has been previously rising within the soundscape.

Just after a minute, where the audience is drawn around by the camera as if orbiting the Blue Ball (01:03), the Space Shuttle *Explorer* comes almost indistinctly into view to increasingly establish its presence orbiting Earth within the frame, taking nearly a minute and a half to do so. In this opening sequence, the presence of orbiting bodies and objects arise through a slow build in their presentation and, with some intricate camera work, the audience experiences a verisimilitude of floating in space. Behind the orbiting objects and people, the image of Earth dominates the framing, so that the camera work and positioning of Earth present a continuous blurring of the subject within the camera frame between Earth and the objects floating in space.

During these moving visuals and the radio communication banter with Houston, we are primed with the information (01:15) that Dr. Stone is not well according to ground control ECG readings. Additionally, her physical presence is not seen until other astronauts have been introduced and completed their own work on the Hubble telescope (03:45), and she is revealed to be struggling to complete her mission to enhance it. Repeated radio communication asks about her health and she insists that she is fine, though the audience is being fed information to the contrary; for example, at one stage, she even drops a screw (06:12–06:19), to be caught by her more experienced colleague, Kowalski. Though it is unlikely to be a conscious decision by Cuarón to present his female protagonist as less competent, as the audience is presented with statements within the film narrative suggesting her genius, Stone is certainly increasingly alienated in the boundaries of space, and her presence as physically competent is far from assured, which echoes some of the concerns raised by Lathers of the ideologies of gender difference coming into play (2010).

The issue of gender and space becomes manifest when Kowalski offers to help the bumbling Stone to complete her task after some meaningless banter about her "pretty blue eyes." The scenography incorporates the view of Earth and suggests that the audience is experiencing Kowalski's gaze of this admirable "view." The camera then pans around from his gaze on the Blue Ball, teasingly reflected on his visor, to a long sweep of a vision of a Gaia-style image, to discover details such as mountains and a lush American continent, to then move back to Kowalski's admiring face (07:56–08:43). The cinematic sweep across Earth like this is an odd one, as we see Kowalski framing the shot on the left of the screen to start the pan and then end up with him on the right, which crosses the usual 180-degree line of presentation in film. The language of gazing and feminine sexuality is certainly implied in this framing, and Laura Mulvey's (1975) idea of the "male gaze" is replicated here and applied to both Stone, who refutes it, and then projected to the passive and beautiful Gaia image of Earth.[75] Here, the vocal and visual feminization of Earth as woman, as seen through Kowalski's eyes, is a form of scopophilic enjoyment that implicates the audience. Both Stone and the image of the Blue Ball are conjoined here, as they are subject to the male gaze of feminization in both the visual scenography and in the narrative that accompanies it, despite Stone's denial. However, the pleasure of looking is at this point directed toward the image of Earth, so, when Kowalski says, "Got to admit one thing— can't beat the view" (08:00–08:05), his comments about Earth as an object of desire are ostensibly said to Stone while she is working, but they are also fundamentally directing and priming the reveal of the Blue Ball to the audience as an appealing, feminine image.

The Blue Ball, presented in this manner as a calm, feminine image of desire, prepares the audience to witness the cinematic presentation of trauma and Stone's general Cassandra-like, chaotic state in the face of disaster. During the catastrophic events (11:53–12:37), Stone's necessary connection to Earth is physically manifested when one colleague is killed and she and Kowalski are rendered detached and drifting in Earth orbit, bound together by an umbilical tether and without a suitable craft to transport them home. The image of the Blue Ball becomes an important signifier that demands such a re-Earthing.

As this first catastrophe strikes, Stone increasingly becomes the main protagonist, and the culmination of this growing image occurs when Stone is spinning away from Earth and out of sight of her colleague Kowalski, who is trying to restore verbal commands to control her crisis. As Kowalski demands that Stone should "focus" to survive and describe where she is by using a visual landmark, Stone retorts repeatedly "I see nothing" (12:56). Here, the audience witnesses a deliberate filmic irony, as Stone's claim that she sees "nothing" to orientate herself is in potent contrast to the visual image that the audience is witnessing. As Stone tumbles away from the Blue Ball, the camera dwells on a long and slowly

---

[75] Mulvey's psychoanalytic argument discusses the image of the woman as other and the dynamic of how an audience is implicated in a "male gaze" in cinematic narrative.

increasing close-up of her face (13:13–14:00). During this close-up sequence, the image of Earth rapidly alternates between framing Stone behind her and being reflected on her visor, suggesting to the audience that the Blue Ball is what Stone is momentarily viewing as she falls. In this cinematic sequence, there is an implicitly repressed reading as the vision of the Blue Ball is clearly present in Stone's visual field, so that Earth becomes a potent symbol which remains unarticulated. The dynamic of this cinematic language directs the audience to desire Stone's return to the planet, for, by physically branding her presence with Earth in this scenographic framing, it sets up the return to the Blue Ball as the ultimate desire for the duration of the film. Additionally, within this visual language, there is a repeat of the distinct merging of figure and ground within this first disaster sequence. The juxtaposition of what is figure and what is ground in witnessing the repeated image of Blue Ball repeatedly blurs the audience's focus for the subject of the film, as the scenography alternates from Stone to Earth's presence and back again in the framing. Thus, Stone's feminine Earth-bound call is thoroughly embedded in the visual language. How her "nature" then performs is directly related to how she is constructed in relation to her framing by the image of the Blue Ball, which suggests throughout that there is still a need to be Earth-bound. The completion of this call at the end of the film satisfies the audience's desire when Stone lands in the lake. In the closing shot, she slips on the mud at the end of the film, as she emerges physically and psychologically reborn from the amniotic waters of the lake (01:18:45). She checks her balance within this final scene, demonstrating to the audience that she is no longer the chaotic, hysterical figure that she was in orbit. Grounded and bound by the call of "gravity," Stone's birth and emergence on Earth are yet again contradictory, embodying the leaden realism of the maturity of womanhood, yet visually suggesting her emergence in an infant state. Her rebirth on Earth, rather than in space, is weighed down with verisimilitude, and it leaves her self-consciousness and the audience's acceptance of her new life devoid of an alternative cinematic imaginary. Stone's child-like emergence is fundamentally denied the modernist transcendence of the truly spiritual through the experience of the "other" of space and via the film's alleged commitment to realism. Her newly born, imago existence is grounded by nothing new and, as an image of the apex of women in space, Stone's female rebirth emerges as far away as possible from the male symbol of space enlightenment that can be seen in the extraterrestrial evolution of Bowman in his becoming the star child.

10.3.15 Transcending realism

The use of the realist aesthetic in this symbolic final act within *Gravity*'s (2013) narrative directs critique toward the insistent claims for realism by the film production team. Despite the immersive and convincing verisimilitude created by the scenography for the audience, the character of Dr. Ryan Stone becomes fundamentally emblematic of a modernist-styled everywoman, connected to Earth and Nature in her rite-of-passage rebirth. Like "Sam" in *Moon*, Ryan's character in *Gravity* completes the modernist crisis of solipsism and existentialism at the end of the films in her return to Earth. Yet, while the clone "Sam 6" publicly and successfully confronts the inhumane corporate practice of cloning on the Moon, Ryan has no obvious post-film story, other than one of individual wholeness and suggested psychological well-being, since she is no longer "neurotic" on landing. Both narratives,

Scenographies and space: reading the gendered tyranny of the "Blue Ball" 245

however, respond to a call from the Blue Ball in the scenography and satisfy this demand as the *dénouement* of the story. In both films, Earth binds the protagonists through their gender, philosophically and spiritually for men, and with a more physical, grounded focus for women. Christopher Nolan's recent film, *Interstellar*, however, contrasts directly with both *Gravity* and *Moon* in terms of its narrative's plot, as it addresses directly the possibility of positive disconnection from Earth. *Interstellar*'s sci-fi narrative hinges on the attempt of mankind to leave an ailing Blue Ball, and much of the action and focus occur within space, rather than simply creating a binary connection with Earth and the modernist boundaries that determine space's "otherness." *Interstellar*'s aesthetics also combine the plausible with the realistic, so that the physical depiction of space allows for an imaginative landscape beyond Earth's boundaries.

In spite of this key narrative drive within the drama, the film is still, however, bound by residual, cinematic conventions toward the Blue Ball. Coop, the male protagonist, is a complex character, and certainly at times transcends the modernist, solipsistic angst of the male astronaut, yet the film does not overall escape the residual narrative of gender essentialism. The connection to Earth in relation to gender is still explicit within the protagonists including Murph (Coop's daughter, who is named after Murphy's law) and Coop's female astronaut colleague, Dr. Amelia Brand. In *Interstellar*, it is also notable that environmental crisis creates the conditions that give rise to identity anxiety and drives the narrative's necessity to leave the Blue Ball, which once again binds the characters' actions though their gender.

10.3.16 *Interstellar* and the residue of Blue-Ball thinking

With a brief narrative that shows a "caretaker" society that has been ravaged by Earthly abuse, Nolan's screenplay sets up the main protagonist Cooper ("Coop") as an Earthbound, underutilized, and overqualified single-parent farmer. His deskilled predicament places strain on his identity and meaningful existence, and echoes a combination of modernist values demonstrating both the conditions of solipsism and self-mastery within his character. He is persuaded by a secret regrouping of NASA to leave his children to altruistically find a new home for the human population using his previously redundant skills as an engineer and NASA pilot.

The audience learns in an early scene that the Luddite cultural values of this near-future society seem to only tolerate functional, Earth-focused science. This myopic position is best demonstrated when Cooper is called in to defend his daughter, Murph, at her school. She is in trouble, having shown her peers Coop's old science textbook which contains the Apollo Moon landings. "You don't believe we went to the moon?" (11:38) Coop incredulously scoffs at the female teacher in the meeting. Her retort is that to not "repeat the excess and wastefulness of the twentieth century … we need to teach our kids about this planet, not tales of leaving it" (11:49–11:57). This exchange set up the plot's sociocultural concerns, which are focused on farming rather than exploration (which lacks public support), and highlight the undesirable constraints of Coop's Earthly existence and the necessary drive for this protagonist to leave it.

There is, however, considerable contradiction in the narrative of the validity of the exploration. The selfish selflessness of Coop's agreement to go on the mission and leave his children behind to find alternative homes for the population of Earth is, later on, revealed as based on a lie from a trusted key figure called Professor Brand. We discover mid-mission (along with Coop and Dr. Brand, the astronaut who is also the professor's daughter) that the "Plan A" of space exploration, which was to transfer the current population of Earth to a suitable new home, was never feasible due to time constraints. Additionally, it transpires that the real motive of "Plan B," to incubate the 5000 fertilized human eggs aboard the ship, also appears fanciful, as finding a suitable new, habitable planet seems increasingly unlikely.

This deliberate complexity within the plot, which does not allow a simple motive of merely returning to Earth as a solution, also translates into the film's scenographic dynamic. The landscape of space in *Interstellar* visually transcends the modernist fixity of seeing space as purely an immobile "other" (Massey 2005), since it constructs an interactive relationship with Earth and a highly complex image of outer space, which can embody multiple dimensions and representations. By way of contrast, there is an immobile dynamic in *Moon* and *Gravity*, which is presented through the modernist framing of space, since the longing for the Blue Ball, which dominates both narratives, presents a simple binary position of Earth as home and space as not-home, "other." Indeed, as *Interstellar*'s screenplay was assisted by the theoretical physicist Kip Thorne, the realism of the project saw itself as a less Earth-bound science-fiction story with "credible science at the heart" (Duncan 2015, pp. 36, 38) and produced the solution of the dilemma of lost time and space as a cinematic spectacle of wonder in the CGI-made tesseract or "hypercube." The tesseract was a physical representation of the fourth dimension of time, which allowed people on Earth to be saved by the actions of Coop and his mature daughter, Murph. They use this hypercube as a conduit between time and space to exchange vital information about gravity that is necessary to save the world. The resulting scenography, which created an Escheresque cubic matrix, visualized physics in a way previously unrealized and additionally entertained cinema goers with a sense of visual wonderment (Thorne 2014). *Interstellar*'s scenography, therefore, embodied both an imaginative and a materialist/realist aesthetic. The sources to create the environments of space used realistic mapping from scientific photographs for some scenes, while also reconstructing Thorne's theoretical imaginings, to create a credible universe. In acknowledgment of the known fabric of understanding the universe, *Interstellar* fused accepted science with material references to construct both a plausible and a realistic scenography within the film.

10.3.17 Creating the reality of space in *Interstellar*

Space imagery was digitally produced by Double Negative and the landscape of space was constructed using as much real photography as possible; the production team referred to scientific material such as NASA's star field maps (Lockely, in Duncan 2015, pp. 38, 40) so that they could reference the correct constellations in the landscape of space.[76] These images of space were then projected live into the set, as well as being used post production,

---

[76] This was a technique first used in "1998 when the company Dream Quest created scientifically accurate starfields for *Armageddon*" *(see* Edwards 2015*)*.

and the effects used a combination of miniatures and sets to convey the scenes involving space travel. The images of the Blue Ball shown as the craft leaves, perhaps never to return (41:32), were composed from "low Earth photography taken primarily from the International Space Station" and there was also a key mediatized image of Man in space:

> "... from Chris Hadfield in this amazing observation dome; he's singing away and you've got the Earth rotating behind him. The sun is hitting the Earth full on and it's totally blown out. It's raw unfiltered sunlight and that's something you generally don't see in science fiction films." (Franklin, in Hogg 2015)

As with both *Moon* and *Gravity*, the production values placed much emphasis on using scientific data and images to recreate the experience of leaving Earth behind. Clearly, this sequence is an inescapable scene in this genre and would be expected within such a film narrative; yet, what is of interest in the scenography of *Interstellar* is that the indulgence of seeing the Blue Ball is relatively measured in its on-screen duration.[77] This visual sequence of leaving Earth in *Interstellar* Interstellar (2014) has additional significance, since it primes the audience for a narrative twist that the potential of space exploration might be a futile exercise that will have devastating personal consequences. This threat of failure is also a significant narrative trope within the plot's overall dramaturgy. The precariousness of the exploration is dramatically highlighted in the sequence of leaving Earth behind as a residual wistfulness in the presentation of the image of the Blue Ball.

10.3.18 Attempting to leave the Blue Ball behind

The image of the "call to the Blue Ball" is present as a visual theme within the cinematic language during the key section of the film in which the crew moves away from Earth (40:33–48:18), yet it functions ultimately as a classic modernist signifier of Earthly detachment and ultimate alienation. During this section of the narrative, there are periods in which the image of Earth is emphasized by significant moments of silence within the soundtrack, so that its image punctures the scenographic reading and draws itself into central vision (e.g. 41:52–41:56 and 41:59–42:03). Here, the presence of the spectacularized Blue Ball is relatively minimal, and its framing is often held for a few seconds. For example, at one stage, a single brief "Oh wow" comment is made by Dr. Brand (45:27–45:30) as the crew reaches the main *Explorer* craft and she sees the image of Earth from its window.

Other frequent presentations of the Blue Ball occur where it is subtly framing the craft as the ship is undergoing disengagement or docking. At such times, particularly when the scenography shows the exterior of the craft, there is certainly a notable merging of figure and ground between the Blue Ball and the ship; yet, unlike the aesthetics of *Gravity*, the overall visual narrative does not allow for excessive dwelling on this fusion of subjects, as the film action rapidly moves the sequence along, so that the craft still holds its presence as a separate moving object.

---

[77] It is of interest that Kubrick's *2001* (1968) also shows similar reserve in its equivalent sequence of leaving Earth behind.

There are two notable dramatic sequences, however, in which the Blue-Ball image does hold the attention of the viewer for considerably longer. The first extended image occurs when Coop begins the initiation of the spin cycle, which rotates the *Endurance* after the successful docking of the shuttle (45:44–48:16). This first long sequence uses the image of the Blue Ball again to reinforce the modernist stance of space and individual solipsism, and quickly establishes the journey of the crew leaving Earth in the film. The idea of grief is also accelerated momentarily within the rapid action, and the distance and separation of the astronauts from Earth are additionally established in a televised communication with Professor Brand, who cites part of the Dylan Thomas poem, *Do Not Go Gentle into that Good Night*, to emphasize the critical level of the mission.

Professor Brand's narration occurs as an overlaid narrative track which frames the visual sequence and changes the pace from the wistful, romantic music that has framed the rotating Earth. The visual scenography also emphasizes concerns with being isolated from family, which has been mentioned in a previous scene during a brief exchange between Coop and Amelia Brand (42:13–42:53).[78] As the *Endurance* craft spins away from the planet, we see its trajectory with Earth underneath as a repeated pattern in the ship's rotation. Following this, the Blue Ball occurs as an incidental image through a window when Professor Brand is communicating his good-luck talk on a video screen. Ultimately, within this first longer sequence, Earth becomes a dominant image, as a large subject framing the smaller image of the *Endurance* ship as it moves away from the Blue Ball's orbit. This shot then fades to black with the orchestral soundtrack by Hans Zimmer finishing abruptly as it does so (48:05–48:14).

Here, these multiple, frequent, and notable presentations of the Blue Ball in the framing of this leaving-home sequence establishes the isolation of the ship, yet it does so in a notably brief and accelerated manner. The on-screen presence and size of the Blue-Ball Earth create both a beautiful and wistfully romantic image, yet also suggest a temporary one, which needs to be quickly discarded to progress the plot. This is an important development in film language, as it subtly contrasts in its meaning in comparison with the films *Gravity* and *Moon*, since this sequence in *Interstellar* is less indulgent in its overall connection to the mediatized Blue Ball. This first Blue-Ball sequence is, however, followed by a second passage that again involves this image of Earth within the visual narrative and again reinforces the isolation of the astronauts from their home planet.

During this second sequence (48:16–51:41), we see the Blue Ball from the point of the view of the spacecraft, through one of its windows (48:16–48:23). The framing then cuts to a long interior shot within the ship, revealing Coop contemplating his relationship with the planet as the others prepare for the necessary long sleep (48:42–48:45). Coop declares to Brand that "it's a perfect planet and we are not going to find another one like her." Brand attempts to bring the discussion from the emotional level to a rational one by focusing on the mission. However, she unwittingly reveals to Coop though her tone and body language that she has feelings for one of the three explorers who have sent back positive data, and

---

[78] Here, Brand and Coop have a verbal exchange and Brand closes down Coop's inquiry into how she is feeling. This is also briefly preceded with an image of the ship silently orientating itself as it leaves the Blue Ball.

she deflects this by talking about the heroic sacrifice of the scientists, lauding the previous mission leader Dr. Mann as "remarkable … the best of us" (49:37) since he persuaded people to follow him "on the loneliest journey in human history" (49:40). As Brand narrates this fact, the camera cuts to the spinning Blue Ball in the window for a few seconds to reinforce the idea of isolation. Here, she wistfully looks at Earth while talking about a lost lover, yet she covers these thoughts by projecting the narrative toward the mission. Her contemplation of Earth in this manner, though brief, reveals her as flawed and potentially untrustworthy, and this is used later on in a key moment in the film to question her decision-making. Here, dwelling on the image of the Blue Ball suggests emotional deception on Dr. Brand's behalf as much as alienation and detachment.

At the end of this sequence, Coop also uses his contemplation on the image of the Blue Ball as part of a deception, using it as a delaying tactic so that he can confirm his suspicions about Dr. Brand with the robot, TARS (49:27–51:04), and discovers that she has had a relationship with the explorer Edmunds. Both Brand and Coop contemplate the Blue Ball in a partially dissembling manner, yet they have different connections to its emotional significance. For Brand at this moment in the film, Earth symbolically represents her lost lover who has traveled into the unknown of outer space. In Cooper's case, there is love for Earth with its dust but, rather more completely, it evokes the domestic through the memory of the family he has left behind.

In the *dénouement* of this sequence, Coop records a message for his family before he goes for the "long nap" (51:17). Here, the Blue Ball is incidentally framed in the window (51:05–51:10) and it spins to the sound of a single pitch in the scenography, which underpins the visual narrative. From the shot inside the ship, the camera then moves to the exterior, with Coop's voice still talking to show the rotating *Endurance* in the cosmos in the star-lit void. Coop notes in his message, "the earth looks amazing from here" (51:23–51:26). A long shot shows the *Endurance* traveling away from the camera, immersed in a starscape, and then the image cuts to the farmstead on Earth (51:30) where a vehicle travels toward the house down the main road leading to it. The scene then changes from the landscape of space, where Coop's voice is talking of Earth, with a single-pitch note underscoring this image, to a final shot of the farm, and creates a potent image of displacement. The scenography transitions from the void of space to the firmness of place in the homestead, and echoes neatly the existential experience of the male astronaut.

This priming of the desire for Earth occurs using the image of the Blue Ball, which represents the wonderment of perspective, and also has a gravitational pull in terms of the emotional meaning in the narrative. Coop plays out the traditional role of the male astronaut, expressing the solipsistic, modernist tones of being detached from his family on Earth. His desire for Earth is, however, also managed by the frontier spirit that is contained in the mission to find another home for the planet, and fundamentally in his ability to see through the crisis of his condition in space to save humanity. So, while the male space-traveling protagonist Coop achieves ultimate freedom in his condition in the environment of space though the recasting of his mission and identity, Earth yet again commands a more significant lure for the female characters of Murph and Dr. Brand as they experience a greater pull to be Earth-bound.

10.3.19 Gender dynamics and the residue of Blue-Ball thinking

While Coop manages to journey forth into space, despite his deep connection with his family on the homestead, the female protagonists are connected or grounded. These female characters are cast in relation to Earth in an ambivalent state of self-consciousness as they, again, like Stone in *Gravity*, need to be finalized by the male protagonist Coop and are ultimately denied a full spiritual act of enlightenment in their grounding to Earth. Furthermore, their attempts to understand the complexity of this mission are both first disbelieved and then completed by the actions of Coop, echoing the Cassandra-like dynamic of both these female characters within the film narrative, who need a patriarchal figure to order the chaos that they evoke.

First, Murph, whom we meet at the beginning of the film, is set up as questionable in terms of her ability to tell the truth. She initially reports the activities of a "poltergeist" in her bedroom, which establishes a childish, female irrationality in her narrating the phenomena of flying books, which are perhaps communicating to her in her bedroom. This repeatedly dismissed by her family, including Coop, who declares, "I just don't think your bookshelf is trying to talk to you" (14:37–14:38). After witnessing an episode in her bedroom, her father rationalizes the occurrences as due to an anomaly in "gravity" (20:11). However, Murph's observations about the "poltergeist" communication is fundamentally correct, as it is later revealed that the books are being pushed in code in from the bookshelf by Coop himself from the hypercube space of the tesseract, where he passes information from another dimension. Decoding the information as an adult, Murph uses her scientific training to solve the problem of harnessing gravity so that a viable amount of human life can get off the planet on a spaceship. It is Murph's reading of the Earth-bound phenomenon that completes the mission, and this is recognized in the name of the craft—*Cooper Station*—that she travels on; yet the success, however, is Coop's role in providing these data in his lateral thinking that becomes the key to saving humankind. The audience reads the discrepancy in this acknowledgment of recognition in the final scenes of the film, when Coop erroneously thinks that *Cooper Station* is named after him, to find it is his daughter who has the credit for the mission when he awakes on the craft after being rescued. Coop innovates, orders, and rationalizes what appears to be, on the surface, chaotic and potentially useless knowledge provided by Murph. Her role is ultimately defined by her ability to understand the universe from the position of Earth, so that sustainable space travel can be achieved, and partially from within the confines of the family home when Coop leaves for the mission. Her character is not seen as having left the planet other than at the end of the film, when she meets Coop in her old age on the simulated Earth spaces of several craft that have left the dying planet.

Similarly, Coop's female colleague in space, Dr. Brand, is also fundamentally connected to a grounded position in her role as an astronaut. Like Murph, there are Cassandra-like traits in Dr. Brand that the audience experiences, which are problematic when considering the relationship of women with Earth. There is, again, an undercurrent of the essentialist gendered thinking that the Blue Ball evokes, which seems to bind the character. Coop's irrationality is accepted as part of the solipsism and consequential derring-do of the male hero in space, yet Brand's feelings, like Murph's, are potentially chaotic.

First, Dr. Brand's actions when she arrives on Miller's planet (the first planet to be explored in the mission) are narratively charged. She ignores Coop's orders to return to the ship, which costs the life of a crewmember, and, additionally, decades of time are lost in terms of space travel (01:08:00–01:10:42). This action impacts on the need to solve the problem of resettling humankind, making it more difficult for the remaining crew. In an angry exchange, she accuses Coop of not focusing on the mission and of trying to "get home" instead, but reveals to him the potential of gravity to be able to theoretically cross other dimensions (01:12:10–01:12:14). Coop, through his quick-wittedness, is able to use this information later on and save humankind in the tesseract, as he draws the potentiality of this knowledge into a viable plan. Brand's articulation of the information is one of dismissive negativity.

Dr. Brand is also disbelieved later in the mission, again at a crucial point in the plot. Her suggestion to visit Edmunds' planet, as the second place to explore for human life, is called into question, even though her scientific training is to determine whether a planet can sustain life or not. Despite this noteworthy "vocational" connection to Earth as a scientist, her judgment is suggested as faulty, due to the earlier revealing of her relationship with Edmunds. However, Brand's predictions for Edmunds's planet are shown to be true to the audience at the end of the film, when Coop meets the aged Murph on *Cooper Station*. He is instructed by his dying daughter to follow the frontier spirit and find Dr. Brand, who is concurrently shown as a lonesome, home-making caretaker on Edmunds's future "Earth" (02:35:16–02:37:14). Notably, Coop's earlier underlying disdain for a scientific caretaker role is also pertinent, as it fundamentally shows Dr. Brand in a less dignified position to the audience, passively waiting to be saved and discovered.[79]

Like Murph, Dr. Brand's potential pioneer spirit, despite her contribution, is grounded by her connection to Earth in both her professional training and her ultimate function in *Interstellar*'s film narrative. The female protagonists in *Interstellar* are denied the full liberation of space exploration and need to find their dark, brooding meaning through their ultimate connection with the grounding on the firmness of Earth, even though the overall narrative of progress in the film *Interstellar* is one that suggests leaving it behind. Again, though the image of Earth is dealt with swiftly within the scenographic narrative, the modernist residue of the symbolism that Blue-Ball thinking evokes still interrupts the cinematic imagination of the characters and defines their agency through their gendered relationship with the image of Earth.

10.3.20 Conclusion

"We used to look up in the sky and wonder at our place in the stars. Now we just look down and worry about our place in the dirt." (Coop, in *Interstellar* 2014, 16:01–16:13)

Despite the groundbreaking engagement with hard science and the positive (because necessary) presentation of space travel in *Interstellar*, the Blue-Ball image of Earth that evokes

---

[79] It is also of interest in considering that Brand's motivation to travel alone to this potential, new "Earth" is ultimately made by Coop when he forces the situation at a critical point near the black hole (02:11:41–02:12:09) so that she cannot. In Gargantuan, Coop encounters the tesseract alone, and solves the critical aspect that allows the mission to ultimately succeed putting knowledge both from Brand and Murph into play in his actions.

the emblematic elements of modernist values is still a significant symbol within the film narrative. The stasis for action on seeing Earth as a symbol is echoed in the still photo from the final scene from the drama *The Blue Ball* (Godfrey 1995) that prefaces this writing. In this image, the writer character, Paul, is rising though the universe, yet is bound by the production conventions of the performance space. Similarly, the philosophical parameters that we have linked to Earth though this Blue-Ball image seem also suspended in the nostalgic parameters of modernist-identity formation that are bound by environmental anxiety. Constructed in this manner, the Blue-Ball image of Earth in current film-making limits the potential for science-fiction thinking (Landon 2002). This mode of dwelling on the spectacular image of the Blue Ball calls an individualist human connection to Earth into an audience's consciousness. It triumphs a heroic narrative that is often bound by solipsistic and environmental pessimism or stasis, or, as Coop says in the quotation above, "our place in the dirt."

There is a retrogressive, distinct back-to-Nature undercurrent in this presentation that both determines and plays out gender dynamics that belong to a previous century, since they fix traditional roles within the character of protagonists. This is demonstrated in the relationship between the image of Earth and the protagonists in both the visual narrative and the narrative dynamic created by this relationship, and echoes in the residue that it leaves behind in its wake.

Blue-Ball thinking is additionally wrapped in the aesthetics of realism that suggest that, in using scientific images in the film, production values creates a form of truth-telling. In this respect, the current presentation of realism in filmic scenography uses mediatized images of an experience of space, perhaps fueled by the current CGI demand for realism, which flatten the distance for an audience and create the idea of knowledge or reality. This reality is, of course, a fiction, but nevertheless a powerful one that presents an existence on Earth which is presently limited to a grand narrative mode of individual thinking in the face of crisis.

The image of the Blue Ball and its framing in scenography needs to be released from its current realist narrative in film and modernist parameters and employed differently. Rethinking this relationship would encourage a more imaginative engagement to our connection and interest in space, and would further the relationship between artistic vision and scientific interest. Hollywood films need to engage with more radical thinking about space, not by necessarily recreating an experience of the cosmos through the cinematic immersion in technological space, but by framing Earth in a way that allows us to imagine an alternative relationship. The cinematic imaginary of Earth needs to be rethought to produce a less territorializing image that escapes the current determinist boundaries of discussion, which construct simplistic binary thinking about Earth and the "otherness" of space in which we are immersed at present.

Not literalizing the narrative using the "Blue Ball" in the way it is currently presented, with its fixed thinking, would be a way forward, and there perhaps needs to be greater space for an audience's contribution to the visual meaning of the scenography, if science-fiction narratives in film (and theater) are to fulfill their potential. Scenographically reconfiguring our "real" relationship with the image of the Blue Ball would liberate science-fiction thinking and reinvigorate humanity's ability to act, discover, gain knowledge, and explore the unknown.

COMBINED REFERENCE LIST FOR CHAPTER 10

E. Armstrong, A scenographic analysis of emergent British "National" identity, on the stages of the national theatre between 1995–2005. Ph.D. thesis, University of Surrey, 2010

M. Benjamin, *Rocket Dreams* (Chatto & Windus, London, 2003)

W. Benjamin, *Illuminations, translated by H. Arendt* (Pimlico, London, 1999)

M. Billington, Michael Billington finds Moon Exploration Shaky Ground on Which to Base Drama: Lost in Space, The Guardian, April 1 (1995) Accessed at the National Theatre archive, Press reviews folder (RNT/1550 PR/4/1/586)

J. Collins, A. Nisbet, *Theatre and Performance Design: A Reader in Scenography* (Routledge, London, 2010)

N. De Jong, The playwright who fell to Earth: The Blue Ball Cottesloe theatre. Evening Standard (1995), Accessed at the National Theatre archive, Press reviews folder (RNT/PR/4/1/586)

M. Dekoven, Gender and Modernism, in *The Cambridge Companion to Modernism*, ed. by M. Levenson, 2nd edn. (Cambridge University Press, Cambridge, 2011), pp. 212–231

J. Duncan, That our feet may leave, Cinefex 140, Jan 2015, pp. 36–73

G. Edwards, Dream landscapes—outer space (2015), http://cinefex.com/blog/outer-space/. Accessed 30 May 2015

J. Fordham, Gravity. Cinefex **136**(January), 42–75 (2014)

P. Godfrey, *The Blue Ball* (Methuen, London, 1995)

C. Hadfield, Space oddity (2013), www.youtube.com/watch?v=KaOC9danxNo. Accessed 9 July 2015

T. Hogg, Paul Franklin talks about the Oscar nominated interstellar (2015), cgsociety.org, www.cgsociety.org/index.php/CGSFeatures/CGSFeatureSpecial/paul_franklin_talks_about_the_oscar_nominated_interstellar. Accessed 3 Mar 2015

C. Innes, *Modern British Drama: The Twentieth Century* (Cambridge University Press, Cambridge, 2002)

G. King, *Spectacular Narratives: Hollywood in the Age of the Blockbuster* (I.B. Tauris, London, 2000)

S. Knapton, Beagle 2 found on surface of Mars after vanishing for 12 years. The Telegraph (2015), www.telegraph.co.uk/news/science/science-news/11349163/Beagle-2-found-on-surface-of-Mars-after-vanishing-for-12-years.html. Accessed 3 Mar 2015

K. Knowles, Rosetta mission: Philae runs out of steam...but all its data is home. The Independent (2014), www.independent.co.uk/news/science/rosetta-mission-philae-runs-out-of-steam-but-all-its-data-is-home-9863480.html. Accessed 3 Mar 2015

B. Landon, *The Aesthetics of Ambivalence: Rethinking Science Fiction Film in the Age of Electronic (Re) Production* (Greenwood Press, Westport, CT, 1992)

B. Landon, *Science Fiction after 1900: From the Steam Man to the Stars* (Routledge, London, 2002)

M. Lathers, *Space Oddities: Women and Outer Space in Popular Film and Culture, 1960–2000* (Continuum International Publishing Group, New York, 2010)

B. Latour, To Modernize or to Ecologize? That's the Question*, in *Remaking Reality: Nature at the Millenium*, ed. by N. Castree, B. Willems-Braun (Routledge, London, 1998), pp. 221–242

P. Lewis, *The Cambridge Introduction to Modernism* (Cambridge University Press, Cambridge, 2007)

R. Luckhurst, Space is the Place. Sight & Sound **23**(12), 24–28 (2013)
A. Macaulay, Lost in space, crippled by irony. Financial Times Weekend (1995), Accessed at National Theatre archive, Press reviews folder (RNT/PR/4/1/586)
D. Massey, *For Space* (Sage, London, 2005)
M. McLuhan, *Understanding the Media: The Extensions of Man* (Routledge Classics, London, 2001)
L. Mulvey, Visual pleasure and narrative cinema. Screen **16**(3), 6–18 (1975), http://iml-portfolio.usc.edu/ctcs505/mulveyVisualPleasureNarrativeCinema.pdf. Accessed 20 June 2015
M. Nijhuis, Is the "Ecomodernist Manifesto" the future of environmentalism. The New Yorker (online), June 2 (2015)
Royal National Theatre (RNT), *The Blue Ball: A New Play by Paul Godfrey* (RNT/PP/1/4/171) (1995)
C. Sagan, *Pale Blue Dot: A Vision of the Human Future in Space* (Random House Publishing Group, New York, 1994)
S. Schwam, The Making of 2001: A Space Odyssey (The Modern Library, New York, 2000)
L.L. Schapira, *The Cassandra Complex—Living with Disbelief: A Modern Perspective on Hysteria* (Inner City Books, Toronto, 1988)
E. Shay, Moon. Cinefex. **118**(July), 11–20 (2009)
K. Shepherd-Barr, *Science on Stage: From Doctor Faustus to Copenhagen* (Princeton University Press, Woodstock, 2006)
V. Sobchack, *Screening Space: The American Science Fiction Film*, 2nd edn. (Rutgers University Press, New Brunswick, 2004)
C. Spargo, Ticket holders set to travel on virgin spaceship show their support for Richard Branson and his program as CEO says he will refund $250,000 tickets after tragic explosion—but will celebrities still take risk? Daily Mail Online November 2, (2014), www.dailymail.co.uk/news/article-2816853/Celebrities-ticket-holders-set-travel--Virgin-spaceship-support-Richard-Branson-program-CEO-says-refund-250-000-tickets-tragic-explosion.html. Accessed 3 Mar 2015
K. Thorne, *The Science of Interstellar* (WW Thornton and Company, New York, 2014)
T. Webber, A. Cuarón, BFI Masterclass, October 11, 2013, published in Leaving the Earth Behind. Sight & Sound **23**(12), 29–30 (2013)

Films

Alien (1979) Directed by Ridley Scott (DVD) UK, Twentieth Century Fox Film Corporation
Elysium (2013) Directed by Neill Blomkamp (DVD) UK, MRC II Distribution Company L.P.
Gravity (2013) Directed by Alfonso Cuarón (DVD) UK, Warner Home Video
Interstellar (2014) Directed by Christopher Nolan (DVD) UK, Warner Home Video
Moon (2009) Directed by Duncan Jones (DVD) UK, Lunar Industries Ltd.
Oblivion (2013) Directed by Joseph Kosinski (DVD) UK, Universal Pictures
Outland (1981) Directed by Peter Hyams (DVD) UK, Warner Home Video
Silent Running (1972) Directed by Douglas Trumbull (DVD) UK, Universal Pictures
2001: A Space Odyssey (1968) Directed by Stanley Kubrick (DVD) UK, Warner Home Video

11

Space ecology

Michael N. Mautner and Simon Park

Two individual authors Mautner and Park discuss different aspects of the nature and quality of living things in interstellar space and their relationship with humanity.

11.1 Saving life itself: life-centered ethics, astroecology, and our cosmological future

Michael N. Mautner, Research Professor at Virginia Commonwealth University.

11.1.1 Biology, ethics, and purpose

It is the human purpose to propagate life. This objective can be realized best in space, where life can have an immense future, giving our existence a cosmic purpose[1].

We are part of life and, as humans, we seek a purpose for our existence. We can derive this purpose from our fundamental unity with all biological self-perpetuating life. Belonging to life then implies a human purpose to safeguard, expand, and propagate life. This purpose can be fulfilled in space, whose resources allow life to expand immensely in time, space, complexity, and diversity.

We share with all cellular organic life the complex structures and processes that we carry in our every cell. At the heart of these processes is self-reproduction through gene/

---

[1] www.astroecology.com contact mmautner@vcu.edu.

Michael N. Mautner (✉)
Research Professor, Department of Chemistry, Virginia Commonwealth University, Richmond, VA 23284-2006
e-mail: mmautner@vcu.edu

Simon Park (✉)
Senior Lecturer in Molecular Biology, Department of Microbial Sciences, Senate House, Stag Hill Campus, Guildford GU2 7XH, United Kingdom
e-mail: s.park@surrey.ac.uk

protein cycles: DNA sequences encode proteins that help to reproduce the DNA code. These cycles propagated life over three billion years. The fragile cell outlasted billions of stars that formed and faded while life continues.

Life is unique in its complexity and because the laws of Nature narrowly allow biology to exist. Biology would not be possible if electro-magnetism, chemistry, thermodynamics, nuclear physics, and gravity were even slightly different. In these respects, the universe came to a special point in life.

Also special to life is active self-propagation. The resulting human purpose to propagate life defines the basic values of life-centered biotic ethics: acts that promote life are good, and acts that destroy life are evil. In particular, threats that can destroy all life are ultimate evils that can never be justified. Conversely, acts that secure life have the highest moral value. Panbiotic ethics apply these principles for extending life space.

Ultimately, in a self-fulfilling future, our destiny will depend on our ethics. Guided by life-centered ethics, space can transform life and life can transform space. Our ethics can then become a controlling force in Nature (Mautner 2005b, 2009).

11.1.2 The human role

"What is life that we aim to propagate?" The definition of "life" is a question of semantics that we can answer using science. Biological life can be defined as a process that involves a flow of mass and energy, which enables DNA sequences to code for proteins which in turn help to replicate the DNA sequences. Life is, then, a self-propagating cycle wrought in molecular structures.

This self-propagation is central to life. All life seeks this outcome actively and therefore, in effect, life pursues a purpose. Where there is life, there is purpose, and a universe that contains life contains purpose. Seeding the universe can be the most profound human enterprise. It can also promote peace by motivating peaceful cooperation for a shared future in the galaxy.

The survival of life will depend on human powers because, without us, life will end with the Red Giant Sun. However, the duration of our spacefaring abilities is unknown and we should pursue a panbiotic program promptly. Life in space can then develop many new branches, some intelligent, who will further expand life in the galaxy. Humankind itself can also diversify into many species adapted to various planetary and space environments. These new species will remain our descendants if their genes build on our human genes. We may also coexist with other species, and possibly with robots and cyborgs. However, control must remain in biological gene/protein brains with a vested interest to propagate life.

For the present, our unique human powers should be guided by life-centered ethics to secure our survival. Our descendants will be there then to expand life and reach for infinity (Mautner 2005b, 2009).

11.1.3 The origins of life and ecology

11.1.3.1 Can we estimate the probability of life?

Expanding life in space is vital if we are alone, with the fate of all life in our hands. There is no scientific evidence for extraterrestrial life presently, but we can end this cosmic isolation by seeding new solar systems with microbial representatives of our gene/protein life form. Tens of billions of habitable solar systems are available for this purpose in the galaxy.

Saving life itself: life-centered ethics, astroecology, and our cosmological future   257

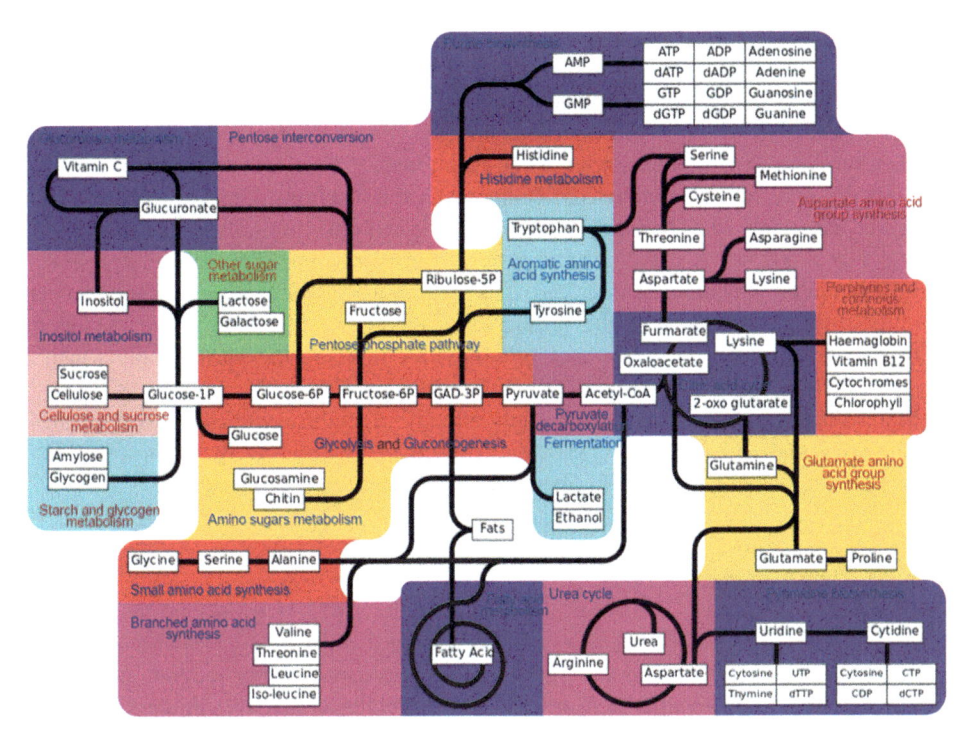

Figure 11.1 The complexity of life. Every cell, including microorganisms, must perform all the chemistry in these metabolic pathways, each containing up to hundreds of complex compounds and reactions. We cannot quantify experimentally the probability that these components should assemble abiotically to form a protocell. Credit: Michael N. Mautner

Ecology concerns the interactions of biota with its environment and, similarly, astroecology concerns interactions of life with space environments. Specifically, astroecology can estimate the amount of life possible using space resources, and cosmo-ecology extends these relations to cosmological timescales (Mautner 2002a, b, 2005a).

Are we alone? The complexity of life is key to this question. To answer, we need to estimate the probability that complex cells can arise and survive. We don't know how much complexity a viable system requires, but every cell contains thousands of different chemicals (DNA/RNA, proteins, energy apparatus, membranes, enzymes, hormones, lipids, saccharides). All have complex specific structures and need to work in unison (Purves et al. 2001). Even bacterial DNA contains over a million base-pairs that can form over $4^{1,000,000}$ different sequences, only one of which may be viable. In this sense, the DNA of one bacterium contains more information than all the 10^{80} particles of the inanimate universe (Fig. 11.1).

11.1.3.2 Chemical kinetics, complexity, and the origin of life

In a protocell, compounds react with each other at rates determined by rate coefficients, which are affected by temperature, pressure, pH, ionic strength, catalysts, and inhibitors. These parameters can give each rate coefficient a continuum range of values. For simplicity, we assume that each rate coefficient can have ten discreet values.

For example, the citric acid cycle to form ATP (middle-right purple cycle in Fig. 11.1) and its precursors involve nine reactions, each catalyzed by an enzyme. If each rate coefficient and enzyme concentration can assume ten discreet values, then this system can have 10^{18} chemical states and a probability $P_{origin} = 10^{-18}$ to form a viable combination abiotically, even for this small subset of cell reactions.

However, cells involve hundreds of components and reactions. If all n compounds in a cell react with each other pairwise, including self-reactions, they can undergo $n(n+1)/2$ bimolecular reactions. A simple protocell of only $n = 100$ compounds can then undergo 5050 different bimolecular reactions and, if each is controlled by a rate coefficient that can have ten different values, then the system can have $10^{5,050}$ different chemical states as defined by these parameters. It is clearly impossible to explore experimentally whether any of these states can lead to life.

Moreover, the first microbial community already needed to synthesize biomolecules, recycle nutrients, and adapt to changing environments, increasing the complexity of a viable ecosystem and reducing P_{origin} even further.

11.1.3.3 The origins of ecology

Beyond its abiotic assembly, the first cell also needed to survive in an environment controlled by the combination of many parameters, only a few of which may allow survival (small $P_{survival}$). The probability to form a minimal viable biosphere of a microorganism in a viable environment as then $P_{biosphere} = P_{origin} \times P_{survival}$ (Eq. 11.1 below).

Consider a simplified environmental system of six processes controlled by six rate coefficients each with ten possible discreet values. This allows 10^6 combinations of rate coefficients. If one combination is viable, then $P_{survival} = 10^{-6}$. Combined with the above citric acid cycle "protobiology," this gives $10^{18} \times 10^6 = 10^{24}$ possible states, even for this greatly simplified ecosystem, leading to a probability of $P_{biosphere} = P_{origin} \times P_{survival} = 10^{-18} \times 10^{-6} = 10^{-24}$ that life will arise there spontaneously (Fig. 11.2).

Experimental testing of the possible states of even such a small subset of a prebiotic biosystem would require 10^{24} experiments to determine whether they can form life. If each state is tested in one day, then this would require 10^{24} days (over 10^{21} human-years—an improbable billion scientists working for a trillion years on this one model). Real

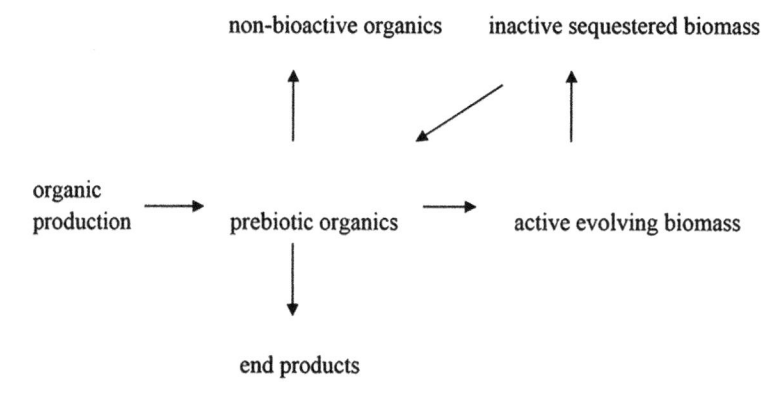

Figure 11.2 Processes in an early ecosystem. Credit: Michael N. Mautner

Saving life itself: life-centered ethics, astroecology, and our cosmological future

biological systems are still much more complex. Therefore, models for the origins of life with realistic complexities cannot be tested experimentally in practice.

Further, exponential growth of the microorganisms would soon exhaust all the nutrients in a finite ecosystem. The first populations then needed both autotrophs to synthesize biomolecules and heterotrophs for recycling. It is unlikely that both types will arise simultaneously in the first ecosystem.

The probabilities P_{origin} and $P_{biosphere}$ for life to arise can then have values from near certainty to vanishingly small probabilities even in (10^{11} galaxies in the universe $\times 10^{10}$ habitable solar systems/galaxy) = 10^{21} habitable solar systems in the universe. This prevents quantitative estimates of the probability that extraterrestrial life started elsewhere.

By formal logics, the statement "there exists extrasolar life" can be falsified only if every habitat in the universe is barren, which cannot be tested. Conversely, the statement "there is no extraterrestrial life" can be falsified by one instance of extrasolar life, but probe missions to even a few extrasolar systems to test this may last for millennia.

If we cannot estimate, prove, or refute extraterrestrial life scientifically, then we must assume that we may be alone. For caution, we must then seed the universe promptly to secure life.

11.1.4 Biological challenges

Beyond space transport, the key to space is biology. Fortunately, physics allows a multitude of life forms beyond what Nature has already realized. Given immense galactic resources and trillions of eons of future time, biology will implement these multitude biological forms naturally or by design. These developments will be limited only by survival: those species that survive and propagate will prevail, and those that don't propagate will perish. The forms of life and the mechanisms of evolution may change, but the logic of life, test by survival, is permanent.

The first space colonizers will likely be microorganisms seeded into unpredictable environments. These starting populations need to contain poly-extremophiles that can be developed with genetic engineering to survive diverse temperatures, pressures, pH, salinity, and radiation (Montague et al. 2012). They need to include anaerobes for probable $N_2/CO/CO_2$ or hydrogen/water/ammonia/methane atmospheres, and both autotrophs and heterotrophs to synthesize biomolecules and to recycle nutrients. The colonizers should also include hardy cysts of simple multi-cellulars such as rotifers to jump-start higher evolution.

Such self-sustaining colonizer populations are relevant both to eco-synthesis for human populations on asteroids and planets here and for the seeding of extrasolar planets. Current levels of biotechnology can develop these microorganisms and test them in model planetary environments.

11.1.5 A roadmap to the galaxy

Bringing life to the universe will be the most profound human enterprise, and establishing this future can yield deep moral rewards. In fact, this life-centered program can be the agent of peace, as society can work united so that our descendants will share the galaxy.

The roadmap to space can follow two strategies: human expansion in our Solar System, and seeding new exoplanets with microbial representatives of our gene/protein life form.

We started to take steps in the Solar System and we can estimate the scope of life possible here. We can also start soon to seed the galaxy, with a panbiotic program of directed panspermia to secure and expand life. This panbiotic program can be accomplished with proven technologies and launched from a space infrastructure by groups or individuals.

The needed basic technologies are advancing, and missions can be developed now in engineering detail. However, pragmatically, the first steps in space must benefit society on Earth, as follows:

1. a space infrastructure to serve human needs: satellite solar power systems for permanent clean energy; a space-based solar shield against climate warming (Seifritz 1989; Mautner 1989, 1991); permanently cold lunar gene banks for endangered species (Mautner 1996); mining asteroid resources (O'Leary 1977);
2. space settlements to accommodate growing populations, toward colonizing the Solar System (O'Neill 1974);
3. astronomy, including lunar telescopes, to identify habitable exoplanets as targets for panspermia and future human colonization;
4. fast interstellar propulsion and in-flight navigation (Matloff 2012);
5. biology and ecology; genetic engineering of hardy poly-extremophile microorganisms and small multicellulars to start higher evolution; studying their survival in model planetary environments and on asteroid/meteorite materials;
6. continued searches for extrasolar life; if found, life is secure and a panbiotic program can help its expansion; however, if all observed extrasolar systems are barren, we may be alone; the fate of life may be then in our hands, and a panbiotic program will be imperative.

11.1.6 Astroecology

Ecology concerns the relationship between life and its environment. Similarly, astroecology concerns the relations between life and resources in space. These relations apply to early life on Earth and to future life that can be based on in-situ biological resources in space.

To study these relations quantitatively, we must first define the amounts of life in an ecosystem. This amount can be defined as the amount of biomass as a function of time multiplied by its duration in time (Eq. 11.2 below).

One subject of experimental astroecology in the Solar System is the capability of planetary (including asteroid/meteorite) materials to sustain microorganisms and plants. For these purposes, we studied nutrients in meteorites and the growth of autotrophs (algae), heterotrophs (bacteria), and plant-tissue cultures, and even brine shrimp on these materials (Mautner et al. 1995; Mautner 2002a, b, 2014). The results showed that a variety of organisms can grow on these substrates. Miniaturized soil tests yielded fertility ratings of meteorites that were similar to terrestrial agricultural soils. Martian meteorites performed well because of high soluble phosphate contents.

For example, we measured recently the contents of soluble bioavailable nutrients in CM2 (Murchison), CV3 (Allende), and several Antarctic carbonaceous chondrite materials, and derived the nutrients in a composite asteroid soil that may be constructed in a

Saving life itself: life-centered ethics, astroecology, and our cosmological future

Table 11.1 Biomass and human populations that can be constructed from the total elemental contents of carbonaceous chondrite asteroids in the Solar System[a]

Meteorite type	Organic C	Organic N	Nitrate N	Phosphate P	K+	Population[b]
Murchison CM2	10.3	2.2	0.1	0.06	7.0	6E15
Allende CV3	2.1	1.9	0.1	0.19	0.8	1E16
Av. soluble in carbonaceous chondrites[c]	2.4	1.5	0.3	0.08	2.5	8E15
Av. total in carbonaceous chondrites[d]	15.9	5.9	–	74	17.8	6E17

[a]Units of gram biomass/kg meteorite. For each element x, maximum biomass (grams) that could be constructed from 1 kg resource material if element x was the limiting nutrient
[b]Population of humans (100 kg/person with average biomass composition, Bowen 1966) allowed by the limiting nutrient in 10^{22} kg carbonaceous asteroid materials
[c]Biomass from soluble bioavailable contents in average of carbonaceous chondrites
[d]Biomass from total elemental contents in average carbonaceous chondrites (Mautner 2014)

space settlement. Comparing these with biomass requirements then allows calculating the amounts of time-integrated biomass, and the human populations that can be constructed in the Solar System from carbonaceous asteroids or later from comets (Table 11.1). The relevant equations are shown below (Fig. 11.3).

11.1.7 Human prospects

11.1.7.1 Human adaptation

Humankind will diversify in space into new post-human species. The following can serve as guiding principles.

Human colonization of space will require new physiologies suited to diverse planetary environments. Conversely, planetary environments will need to be modified to suit these post-human physiologies. However, full adjustment entails life in open space, with self-contained physiology and full recycling of materials. Direct conversion of stellar radiation to bioenergetics is also needed, possibly through symbiosis with cyanobacteria contained in photosynthetic organs (Tsiolkovsky 1903). This space-based physiology will also need adjustments to extreme cold or heat (insulation); zero gravity (altered bones/muscles, solar sail organs for transport); communication in vacuum (optical and radio organs); and probably asexual reproduction.

Further, social adjustment will be needed, ranging from solitary life to crowded habitats, and coexistence with other species, even hardy robots and cyborgs. However, to continue biological life, control must always remain with biological gene/protein brains with a vested interest to propagate life. These space-adapted species will remain our descendants if, as in natural evolution, their genes build on our human genes.

11.1.7.2 Collective organisms and interstellar human travel cell by cell

Interstellar travel appears feasible for microorganisms but not for humans (Mauldin 1992). Nevertheless, we aspire to spread intelligent life in the galaxy, preferably our descendants.

Complex human organs can be damaged irreversibly during prolonged interstellar travel. However, individual cells could self-repair, for example, by mechanisms borrowed

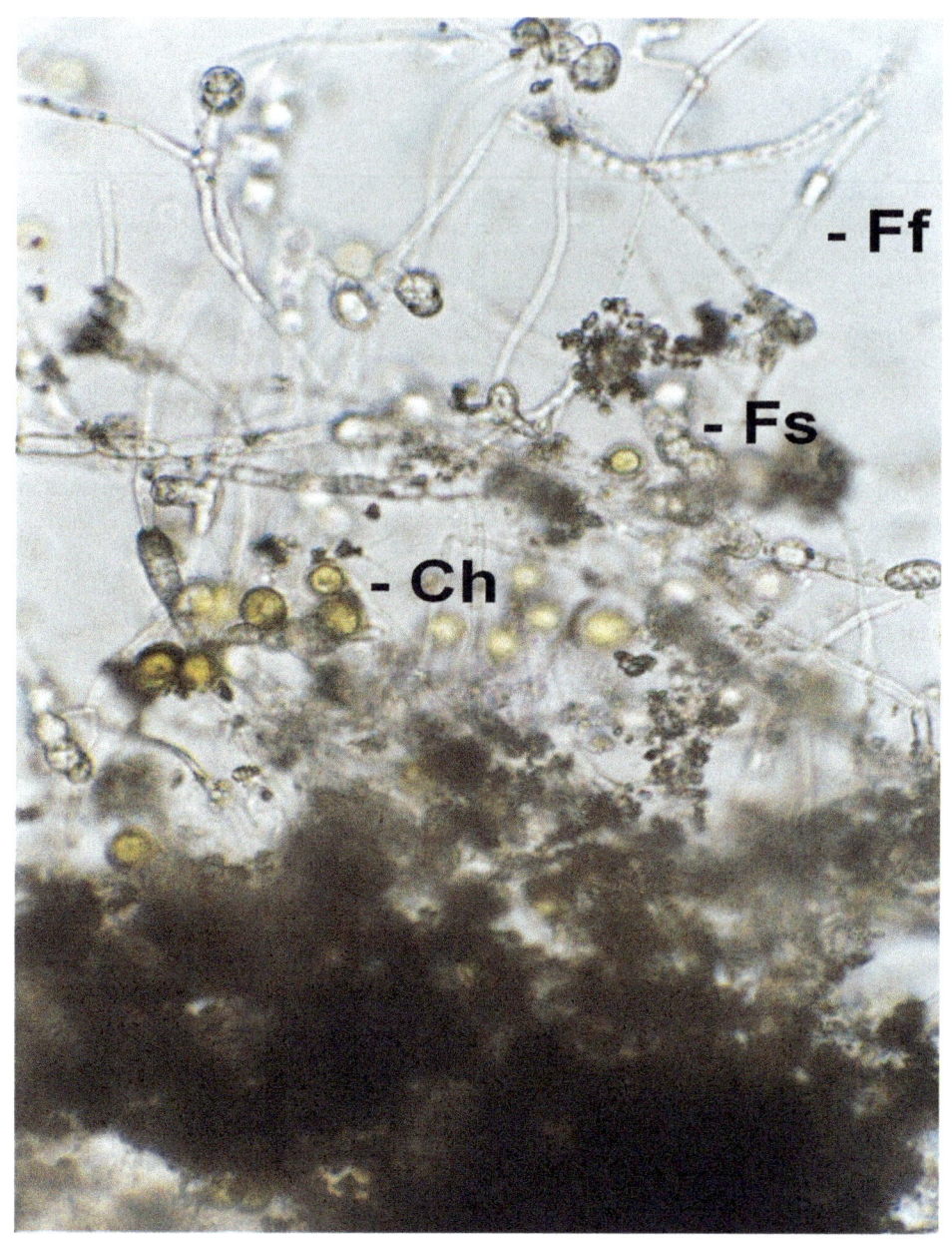

Figure 11.3 Planetary ecosystems: microorganisms growing on meteorite soils. Similar populations of photosynthetic algae and recycling fungi can start self-sustaining ecosystems on new planets (*Ch* chlorella, *Ff* fungal filaments, *Fs* fungal spores) Credit: Michael N. Mautner (2002a)

Saving life itself: life-centered ethics, astroecology, and our cosmological future

from *Deinococcus radiodurans* into human stem cells. Once arrived in suitable environments, they can reproduce and assemble into organs and humans, similarly to microorganisms that can live single-celled or in colonies. They can establish evolving ecosystems supported by environmental microorganisms that were included in the missions. Guided by panbiotic ethics, they can develop spacefaring societies that expand further.

Life can unfold in space in innumerable ways. Such collective organisms (Clarke 1956) may be one way for sentient life to expand in the galaxy.

11.1.8 Cosmo-ecology and the long-term future

We may wonder about the long-term future of life that we strive to propagate. By cosmological predictions, biological life may remain possible for over 10^{20} years about stars, and possibly over 10^{35} years until baryonic matter decays (Adams and Laughlin 1999). What is the ultimate amount of life that can form in the universe, and is it finite?

In terms of time-integrated biomass ($BIOTA_{int}$), life can grow until all matter is turned into life and its supporting materials, sustained eventually by stellar energy about White and Red Dwarf stars.

Table 11.2 summarizes the maximum amounts of life allowed by limiting resources that are carbonaceous asteroid materials in the Solar System or the power output of future stars (column 2) that allow the biomass (column 3). This biomass and its potential lifetime (column 4) yields the time-integrated biomass allowed about each star. In turn, the number of these stars in the galaxy (10^{11} solar-like systems, 10^{12} White and Red Dwarfs) yields the $BIOTA_{int}$ possible in the galaxy (column 5). The calculations assume using the total elemental contents of the asteroids, and a power requirement of 100 W/kg biomass.

We may also estimate the maximum possible amount of life in the universe. This would be realized if all the ordinary matter is converted to biomass, and then some of it converted relativistically ($e=mc^2$) to energy (see equations below). The last two rows of Table 11.2 list the maximum amounts of biological life that can then exist in the galaxy and in the universe (Mautner 2005a). Abstract organized entities but not biological life may or may not be possible indefinitely (Dyson 1979).

The ultimate future depends on the long-term behavior of dark matter and energy that may require observations for eons. Our descendants may then understand Nature more deeply and transform it to serve life permanently.

Table 11.2 Resources, potential biomass, and time-integrated biomass ($BIOTA_{int}$) in space

Location	Materials (kg) or energy (W)	Biomass (kg)	Future lifetime (years)	$BIOTA_{int}$ in the galaxy (kg years)
Earth to present		10^{15}	4×10^9	4×10^{24}
Solar system	Asteroids, 10^{22} kg	6×10^{20}	5×10^9	3×10^{41}
White dwarfs	10^{15} W	10^{13}	10^{20}	10^{45}
Red dwarfs	10^{23} W	10^{21}	10^{13}	10^{46}
Galaxy (baryons, 10^{41} kg)	mc^2/t	$<10^{41}$	10^{37}	3×10^{48}
Universe (baryons, 10^{52} kg)	mc^2/t	$<10^{52}$	10^{37}	3×10^{59}

11.1.9 Quantitative astroecology

11.1.9.1 Origins of life and ecology

The first protocell had to arise in a supporting environment that needed to fulfill many conditions with many variables to allow a complex cell to originate, and the environmental variables that support its survival (small $P_{survival}$). The probability to form a viable biosphere can be then given then in Eq. 11.1:

$$P_{biosphere} = P_{origin} \times P_{survival} \tag{11.1}$$

Both P_{origin} and $P_{survival}$ may be products of the probabilities of other variables that need to coincide into viable combinations.

11.1.9.2 The amounts of life in finite ecosystems: time-integrated biomass

To analyze future life, its amounts must be quantified. Any finite ecosystem is limited in mass and time, and the amount of life that it harbors through its duration may be quantified as time-integrated biomass $BIOTA_{int}$ (Biomass Integrated Over Time Available):

$$BIOTA_{int} = \int m_{biomass}(t) dt. \tag{11.2}$$

Here, $m_{biomass}(t)$ is the biomass at time t and integration is from the start of life t_o to time t. For constant steady-state biomass lasting for time t, then $BIOTA_{int} = m_{biomass,steady-state} t$ (in units, for example, of kg-years).

If a fraction of an initial biomass m_o is used up or lost as wastage per unit time without replacement, then the mass decays exponentially and the remaining biomass m_t at time t is given by Eq. 11.3, where k is the decay rate coefficient:

$$m_t = m_o e^{-kt}. \tag{11.3}$$

The time-integrated biomass from time t_o (time 0) to t is then given by integrating Eq. 11.4:

$$BIOTA_{int} = (m_o / k)(1 - e^{-kt}) \text{ and for } t = \text{infinity } BIOTA_{int} = m_o / k. \tag{11.4}$$

Equation 11.4 applies to any unit biomass constructed. The total $BIOTA_{int}$ in the ecosystem depends only on the total biomass constructed from the resources (Eq. 11.5 below) and the decay coefficient k, but not on the rate of construction. Therefore, the duration of life in an ecosystem can be extended by constructing and maintaining a smaller steady-state biomass, rather than converting and then depleting all the resources quickly. In the Solar System, this can continue life throughout its habitable lifetime.

11.1.9.3 Relations between biomass and resources

The potential total biomass in an ecosystem is determined by the amount of resource materials and their nutrient contents versus the elemental contents of the biomass:

Saving life itself: life-centered ethics, astroecology, and our cosmological future

$$m_{x,biomass} = m_{resource} c_{x,resource} / c_{x,biomass}. \quad (11.5)$$

Here, $m_{resource}$ is the amount of resource materials (e.g. soils or meteorite solids) and $c_{x,resource}$ and $c_{x,\,biomass}$ are the concentrations of element x in the resource materials and in biomass, respectively, that could be constructed if x was the limiting nutrient.

Alternatively, if energy is the limiting factor, then Eq. 11.6 applies:

$$m_{biomass} = P_{source} \times c_{eff} / P_{use}. \quad (11.6)$$

Here, P_{source} is the power output of the energy source (e.g. star), c_{eff} its conversion rate, and P_{use} is the power use (energy/(time × biomass)).

11.1.9.4 Ultimate life in the galaxy and universe

The ultimate biomass in the galaxy or universe would be obtained by converting all ordinary matter to biomass and then converting a fraction to energy continually to power the biomass. If a fraction of the biomass is converted relativistically to energy ($e = mc^2$), then $m_{biomass}$ will decay exponentially as in Eq. 11.2 with $k = P_{biomass}/E_{yield} = P_{use}/c^2$:

$$m_t = m_o e^{-(P(use)/c^2)t}. \quad (11.7)$$

Starting with biomass m_o, then the total time-integrated biomass from t_o to infinity is then given:

$$BIOTA_{int} = m_o c^2 / (P_{use}). \quad (11.8)$$

The maximum theoretical $BIOTA_{int}$ in the galaxy and the universe was derived by Eq. 11.8 assuming a galactic mass of ordinary matter of 10^{41} kg and universe mass of 10^{52} kg and power use of 100 W/kg biomass.

11.1.10 Conclusions

We are entering an era of self-determination empowered by science. To survive, we must then pursue survival deliberately, guided by the human purpose to propagate life. These life-centered ethics must themselves be propagated, because life will always be tested by survival.

We shall probably develop adequate methods of space transport and in-situ resource utilization in this century. Beyond that, the key to space lies in biology. Planetary habitats or space-adapted individuals will need supporting ecosystems of new extremophile microorganisms. The space colonists will also have to evolve in these environments, and possibly coexist with other species, and with robots and cyborgs. However, control must always remain in biological brains with vested interest to continue our gene/protein biological life form.

As for space resources, astroecology experiments found that microorganisms and plant cultures can grow well on asteroid/meteorite materials, whose fertilities are similar to terrestrial soils. Similar materials should exist in other solar systems with far-reaching implications: if life can grow on Earth, then life can flourish throughout the universe.

Given galactic resources and trillions of future eons, life can then reach its full potentials. In that future, our descendants may understand Nature more deeply and try to extend life indefinitely. Founding this immense future can give our human existence a cosmic purpose.

11.2 Saving life itself: science and strategies for seeding the universe

Michael N. Mautner, Research Professor at Virginia Commonwealth University.

11.2.1 Synopsis

We belong to life, and we can derive our human purpose from this identity. The unity with all self-perpetuating biological life then implies a human purpose to safeguard, expand, and propagate life[2]. This purpose is best fulfilled in space, where life can have an immense future. We can secure this future by seeding new near solar systems with microbial representatives of our organic gene/protein family of life. This panbiotic program can be accomplished with current technologies, to start life on exoplanets nearby and in young solar systems in star-forming interstellar clouds. Life in space can then create diverse new species, some intelligent who can expand life further in the galaxy. In these descendants, our life can fulfill a cosmic purpose.

11.2.2 Seeding the universe: motivations for directed panspermia

It is the human purpose to propagate life. This is best achieved in space where life has an immense future. Founding that future gives our existence a cosmic purpose. Seeding the universe will be the most enduring human enterprise, which can also give us deep moral satisfaction presently. We can start soon this panbiotic program with proven technologies. The missions can be also launched readily by motivated groups and organizations from an emerging space infrastructure. This chapter outlines the main strategies of this program.

Searches for extrasolar life are continuing. If found, life is secure and a panbiotic program can help its expansion. However, if all the explored solar systems are barren, we must assume that we may be alone. The fate of life is then in our hands and a panbiotic program to secure life is imperative.

Although Earth is the cradle of life, "one cannot live in the cradle forever" (Tsiolkovsky 1903). We are already pressing the carrying capacity of Earth, and life on Earth will end with the Red Giant Sun or sooner.

In contrast, immense scopes of life can last for trillions of eons in the galaxy. Interstellar human colonization is impractical presently (Mauldin 1992) but microorganisms also carry the essential features of our organic gene/protein life form. Directed microbial panspermia can therefore secure and expand biological life and allow it to realize its full potentials. Ultimately, life will pervade space and space will transform life (Tsiolkovsky 1903; Shklovskii and Sagan 1966; Crick and Orgel 1973; Montague et al. 2012).

For these reasons, we examined the ethics, technology, and strategies directed at panspermia for seeding exoplanets and star-forming interstellar clouds as summarized below (Mautner and Matloff 1979; Mautner 1995, 1997; 2005a, b).

---

[2] www.panspermia-society.com contact mmautner@vcu.edu

Saving life Itself: science and strategies for seeding the universe

11.2.3 Evaluating potential targets

The purpose of directed panspermia missions is to secure and extend life in space. The desirable features of the targets include:

- location nearby, for accurate targeting and survivable transit times;
- accurate knowledge of the target's proper motion, for precise aiming;
- large, easy targets;
- Earth-like planets in habitable zones, with moderate temperatures, liquid water, shielding atmosphere, and light for photosynthesis;
- young solar systems, where local life, and especially advanced sentient life, could not have formed yet;
- long habitability to extend the duration of life and allow its advanced evolution.

We can use two strategies: to start life in nearby solar systems or in clusters of new stars in interstellar clouds. On Earth, a single primordial cell may have started life but, to be sure, we assume that delivering a critical biomass with a self-sustaining mix of 100,000 microorganisms can start a viable ecosystem. Evolution toward intelligent life can be jump-started also by including small hardy multicellulars such as rotifer cysts.

11.2.3.1 Exoplanets

Nearby exoplanets can be reached after relatively short transit times, which reduces their positional uncertainty δy when they arrive. This uncertainty is strongly sensitive to the distance, as $1/(\text{distance})^4$ (see Eq. 11.10 below) (Fig. 11.4).

Seeding planets about Sun-like stars may seem obvious but, with their Sun-like lifetimes, this will not extend life significantly. Life seeded on exoplanets about Red Dwarf stars that last trillions of eons can harbor life for much longer. Ultimately, life may reach the estimated 30–80 billion Red Dwarfs in the galaxy (van Dokkum and Conroy 2010).

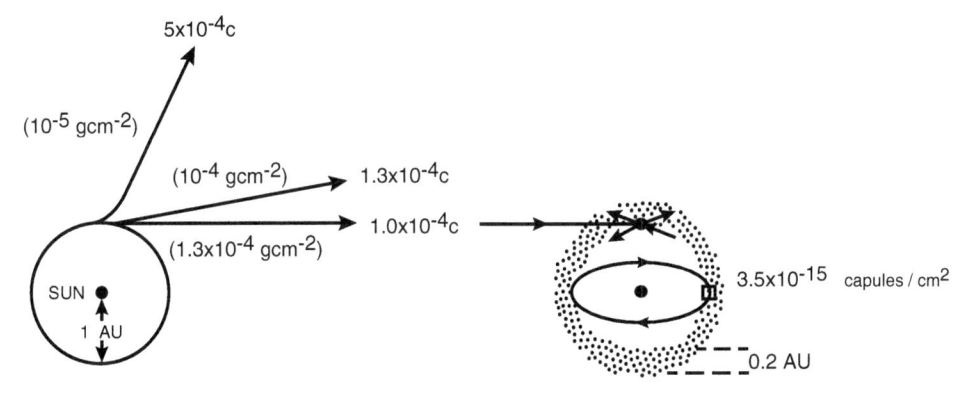

Figure 11.4 Directed panspermia solar sail missions to seed an exoplanet. The transit velocity, as shown by the *left arrows*, depends on the surface density of the sails. The captured microbial capsules are dispersed with the density shown in a ring in the habitable zone of a target star, where capsules are then swept up by a planet (Mautner and Matloff 1979). Credit: Michael N. Mautner

Some Red Dwarfs have habitable planets with temperatures of up to 280 K that allow liquid water. Some also have debris disks and comets that can capture and preserve microbial capsules, and deliver them in meteorites or mixed with dust to exoplanets later, when their environments can support life. Ejected comets can carry the microorganisms also to other solar systems.

The nearest extrasolar planets about Red Dwarfs include Groombridge 34, at 11.6 light years; Kapteyn b, 13 light years; GJ832 c (temperature 233–280 K), 16 light years; GJ581 g (debris disk with comets), 20 light years (Schneider 2015). Young Red Dwarf solar systems are preferable because they would not yet contain local life.

11.2.3.2 Star-forming clouds

The second strategy aims at clusters of new young solar systems in interstellar clouds. One mission can seed there hundreds of new solar systems, including Red Dwarfs that remain habitable for trillions of years (Mezger 1994; Palla and Stahler 2002).

Suitable star-forming clouds are hundreds of light years away but are still easy large targets. The microbial capsules will mix with dust, some of which falls on planets, and some form asteroids and comets whose meteorite fragments will deliver capsules to planets. A modest 1 kg payload of 10^{15} microorganisms can then in principle seed up to 1000 young solar systems in a star cluster with 10^{12} microorganisms each. Dilution of the capsules in the dust will reduce these numbers, but this can be minimized by targeting small dense cloud fragments.

Suitable targets may be Pleiades Cluster, 440 light years away, especially its 8-light-year dense core with over 1000 young stars aged 75–150 million years; and the Rho Ophiuchus Cloud, 427 light years away, with dense star-forming regions and filaments that contain over 600 young stars and protostars that can be seeded by one mission (Fig. 11.5).

The following mission parameters apply to both missions to nearby exoplanets and star-forming clouds.

11.2.4 Propulsion and navigation

Propulsion should produce fast interstellar transit to moving targets to minimize radiation exposure. Thin sails propelled by solar radiation pressure are a proven method of acceleration. The sails can be deployed in Earth orbit and accelerated to about 0.0001–0.0005 times the speed of light (10^{-4}–5×10^{-4} c (speed of light)). Faster speeds may be achieved by lighter graphene sails (Matloff 2012) and by sails pushed by laser beams. The payload of microbial capsules may be bundled and protected, or capsules can be covered in thin reflective covers that act as solar sails, and launched in a fleet.

Advanced technology will allow in-flight navigation, but presently the missions need to be aimed at launch to the future positions of the moving targets. These positions must be known accurately from precise astrometric data of their current locations, directions, and speed (Eq. 11.10 below). The required precise astrometry is advancing.

Saving life Itself: science and strategies for seeding the universe   269

Figure 11.5 Clusters of new stars in the Rho Ophiuchus Cloud. The dark dense star-forming cores and filaments extend for tens of light years, making them easy targets. Credit: NASA/JPL-Caltech/Harvard-Smithsonian CfA

11.2.5 Capture in target zones

The first strategy above aims at nearby extrasolar planets. A protected payload of 10 kg of microorganisms is launched, then decelerated by radiation pressure of the target planet or by dust, and captured into orbit, dispersed, and swept up by planets.

In the second strategy, a shielded payload or a fleet of capsules are sent, captured, and dispersed in an interstellar cloud. Capsule masses and sizes may be designed for deceleration in various density zones, from the whole cloud to dense protostellar condensations and accretion disks. The capsules fall onto planets or, mixed with dust, condense into asteroids and comets.

The fraction of the capsules that are mixed homogenously with the dust and delivered to planets will be equal to the fraction of the dust that is delivered to planets (equal f_{infall};

see below). Small planet-forming zones are harder to hit but the capsules will mix there with fewer solids and a larger fraction will eventually land on planets.

11.2.6 The microbial payload

The captured microbial capsules are scattered in a ring about a star or mix with dust. The capsules should be similar to dust particles with a median radius of 50 µm and a range of 10–200 µm, and weighing about 10^{-9} kg. Assuming 1 µm of microorganisms of 10^{-15} kg, 1000 microorganisms per capsule will occupy a sphere with a radius of 2.4 µm. Protective and reflecting organic polymer walls can increase this size to 50 µm to decrease the Poynting-Robertson drag into the star, allow intact infall through planetary atmospheres, and provide the first organic nutrients after landing. Capsules that mix with the dust can be smaller because turbulence will prevent them from falling into the star or from expulsion by radiation.

Missions with speeds of 0.0005 c can reach star-forming clouds at 400–500 light years in about a million years. The payload must survive this, but microorganisms can survive even longer, over 25–40 million years, in amber (Cano and Borucki 1995). Survival in transit can be helped by chemical preservatives and interstellar deep-freeze at 3 K. Cosmic radiation damage may be self-repaired after arrival by mechanisms borrowed from *Deinococcus radiodurans*.

11.2.7 Inducing a biosphere

We assume that 1000 microorganisms delivered to a planet can start life there, or 100,000 could be delivered for safety. A diverse ecosystem will require a mix or autotrophs for biosynthesis and heterotrophs for recycling, and of prokaryotes and eukaryotes and hardy cysts of small multicellulars to start higher evolution (Mautner 2005a, b).

11.2.8 Quantitative analysis

11.2.8.1 Requirements for success

The following sections consider targets surrounded by interplanetary or interstellar dust. The captured microbial capsules mix and fall with dust on planets. The probability of success is estimated as a composite of several stages.

The probability P_{target} that the probe will arrive in the target cloud is a function of the resolution of target star proper motion (α_p) needed for aiming; the distance to the target (d); velocity (v) and time span of interstellar transit (d/v); and the size (radius r_{target}) of a circular IDP cloud about the target star. Every probe or capsule arriving in the cloud is assumed to be captured and mixed in the target cloud and eventually fall on planets. $P_{seeding}$ is the probability that enough microorganisms will fall and accumulate on the planet during a short enough survival period and start an evolving ecosystem with probability $P_{success}$ (if > 1, we use $P_{success} = 1$):

$$P_{success} = P_{target} \times P_{seeding} \geq 1. \tag{11.9}$$

Saving life Itself: science and strategies for seeding the universe

For example, 10^5 microorganisms with 10^{-10} kg biomass accumulating in 1000 years may seed the planet. For clusters of n new stars, we can calculate the probability that at least one or all are seeded.

11.2.8.2 Positional uncertainty and capture in the target zone

The microbial payload may be launched with velocity v toward a moving target star at a distance d and transit time d/v. A critical parameter for acceleration is the areal density of (sail area)/(mass of the sails + payload).

The position of the probe with respect to the target at arrival time has an uncertainty δy depending on the resolution of the target proper motion α_p (in arcsec/year):

$$\delta y = 1.5 \times 10^{-13} \alpha_p (\text{arcsec}/\text{year})(d^2/v). \tag{11.10}$$

Here, α_p is in arcsec/year (2.4×10^{-14} fraction of a circle/s); all other values in SI units (δy and d in m, v in m/s). The uncertainty in the distance of the sail-ship when it arrives from the target star is then δy (m) $= [2.4 \times 10^{-14} \times 2\pi d \times \alpha_p$ (arcsec/year)], namely $1.5 \times 10^{-13} \alpha_p$ (arcsec/year) (d)m for each sec of the $t = d/v$ interstellar travel time. (Note: the mission parameters in Mautner (1997) were calculated correctly but Eq. 11.10 was misprinted.)

The sail-ships or capsule fleet will be then scattered in a circle with radius δy about the target star with a cross-sectional area $A_{arrival} = \pi(\delta y)^2 = 7.4 \times 10^{-26} \alpha_p^2(d^4/v^2)$. For example, with a resolution $\alpha_p = 10^{-5}$ arcsec/year, for a mission to an exoplanet at $d = 20$ light years (3.9×10^{17} m) and transit velocity 0.0005 c (1.5×10^5 m/s), the area of this circular zone about the target star is $A_{arrival} = \pi (\delta y)^2 = 7.6 \times 10^{24}$ m^2.

In comparison, an IDP dust ring about the star has a radius r_{target} and area $A_{target} = \pi r_{target}^2$. The probability of capture of the capsule fleet in the dust disk (P_{target}) is estimated here as a function of the ratio of $A_{target}/A_{arrival}$ but a full analysis is needed (units are as in Eq. 11.10):

$$P_{target} = A_{target}/A_{arrival} = r_{target}^2/\delta y^2 = 4.2 \times 10^{25} \left(r_{target}^2 v^2/\alpha_p^2 d^4\right). \tag{11.11}$$

11.2.8.3 Arrival at, and seeding of, the target planet

A fraction of the dust cloud f_{infall}/year falls onto a planet; therefore, the infall rate of the dust mass is $f_{infall} \times m_{\text{(IDP cloud)}}$ kg/year. For example, the fraction of our IDP dust cloud falling on Earth is $f_{infall} = 10^{-9}$/year.

The fraction of the capsules that are mixed homogenously with the dust and delivered to planets will be equal to this f_{infall}. The mass infall rate of the microorganisms is then $f_{infall} \times m_{microorganisms}$ (kg/year) where $m_{microorganisms}$ is the mass of the captured microorganisms in the dust cloud. These relations yield the general Eq. 11.12 for the probability $P_{seeding}$ of seeding the planet successfully by this infall:

$$P_{seeding} = \frac{m_{microorganisms}}{m_{\text{(IDP cloud)}}} \times \left(f_{infall} \times m_{\text{(IDP cloud)}} \times t_{infall}\right)/m_{\text{(seeding)}}. \tag{11.12}$$

272 Space ecology

Here, $m_{microorganisms}$ is the mass of microorganisms mixed in the IDP cloud whose mass is $m_{(IDP\ cloud)}$, the IDP infall rate is the mass of the infalling IDP dust per year (kg/year), t_{infall} is the time span of the infall, and $m_{seeding}$ is the mass of microorganisms needed seed to the ecosystem. Note that $m_{(IDP\ cloud)}$ cancels, leading to:

$$P_{seeding} = m_{microorganisms} \times f_{infall} \times t_{infall} / m_{seeding}. \quad (11.13)$$

11.2.8.4 Biomass requirements

Combining Eqs. 11.11 and 11.12 yields Eq. 11.14 for the biomass that needs to be launched for seeding a planet.

We assume sending and scattering in the target zones swarms of microbial capsules, and that the arrival of 10^5 microorganisms with $m_{seeding} = 10^{-10}$ kg in a survivable 1000-year period will seed the planet with probability $P_{seeding} = 1$. Then, $P_{success} = P_{target} \times P_{seeding} = 1$ and, combining Eqs. 11.11 and 11.13 yields Eq. 11.14 for $m_{microorganisms}$ that needs to be captured in the cloud if all the launched biomass is all captured in the cloud as assumed here:

$$m_{microorganisms} = \frac{m_{seeding} \alpha_p^2 d^4}{4.2 \times 10^{25} r_{target}^2 v^2 f_{infall} t_{infall}} \quad (11.14)$$

Alternatively, Eqs. (11.11) and (11.13) can be combined to yield Eq. (11.15).

$$P_{success} = P_{target} P_{seeding} = \left(r_{target}^2 / \delta y^2\right)\left(m_{microorganisms} \times f_{infall} \times t_{infall} / m_{seeding}\right). \quad (11.15)$$

Setting $P_{success} = 1$ then yields Eq. (11.16) for $m_{microorganisms}$, the microbial mass that we need to launch for success (symbols and units as above).

$$m_{microorganisms} = \left(\left(\delta y^2 / r_{target}^2\right) m_{seeding}\right) / \left(f_{infall} \times t_{infall}\right). \quad (11.16)$$

For missions to interstellar clouds containing n_{stars} stars, these probabilities apply to each star.

11.2.8.5 Sample targets and mission parameters

The missions aim at targets about Red Dwarf stars where life can last for a long time. Such targets may be the GJ832 c exoplanet 16 light years away with an equilibrium temperature of 253 K and possible liquid-water hotspots; and GJ581 g that has a mass 3 M_{earth} and a host star with a circumstellar debris disk and comets. We assume a surrounding IDP cloud with a radius of 5 au, with a steady-state mass of 10^{16} kg of which 10^7 kg per year falls on the planet—that is, a probability of 10^{-9} per IDP particle, similar to Earth.

A star-forming cloud target may be the Pleiades Cluster, 440 light years away, especially its 4-light-year radius (area 1.8×10^{34} m²) dense core containing over 1000 young stars; or the Rho Ophiuchus Cloud, 427 light years away, with dense star-forming regions and filaments, with a cross-sectional capture area of about 40 light years² (3.6×10^{33} m²) containing a mass of 3000 M_{sun} including 641 young stars with circumstellar disks. A typical dense star-forming core would then have a capture cross-section area of 10^{34} m² and dust mass in the order of 10^{34} kg, in which the capsules can mix and then fall on planets with f_{infall} 10^{-9} per dust particle per year. Microbial capsules scattered there by one mission can then seed hundreds of planets and secure life for trillions of eons.

The amount of biomass needed to seed these targets can be calculated from Eq. (11.16) using these parameters.

11.2.9   Designing the biological payload

Colonizing microorganisms can be damaged by radiation during interstellar transit. This may be corrected by colonizers with mechanisms for self-repair borrowed from radiation-resistant *D. radiodurans* that survives 5000 Gy of ionizing radiation. It repairs DNA using multiple copies of its genome and by exchanges between damaged and normal DNA. It can also survive extreme cold, dehydration, vacuum, and acid, making it a polyextremophile.

The messenger organisms need to combine the resiliency of *D. radiodurans* with photosynthesis and anaerobic metabolism, in one organism or a mixture in the colonizing populations. Once microorganisms were captured in new habitats, can they survive and evolve there?

Astroecology experiments showed that algae, bacteria, fungi, and plant cultures can grow on nutrients in asteroid/meteorite materials of our Solar System, and therefore also on similar materials in billions of other similar planetary systems. What complexities do biota and ecosystems need to survive there? Astroecology experiments with microbes and meteorites can help us to design viable panspermia payloads to advance life in the galaxy.

11.2.10   Conclusions

Biota planted about Red and White Dwarf stars can secure life for more than 10^{21} years (a trillion eons), in tens of billions of solar systems in the galaxy.

We can secure this future for our family of biological gene/protein life through a directed panspermia panbiotic program using current level technologies. Engineering technology developments include high-precision astrometry, lighter sails, and analysis of the capture process.

Biology is also a key to adapting life to space. The bioengineering of hardy polyextremophile microorganisms for diverse environments can produce a mix of organisms that can start self-sustaining ecosystems is a key to directed panspermia.

Also critical is a wide public acceptance of life-centered ethics focusing on a human purpose to propagate life. In future eons, life can then assume many new forms, some intelligent, who can secure and expand life further. Establishing this immense future for life can give us deep moral satisfaction and a cosmic purpose.

11.3   Silent running: the bacteriology of spaceflight

Simon F. Park, Senior Lecturer in the Faculty of Health and Medical Sciences at the University of Surrey.

11.3.1 Introduction

Science fiction is an extremely powerful means through which we can predict and gain an understanding of the possibilities of our future existence, through reasoned speculation and, usually, profound storytelling. In the context of the Star Ark project, this literary genre has explored, almost to the limits of imagination, numerous concepts in terms of the mechanisms of future space travel, the discovery of extraterrestrial planets, contact with alien life forms, and the colonization of distant worlds by humankind. While we are the only life form on planet Earth capable of space travel and dreaming about its future possibilities, we are a long way from being the dominant form of life on our world. That title has to go to Earth's bacteria, in terms of their numbers, diversity, and activity, and, as we explore space, these microscopic and very independent organisms will inevitably travel with us, whether this is our intention or not. If we do ever find complex, and even intelligent, extraterrestrial life, then it too is very likely to be underpinned by a dominant and alien microbiology.

In the context of the above, we all live out our lives on a small planet, which is dominated by forms of life that we usually cannot see, yet without which we would not, or could not, continue to exist. Yet, in all of the speculation in science fiction about the future of spacefaring, imagined galactic civilizations, and alien contact, rarely has the role of the apparently humble bacterium been considered in any detail. Yes, science fiction has occasionally explored the science of microbiology, but usually in contexts much closer to home, and with a focus on some of its more negative aspects, such as in Michael Crichton's *The Andromeda Strain* (1969), which charts a deadly epidemic caused by a crystal structured extraterrestrial microbe. Strangely, we have to go back in time, and nearly to the beginnings of science fiction itself, to find an account that predicts what might happen when microbiological worlds that are completely alien to each other interact. In this respect, *The War of the Worlds* (1898) by H.G. Wells is a remarkably prescient work that is bookended by vivid and powerful microbiological prose. For example, the book begins by comparing our activities on Earth from the perspective of a distant and superior extraterrestrial observer to the way in which an earthly microbiologist might today observe unwitting microbes under a microscope:

> "... as men busied themselves about their various concerns they were scrutinised and studied, perhaps almost as narrowly as a man with a microscope might scrutinise the transient creatures that swarm and multiply in a drop of water. With infinite complacency men went to and fro over this globe about their little affairs, serene in their assurance of their empire over matter. It is possible that the infusoria under the microscope do the same." (Wells 1898)

The book also ends with an exploration of how an immunologically naïve alien race might interact with Earth's native bacteria, so that, in the end, microbes unwittingly become our allies in the war against the Martians:

> "From the moment the invaders arrived, breathed our air, ate and drank, they were doomed. They were undone, destroyed, after all of man's weapons and devices had failed, by the tiniest creatures that God in his wisdom put upon this earth. By the toll of a billion deaths, man had earned his immunity, his right to survive among this planet's infinite organisms. And that right is ours against all challenges. For neither do men live nor die in vain." (Wells 1898)

Obviously, in the context of future space travel, this concept might be juxtaposed, so it is us who are immunologically naïve, and the pathogenic bacteria are from another alien world.

As long ago as 1897, then, science-fiction writers were exploring at least some of the important implications that microbiology might have in terms of space travel and extraterrestrial encounters. With our modern understanding of bacteria, and consistently with their manifold and essential activities here on our home planet, any reasonable consideration of our spacefaring future has to include the likely role and impact that bacteria will have on this endeavor. This is the overall aim of this chapter, which will consider the importance of Earth's bacteria, the prospect that they may have already traveled to extraterrestrial bodies, and why we might also choose to deliberately take them with us on our spacecraft. Finally, if we do find extraterrestrial life, the chances are that it will be more likely to be microbiological than complex and, even if it is complex, then it will have to be underpinned by the activity of alien microbes, so this chapter will also explore what might happen when terrestrial and alien microbiologies meet.

In the context of the above, it is interesting to note here that bacteria have already experienced space travel in numbers far greater than us. Very few humans have experienced space travel, and even fewer have actually walked on the surface of a celestial body other than Earth. Twelve astronauts did walk on the Moon between 1969 and 1972, but each of these individuals would have been carrying between 200 and 1400 g of bacteria, in the form of their normal bacterial flora. This is a small weight relative to the host (1–3 % total body mass) but, because bacteria are very much smaller than us, this represents around 10^{14} bacterial cells, and so the numbers of bacteria that have ventured into space already dwarfs that of its human explorers.

11.3.2 Earth: planet of the bacteria

Before we move on to the important role of bacteria, let us briefly consider their nature. Bacteria are a group of microscopic, single-celled organisms (most bacteria are around 0.2 µm in diameter and 2–8 µm in length) that lack a membrane-bound nucleus and other internal organelles (e.g. mitochondria) found in more complicated life forms like ourselves. Unlike the situation in complex multicellular organisms, their reproduction is rapid, asexual, and simple, in that they generally grow to a fixed size and then reproduce through binary fission to produce two daughter cells. In appropriate conditions, certain types of bacteria can grow and divide so rapidly that a population might double in size every 9.8 min.

As a group of organisms, bacteria also display an exceedingly diverse range of metabolic capabilities, from being able to use and survive on almost any known carbon-containing compound, no matter how complex, to being capable of using just simple inorganic compounds. Different types of bacteria can obtain the energy needed for their growth from a number of different sources, such as from just light (photosynthesis), from inorganic compounds (e.g. simple iron-, sulfur-, and nitrogen-containing chemicals), and more commonly from organic compounds (bacteria that obtain energy from organic compounds are known as organotrophs and are the most commonly encountered bacteria in most environments). Bacteria also need a source of carbon to grow, as this element forms the basis of their cellular structure and enzymes. Most of them obtain this carbon from the breakdown of organic compounds like ourselves, but some can obtain it from carbon dioxide via photosynthesis. Photosynthetic bacteria, like cyanobacteria, which obtain energy from light and carbon from carbon dioxide, have the simplest growth requirements known, and can grow on little more than sunlight, water, and air.

For many years, bacteria were studied in the laboratory as isolated species, in isolated monoculture, and under artificial conditions, and this led to a misunderstanding that has perpetuated throughout most of the history of microbiology that bacteria were simple, solitary, and ultimately selfish creatures. We now know this to be untrue, and discoveries in the past 40 years have shown bacteria to be capable of communication with each other and with most of the other life forms with which they form associations, so that, under the demands of the wild environment, these highly versatile life forms are capable of working in teams and in dynamic communication. In the manifold environments that we find bacteria in on Earth, they form productive and powerful interspecies communities, in which they use cell-to-cell communication to share information, to adjust and coordinate their activities and metabolic functions, and to collaborate (Straight and Kolter 2009).

On our home planet (and presumably on those extraterrestrial planets that we might colonize in future), bacteria (will) play critical roles in sustaining more complex life like ourselves. Despite this, we are all too often macroscopically blinkered and usually ignore these organisms but, as has been revealed by modern technologies, the scale and activity of bacterial life are staggering. Bacteria have been present for perhaps three-quarters of Earth's history and have adapted to almost all of our planet's available ecological niches, and my own feeling is that, if we do find extraterrestrial life on other planets, even if it has not evolved beyond the microbiological, then, given what we find on Earth today (and as highlighted above) and the rate at which this type of life can replicate, evolve, and adapt, the scale and diversity of an extraterrestrial microbiology will be vast. Some of the environments offered by extraterrestrial planets may be hostile, but we should remember that, when the first bacteria colonized Earth, almost four billion years ago, it was not habitable by our standards. There was no free oxygen, no ozone to block the Sun's deadly ultraviolet radiation, and the atmosphere was hot and full of toxic chemicals; and so, if we do not, then bacteria certainly have the ability to colonize hostile planets. Even our own habitable planet of today contains many extreme environments, and in many of these we find microbiological life—in the Atacama desert in Chile, for example, perhaps the driest environment on Earth, at temperatures ranging from −15 to 121 °C, and in environments where acidity and alkalinity range from pH 0 to pH 13. Bacterial life has been found in all of these extreme environments, and there are very few natural environments on Earth where microbiological life is absent (Anon 2007).

Traditional estimates for the amount of bacterial life on Earth were based on numbers acquired via laboratory culturing techniques, but we know now that most bacteria cannot be grown in laboratory culture, despite the best efforts of over 170 years of microbiological research. The true extent of Earth's bacterial presence, and its diversity, has only recently been revealed through metagenomics and modern sequencing technologies, and today we recognize that the number of different bacterial species probably extends into the millions (Rinke et al. 2013), while the number of individual bacterial cells with which we share our world is estimated to be $4-6 \times 10^{30}$ (Whitman et al. 1998). In comparison, there are "only" about 10^{22} to 10^{24} stars in the known universe. In 2004, scientists developed a cantilever device able to measure the weight of a single bacterial cell and, when this device was used to empirically determine the mass of a single cell of the bacterium *Escherichia coli*, it was found to be 665 femtograms (a femtogram is a billionth of a gram) (Craighead 2007). While the weight of a single bacterial cell is minute, there are a great number of bacteria on our planet, and so bacteria account for at least half of the total live biomass on Earth (Whitman et al. 1998). The remaining non-bacterial biomass is the stuff that we can see—that is, all the insects, plants, and animals.

The Sun was once thought to provide energy for all life on Earth and, because of this, it was thought that life on Earth had to be limited to close to its surface, and thus to the Sun's energetic reach. We know now, however, that bacteria are able to obtain the energy they need to live through chemical reactions that do not involve sunlight, and that these organisms thrive in many deep subsurface environments, from sediments below the deepest ocean floors to microscopic pockets of water inside deep solid rock. A pioneering study of global biomass by Whitman and colleagues (Whitman et al. 1998) proposed that subsurface microbes comprise 35–47 % of Earth's total biomass, nearly equal to plants in their total carbon content. These findings have important implications for space exploration, because they mean that, even if there is no visible life on the surface of an extraterrestrial body, there may be a complex and extensive subsurface ecology similar in scope and diversity to the one that we find on Earth hidden beneath the surface.

Where we find them, bacteria form myriad associations with other organisms and carry out vital biogeochemical processes on a global scale, and so they are not just the most important form of life on Earth in terms of numbers, but they are also its most important life form in terms of their ability to sustain all other living organisms. Complex communities of bacteria can no longer be viewed in isolation anymore, and must now be considered to be integral and functionally important components of the hugely diverse range of earthly ecologies, ranging from the collectives present in soil, for example, to our own personal human microbiomes (Straight and Kolter 2009). Below is a discussion of just some of the important roles that bacteria play in sustaining life on our home planet and that may also be needed should we ever colonize extraterrestrial bodies.

Of all the many bacterial "inventions" that have arisen during the history of life on Earth, the machinery to oxidize water, using just sunlight (photosystem II) must surely be one of the greatest. This gave life the ability to use water and light as an almost limitless supply of energy. Moreover, the production of oxygen, as an apparent waste product of this, also profoundly changed the composition of the world's oceans, continents, and atmosphere. Photosynthesis arose in oxygenic photosynthetic bacteria, called cyanobacteria, and using this process these organisms transformed Earth's atmosphere from an anoxic state, containing oxygen at concentrations 100,000 times less that today's atmosphere, to an oxygenic one during the great oxygenation event, beginning approximately 2.45 billion years ago (Sessions et al. 2009). Today, cyanobacteria such as *Prochlorococcus* and *Synechoccocus* still exist in a similar form and, being two of the most abundant cellular life forms on Earth, these two cyanobacterial species alone still generate globally significant amounts of atmospheric oxygen. Looking toward the future, such photosynthetic and oxygen-producing bacteria are likely to have utility in providing a source of oxygen during prolonged journeys in space and in the global oxygenation of other worlds.

Another vital, but usually unheralded, biological role unique to Earth's bacteria is their ability to fix atmospheric nitrogen—that is, to convert the inert and gaseous elemental form present in air into a series of nitrogen-containing compounds that can be used by other life and plants. Plants have an absolute requirement for nitrogen in these fixed forms and, without the bacterially powered fixation process, plants and other forms of life that rely on plants and photosynthesis could not exist.

Bacteria also form many vital associations with animals and, in one in particular, are essential for mediating the transfer of the carbon fixed by plants during photosynthesis into the bodies of many kinds of animals. Numerous animals that are a source of food for carnivorous ones or for ourselves (e.g. cows, sheep, etc.) subsist on diets that are very high in the plant structural polysaccharide cellulose. In plants, cellulose gives rigidity to their cells and, because of the vast amount of plant life, it is Earth's most abundant organic compound and source of carbon. Consequently, the hydrolysis of cellulose in plant cell walls is a key step in the global carbon cycle. However, most animals cannot digest cellulose alone, and those that are able to rely on a symbiosis formed with cellulose-degrading bacteria.

The half-life of cellulose at neutral pH in the absence of biological enzymes is estimated to be several million years and, in light of this, it is easy to see the importance of bacterial activity, which is responsible for most of the turnover of the carbon in cellulose in the environment (Wilson 2011). In addition, without this vital contribution made by cellulose-degrading bacteria, many of the ruminants that underpin much of our farming (e.g. cows, sheep, and goats) would not be able to derive any benefit from their cellulose-rich diets and would not survive.

In the context of the importance of bacteria discussed above and the possibilities of colonizing other planets, it is interesting to speculate as to how long we would survive if someone waved a magical and hypothetical wand that instantaneously made all bacteria disappear from planet Earth, or how long we would survive on an extraterrestrial world that lacked microbiological life. Such speculation forms the basis of an article by Gilbert and Neufeld (2014), who predict that complete societal collapse would occur within a year, and this would be linked to the rapid and catastrophic failure of the food supply chain. It is clear, from the discussion above, that most global biogeochemical cycling would cease in the absence of bacteria, and that human and animal waste would accumulate rapidly, as there would be very little decomposition of it. Most ruminant livestock would rapidly starve without microbial symbionts and, without bacterial nitrogen fixation, plants would rapidly deplete nitrogen, cease photosynthesis, and then die. Indeed, it has been estimated that, without bacterial activity, most available nitrogen in soil would be used up in about a week, shortly after which photosynthesis would cease. It is possible that small pockets of humans would survive for a time, especially through the use of any technology that remained, possibly for decades or centuries, but our long-term survival would be very doubtful, and it would be a very unpleasant existence. In this context, it seems very unlikely that we would be able to venture very far into space or colonize other extraterrestrial worlds in the absence of an underpinning microbiology. The implications of bacteria and space travel are considered in the following section.

11.3.3 Bacteria in space

As and when we venture into space and onto other planets, bacteria will inevitably accompany us on these journeys; such is their ubiquity and dominance here on our home planet. We may deliberately choose to carry them aboard our spacecraft, for reasons outlined above and below, but, even if not, the many billions that are present as part of our own human microbiomes will accompany us. In addition, those that have contaminated any parts of our spaceships will be accidently transported to other worlds and, even if they are present on the outside of the craft, such is the resistance and persistence of these life forms that they are likely to survive the journey.

Today, where we await possible discoveries and scientific confirmation of extraterrestrial life, we have to ensure that any life that we might detect on our neighboring celestial bodies has not accidently been introduced there from Earth, having been carried there on our unmanned spacecraft. Consequently, to ensure that space missions do not accidentally transfer bacteria to other cosmic bodies, spacecraft are currently allowed to harbor only a certain level of microbial life (called the bioburden), and most spacefaring countries have agreed to follow guidelines for planetary protection from the International Council for Science's Committee on Space Research to reduce the chances of their vehicles carrying terrestrial life to other planets.

To ensure that this bioburden is kept to a minimal level, spacecraft are kept in highly specialized clean rooms where the incoming air is stringently filtered, and surfaces are cleansed regularly with powerful disinfectants. Obviously, humans are a huge source of bacterial contamination, not because we are inherently dirty, but because of the microbiome that we carry with us; so any human who enters the rooms containing spacecraft must be clad head to foot in a protective suit. As a consequence of these stringent measures, these environments are some of the cleanest places on Earth. Nevertheless, as an additional precaution, spacecraft are usually heat treated prior to launch, to reduce the bioburden even further, although the delicate electronics of a spacecraft limit the range of temperatures that can be used and, thus, this process cannot be relied on to provide complete sterility. To ensure absolute sterility in a microbiology laboratory, either to prepare sterile media for experiments or to kill bacteria to enable the safe disposal of biological waste, samples are routinely heated under pressure to 121 °C for 20 min. For comparison and as an example, the Viking 1 lander was heated at 111.7 °C for 30.23 h prior to launch.

Despite humankind's best efforts, some bacteria inevitably survive in these environments, but the powerful bactericidal processes used do select some very unusual organisms that can survive the repeated sterilization procedures used in clean rooms and also cope with the severe lack of nutrients and water available in and on spacecraft. One such bacterial spacecraft clean-room specialist, which has been isolated independently from a number of different clean rooms, is *Tersicoccus phoenicis* (*Tersi* is Latin for clean). The remarkable ability of this bacterium to survive in harsh, clean-room environments, with virtually no access to water and nutrients, raises the possibility that it may have traveled to Mars with the Phoenix Mars lander, and that it could have potentially contaminated the red planet with terrestrial bacteria (Vaishampayan et al. 2013), ironically bringing about something that we were trying so hard to avoid.

A number of studies have also suggested that bacteria could survive the trip from Earth to Mars, and perhaps even beyond, and that they might survive and grown on extraterrestrial bodies once they get there. Spores of the bacteria *Bacillus subtilis* and *Bacillus pumilus*, for example, have survived on surfaces mounted outside the International Space Station (ISS) for 18 months, confirming that dormant and highly resistant bacterial spores, at least, could survive the vacuum of space, its extreme temperature fluctuations, and the barrage of radiation to which they would be exposed (Onofri et al. 2012; Vaishampayan et al. 2012). There is also some evidence that bacteria accidently carried from Earth may have survived on the Moon for an extended period of time. On November 19, 1969, Apollo 12 astronauts Pete Conrad and Alan Bean made their touchdown on the Moon just 163 m away from the site where the unmanned Surveyor 3 lander had landed on lunar soil previously, on April 20, 1967. The Surveyor 3 camera was brought back to Earth under sterile conditions by the Apollo 12 crew and, when analyzed microbiologically, living cells of the bacterium *Streptococcus mitis* were recovered, with the suggestion that this bacterium had survived launch, the harsh vacuum of space, and an average lunar temperature of −253 °C for over 2 years. However, this recovery remains controversial, and it is also possible that the contamination may have happened after the camera was returned to Earth. In any case, given the results of the studies by Onofri et al. (2012) and Vaishampayan et al. (2012) above, it seems likely that Earth's microbes could survive the perilous journey to other planets and moons, making it difficult to determine whether any microbial life discovered on those bodies originated there or, alternatively, was accidently introduced from Earth.

When we do send manned spacecraft to other worlds or moons, then it will be impossible to prevent the contamination of these, and their living spaces, by bacteria, given the huge numbers present on and in our own bodies. Indeed, such contamination has already been described during Space Shuttle missions (Koenig and Pierson 1997) on the ISS (Castro et al. 2004), and on the Mir space station (Song and Leff 2005). As is the case here on Earth, not all of the bacteria will be problematic, but some may be so, and especially so on extended journeys where their activities have time to take effect. Aside from being unattractive, bacterial growth may attack the structure of a spacecraft itself as, given appropriate conditions, these organisms can degrade steel, etch glass, grow in fuel, make rubber brittle, and also foul air and water filters. Wherever there is water or moisture on a spacecraft, bacteria are likely to grow and form biofilms, which are complex, 3D microbial communities commonly found in Nature. Bacterial biofilm formation was found to be extensive on the Mir space station and continues to be a challenge on the ISS. It is also possible that the effects of space travel itself, and particularly the absence of gravity, may exacerbate the problem of bacterial biofilms, as space-borne biofilms have a greater biomass and increased thickness when compared with ground-formed controls (Kim et al. 2013a, b). Consequently, the altered biofilms that form during spaceflight may have a detrimental impact on long-term spaceflight missions, and lead to increases in biofouling and bacterially induced corrosion.

Among the bacteria that have been isolated from manned spacecraft are *E. coli*, *Serratia marscens*, and a presumptive species of *Legionella* (Ott et al. 2004), all of which can be opportunistic pathogens. While these bacteria will not cause illness in otherwise healthy individuals, they may pose a particular threat to astronauts, whose immune systems may be affected by microgravity, closed environments, and radiation. Of particular concern is a recent study which suggests that space travel might have dramatic effects on the very nature of the bacteria themselves. In this study, *Salmonella* grown on board the Space Shuttle were more virulent than their terrestrially grown counterparts, raising the concern that long manned missions will increase the virulence of microorganisms that astronauts inevitably carry with them (Wilson et al. 2007). The possible combination of pathogenic bacteria with increased virulence and the presence of individuals with immune systems attenuated by the rigors of spaceflight is a problem that may need to be addressed on long-term spaceflight missions. It is becoming increasingly apparent that spaceflight affects community-level behavior in bacteria, and it is vital that we gain an understanding of ways in which both the harmful and beneficial human–microbe interactions may be altered during spaceflight, as this is likely to have profound impacts on the success of future manned missions.

Illness in space is a big problem, not only for the health of astronauts in terms of being distant from sophisticated health care facilities, but also because of the unintentional liberation of bodily material in a zero-gravity environment, especially with regard to vomit and diarrhea, and the ensuing damage and problems that this might cause. With this in mind, food safety in orbit was a significant problem that NASA had to address during its early spaceflights. The organization could not risk having its astronauts exposed to foodborne pathogens in space, which was a major problem because this could not be guaranteed under manufacturing practices of the time. As a consequence of this, a process called hazard analysis and critical control points (HACCP) was conceived when NASA collaborated with Pillsbury to design and manufacture the first foods specifically for spaceflight in the 1960s. What emerged was a modern, science-based, and systematic preventive approach for ensuring food safety that has since been adopted internationally back on Earth as an important tool for reducing risk. HACCP is likely to be crucial for the design of safe food to sustain larger populations on much longer and future journeys into space.

The consumption of food by astronauts and other spacefarers may also give rise to less obvious bacteriological problems during spaceflight beyond the immediate concerns regarding food safety. Humans produce two highly flammable gases—hydrogen and methane—via the activity of their normal gut microflora on the food that they consume. To most of us, this is not an issue, beyond the embarrassment of flatulence, but, to those in enclosed spaces where the gases can accumulate, this may be a problem because the gases could build up to sufficient concentrations to ignite. In the early days of spaceflight, this was considered to be a serious problem by NASA and, because of this, they embarked on a number of studies to explore the potential buildup of hydrogen and methane in relation to astronaut diet (Calloway and Murphy 1969). With larger numbers of people traveling into space in enclosed environments, the production of these flammable gases in relation to diet may warrant further investigation.

As discussed above, bacteria may be detrimental to human health and also to the structure of spacecraft, and thus there are a number of critical bacteriological issues that must be taken into account before we can safely venture farther out into space. Enclosed cabin or habitat conditions present many challenges to their design with regard to the health of the spacefarers that occupy them due to the potential buildup of bacteria in the atmosphere of the spacecraft, in its water systems, and in other moist environments. On Earth, the bacterial load in the atmosphere, and in the air that we breathe, is reduced by the natural bactericidal effects of sunlight and, thus, it may be possible to design artificial lighting systems to reduce the general bacterial load of the habitat.

It is also very likely that we will choose to take bacteria with us on our starships, and that they will find more useful roles in life-support functions. These are likely to take the form of bacterially powered living systems that will exploit the well-established ecological pathways that have been characterized here on Earth. These have been discussed in the above section, so will only briefly be considered here. Such systems may take the form of photosynthetic systems for generating oxygen, systems for recycling waste and generating heat, and microbial factories for generating food and nutrition. Such bacterial systems are already being developed, and one prototype anaerobic digester produces biogas, a mixture of methane and carbon dioxide, from human waste at a rate of 290 L of methane per crew per day (Dhoble and Pullammanappallil 2014). In addition, microbial fuel cells are an emerging technology that uses bacteria to generate electricity from a variety of sources that might be waste, soil, or mud and, in future, these might provide clean and sustainable electricity for future spaceships. It is very likely, then, that living bacterial systems will become an integral part of the spacecraft or habitat itself.

11.3.4 When alien microbiologies meet

H.G. Wells raised a very important issue regarding contact with extraterrestrial life, which, even today when its possibility comes ever closer, does not seem to be considered much in his book, *The War of the Worlds* (1898), in the sense of what will happen if we do encounter extraterrestrial microorganisms. There are two key considerations that arise from this. What will happen to naïve extraterrestrial ecologies when they encounter the Earth-derived bacteria that we will either deliberately or unintentionally bring with us, and what will happen when our own Earth-evolved immune systems encounter unique extraterrestrial microbes which they have never encountered previously? Obviously, this has not happened yet so we can only speculate, but our own history might have some very important lessons for us in this context, if we travel back to a time when Earth's own distant and separate populations began to meet and interact. Until the coming of the Europeans, the New World was free of many dangerous and devastating diseases like smallpox, typhus, cholera, and measles. Smallpox is believed to have arrived in the Americas in 1520 via a Spanish ship sailing from Cuba, carrying an infected African slave, and so, when the Europeans invaded the New World, they were aided by an adventitious and insidious ally more potent than any of their armies: the smallpox virus caused an infectious disease to which the native South Americans had no natural immunity and, within just a few generations, it is estimated that approximately 20 million people may have died in the years following the European invasion and this represented up to 95% of the population of the Americas and the indigenous population was "undone, destroyed … by the tiniest creatures that God in his wisdom put upon this earth."

Whether extraterrestrial bacteria will be dangerous to us when we encounter them on alien worlds cannot be empirically known at present. The good news is that the pathogenic microorganisms that cause illness in humans on Earth have had millions of years of competitive evolutionary interchange with our immune systems, so they are both sophisticated and highly adapted to their host. For example, most pathogenic bacteria initiate an infection by binding onto a specific molecule on their host called a receptor, and lock onto this via a protein that they produce called an adhesin. These two components act in a similar way to a lock and a key and, without the intimate and highly specific interaction between the adhesin and receptor, the pathogen cannot initiate an infection. It seems highly unlikely that alien microorganisms would have this ability to recognize us as hosts and be able to bind to our tissue to establish an infection. In a general sense, then, we may be immune to bacteria on other worlds. However, if the human body is recognized as a source of nutrition by them, and alien bacteria have the capacity to harvest this, this is a situation that might give rise to problems. For example, if alien bacteria required amino acid building blocks, similar to those that make up our own bodies, then we might be "seen" as a potential source of nutrition. To be destroyed by our immune systems, and thus be prevented from using our amino acids and thus causing damage to the human body, the alien bacteria would have to be recognized by it but, if their surface features are so different from earthly bacteria, then this might not happen and such bacteria may grow in the body unchecked with likely a disastrous outcome for the human host. Similarly, bacteria from extraterrestrial worlds might be inherently resistant to our immune systems, especially if they lack the components that its bactericidal activities target. Finally, Earth's own bacteria produce some incredibly toxic molecules, many for reasons that are not well known, and in some cases their potent action on the human body may just be accidental. Botulinum toxin, for example, is a neurotoxin produced by a bacterium that is the most acutely lethal toxin known, with an estimated median lethal dose to humans of 10–13 ng/kg when ingested. By sheer chance, then, it is possible that alien microbes may also produce chemicals that are highly toxic to humans. Interestingly, this was the case for the crystalline life form in Michael Crichton's fictional *Andromeda Strain* (1969), and so, as we come to the end of this chapter, we end up where it started: a discussion of the coverage of microbiology in science fiction.

COMBINED REFERENCE LIST FOR CHAPTER 11

F. Adams, G. Laughlin, *The Five Ages of the Universe* (Touchstone Books, New York, 1999)

Anon, *The Limits of Organic Life in Planetary Systems*. Committee on the Limits of Organic Life in Planetary Systems, Committee on the Origins and Evolution of Life, National Research Council (2007)

H.J.M. Bowen, *Trace Elements in Biochemistry* (Academic, New York, 1966)

D.H. Calloway, E.L. Murphy, Intestinal hydrogen and methane of men fed space diet. Life Science Space Res. **7**, 102–109 (1969)

R.J. Cano, M.K. Borucki, Revival and identification of bacterial spores in 25- to 40-million-year-old Dominican Amber. Science, New Series **268**(5213), 1060–1064 (1995)

A. Castro, A.N. Thrasher, M. Healy, C.M. Ott, D.L. Pierson, Microbial characterization during the early habitation of the international space station. Microb. Ecol. **47**, 119–126 (2004)

A.C. Clarke, *The City and the Stars* (Frederick Muller, London, 1956)

H. Craighead, Nanomechanical systems: measuring more than mass. Nat. Nanotechnol. **2**, 18–19 (2007)

M. Crichton, *The Andromeda Strain* (Knopf, New York, 1969)

F.H. Crick, L.E. Orgel, Directed panspermia. Icarus **19**, 341–344 (1973)

A.S. Dhoble, P.C. Pullammanappallil, Design and operation of an anaerobic digester for waste management and fuel generation during long term lunar mission. Adv. Space Res. **54**, 1502–1512 (2014)

F. Dyson, Time without end: physics and biology in an open universe. Rev. Mod. Phys. **51**, 447–468 (1979)

J.A. Gilbert, J.D. Neufeld, Life in a world without microbes. PLoS Biol. **12**(12), e1002020 (2014). doi:10.1371/journal.pbio.1002020

W. Kim, F.K. Tengra, Z. Young, J. Shong, N. Marchand, H.K. Chan, R.C. Pangule, M.P. Parra, J.S. Dordick, J.L. Plawsky, C.H. Collins, Spaceflight promotes biofilm formation by *Pseudomonas aeruginosa*. PLoS One **8**(4), e62437 (2013a). doi:10.1371/journal.pone.0062437

W. Kim, F.K. Tengra, J. Shong, N. Marchand, H.K. Chan, Z. Young, R.C. Pangule, M.P. Parra, J.S. Dordick, J.L. Plawsky, C.H. Collins, Effect of spaceflight on *Pseudomonas aeruginosa* final cell density is modulated by nutrient and oxygen availability. BMC Microbiol. **13**, 241 (2013a). doi:10.1186/1471-2180-13-241, PMID: 24192060

D.W. Koenig, D.L. Pierson, Microbiology of the space shuttle water system. Water Sci. Technol. **35**, 59–64 (1997)

G.L. Matloff, Graphene: the ultimate solar sail material. J. Br. Interplanet. Soc. **65**, 378–381 (2012)

J.H. Mauldin, *Prospects for Interstellar Travel* (AAS Publications, Univelt, San Diego, 1992)

M.N. Mautner, Deep-space screens against climatic warming: technical and research requirements, in *Space Utilization and Applications in the Pacific*, ed. by P.M. Bainum et al. Paper AAS 89-668, American Astronautical Society, Advances in the Astronautical Sciences, vol. 73, (1989), p. 771, www.astro-ecology.com/PDFSolarScreensAmericanAstronauticalSociety1989Paper.pdf

M.N. Mautner, A space-based solar screen against climatic warming. J. Br. Interplanet. Soc. **44**, 135–138 (1991). www.astro-ecology.com/PDFSolarScreenJBIS1991Paper.pdf

M.N. Mautner, Directed Panspermia. 2. Technological advances toward seeding other solar systems, and the foundations of panbiotic ethics. J. Br. Interplanet. Soc. **48**, 435–440 (1995)

M.N. Mautner, Space-based cryoconservation of endangered species. J. Br. Interplanet. Soc. **49**, 319–320 (1996)

M.N. Mautner, Directed Panspermia. 3. Strategies and motivation for seeding star-forming clouds. J. Br. Interplanet. Soc. **50**, 93–102 (1997). www.astro-ecology.com/PDFDirectedPanspermia3JBIS1997Paper.pdf

M.N. Mautner, Planetary bioresources and astroecology. 1. Planetary microcosm bioessays of martian and carbonaceous chondrite materials: nutrients, electrolyte solutions, and algal and plant responses. Icarus **158**, 78–86 (2002a), doi:10.1006.icar.2002.6841, www.astro-ecology.com/PDFBioresourcesIcarus2002Paper.pdf

M.N. Mautner, Planetary resources and astroecology. Planetary microcosm models of asteroid and meteorite interiors: electrolyte solutions and microbial growth: implications for space populations and panspermia. Astrobiology **2**, 59–76 (2002b). www.astro-ecology.com/PDFAsteroidAstrobiology2002Paper.pdf

M.N. Mautner, Life in the cosmological future: resources, biomass and populations. J. Br. Interplanet. Soc. **58**, 167–180 (2005a). www.astro-ecology.com/PDFCosmologyJBIS2005Paper.pdf

M.N. Mautner, *Seeding the Universe with Life: Securing Our Cosmological Future* (Legacy Books, Washington, DC, 2005b). www.astro-ecology.com/PDFSeedingtheUniverse2005Book.pdf

M.N. Mautner, Life-centered ethics, and the human future in space. Bioethics **23**, 433–440 (2009). doi:10.1111/j.1467-8519.2008.00688.x. www.astro-ecology.com/PDFLifeCenteredBioethics2009Paper.pdf

M.N. Mautner, In situ biological resources: soluble nutrients and electrolytes in carbonaceous asteroids/meteorites: implication for astroecology and human space populations. Planet. Space Sci. **104**, 234–243 (2014). http://dx.doi.org/10.1016/j.pss.2014.10.001 www.astro-ecology.com/PDFNutrientsPopulationsPlanetarySpaceSciencePaper.pdf

M. Mautner, G.L. Matloff, Directed Panspermia: a technical and ethical evaluation of seeding nearby solar systems. J. Br. Interplanet. Soc. **32**, 419–423 (1979)

M.N. Mautner, R.L. Leonard, D.W. Deamer, Meteorite organics in planetary environments: hydrothermal release, surface activity, and microbial utilization. Planet. Space Sci. **43**, 139–147 (1995)

P.G. Mezger, The search for protostars using millimeter/submillimeter dust emission as tracer, in *Planetary System: Formation, Evolution and Detection*, ed. by B.F. Burke, J.H. Rahe, E.E. Roettger (Kluwer, Dordrecht, 1994), pp. 197–214

M. Montague, G.H. McArthur IV, C.S. Cockell, J. Held, W. Marshall, L.A. Sherman, N. Wang, W.L. Nicholson, D.R. Tarjan, J. Cumbers, The role of synthetic biology for in situ resource utilization (ISRU). Astrobiology **12**, 1135–1142 (2012)

B.T. O'Leary, Mining the Apollo and Amor asteroids. Science **197**, 363–364 (1977)

G.K. O'Neill, The colonization of space. Phys. Today **27**, 32–38 (1974)

S. Onofri, R. de la Torre, J.-P. de Vera, S. Ott, L. Zucconi, L. Selbmann, G. Scalzi, K.J. Venkateswaran, R. Rabbow, F.J. Sánchez Iñigo, G. Horneck, Survival of rock-colonizing organisms after 1.5 years in outer space. Astrobiology **12**, 508–516 (2012)

C.M. Ott, R.J. Bruce, D.L. Pierson, Microbial characterization of free floating condensate aboard the Mir space station. Microb. Ecol. **47**, 133–136 (2004)

F. Palla, S.W. Stahler, Star formation in space and time: Taurus-Auriga. Astrophys J **581**, 1194–1203 (2002)

W.K. Purves, D. Sadava, G.H. Orians, C. Heller, *Biology: the science of life* (Sinauer Associates, Sunderland, MA, 2001)

C. Rinke, P. Schwientek, A. Sczyrba, N.N. Ivanova, I.J. Anderson, J.F. Cheng, A. Darling, S. Malfatti, B.K. Swan, E.A. Gies, J.A. Dodsworth, B.P. Hedlund, G. Tsiamis, S.M. Sievert, W.T. Liu, J.A. Eisen, S.J. Hallam, N.C. Kyrpides, R. Stepanauskas, E.M. Rubin, P. Hugenholtz, T. Woyke, Insights into the phylogeny and coding potential of microbial dark matter. Nature **499**, 431–437 (2013)

J. Schneider, Interactive extra-solar planets catalog. The Extrasolar Planets Encyclopedia 9 February 2015 edn (2015)

W. Seifritz, Mirrors to halt global warming? Nature **340**, 603 (1989)

A.L. Sessions, D.M. Doughty, P.V. Welander, R.E. Summons, D.K. Newman, The continuing puzzle of the great oxidation event. Curr. Biol. **19**, R567–R574 (2009)

I.S. Shklovskii, C. Sagan, *Intelligent Life in the Universe* (Holden-Day, San Francisco, 1966)

B. Song, L.G. Leff, Identification and characterization of bacterial isolates from the Mir space station. Microbiol. Res. **160**, 111–117 (2005)

P.D. Straight, R. Kolter, Interspecies chemical communication in bacterial development. Annu. Rev. Microbiol. **63**, 99–118 (2009)

K.E. Tsiolkovsky, cited in N.A. Rynin (1971) *K.E. Tsiolkovskii: life, writings and rockets*. Leningrad Academy of the Sciences of the USSR (1903)

P.A. Vaishampayan, E. Rabbow, G. Horneck, K.J. Venkateswaran, Survival of *Bacillus pumilus* spores for a prolonged period of time in real space conditions. Astrobiology **12**, 487–497 (2012)

P.A. Vaishampayan, C. Moissl-Eichinger, R. Pukall, P. Schumann, C. Spröer, A. Augustus, A.H. Roberts, G. Namba, J. Cisneros, T. Salmassi, K. Venkateswaran, Description of *Tersicoccus phoenicis* gen. nov., sp. nov. isolated from spacecraft assembly clean room environments. Int. J. Syst. Evol. Microbiol. **63**, 2463–2471 (2013)

P.G. van Dokkum, C. Conroy, A substantial population of low-mass stars in luminous elliptical galaxies. *Nature* **468** (7326), 940 (2010)

H.G. Wells, *The War of the Worlds* (William Heinemann, London, 1898)

W.B. Whitman, D.C. Coleman, W.J. Wiebe, Prokaryotes: the unseen majority. PNAS **95**, 6578–6583 (1998)

D.B. Wilson, Microbial diversity of cellulose hydrolysis. Curr. Opin. Microbiol. **14**, 1–5 (2011)

J.W. Wilson, C.M. Ott, K. Höner zu Bentrup, R. Ramamurthy, L. Quick, S. Porwollik, P. Cheng, M. McClelland, G. Tsaprailis, T. Radabaugh, A. Hunt, D. Fernandez, E. Richter, M. Shah, M. Kilcoyne, L. Joshi, M. Nelman-Gonzalez, S. Hing, M. Parra, P. Dumars, K. Norwood, R. Bober, J. Devich, A. Ruggles, C. Goulart, M. Rupert, L. Stodieck, P. Stafford, L. Catella, M.J. Schurr, K. Buchanan, L. Morici, J. McCracken, P. Allen, C. Baker-Coleman, T. Hammond, J. Vogel, R. Nelson, D.L. Pierson, H.M. Stefanyshyn-Piper, C.A. Nickerson, Space flight alters bacterial gene expression and virulence and reveals a role for global regulator Hfq. PNAS **104**, 16299–16304 (2007)

12

Space architectures

Barbara Imhof, Peter Weiss, Angelo Vermeulen, Emma Flynn,
Richard Hyams, Christian Kerrigan, Max Rengifo, Susmita Mohanty
and Sue Fairburn

Three author groups namely: Imhof, Weiss, and Vermeulen; followed by Astudio (Flynn, Hyams, Kerrigan, and Rengifo); and finally, Mohanty, Imhof and Fairburn, discuss different ways of making cities, architectures, habitats, and environments in space

12.1   The world in one small habitat

Barbara Imhof, co-founder LIQUIFER Systems Group,
Peter Weiss, Head of the Space and Innovations, Department at COMEX, and
Angelo Vermeulen, space systems researcher, biologist, artist, and community architect at the Participatory Systems Initiative, Delft University of Technology.

When an article about one of the largest biospheric projects entitled "Noah's Ark—The Sequel" (Reingold 1990) appeared in September 1990, the author Edwin Reingold could not anticipate that Biosphere 2 would become the proof that constructing a closed ecology is one of the toughest challenges for humankind and for long-term spaceflight. It is not

---

Barbara Imhof (✉) • Peter Weiss (✉) • Angelo Vermeulen (✉)
LIQUIFER Systems Group Gmbh, Vienna, Austria
e-mail: barbara.imhof@liquifer.com

Emma Flynn (✉) • Richard Hyams (✉) • Christian Kerrigan (✉) • Max Rengifo (✉)
Astudio, London, United Kingdom
e-mail: emma.flynn@astudio.co.uk

Susmita Mohanty (✉)
CEO Earth2Orbit Consulting Private Limited, Maker Tower, Mumbai, Maharashtra 400005, India
e-mail: susmita@earth2orbit.com

Sue Fairburn (✉)
Co-founder, Fibre Design Inc., QC, Canada

commonly known that it is in fact a prerequisite for future space exploration to have a reliable closed-loop life-support system in place. To date, nobody in the world has such a system ready. We now know that the Biosphere 2 experiment failed to a certain extent because of soil bacteria respiring and producing a lot of carbon dioxide. This means that the oxygen levels decreased (Fisher Smith 2010) quickly and dramatically. It seemed that the wrong choice of soil was made for the climate in Arizona. Soon after the door was sealed, the crew already felt the change in the atmospheric composition, although they were only 3 months into the mission (Broad 1993).

Further, the crew was hit by bad harvests and, as a result, needed to slaughter the livestock because the animals consumed too much food. Finally, the crew ended up sharing all the remaining food while oxygen levels decreased further. The psychological atmosphere among the crew got tense because the eight biospherians split into two groups who had opposite views on how to direct their experiment after it was clear that oxygen and food needed to be added from the outside to guarantee their survival. The two groups stopped speaking to each other, even after the experiment was terminated. They remained apart even after the end of the experiment, until one of their fellow crewmembers died and they all met at his funeral.

Rebuilding Earth with its closed ecology requires a lot more experimentation, knowledge, and skills. Only if we master closed-loop biological life-support systems will we be able to leave our terrestrial cradle to settle somewhere else or travel out of our Solar System.

12.1.1 MELiSSA: the agency's perspective

The European Space Agency (ESA) is currently conducting a study[1] to identify the most pressing needs to be able to leave Earth's orbit with human-tended spaceships and to close the technology gaps in the creation of closed-loop life-support systems, ideally including living ecologies.

One step in this direction of paving our way further into the Solar System is ESA's MELiSSA project. It is conceptualized as a closed-loop system, and is conceived as a network of five connected biological compartments, each of them with a different biochemical function which allows the exchange of gases and fluids (Fig. 6.1). The astronauts produce biowaste that is fed into the liquefying compartment where anoxygenic fermentation takes place using bacteria—a process which sets free volatile fatty acids, ammonia, and minerals. These reach the second compartment, which allows a photoheterotrophic process—basically, the decomposition of the materials through another type of bacteria. Both compartments and the astronauts set free carbon dioxide which is fed into the loop to the algae compartment. In this compartment, the algae transform the carbon dioxide into oxygen through a photosynthetic process, and the astronauts can utilize water as a by-product. After the second bacterial decomposition, ammonia and minerals go to the nitrifying compartment where oxygen is added. The resulting nitrates are used for the plant compartment. The crew tends the plants and harvest vegetables. MELiSSA has been in development for 25 years in different parts of Europe and, in 2009, its different compartments were brought

---

MELiSSA is an abbreviation of "Micro Ecological Life Support System Alternative."

[1] ESA study currently conducted by Space Applications Services, Comex, and LIQUIFER Systems Group; LUNA—Analogues for Preparing Robotic and Human Exploration on the Moon—Needs and Concepts. Available online at www.liquifer.com/?p=2207.

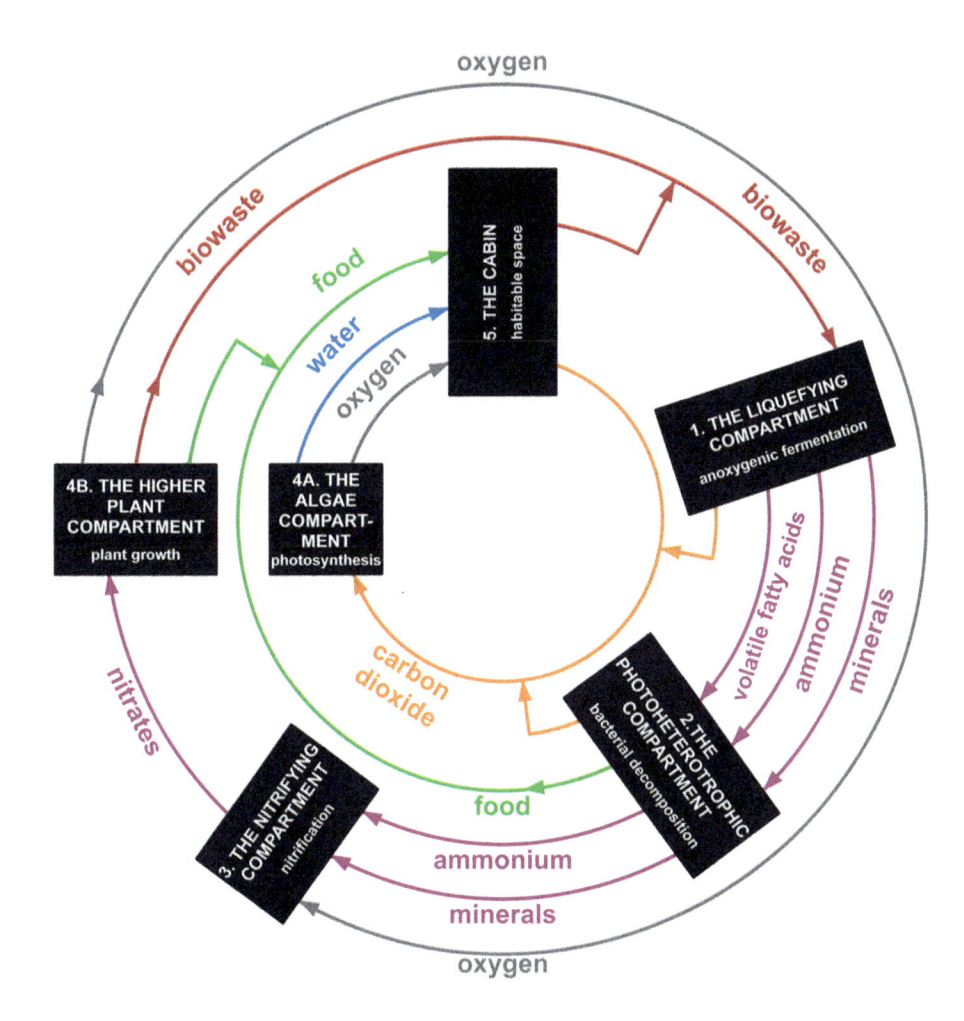

Figure 12.1 MELiSSA support systems. Credit: LIQUIFER Systems Group, schematics: Anne Marlene Rüede (2014)

together at the Autonomous University of Barcelona. In this set-up, 25 rats equal the metabolism of a human, and function as substitutes for various tests of system parts. However, the different compartments have not yet been connected, and the entire system needs a lot of miniaturization before it can be sent off into space (Fig. 12.1).

Competitors are NASA, the Russian space agency, and, foremost, the Chinese space agency, who have established a nearly closed-loop cycle in their Lunar Palace project. Chinese inhabitants serve also insects for dinner. On the International Space Station (ISS), however, the crew eats what we cook at home and, to a certain extent, also lives on the earthly supplies of water and oxygen.

Therefore, the likelihood of waving goodbye to a Chinese crew leaving Earth's orbit seems higher than the other nations mentioned. On our spaceship, Earth, we are more and more concerned with the limitation of life-essential resources such as water, good air quality, and clean energy. Looking at current tendencies and believing the extrapolations made by renowned researchers, our future lies within megacities. With the decrease of vital resources and the growth of urban areas, concepts and technologies from space create inspiration, such as the CAAS (City As A Spaceship) project (see Part II, Chap. 8). Further, there are technologically transferred applications (spin-offs from space), such as business products of IP-Star, a Dutch company which took the water regeneration system from MELiSSA and created a terrestrial application for water purification for large hotel structures.

12.1.2 MEDUSA: designing for living in outer space—underwater! (Fig. 12.2)

One design project in the realm of the topic's focus is MEDUSA—from subsea to Moon and Mars. It is anticipated that humans will return to the Moon in the time frame around 2025 (ISECG 2013). Future lunar explorers will, however, not visit our celestial neighbor for some days—as during the time of the Apollo program—but will work to build a permanent base on the lunar surface for extended crewed missions. This vision relies on the development of habitation concepts that, while being functional and livable, can be transported to the surface of the Moon. Inflatable structures are identified today as a technically feasible concept to build a habitat on the lunar surface: when folded, those living spaces can be more easily transported in space due to their low volume and weight.

But how can the internal design be combined with the concept of a foldable habitat structure?

And how can such foldable structures be installed on the lunar surface in reduced gravity?

Before one starts building the real hardware to fly into space, one needs to test, validate, and train in simulated environments. And this is what the concept of MEDUSA delivers: an answer to these design questions. Further, since it has been conceived by architects, engineers, artists, and scientists, the MEDUSA project presents a holistic approach, incorporating an interesting concept of life-support systems in combination with human factors considerations.

MEDUSA's concept is based on an inflatable volume that can be built around any type of lander: for its final destination to the Moon, a smaller version of the Altair or LEM[2]-type lander can be imagined. MEDUSA could be wrapped around the newly designed NASA Altair.

On MEDUSA's road map of development, it is foreseen to test and validate MEDUSA underwater and in orbit (Fig. 12.3). For the underwater simulation, the lander in the center of the habitat will be replaced by a classic diving bell (Fig. 12.4), leading to reduced costs in testing the first prototype of MEDUSA. Then, in a second stage, MEDUSA can be fixed outside a Soyuz or Progress module and brought up to the ISS for space validation (Figs. 12.3 and 12.4).

The subsea simulation of an underwater version of the habitat incorporates the idea of combining neutral buoyancy simulations, such as are conducted for astronaut training in water pools by ESA or NASA, with the use of analog sites on the seabed (ISECG 2013). As an inflatable structure, it can be deployed on a site of interest and recovered for storage and maintenance, which will significantly reduce the running costs of the facility compared with other underwater laboratories. Subsea simulations will allow testing the installation of the habitat in reduced gravity. Inside the base, the astronauts, or aquanauts, train

---

[2] Lunar Excursion Module.

The world is one small habitat 291

Figure 12.2 MEDUSA from a subsea habitat for astronaut training toward a lunar base. Credit: LIQUIFER Systems Group, visualization: René Waclavicek (2013)

PHASE A Subsea Simulation:
MEDUSA as simulation base for subsea training in neutral buoyancy.
- Surface operations (EVA)
- Habitability
- Life support systems

MEDUSA is fitted to a diving bell as "lander".

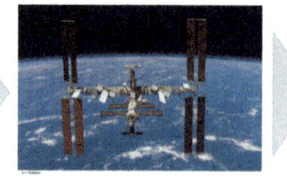

PHASE B Orbit Validation:
MEDUSA undergoes orbital validation at ISS.
- Hull material resistance
- Radiation shielding
- Micro-meteoroids
- Life support system

-MEDUSA is fitted to a Soyuz or Progress vehicle for transport.

PHASE C Moon Base:
MEDUSA's final destination is the Moon where it would serve as living base for future astronauts.

MEDUSA is fitted to a lunar landing vehicle.

Figure 12.3 MEDUSA's road map of development from underwater trials and validations to a lunar base. Credit: LIQUIFER Systems Group and COMEX, 2012, ISS photo courtesy of NASA

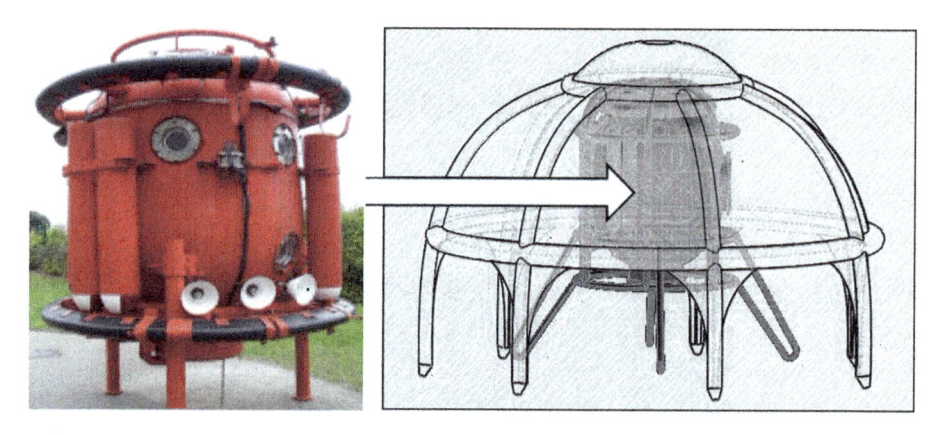

Figure 12.4 MEDUSA's concept is based on an inflatable habitat being installed around a landing vehicle; for the underwater trials, a diving bell will be used. Credit: COMEX (2012)

to live in confined space for a longer period of time. Functionality, as well as aspects of habitability and usability of the interior design, can be tested and improved for the next steps in development. In this simulation phase, the aquanauts already experience living and working in an extreme environment. They leave the habitat to perform field explorations or scientific missions underwater in a lunar-specific reduced-gravity simulation at one-sixth G through adequate buoyancy adjustment (Fig. 12.5).

An orbit simulation would be the second stage of MEDUSA's way to the Moon or farther destinations. A space-fit version of the habitat, built of resistant materials, is fixed onto a

Figure 12.5 Astronaut training on the seabed. Credit: COMEX (2013)

transfer craft toward the ISS, where it will deliver additional living and working space to the crew. In orbit, the habitat can be tested for its resistance to and use in the space environment.

The MEDUSA concept integrates the landing vehicle as a central unit into the layout of the habitat which serves as shelter, sluice, or storage space. It is surrounded by the rigid service ring containing all the necessary working, living, and life-support elements in folded or stored state during transport. After deploying the general living and working space through inflation and creating the stabilizing interior pressure of the hull compartments, the rigid floor will be folded from a vertical into a horizontal position (Fig. 12.6).

The configuration of the service ring is designed in such a way that the habitat is ready for use immediately after the completion of the deployment process. The transformation capacity of the ring-based elements also ensures maximum flexibility during operation. Furniture elements not in use can be folded away, so that the confined living and working space can be individually configured and optimized.

Life support is, strictly speaking, the management of products to sustain the astronaut's life in the extreme environment of space. However, in the MEDUSA concept, this aspect is seen on a larger scale: a life-support system shall not only handle the necessary living

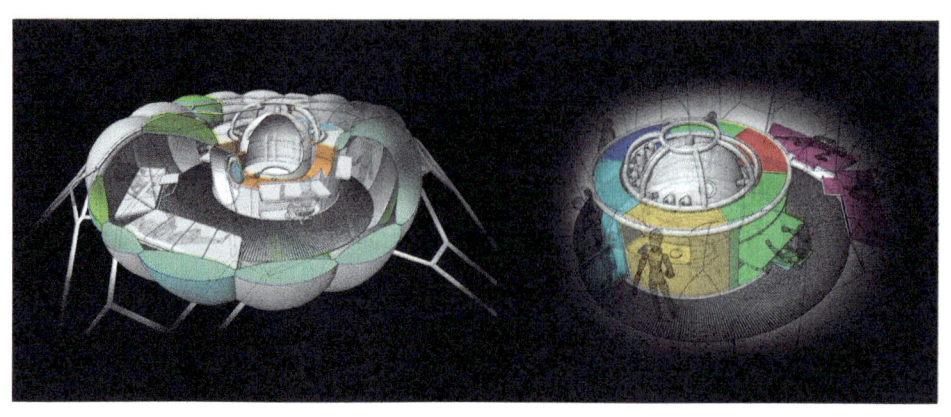

Figure 12.6 MEDUSA's internal design. Credit: LIQUIFER Systems Group, visualization: René Waclavicek (2012)

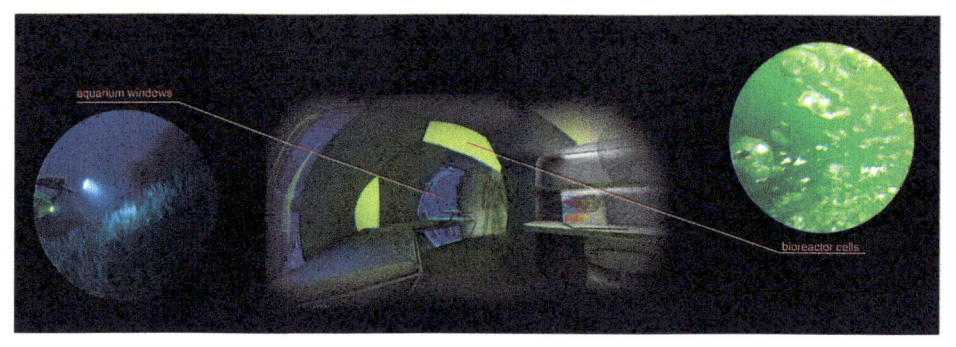

Figure 12.7 Aquarium windows and bioreactor cells inside MEDUSA. Credit: LIQUIFER Systems Group and COMEX (2012)

products (air, food, water, and waste), but also contribute to the psychological well-being of the astronauts in their lunar home. The MEDUSA life-support system is an integral part of the inflated structure of the habitat. The outer hull of the living spaces will be composed of cells that not only protect the inhabitants from the extreme environment, but will also add a comfortable atmosphere inside the living volume.

These cells will be filled with water for structural purposes. The majority of cells will also serve to protect the astronauts against ionizing radiation (water is a good shielding element for solar energetic particles). These water-filled cells can also be used as windows to the outside, thus increasing the perceived living space and adding a blue light to the internal environment (Fig. 12.7).

Other cells will include bioreactors filled with algae that can assist in recycling carbon dioxide to oxygen and deliver food products. These cells are illuminated (by the Sun or by artificial light integrated into the cell), and thus add a green, natural environment to the home of astronauts. They can be exchanged for maintenance or harvesting purposes (Fig. 12.7).

MEDUSA is in some way designed as our natural "life-support system" on Earth—Nature—not only sustaining life, but also necessary for the well-being of its habitants.

12.1.3 Water Walls life support: the next step in evolution

The Water Walls concept, derived by space architects Marc Cohen, Renee Matossian, and engineer Michael Flynn (NASA Ames),[3] explores more concretely the step into a fully fledged MEDUSA-like life-support system. The Water Wall takes Nature's approach of being a passive system: chemically passive and massively redundant. This opposes most space agencies' approaches of using electromechanical systems which tend to be failure-prone.

"The core processing technology of Water Walls is forward osmosis. Each cell of the Water Wall system consists of a polyethylene bag or tank with one or more forward osmosis membranes to provide the chemical processing of waste. Water Walls provides four principal functions of processing cells in five different types plus the common function of radiation shielding:

1. Gray water processing for urine and wash water,
2. Black water processing for solid waste,
3. Air processing for carbon dioxide removal and oxygen revitalization,
4. Thermal and humidity control,
5. Food growth using green algae,
6. Provide radiation protection to the crew habitat.

"Forward Osmosis uses a semi-permeable membrane to effect separation of water from dissolved solutes.[4] In effect black water can be purified this way and exhausted forward osmosis bag elements can be delivered to other functions of the system. These bag elements are drained, fluids are then mixed with faeces, solid organic wastes, and advanced water treatment residuals are either re-injected for sludge treatment, or simply cured in place to a stable solid." (Cohen 2012)

This life-support system could be filled with water from extraterrestrial surfaces, and the crew would also contribute to the water resources through their urine. Further, the bags are assembled such that they are easy to maintain, replace, and repair the system if problems occur. Thus, it can be considered as a sustainable system, or even resilient, because there is always one part which can serve as a backup for another part which is malfunctioning. In the course of a long journey, the crew can even exchange the forward osmosis bags. In this way, this concept is applicable not only to a small habitat, but also to a large star ark. Currently, the team is working on breadboarding parts of the system with a view to completing a full prototype sometime in the near future.

12.1.4 Toward a robust and resilient future in outer space

It has been more than 60 years of human spaceflight and we still have not gone farther than the Moon, and are only present in a low Earth orbit on a regular basis. The pressing issue of a system that supports human life in harsh and unforgiving environments, in outer

---

[3] The full team also incorporates Sherwin Gormly and Rosso Mancinelli.
[4] https://en.wikipedia.org/wiki/Forward_osmosis.

space, has been with us since the dawn of space travel. Today, with renewed public and private interest, we will get closer to the Moon again and farther, onto Mars and beyond.

The project MEDUSA and its habitat concept present a step into a direction where we could speak of a living *space* architecture, where organisms, humans, and technology are equal parts of a biospheric and holistic approach. MELiSSA and the Water Walls life-support system are leading with other life-support system approaches, such as the Chinese Lunar Palace simulator, in the right direction of getting us farther away from Earth.

Research in this direction also helps us to understand our spaceship Earth and its biospherics much better, so that we can learn to improve the environment on our home planet. From Earth, we learn that a useful life-support system needs to be a closed ecology and a living system, not a purely mechanical system—at least, something hybrid.

Will we be able to transform our dreams into reality, and explore and leave this planet to go farther and beyond our Solar System?

Will we be able to create even larger structures, such as a generation spaceship, which represents *the whole world*, steered by crews that set out to meet the unknown?

These questions cannot be answered today, though they leave space for speculation, and the recent DARPA (Defense Advanced Research Projects Agency) initiative to create a 100-year spaceship is only one of such programs in this direction. It may look as thought, after 60 years of human spaceflight, we have not come farther than a continuous presence in orbit and a short trip to the Moon, but from another perspective we can proudly say that we have come this far already—that, as a species, we have developed the intellectual capacities to develop technologies intertwined with living ecologies and that we will also pose the right ethical questions which will demand that we create our values in these matters. It is imperative that we make sure that recreating a biosphere and living ecologies in convergence with technology follows a strong system of values and an ethical component.

12.2 Starship cities: a living architecture

Emma Flynn, Research and Development lead, Astudio,
Richard Hyams, Co-Director, Astudio,
Max Rengifo, Co-Director Astudio and
Christian Kerrigan, architect, Astudio.

12.2.1 Introduction

Astudio are world-class architects focused on reducing our buildings' impact on the environment. Our courage in the pursuit of new ways to reduce carbon and change behaviors gained us the coveted title of Architectural Practice of the Year in 2012. This unprecedented achievement led us to establish our research and development group, whose role is to investigate new ways of seeing and thinking about the challenges we face in the industry. With this group, we explore visionary concepts that enable us to consider how we may cross boundaries taking us out of architecture and work with different fields of expertise that can help us solve sustainable construction in new ways. We look to the future to propose buildable solutions today.

12.2.2 Tomorrow's city

12.2.2.1 Today's challenge

In the 30 min in takes you to read this article, the world's population will have risen by 5000 people.[5]

By 2050, with population growth rising by 74 million people a year, two-thirds of us will be living in cities, and the environmental impact of our increased demand for resources and services will be intensified by the collective impact of industrial processes.

Within the next few decades, we will face life-threatening issues within our city, such as resource shortages, waste-disposal issues, pollution, traffic congestion, crime, and homelessness. Despite our best efforts to reduce our consumption of the natural resources that we rely on, our current consumer-led and resource-intensive lifestyles and dependence on current "take, make, dispose" industrial processes will not be able to contain the scale of the problem that we will face within our lifetimes.

Climate change caused by increasing carbon dioxide levels and the resultant global warming is a precise result of our growing population and our wasteful behavior. Extreme weather conditions and acts of Nature are on the rise, and are progressively threatening our existence in both rural and urban surrounds.

In response to this environmental crisis, the UK has committed to a carbon emission reduction of 80% by 2050. This places responsibility on all members of society, from governments and industry to individual citizens, to make sustainable change. How we govern, design, and inhabit our cities will have a fundamental impact on our collective sustainable future.

The construction industry will need to play a key role in this sustainable vision if we are to meet the 2050 carbon-reduction target. In the UK, the construction industry accounts for approximately half of the UK's emissions of carbon dioxide, generated from depleting fossil fuels consumed in the construction and operation of buildings. This commitment, translated in real terms for the construction industry, means that all buildings whether new or existing need to be carbon-positive by around 2040 (the remaining challenges being in relation to infrastructure and non-regulated uses). This is a considerable challenge for the industry.

The built environment—the city: its architecture, infrastructure, and urban realm— needs to be reassessed in the light of these new demands. With global warming increasing, amplifying extreme weather conditions and frequency, buildings can no longer remain static and unresponsive to their surrounding environment. Architecture must become adaptive, reactive to changes in climate, and reducing CO_2, not generating it.

Incremental thinking is not going to help us achieve this. We need to adopt strategies that are visionary so that we can raise the game beyond what we presume possible—and explore new ideas and approaches that may lead to radical shifts in thinking, practice, and the way we live.

---

[5] Worldometer World statistics, available online at www.worldometers.info (accessed May 20, 2015).

12.2.2.2 Starship cities

"I've often heard people say: 'I wonder what it would feel like to be on board a spaceship,' and the answer is very simple. What does it feel like? That's all we have ever experienced. We are all astronauts on a little spaceship called Earth."[6]

By the middle of the century, we will have entered the next phase of human expansion. With the impact of an increased population threatening life on Earth, space colonization is becoming a serious reality as the concern for planetary-scale disaster grows. Human space exploration, driven by our innate intellectual curiosity to discover, is divulging new possibilities for the human race beyond simply mere survival. Earlier this year, the Mars One mission selected a shortlist of 100 candidates for a one-way trip to Mars, as they hope to become the first humans to set foot on the Red Planet. In March, two astronauts embarked on the first one-year mission to the International Space Station (ISS), twice as long as any previous US missions.

Long-duration spaceflight is now a regular occurrence, but interplanetary travel is one thing; designing an interstellar spaceship capable of reaching our nearest star system is a different matter. It requires enormous amounts of energy and a starship design that will endure great speeds and vast distances, which could be hundreds, potentially thousands, of years. A worldship, capable of sustaining human inhabitants for the duration of the journey, would require the design of a life-supporting interior that provides food and water, recycles waste, protects against radiation, and generates significant amounts of energy to run those systems, not to mention catering for the social and physiological issues faced by the crew on board.

While a worldship is some time off being reality, the question of how we may live in space is important, as it asks us questions about sustaining life on Earth right now. Thinking about how we may design a whole city that self-perpetuates, with a limited set of resources and conditions for existence, is exactly the kind of challenge we are trying to address in cities today.

"Spaceship Earth," the worldview term, popularized in the mid-1960s by R. Buckminster Fuller, conceptualizes the finite amount of resources available on Earth, and encourages humanity to live harmoniously without negatively impacting Earth's ecosystems and regenerative capacity. This capacity is now at threat—the rapid development and industrialization of "Spaceship Earth" is triggering a new stage of "metabolic evolution" on Earth as we alter our environment through the burning of non-renewable fossil fuels.[7] This transformation preempted the 1966 essay, *The Economics of the Coming Spaceship Earth*. In this essay, Kenneth E. Boulding described the past open economy of apparently illimitable resources, which he said he was tempted to call the "cowboy economy," and continued:

"The closed economy of the future might similarly be called the 'spaceman' economy, in which the Earth has become a single spaceship, without unlimited reservoirs

---

[6] Fuller, B. *Operating Manual for Spaceship Earth*, pp. 55–56. Southern Illinois University Press, Carbondale, IL (1969).

[7] Armstrong, R. *Starships on Earth*. Available online at www.centauri-dreams.org/ (accessed June 15, 2015) (2014).

Figure 12.8 Spaceship Earth. Credit: "Earth From Space," image by Luigi, from openwalls.com, https://poesypluspolemics.files.wordpress.com/2013/12/earth_from_space_5_1920x1080.jpg

of anything, either for extraction or for pollution, and in which, therefore, man must find his place in a cyclical ecological system."(Fig. 12.8)[8]

This "spaceman economy" is a new age in which resources are now limited, and the pressure of human development is compromising our planet's native life-support systems. If we are to ensure a sustainable future, whether in space or on Earth, we need to start thinking of our habitats ecologically, and find inventive ways to design how it works as an autopoietic

---

[8] Boulding, K. *The Economics of the Coming Spaceship Earth.* Available online at http://dieoff.org/page160.htm (accessed June 15, 2015) (1966).

system as a priority. If Buckminster Fuller called Earth a "spaceship," then we can consider all our cities as being starship environments (although they have not broken free of their orbital flight path around our Sun). In other words, we can consider our cities as megastructures that cannot simply draw on some untapped terrestrial resource to supplement its metabolism, but must become independent ecological systems that are self-sustaining.

The challenge of starship design enables us to deal with the complex scarcity of issues on living in a resource-constrained world where we have to rethink traditional practices and relationships for a more humane and habitable world. Starships allow us to transcend the technical challenges of space travel alone and propose a visionary architecture through which tomorrow's cities can evolve and flourish—not merely survive through austerity. They provide a realm in which we can explore new forms of value production, facilitate different kinds of economy, generate the conditions in which alternative social structures may evolve, and facilitate notions of community that are not dictatorial and prescriptive, but enabling, creative, and responsive to contingencies and change. This is beyond the demands of a colony on Mars—it is to embrace that visionary potential and engage the imagination with what may be possible if we share value systems and ambitions for a new way of living.

12.2.2.3 Future Nature

From the 1960s, the imagined and real environments in outer space were, to environmentally concerned designers, models for how to handle the ecological crisis on Earth. The construction of space-cabin environments for astronauts linked space exploration to city design, due to the fundamental nature of these life-supporting infrastructures. These concerns were centered on the issue of "sustainability."

The importance of space research to "sustainable design" was first realized by architects Serge Chermayeff and Christopher Alexander. Their *Community and Privacy* (1963) noted the environmental destruction of farmland and wilderness by increasing suburban development and industrialization. In response to this invasion, they believed that buildings should be designed as ecologically autonomous rather than exploiting the natural environment around them. Looking to space research for inspiration, they believed there was an urgent "need to design fully functioning self-contained environments, capable of sustaining human life over long periods." They recognized that "the nuclear submarine and the space capsule have been designed to support life over protracted periods without the possibility of escape," although they acknowledged the danger of occupying closed environments for long durations.[9]

Following *Community and Privacy*, Buckminster Fuller was soon to adopt "space ecology" to develop his own unique set of "design-science" principles.[10] He envisioned utopian cities with their own closed ecological systems, free of politics, and designed for all of humanity to live in freedom and harmony. Fuller, aware of the finite resources the planet has to offer, promoted a principle that he termed "ephemeralization," which refers to the ability of technological advancement to do "more and more with less and less until eventually you

---

[9] Alexander, C.; Chermayeff, S. *Community and Privacy*, pp. 46–47. Penguin, Harmondsworth (1963).

[10] "Design science" term coined by R. Buckminster Fuller. This subject is covered in both "Critical Path" and "Utopia or Oblivion."

can do everything with nothing."[11] It was the idea that resources and waste material could be continually recycled into more valuable products, increasing the efficiency of the entire process. The intention was to provide an increasing quality of life for a growing population despite finite resources. This concept was exemplified with his Dymaxion house prototype, which represented the first attempt to build an autonomous building. Incorporating passive cooling strategies through the "dome effect" of the architecture, and by integrating devices such as "fog showers," it was designed to reduce water and energy use.

While these early ideas of "sustainability" pioneered the notion of closed ecosystems, they operated solely on the principles of resource conservation—essentially demonstrating a "better" kind of industrialization. This remains the prevalent paradigm of sustainability today. Today's sustainable solutions are focused on reducing carbon emissions, conserving energy and water, and recycling our waste. Indeed, this strategy succeeds in slowing down the depletion of the planet's resources, but it does not address fundamental issues regarding consumption and population growth—nor does it promote the "liveliness" of spaces. Our current idea of "sustainability" is based on entrenched industrial principles that rely on machine-inspired solutions to continually modify what already exists, rather than exploring fundamentally different ways of thinking. This is a top-down approach to sustainability, which aligns with our Victorian methods of construction, simply imposing structure on the environment.

For generational starships and resilient future cities, we need a different approach. We need to take a long-term, radically creative view of our ambitions for humankind, do more with less, embrace change—and not look to "sustainability" as a means of justifying established practices. We need to remove ourselves from this particular kind of industrial thinking, and reassess architecture's environmental role from a much wider perspective. We need to build sustainable environments that promote life—that are self-sufficient and productive—actively responding to environmental conditions, generating energy, and reducing pollutants. We must move away from the 1970s visions of the future—ideas of sustainability that think solely in terms of closed-loop systems that assume life runs like a machine—but nourish open systems that can flourish under the right conditions, that can adapt and change to their differing circumstances, just like Nature itself.

12.2.2.4 The Tomorrow's City vision

"Tomorrow's City" is a collaborative vision conceived by Rachel Armstrong, professor of experimental architecture at Newcastle University, Tomorrow's Company, and Astudio to move beyond the current model of sustainability and propose practice-based solutions to our future cities. It is a bottom-up approach in which the future city is co-designed by communities, businesses, government, and our ecology—enabled by forms of governance and leadership that can inspire action and participation across stakeholders and generations. It involves communities to become active in the process of change by encouraging projects that can be contributed to through maker movements and schools, as well as multidisciplinary professional and commercial partnerships. It is a vision that values cross-disciplinary collaboration as key to unlocking the answers to future sustainability.

---

[11] Buckminster, F. *Nine Chains to the Moon*, pp. 252–259. Cape, London (1973).

Figure 12.9 Tomorrow's City vision: a vision of a city that exists within its ecological limits. Credit: Astudio, rendering by Christian Kerrigan, 2015

This collective vision is a different way of living. It explores how we can best create long-term value in how we live and work: our energy, food, water, waste, transport, and infrastructure. It looks to invert the current relationship between making a building and its environmental impact. Right now, we assume that creating an architecture inevitably damages the environment but, for us to persist on this planet, then the production of architecture itself must enhance not just our own survival, but the infrastructures and ecosystems on which our prosperity depends. This means we need architectures that do not just take from our environment, but also give back to it, in ways that are ecologically meaningful. In this vision, the city is a living system of exchange between people and their surroundings, so that they are no longer simply consumers of their biosphere, but contribute to it—and become a productive part of it (Fig. 12.9).

This is a vision of a city that exists within its ecological limits. The city is self-sufficient and sustainable, with a circular metabolism that finds opportunity and value in waste for the generation of food, economy, and power. It can be thought of as a "circular economy" in which all exchanges promote welfare throughout the whole value chain underpinning cities: social, economic, or environmental in nature.

Such a transition will be possible by adopting a visionary approach to architectural practice that may be enabled by new ways of thinking about the relationship between architecture and Nature, made tangible through new kinds of technologies that possess lifelike qualities and a broad range of associated cultural and social changes that will result from the new opportunities to which they provide access.

12.2.2.5 An Ecological framework

To develop new thinking, we need to establish new frameworks, which allow us to envision the world differently. At the start of the third millennium, we are already finding ourselves establishing new ways of relating to the city and the natural world. Major cultural and technological developments in our cities are impacting a worldview that is less defined by objects, but increasingly shaped by connectivity. Our cities are becoming progressively networked, connected by the Internet of everything, sensor technologies, data sharing: the ubiquitous smartphone.

We are increasingly understanding more about Nature, and how to manipulate it, through the advances in biotechnology and the emergence of synthetic biology. We have emerged from the machine age of the Industrial Revolution, and the twentieth-century information revolution, to a new era in which biological engineering will be the foremost driver of change (Fig. 12.10).

Our cultural notion of Nature as wilderness and unspoiled landscape is transforming. Whereas, traditionally, Nature and technology were regarded as binary oppositions, today Nature and technology appear increasingly unified and collaborative. Architect Liam

Figure 12.10 Ecological infrastructure: this view of the River Thames depicts the site under the microscope, and also as an aerial plan of a city as the boundary between architecture and environment blurs. These multi-scalar perspectives of where we live are relevant to both Earth or in space. Credit: Astudio, rendering by Christian Kerrigan, 2015

Young of Tomorrow's Thoughts Today explains: "What we've realised is that there's no nature anymore—at least not in the sense that we culturally define it." He concludes: "What there is, is technology. Engineered networks, augmented environments, invisible fields—infrastructure has exploded into bits, to roam the Earth in an architecture of everywhere."[12]

With the emerging landscapes of robotics, biotechnology, and pervasive computing, our cities are becoming a wilderness of complex connections and dependencies in which the traditional definitions of Nature and technology are becoming blurred. This is driving an intricate and fluid understanding of the world, in which the term "Nature" is being redefined by complexity and connectivity. These complexities have political, social, economic, cultural, environmental, and aesthetic implications, which underpin the performance of our cities. By establishing an ecological framework, we can start to imagine these complexities as architectural ecologies of the city, and start to design sustainable solutions to future living.

The following chapter explores the idea of architectural ecologies as networks and technology, which underpin the complexity of sustaining city systems. These architectural ecologies examine the complex relationships that constitute our culture and environment. In contrast to the architectural professions leading interpretations of ecology as the basis for building designs that achieve environmental efficiency, architectural ecologies extend architectural thinking beyond "green building and sustainable architecture," incorporating sociopolitical understandings of ecological thinking and practice. These ecologies impact the complex material, spatial, social, political, and economic concerns which will inform the design of our future habitats, and vice versa.

By thinking ecologically, the future city has the potential to develop new forms of economy, value, and social organization, working symbiotically with the natural environment. It is able to venture beyond the closed loops of sustainability and become self-sustaining, while remaining open and responsive to changing contexts. This future city may be called a starship city, and could apply to inhabitation of terrains that are either terrestrial or non-terrestrial in nature.

The second part of this chapter proposes strategies for implementing this vision today and in the near future. Through projects, proposals, collaborations, and ideas, it looks to propose workable solutions to explore how our machine-like cities could behave more like ecologies.

12.2.3 Architectural ecologies

12.2.3.1 The Future Timeline: implementing change (Fig. 12.11)

While notions of starships are some time away, the challenges that we are facing with "sustainability" in today's cities are not. Our current predictions indicate that we are less than five "building development times" away from finding ourselves at the heart of a

---

[12] Young, L. *Unknown Fields*. Available online at www.dezeen.com/2013/02/04/liam-young-tomorrows-thoughts-today/ (accessed June 15, 2015) (2013).

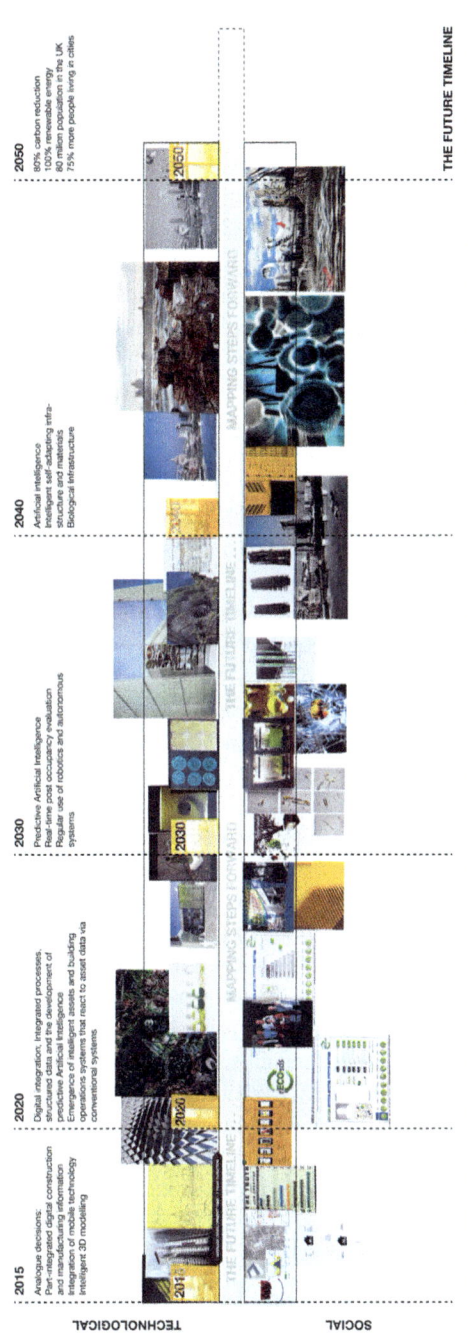

Figure 12.11 The Future Timeline, depicting both the technological and social step changes required in confronting the huge challenges facing the planet. Credit: Astudio, installation and drawing by Emma Flynn, 2015

worldwide crisis on a scale that we have previously never encountered. This crisis will not just be ecological, but also humanitarian.

The average civic-scale building project can take 5 years from inception to completion. This means that we only have five chances, if you consider building projects running end to end, to implement change before 2040, when all buildings need to be carbon-neutral. While this statement is not quite accurate—building projects rarely run in succession—it highlights the enormous challenge that the building industry faces. We need to take big steps forward on a project-by-project basis, which will require substantial advances in thinking and technology if we are to meet our 2050 goals. While, in the past, we have been content with incremental thinking and its implementation, we now need visionary experimentation and prototyping within urban environments to generate these giant jumps forward.

Space exploration has historically driven radical leaps in knowledge and technology, pioneering and transforming breakthrough applications that have enhanced the quality of life on Earth for all. Technologies created for and made possible by space exploration permeate, shape, and are an integral part of our world, providing GPS location devices, lightweight materials, revolutionary medical procedures, television signals to remote parts of the world—the list goes on. While some, like futurist Juan Enriquez, consider that starships may be 300 generations away, others such as the international organization Icarus Interstellar aim to catalyze interstellar flight by the year 2100. These radical goals will be fundamental to driving sustainable cities here on Earth.

The "Future Timeline," developed by the research group at Astudio, sets out both the technological and social step changes required in confronting the huge challenges facing the planet. These sociological and technological advances are underpinned by motions in current legislation—issues of future governance, environment, economy, urban development, and infrastructure. Recognizing the urgency in response to the crisis, the timeline highlights just how little actual conceptual and visionary thinking has permeated into our urban environment and emphasizes how much more ambitious we could be.

Journeying from 2015, through the depletion of fossil fuels to the date coal runs out in 2427, the timeline sets out a pathway to establishing a sustainable urban model for our future cities on Earth and in space. The timeline proposes real solutions. It offers a buildable vision of our future, not simply a critique of it. We are aware of the dangers of "apocalypse fatigue"—continually saying how bad things will be if we do not do things differently offers no solution or route map for change. Moreover, it does not inspire the new insights from which new technologies or behaviors will arise, or engage the different communities who must come together to create those insights.

Developed initially as a digital, interactive holder for our "Future Cities" research at Astudio, the Future Timeline has manifested into a guide, and a warning, plotting our prospective journey toward a sustainable urban future, outlining possible routes and solutions to change, and challenging our current urban models and strategies along the way.

12.2.3.2 Architectural ecosystems

At Astudio, we realize that the Tomorrow's City vision will rely on both technological and social step change, facilitated by adjustments in governance, to achieve realization. While technological developments are rapidly transforming our cities and our experience of them, social change is less quick to materialize. Society still has far to go in adjusting basic attitudes and behaviors regarding waste and consumption.

Currently, we live in a world where the impact of our actions still remains intangible. If you ride a bicycle, when you pedal faster, you experience an increase in velocity. There is a direct relationship between action and impact. So why is it that, in a world where resources are short and finite, we do not understand the impact of our actions? If we charge a laptop overnight, how many aluminum cans do we need to recycle to negate its impact? We still do not know how much energy is used, and carbon generated, by our daily activities and purchases.

At Astudio, we aim to move the issue from wasteful to efficient. Our work proposes solutions to issues of sustainability that simultaneously encourage awareness and behavior change as a starting point. Our goal is to create a range of toolsets that enable people to create new kinds of value, meaning, and ways of living, which will impact the future resilience of our cities.

Cities are complex systems and are difficult to understand and adapt without the right information. We are currently working with the Carbon Trust to actively monitor our buildings performance in use. From the collected data, we can help the occupants use and manage the building more efficiently, reducing wastewater and power, and increasing comfort. To communicate this information to the occupants, we are developing an intelligent building management system (BMS) and user interface to allow our building to "talk" to the users, and vice versa. This is a real-time transfer of information and has the potential to be truly dynamic and responsive. By incorporating a network of sensors, our buildings can actively adjust to how many people are in them to control climate change and conserve resources.

In this way, our cities are becoming increasingly networked. Every chair, wall, and pavement will be filled with sensor technology, and we are seeing it emerging right now. A number of cities in the US are trialing "smart parking," which accurately senses vehicle occupancy in real time and guides drivers to the nearest available parking. Sensors installed in bus stops are optimizing public transport, directing buses to where people are waiting, reducing fuel usage. These enhanced networks of information enable us to monitor, measure, and manage cities to ensure they are happy and healthy places for their inhabitants. They provide interfaces that highlight the cost of our actions in energy and resources terms, where currently we are blind to them. It is a form of communication that develops a new understanding between architecture, humans, and the environment so we are able to tune and adjust our surroundings.

We envision a starship city that exists as an ecological framework of highly sensitive and intuitive feedback networks. It is a system of self-regulating architectural ecologies that function at the scale of an individual building, while simultaneously integrating with the surrounding environment and urban infrastructure. Informed by data collected from factors such as energy and water consumption, transportation, weather, crowds, or occupancy requirements, the built infrastructure is able to execute informed and calculated decisions to optimize environments.

Within buildings, the monitoring of internal conditions allows for the modification of spaces for specific user requirements and programs, adjusting climate, lighting, and acoustics. Externally, building systems can transform their facades to reduce reflectivity and heat adsorption, and minimize the urban heat-island effect. Daylight sensing can manipulate facade opacity to regulate internal temperature and absorb energy, and the composition of structures and use of resources transform according to external conditions and stresses. These nurturing infrastructures, operating at both the building and city scales, remain open and adaptable to changing contexts. Informed by the current conditions of the occupants, the environment, and the city, they respond actively to the challenges of climate change, population, and resources.

Right now, these sensing and computational technologies are driven by traditional digital computing systems, relaying information between humans and the built environment. In the starship cities of the future, we may imagine that levels of communication may extend to the natural world, evolving more "vital" engagements. Data could be used in alternative ways to better manage and engage with complex systems.

In the emerging field of synthetic biology, scientists and artists are exploring ways to manipulate and construct biological systems for useful purposes. A project by designer Alexandra Daisy Ginsberg titled *E. chromi* proposes that bacteria could, one day, be developed to excrete brightly colored pigments when they detect disease inside your body, alerting you via vividly colored poo. It is a concept that develops a new kind of interface for biological computing. Manipulated "natures" that become communication systems embedded at a microscopic level in our environment—a new Nature that could actually "talk" to us.

This is not unrealistic. We already have plants that "tweet" how they are doing. Imagine a garden as a data display; the BMS of today could be reinvented as soil infrastructure, using manipulated bacteria that changes color in relation to the carbon content of the soil, providing an indication of fertility. Alternatively, we can imagine plant life that changes scent or height in response to drought or frost, or algae blooms that indicate contaminated water in rivers or ponds through distinctive growth patterns. Applied to building facades, engineered bacteria could be used to detect toxins or pollutants in the atmosphere. Pigeons, the pests of the city, could be enabled with CO_2-monitoring tags to measure the quality of the air and become positive players in city life, with the skins of buildings able to reflect the findings.

12.2.3.3 Disruptive innovation

The interconnected city landscape, facilitated by the biotechnological revolution and the ever-expanding Internet of everything, has the potential to discover new forms of economy, value, and social structure, which work symbiotically with the natural world to develop new solutions to issues of future sustainability.

The increased power of communication and knowledge exchange over networks and mobile devices is bridging social and cultural boundaries. It has provided a space for the individual voice to be heard and be meaningful, providing citizens with the means to support and enhance collective initiatives concerning public issues, from shared childcare to health services, transport, and safety. Public communities are increasingly shaping and being shaped by online platforms which allow individuals to participate more directly in the design of their communities and cities. Today, our cities are no longer shaped solely by town planners and authorities, but are transformed by the interplay of a multitude of voices. This is a bottom-up approach to city design, which has the potential to radically re-invent the cityscape.

This new form of social organization and empowerment is reflected in the emerging world of "disruptive innovation," which is starting to create new patterns of value creation and economy. Connectivity is generating a new sharing economy, where rooms, homes, cars, and even dogs can be made available to others for rent or share. These disruptive Internet-based services, such as AirBnB, which allows people to locate strangers' spare rooms at a click of a button, and Uber, the Internet taxi service that allows tens of thousands of people to answer ride requests with their own cars, are creating new networks of exchange. These disruptive innovations are altering the requirements of the city and our experience of it.

In the design world, the proliferation of low-cost manufacturing tools and systems is disrupting traditional processes, allowing us all to become creators. Techshop, a start-up from the US, offers a kind of gym membership for designing stuff — providing access to tools, software, and space. The rise of Building Integrated Modeling (BIM) is permitting the pursuit of methods of true design collaboration, where even risk begins to be shared, allowing us to work better, produce better outcomes, and reduce waste. In Amsterdam, the world's first 3D printed house is underway. The house will grow from a single printer, producing zero waste, lowering transport costs, and will be fully recyclable.

This is design democratization, a bottom-up approach that is allowing new voices to speak in the design and construction of our cities. With increasing advancements in rapid-prototyping, 3D printing, and organic growth structures, starship cities of the future may be designed in a fundamentally different way. Synthetic biology has the potential to transform the fundamental process of architecture, growing our surroundings directly from Nature itself.

At Astudio, we are developing "disruptive innovations" that encourage new value systems and networks of exchange in line with our vision of sustainable future cities. One Internet-based example is rECOrds, a carbon- and cost-reduction tool for businesses. It is an online platform that helps businesses to improve their environmental performance. Through the benchmarking of businesses' environmental profiles, the platform enables a highly specialized resources industry to be made aware of a market near them that could benefit from their expertise. Businesses are connected to an online marketplace of professionals who can offer services to improve their sustainability rating.

In a similar way, our "environmental Oyster card" proposal aims to create a new value and economic system around carbon. The 2011 White Paper report, conducted by the Carbon Trust working alongside the Coca-Cola Company, revealed that the average individual consumer is responsible for 23.1 kg of embodied CO_2e emissions. By calculating the carbon reduction needed to ensure we are on trajectory to deliver the UK's 2050 target, the report determined a target guideline daily personal carbon allowance (PCA) of 20 kg CO_2e per day.[13] Our environmental Oyster card proposal takes the concept of PCA and develops it, providing a means to monitor energy and water usage, along with carbon footprint, in a simple, digital way. The environmental Oyster card powered by "carbon currency" is intended to monitor and limit "carbon spend" over the course of the year.

---

[13] Carbon Trust *Personal Carbon Allowances White Paper* (2011).

While we want to avoid a dictatorial system, in which citizens are given allowances and rations, carbon is an interesting proposal when it comes to creating new value systems and new kinds of economy within an ecological framework. What if the main economic system of starship cities worked on carbon exchange? As the element present in all living organisms, the flow of carbon through a city can be thought as of measurement of "life." The measurement of carbon can be considered as an indicator of a biological system called the "carbon cycle" that comprises a sequence of events that are key to making Earth capable of sustaining life. All life is connected through this cycle that provides a complex pathway through which sunlight is converted into biomass and then decomposed into a metabolic substrate, a gas called carbon dioxide, to be turned into "life" again through the action of the Sun and green plants. When part of a balanced system of exchange between humans and the environment, it is sustaining; when unbalanced, it threatens the very life-support systems on which we rely. Understanding the role of carbon in everyday exchanges can help us form communities and develop new ways of living that are mutually beneficial. These new kinds of value systems can underpin exchanges in very practical ways, and could even be used in cities today.

On a starship, or working in a resource-constrained city, we will need to be able to find systems of exchange that inform us of our impact, as indicators of success or failure. They may be carbon-related, but could also be driven by resource use, such as the consumption of water or the minerals of life, such as phosphate and nitrogen. These systems should be driven by our overarching ecological needs, but simultaneously complement individual skill sets, needs, and practices. In very simple terms, we need to build ecosystem exchanges, not only in environmental terms, but also in economic and social terms. This concept underpins Ellen MacArthur's notion of circular economy. The circular approach is one that takes insights from living systems. It considers that our systems should work like organisms, processing nutrients that can be fed back into the cycle—whether biological or technical—and designing out waste. These "closed-loop" or "regenerative" principles are reflected in the notion of closed ecosystems pioneered by Christopher Alexander and Buckminster Fuller, and further augmented by the concept of "Cradle to Cradle" and notions of "upcycling." However, as we have established in the first part of this chapter, there are limits to these approaches; an external energy source like the Sun is needed to keep the ecological metabolism flowing and, even within natural systems, there are imperfections and wastage, and it will never be possible (since the world is complex and interconnected) to completely separate out industrial and biological metabolisms (which is what Cradle to Cradle proposes). A true ecological design must remain partially open to allow for adaptability and diversity that ensures resilience.

Aligning with the broad vision of the Ellen MacArthur Foundation, at Astudio, we are exploring the concept of circular economy systems as interventions in the city today. This idea is explored in two projects—"Social Energy Clusters" and "Wasted Energy Networks"—that allow communities to take ownership of their energy use and waste generation, and collectively reap the benefits from localized processing, energy generation, and urban agricultures (Fig. 12.12).

The "Wasted Energy Network" is an online platform for encouraging inter-business recycling, triggering waste-based economies, and identifying areas of opportunity for sustainable waste management and energy-generation systems. It will be the first mapping

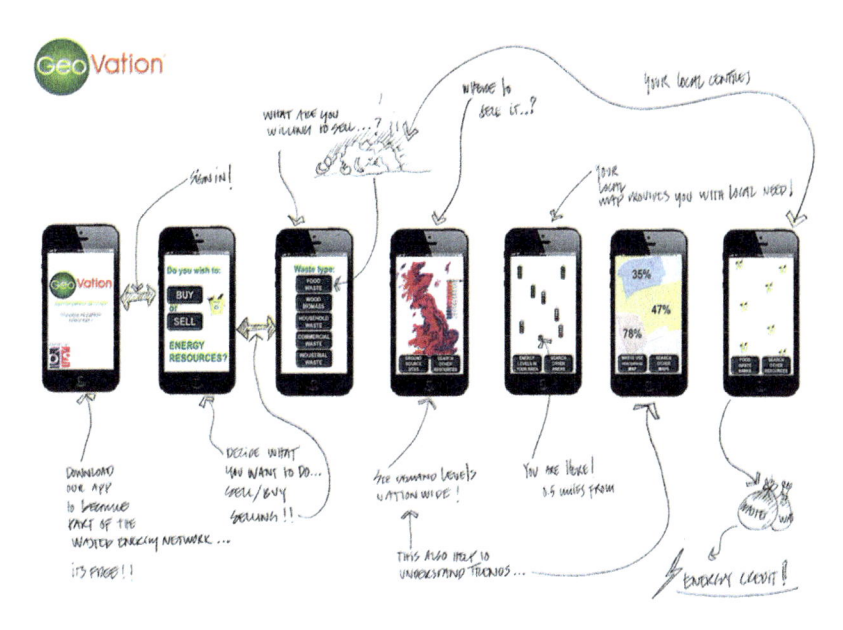

Figure 12.12 The Wasted Energy Network: an online platform for encouraging inter-business recycling, triggering waste-based economies, and identifying areas of opportunity for sustainable waste management and energy-generation systems. Credit: Astudio, drawing by Richard Hyams, 2014

system to display and make connections between waste production and the resultant potential for localized energy generation. One major aim of the network will be to highlight the potential for localized waste-to-energy projects, with the intention of generating business communities capable of producing energy from their own garbage. Astudio is currently working alongside Brunel University and Local Energy Adventure Partnership (LEAP) to develop urban community food waste management through the development of small-scale anaerobic digestion and community composting initiatives. Through the breakdown of organic matter, food, and human waste, methane gas is released, which can power elements of the building. A valuable side product is a nutrient-rich broth that can be used as fertilizer to improve community allotments and gardens, encouraging urban food production. Composting is an alternative way of processing organic matter, which equally produces nutrient-rich compost and heat, which can be utilized in the warming of greenhouses.

"Social Energy Clusters" are the idea of communities working as social units rather than as individuals toward reaching government carbon-saving targets and reaping benefits through partnerships with local councils. There is a very real problem, in that individuals, small businesses, and even councils are not currently able to tap into energy-efficiency initiatives such as carbon credits, as such schemes are aimed at large businesses. Significant issues arising from this include lack of motivation of individuals and small communities to save energy and spiraling energy bills, which is unsustainable given current economic conditions and climatic damage. We intend to create a space in which a community can build value through combined efforts to be energy-smart. We wish to seed cultural change, placing a social value on energy conservation. The introduction of Social Energy Clusters requires simultaneously addressing several interconnected challenges, including motivation and education of individuals, provision of energy usage information in appropriate formats, and promoting joined-up thinking at the levels of councils and government.

This is a vision in which rich sources of abundance are found within the city, ending the austerity measures of our current sustainable practices. It is a vision which forms new value systems that encourage sharing, collaboration, and equality to collectively prosper. There are numerous shared opportunities associated with urban energy and food networks that exploit closed-loop waste-to-energy systems to transform waste into valuable end products that can stimulate localized economies, including urban agriculture.

Urban agriculture can play an important role in the environmental sustainability of a city and its resilience to climate change and other uncertainties. It is increasingly being considered a solution for future economic and food security, ensuring healthy and adequate food access in cities, but equally providing economic, social, and environmental benefits. Developing agricultural capacity within the city not provides a local source of healthful sustenance, but it can also contribute to a household's income, offset food expenditures, and create jobs. Local production reduces food transportation costs and environmental impacts, provides economic development opportunities, and increases access to affordable, fresh, healthy food which may previously have contributed to epidemic rates of obesity and diabetes, especially among low-income populations. There is also an important social facet to urban agriculture — the potential engagement of local communities, who come together for mutual benefit, enhancing the common social and cultural identity of local residents.

These concepts reveal how we can redesign the city's infrastructure and architecture to process waste and generate energy and food. With populations around the world growing and becoming predominately urban, this is driving a need to re-examine how urban inhabitants are fed, and how we develop urban infrastructures to cater for increasing energy, water, and waste demands. The sustainable future city will be one that is autonomous — a city that is productive, able to generate energy, process waste and water, generate soil, and grow food — just like we would need to do in space.

As architects, we are exploring ways that this "starship infrastructure" can be integrated into the fabric of our cities in an ecological way. Moving beyond closed-loop systems, we look to design cities that are open to changing context, not only in terms of their social and economic systems, but also in the materiality of the physical environment. By designing with Nature in mind, we look to integrate architecture with living processes to increase the adaptability and resilience of the urban environment:

$$NATURE + TECHNOLOGY = RESILIENCE$$

12.2.3.4 Designing living architecture

Nature's brilliance is its astonishing ability to survive extreme changes in environment. This resilience is delivered through an capacity to adjust to climate change, moderate potential damages, take advantage of opportunities, or cope with consequences — in short,

Figure 12.13 Algae plays a significant role in aquatic ecology. Credit: Istock image sourced from Google, www.dreamstime.com/royalty-free-stock-photography-algae-image5741367

the capacity to adapt. By incorporating fundamental dynamic processes found in living systems, architecture has the potential to adapt or respond to changes in climate, seasons, or extreme acts of Nature, just as our natural green landscapes respond to the availability of water, sunlight, and wind. As expert moderators of environment, Nature can become a key component in developing technological systems and processes to help reduce the impact of climate change and propose ecological solutions to resource-limited environments. Instead of our buildings remaining inert, they could employ the design of responsive surfaces and building systems to actively moderate and contribute to the surrounding environment (Fig. 12.13).

Historically, Nature and biology have often been inspirations for architecture. The study of Nature as organic form has informed the work of many great architects; take, singularly, Antoni Gaudí, purveyor of Catalan modernism, who adapted the language of Nature into structural form. However, these great works only emulate natural patterns and strategies, and they do not fundamentally alter the way the buildings are made or perform. Today, there is growing acknowledgment of the importance of ecological or "green" urban strategies as a method to ensure the health and sustainability of our cities. The healing potential of green spaces is readily exploited; Nature incorporated into pop-up gardens, green walls, balconies, and roofs is increasingly commonplace. Sometimes, these are thoughtfully integrated; too often, these "greenings" offer nothing more than visual benefit, requiring excessive water irrigation and human management. But, the incorporation of green landscapes can offer great transformative value when integrated correctly.

Trees, grass, and vegetation can absorb carbon dioxide and improve air quality, reduce the heat-island effect, slow rainwater runoff to prevent the flooding of the sewers, even filter and recycle gray water, as well as providing the important facility of beautifying our urban environments to make them more enjoyable and safer environments in which to live and work.

By incorporating Nature into fundamental infrastructure and building systems, we are able to find solutions to the increasingly complex urban and climatic challenges of the future city. Through a natural process called "succession," we work with the inherent complexification of successive plant populations to reach mature and stable ecological systems that can help enhance the resilience of our urban environments. This is an ecological approach to urban and architectural design that looks to work symbiotically with the natural environment, protecting landscapes and ecologies, and harnessing or replicating natural processes to increase the resilience and sustainability of the urban realm. These "living systems" can provide important services for urban communities: protecting against flooding and excessive heat; improving air, soil, and water quality; generating sustainable energy; and, importantly, ensuring psychological well-being. Together, these systems can form an ecological framework for the social, economic, and environmental health of the starship city (Fig. 12.14).

At Astudio Research, we have been working on a number of projects which explore the relationship between Nature and the city, and look to integrate natural processes into building-, landscape-, and city-scale design to increase the resilience and autonomy of our urban environments. These projects have been broadly categorized into "Living Skins," "Living Structures," and "Living Landscapes."

Figure 12.14 Future London: envisioning a new relationship with Nature. Credit: Astudio, rendering by Christian Kerrigan, 2011

12.2.3.5 Designing living skins

Building envelopes or "skins," as the crucial interface between interior and exterior space — between human habitation and the "natural" world — play a fundamental role in the vision of a starship city. As an important environmental moderator, a building's skin impacts transition between environments, affecting heat, light, and air quality inside and outside a building. Successful performance of this interface can regulate and reduce energy consumption used in the heating, cooling, ventilating, and lighting of buildings, affect occupant comfort and productivity, as well as impact the external surrounding environment and climate. Optimizing the responsiveness of this interface will be key to designing for resource-constrained environments and increasing the resilience of our new and existing building stock (Fig. 12.15).

Rather than constructing envelopes with traditional inert surfaces, a "living skin" incorporates dynamic physical and chemical processes found in living and natural systems. Utilizing a range of biologically or synthetic-biology-based technologies, this living cladding would have the ability to transform the performance of buildings within their environmental context. Just as a tree, or our own skin, responds to and moderates its internal and surrounding environment, a living building envelope would be active and adaptable to changes in climate and extreme weather conditions. This envelope would possess the capacity to transform buildings into positive generators, producing heat and providing cooling, capturing carbon dioxide and other pollutants from the atmosphere, and contributing to a new urban ecology. While the "skin" of a spaceship may not necessarily mediate between the interior and outer space in the same way, the architecture of

Figure 12.15 Algae Photobioreactor Façade, Hong Kong Science Park proposal. Credit: Astudio, rendering and photoshop drawing, 2013

Figure 12.16 Algae experiments with Sustainable Now Technologies and Brunel University. Credit: Astudio with Sustainable Now Technologies and Raymond Wilkes from Brunel University, 2015

the envelope could employ similar processes to ensure the optimization of the "internal" environment through the provision of life-supporting infrastructures. Such interfaces with habitat may be likened to the function of biological membranes through which vital exchanges take place. Increasingly, we are likely to see these structures become more permeable to their environment to minimize their reliance on centrally driven utilities — for example, sewage processing plants — and rely more on local resources, such as home composting systems (Fig. 12.16).

Astudio, in collaboration with Brunel University and the Centre for Process Innovation (CPI), is developing a living-skin technology with the intention of optimizing the performance of building envelopes. The team is developing a marketable photobioreactor (PBR) facade system for the production of biofuel and has recently commissioned the first operational industrial algae production unit in the UK. The PBR facade design uses the process of photosynthesis to cultivate algae in uniquely designed photobioreactors, capturing carbon from the atmosphere. Using certain strains of algae with reduced hydrocarbon chains for quick growth, the intention is to produce biofuel as an output to fuel the building, offsetting the building's carbon footprint and energy needs. As a multifaceted renewable energy source, the photobioreactors act as solar thermal collectors providing heat for the building and shading the interior from the Sun. As levels of solar radiation increase, the algae increase their growth rate, and consequently their density and shading potential, adapting in real time to changes in external environmental. This process can be manipulated to accurately regulate the internal temperature of the building. When designed within a holistic building engineering strategy, this living architecture can impact all areas: water filtration, power, heat, and light—a truly dynamic element.[14]

This living, or "bioresponsive," facade system combines both biological and technical systems, which adapt to, and harvest from, the immediate climatic environment. Another example of the application of dynamic skin technology is GLASSX, a 5-cm-thick glazing unit licensed to Saint Gobain that provides thermal storage through the use of a salt hydrate phase change material (PCM). The glass provides thermal storage equivalent to a concrete wall of 20-cm thickness, passively providing heat and cooling over the course of a day. Solar heat is stored in the PCM during the daytime by means of a melting process. During the night-time, the stored heat is delivered to the interior during salt recrystalization, a process of active night purging. While not a "living" material, it demonstrates the dynamic properties of living systems in response to changes in external environment. The building fabric provides an active response to temperature changes, moderating heating and cooling within a building, and providing thermal storage for energy conservation (Fig. 12.17).[15]

Figure 12.17 Star Ark City: future Nature. This view of the city blurs the distinction between the natural realm and architecture that could exist on either Earth or in space. Credit: Astudio, rendering by Christian Kerrigan, 2015

---

[14] Astudio Architecture, *Astudio Research Project: Living Skins* (2014–2015).

[15] Laros Technologies, available online at www.laros.com.au/phase-change-glazing.html (accesses May 20, 2015).

12.2.3.6 Living structures

Synthetic biology techniques offer the ability to transform not only the resilience of the skin of the building, but the structural bones as well. A collaborative work between Astudio Architecture and Rachel Armstrong, professor of experimental architecture at Newcastle University, explores the potential of synthetic biology to regenerate buildings in situ.[16] With 50–70 % of existing buildings expected still to be in use in 2050, this represents the biggest challenge facing architects today. The challenge of upgrading the existing building stock to meet increasing environmental targets is still relatively unknown territory, compared with the new construction of carbon-zero buildings. The challenge will be to reduce energy demand by retrofitting energy-efficient building envelopes, systems, and appliances (lighting alone accounts for up to 10 % of a city's energy demand). Additionally, the ongoing maintenance and repair of these buildings can be 2–3 % of the original cost of a new building every year. With maintenance costs increasing with building age, the work of the collaboration looked to explore ways of introducing self-healing systems into an architectural structure. The work looks to upgrade existing concrete frames of buildings using carbon-nanotube-based materials to improve the mechanical strength of the concrete and extend the framework to accommodate a new dynamic skin. This "re-skin" proposal aims to transform underperforming buildings into energy-efficient, fit-for-purpose spaces that make a positive environmental impact. This regenerative capacity not only has the potential to improve structural stability and energy consumption, but has the means to impact the commercial value of a building, renew an image, and, through doing so, regenerate areas of the city without the unsustainable process of demolition (Fig. 12.18).

Armstrong has been developing an approach to synthetic biology that explores the potential of "lifelike" systems called protocells. These non-genetic molecules, which are not technically alive as they possess no DNA, can be chemically programmed to display the characteristics and behaviors of living systems. Through the interaction of oil and alkaline solutions at a molecular level, new features arise through the protocells' capability for self-organization which exhibit "a range of life-like behaviours such as movement, sensitivity and the growth of micro structures."[17] The characteristics of protocells have allowed Armstrong to speculate on a range of uses, including the development of protocell-based coatings or paints. These water-based protocell coatings, once exposed to carbon dioxide, form a crystalline microstructure akin to limestone over the surface of which it was painted. While the reality of creating a smart, self-regenerating protocell paint is years away, protocell paints offer "the capacity for a unique growth of materials but also potential applications in healing 'broken' buildings, by which molecular interactions detect and deposit material into stress fractures to from 'scar tissue' at the microscale."[18] Protocells harbor the

---

[16] Armstong, R.; Astudio Architecture. AVATAR collaboration, Living Architecture.

[17] Armstong, R. *Designing with Protocells: Applications of a Novel Technical Platform.* Available online at www.ncbi.nlm.nih.gov/pmc/articles/PMC4206855/ (2014).

[18] Armstrong, R. *Living Architecture: How Synthetic Biology Can Remake Cities and Reshape Our Lives.* Published Online, p. Location 568 (2012).

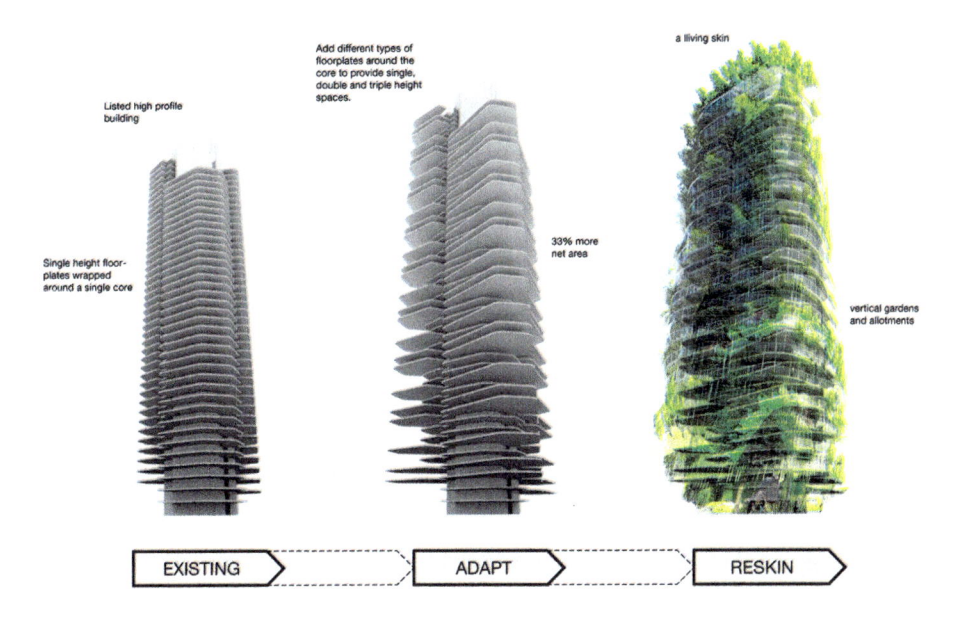

Figure 12.18 Tower 42: Re-Skin Project. Credit: Astudio, render and photoshop drawing by Richard Hyams and Emma Flynn, 2015

potential to grow structures to stabilize unsafe buildings devastated by seismic activity, or, as exemplified by Armstrong, reinforce the sinking timber foundations of Venice disintegrating from centuries of contact with salt water and water-borne organisms.[19] Applied to existing housing stocks as a painted coating, it could offer a method of continued carbon sequestration. Applied to starship design, spaceships could self-repair, or even be grown out of the very life-supporting infrastructures the inhabitants on board require to live.

The possibility of synthetic biology to grow resilience structures extends beyond the futuristic vision of protocell technology. New sustainable construction materials are being developed by harnessing biological organisms, such as fungi, as alternatives to plastics or bricks. In 2014, New York studio The Living completed the MoMA PS1 gallery pavilion, a cluster of circular towers built from bricks that had been grown from corn stalks and mushrooms. The pavilion was entirely biodegradable, creating no waste, no energy needs, nor carbon emissions.[20] Another design studio, Officina Corpuscoli, is researching how fungal organisms can be used to produce alternatives to plastics, eliminating the pollution generated from a petroleum-based plastics supply chain. Biologically produced materials can be "completely non-harmful," they claim. "Once disposed of they just become new nutrients

---

[19] Armstrong, R.; Spiller, N. *Future Venice*. Greenwich University, available online at www2.gre.ac.uk/about/faculty/ach/research/centres/avatar/research/fv (accessed May 20, 2015).

[20] MOMA PS1. Hy-Fi by the Living. Available online at http://momaps1.org/yap/view/17 (accessed May 20, 2015).

for new life."[21] Mushroom mycelium can be grown in a range of organic material, most commonly agricultural by-products. It is a natural, self-assembling glue, which digests crop waste to make environmentally responsible materials. When baked, the grown network of mycelium filaments is transformed into a durable, structural, and waterproof material.

The growth of waterproof or water-responsive structures can provide resilience in areas prone to drought or flooding. Protocell technology has the ability to transform the properties of materials in the presence of water, allowing them to float or expand. Materials can be engineered to absorb and store water in the advent of excessive rainfall or flooding, to expand into cracks and gaps to act as barriers to water flow, and raise bridges over flowing watercourses. In countries experiencing the effects of water shortages, building skins can be designed to conserve and recycle water. Living technologies embedded in the fabric of the building can recycle the gray and black water used within the building, filtering through living materials that purify the water. Water could be sourced, stored, and processed locally instead of increasing demand on the district sewer system. Reservoirs built into the building fabric could be used to cool interior and exterior spaces through both conventional cooling and water misting.

12.2.3.7 Living landscapes

During floods, it is not just buildings that need protecting; so do our landscapes. Synthetic biology could be used to prevent soil erosion by reengineering bacteria in soils. The engineered bacteria would produce a plant hormone that promotes root growth to protect plants from being uprooted and washed away in flash floods.[22]

Soils provide ecosystem services critical for life. Maintaining the health of soils is essential if we are to safeguard our future, retain food security, and mitigate soil, water, and air pollution. With the pressure of an increasing population, new technologies and techniques are continually being developed to optimize soil performance—to produce more food and fuel with less: less land, less water, less energy, and fewer nutrients. New innovative soil management strategies that can mitigate pollution, while also enhancing ecosystem performance, are being developed. One example is Carbon Prophet, a government-funded project, which provides farmers and landowners with a new income stream by selling captured carbon to companies that want to offset their emissions. By increasing the soil fertility of land, soil improves its capacity to absorb and fix carbon, acting as a carbon sink. Working with Astudio, Carbon Prophet have been exploring the potential of manipulating soil fertility in both rural and urban environments as part of building development landscape strategies to offset carbon generated during construction.

---

[21] Montalti, M. *Growing Products from Fungus Could Be the Start of a "Biotechnological Revolution"*, *Dezeen and Mini Frontiers*. Available online at www.dezeen.com/2015/01/21/movie-officina-corpus-coli-growing-products-materials-fungus-biotechnological-revolution/ (accessed May 20, 2015).

[22] The Future of Synthetic Biology. In Imperial College London Fringe event, London, March 22, 2013.

Starship cities: a living architecture 321

The engineering of bacteria in the breakdown of organic waste offers an exciting potential for the production of soil in cities and, indeed, in space. Living skins and building systems could possess the potential for the on-site processing of organic waste, in turn generating nutrient-rich fertilizer and biogas, or compost and heat. Through the manipulation of bacteria in the organic waste makeup, direct heat or biogas output can be controlled and fed directly into the building systems. The subsequent by-products of compost or fertilizer can then be used in the local production of food and biomass.

In the vision of a starship city, we can imagine that soil may be generated and stored by the architecture and infrastructure of the city. Soil could provide important life-sustaining services, where it acts in a number of capacities such as a biofilter to clean exhaust air, carbon sink, waste-disposal system, heat source, or a site of food production. Engineered bacteria could be computationally instructed to control the required output or used as a "natural BMS" to indicate the health of the system.

Bacteria offer further potential for infrastructural-scale application of living technologies. Research is being conducted into the use of bacteria to create zero-electricity light sources, exploiting the natural phenomena of bioluminescence in octopuses, jellyfish, or mushrooms. Bacteria lighting has the potential to redefine streetlight infrastructure, light building facades, and interior spaces without impacting the environment, and provide lighting solutions for areas without electricity. Natural bioluminescence not only provides a light source with zero carbon generation, but can also be used to detect pollution levels of the city—with lights changing color to relay warnings (Fig. 12.19).

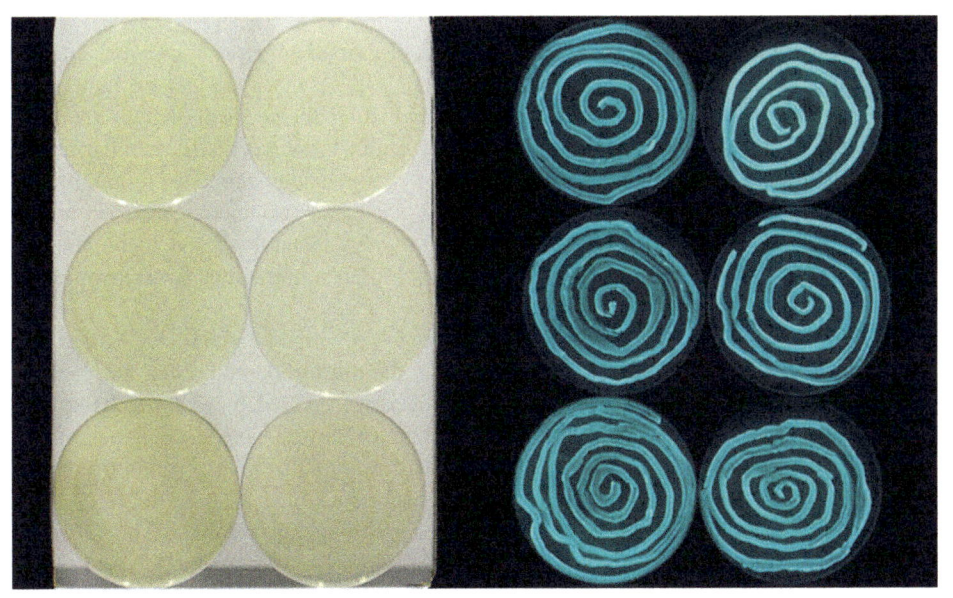

Figure 12.19 Bacteria bioluminescence research with biologist Dr. Simon F. Park and Brunel University. Credit: Astudio with Dr. Simon F. Park, Surrey University, 2014

12.2.4 Near-future architecture

Nature can deliver the means to transform the infrastructure of our cities. If we can learn to harness the properties of living systems in the technology of our cities, living architecture could play an important role in the future resilience of our built environment. Understanding how architecture and technology may possess the characteristics of living systems, design professionals have the potential to transform the built environment from its static, polluting state to an active, responsive, ecologically connected environment that makes a positive impact on the planet. The incorporation of "living technologies" promises an implementable solution for issues of environmental responsiveness and sustainability in our cities. From the futuristic visions of protocell technology to the current manipulations of algae, salt, soil, fungi, and bacteria, living technologies have the ability to confer lifelike qualities on our buildings, to induce regenerative and responsive capacities to counter the negative impacts of climate change and resource depletion.

As "architectural ecologies," they can provide renewable energy sources, generate heat, stimulate plant and food growth, absorb carbon dioxide and other pollutants, and remove waste. Beyond that, and when paired with advances in digital communication, they have the potential to propose radically new value systems within society to transform the economic systems and social organizations that encourage industrialized practices. These ecologies have the means to transform our current consumption habits and wasteful behavior, and to propose new ways of living symbiotically with the natural world. Buildings are to become positive contributors to the city, counted for their benefit, not their cost.

"Living architecture" in the transformation of the fabric of our cities is only one step on the path to future autonomy. It needs to be a social vision that does not pit humanism against environmentalism, but finds common ground where the needs of humans and nonhumans can be met through urban development. It requires flexible, resilient, and efficient urban models that can respond to issues of climate change and resource depletion, urbanization and population growth, by balancing the social, environmental, and economic needs of happy and sustainable cities. Facilitated by new technology and social change, these environments can become thriving, productive urban communities where people and Nature can flourish alongside each other.

This vision of the city as a literal living organism, in which mankind and Nature live symbiotically side by side, is a model for living in the third millennium that will help us establish best practice architectural tactics, which are grounded in ecological thought. These approaches are transferable concepts that give us a chance to construct highly responsive living spaces that work in partnership with Nature that may ensure our survival for generations to come, whether that is on Earth or in space.

The future architect is the conductor of an ever-increasing orchestra. The seats are taken up by unknown specialists in every field (Fig. 12.20).

Figure 12.20 Starship cities: a living architecture. Nature becomes architecture in the design of a starship city. Credit: Astudio rendering by Christian Kerrigan, 2015

12.3 Terraforming our cities

Susmita Mohanty, CEO, Earth2Orbit,
Sue Fairburn, Co-founder, Fibre Design Inc.
Barbara Imhof, co-founder LIQUIFER Systems Group.

12.3.1 Introduction

Space is closer than you think. Elon Musk, the billionaire entrepreneur behind the private spaceflight company SpaceX, claims that SpaceX's main mission is to "make humans an interplanetary species." His ambitious Mars settlement program would start with a pioneering group of fewer than 10 people, who would journey to the Red Planet aboard a huge reusable rocket powered by liquid oxygen and methane. Musk dreams of eventually establishing a Mars colony of up to 80,000 people by ferrying paying explorers.[23] Musk says that, this year, he will reveal the details of his Mars colonial transporter and spacesuits.[24] To make such an ambitious mission feasible, Musk and fellow entrepreneurs such

---

[23] www.space.com/18596-mars-colony-spacex-elon-musk.html.
[24] www.space.com/28215-elon-musk-spacex-mars-colony-idea.html.

Figure 12.21 Mars One's vision of a future Martian base. Credit: Mars One, www.mars-one.com

as Bas Lansdorp (founder of Mars One)[25] will not only need a robust rocket, Mars lander, well-designed spacesuits, advances in Mars in-situ resource utilization (ISRU) technologies, and a well-thought-out and implementable spare-part resupply strategy,[26] but also master soft precision landing on the Red Planet, which is still quite a challenge. On top of all this, we will also need a super-reliable "life-support system" that can sustain the debut crew and then the eventual colony (Fig. 12.21).

Life support will include advanced food-growth systems and a means to create a breathable atmosphere for visitors from planet Earth. It is not that Mars does not have an atmosphere—it is actually one of four terrestrial bodies in the Solar System that does, a short list that also includes Venus and Saturn's moon, Titan.[27] The problem is that the atmosphere of Mars differs radically from that of Earth. For one thing, the atmosphere of Mars is 100 times lighter than that of Earth, making the air too thin to breathe. The low atmospheric pressure is also partially responsible for Mars's frigid average surface temperature of −81 °F (compared with 57° on Earth). And, even if we could breathe the Martian atmosphere, it is composed almost completely of carbon dioxide. Table 12.1 shows at a glance how the atmospheric composition of the two planets compare, according to NASA.[28]

---

[25] www.mars-one.com/.

[26] Do, S.; Ho, K.; Schreiner, S. S.; Owens, A. C.; de Weck, O. L. An Independent Assessment of the Technical Feasibility of the Mars One Mission Plan. 65th International Astronautical Congress, Toronto, Canada, September 29–October 3 (2014).

[27] www.pbs.org/wgbh/nova/space/how-to-get-an-atmosphere.html.

[28] www.srh.noaa.gov/jetstream/atmos/atmos_intro.htm.

Table 12.1 Atmospheric composition of Earth and Mars

	Earth (%)	Mars (%)
Oxygen	20.95	0
CO_2	0.33	95.32
Nitrogen	78.10	2.70
Argon	0.93	1.60
Others	Trace amounts	Trace amounts

Source: NOAA/NASA

For a true Martian metropolis to exist, in which humans and other Earthly life would not merely survive, but thrive, we would need to somehow replace Mars's inhospitable atmosphere with one that mirrors our own. Some scientists and advocates of future colonies on Mars believe that Mars can be made to have an Earth-like atmosphere via "terraforming." Terraforming (literally, "Earth-shaping") of a planet, moon, or other body is the theoretical process of deliberately modifying the atmosphere, temperature, surface topography, and/or ecology to be similar to the biosphere of Earth to make it habitable by Earth-like life.

Robert Zubrin, president of the Mars Society, a non-profit organization dedicated to Martian exploration, suggests that adding fluorocarbons to the Martian atmosphere would increase the greenhouse effect. Unlike the notorious chlorofluorocarbons that have contributed to the destruction of Earth's ozone layer, Zubrin proposes tetrafluoromethane as a refrigerant that would not backfire and destroy the very atmosphere we are trying to create. From there, Zubrin speculates[29]: as Mars warms, its frozen soil would thaw enough to release carbon dioxide, and more carbon in the atmosphere would further accelerate the greenhouse effect, bringing the average temperature up to 32 °C and causing Mars's frozen underground water supply to melt and flow back into ancient riverbeds. And, when the water reaches Martian soil, it would break down latent peroxides, releasing oxygen into the atmosphere—not yet enough to sustain human life, but enough to grow plants—which would further increase the supply of oxygen. Once the plants take root, we could just wait for oxygen to accumulate. At this point, Mars colonizers, who Zubrin imagines would work out of a research base camp and wear something akin to scuba gear to supplement their oxygen intake, would also grow algae and seaweed in ponds, which could anchor a growing food chain. "You could have fish farms on Mars," he says. "Water would become the first environment that would be habitable by higher animals without any kind of artificial assistance." And, once aquatic creatures could live happily, we could move onto insects, mammals, and, of course, humans. With an atmosphere established, we would have breathable air and comfortable temperatures, and could move freely, without bulky suits and supplemental oxygen.[30]

---

[29] http://finance.yahoo.com/news/elon-musks-martian-city-solve-163000855.html.

[30] www.popsci.com/science/article/2011-04/fyi-can-we-make-mars-habitable-pumping-atmosphere-full-oxygen.

Further, Robert Zubrin points out that "We know how to warm planets; we're doing it right now," albeit to our own planet, by emitting greenhouse gases into the atmosphere. So, even before we go and try to terraform Mars, we can use Earth as a test bed to experiment with "ecosystem reconstruction."

12.3.2 Urban atmospheric reconstruction

We do not have to look far for atmospheres needing reconstruction. As experimental precursors to Martian terraforming, hypercities, such as Beijing or Delhi, require improved air quality to ensure continued habitability for humans and other Earthly species. The atmosphere in these megacities needs urgent reconstruction to make it breathable and fit for human habitation. Let us take Delhi as an example, because, in 2015, it surpassed Beijing as the world's most polluted city.[31] Delhi has the world's highest levels of PM2.5 — tiny, toxic particles that lead to respiratory diseases, lung cancer, and heart attacks. The Indian capital averaged 153 µg/m³ in 2013, the World Health Organization (WHO) said, citing government data. That is 15 times more than the average annual exposure recommended by the WHO (Fig. 12.22).

Figure 12.22 Accumulated particulate matter on Delhi foliage in Khirkee village. Credit: Sue Fairburn (2015)

---

[31] www.bbc.com/news/magazine-32352722.

When President Obama visited Delhi in January 2015, the media put the spotlight back on Delhi's abysmal air quality and one of them even went on to claim that "U.S. President Barack Obama could lose roughly 6 h from his expected lifespan after spending three days in India's capital inhaling the world's most toxic air."[32] The US Embassy in New Delhi purchased more than 1800 Swedish air purifiers ahead of Obama's visit, as per a press release by Stockholm-based Blueair AB.[33] Like embassy officials, many of Delhi's residents have actively started using countermeasures, such as installing air filters and growing certain varieties of houseplants that serve as excellent natural air scrubbers. Unlike the visiting US president, the residents have long-term health repercussions to deal with. A Delhi businessman and environmental activist Kamal Meattle,[34] in his TED talk[35] "How to Grow Fresh Air," shows how an arrangement of three common houseplants—the areca palm, the money plant, and a spiky plant with the unappealing moniker of mother-in-law's tongue—used in specific spots in a home or office, can result in measurably cleaner indoor air.

India as a whole is home to 11 of the top 20 cities on the planet with the worst air quality, according to data[36] from the WHO, which collected pollution levels from 1600 metropolitan areas between 2008 and 2013. At the 2014 Biennale in Dharavi, Mumbai's largest slum, an intrepid urban gardener duo, Nicola Antaki and Adrienne Thadani, brought their botanical expertise and enterprise into the Kumbharwada (aka potters' colony) neighborhood of Dharavi. Like Meattle, they too relied on the three most effective houseplants—the areca palm, the snake plant (also known as mother-in-law's tongue), and the money plant—to purify the air polluted by the city pollutants and the smoke from the potters' kilns. These plants filter toxins such as benzene, formaldehyde, trichloroethylene, xylene, and toluene from the air—chemicals known to cause eye, nose, and throat irritation, nausea, and headaches in the short term and cancer in the long term. A person would need about four areca palms, four snake plants, and two money plants to have fresh, clean indoor air. In Kumbharwada homes, however, space is a luxury—as it is in many other parts of Mumbai or, for that matter, on our spaceships. The urban gardeners devised three kinds of gardens that would utilize minimum space and resources. The "Living Screen" and "Air Cleaning Tiles" make use of earthenware and ceramics to form vertical gardens. The Living Screen uses money plants as a living air filter and space divider. Air Cleaning Tiles are a plastic-free garden wall that maximizes vertical wall space to provide an additional layer of insulation to keep buildings cool. The tiles are of glazed earthenware with a pocket for planting, with drip irrigation and a simple drainage system, and can be arranged in any design. Discussions with locals resulted in "Fresh Air for One," an indoor garden designed to fit a small home (Figs. 12.23 and 12.24).[37]

---

[32] www.prweb.com/releases/2015/01/prweb12469026.htm.

[33] www.bloomberg.com/news/articles/2015-01-26/mr-president-world-s-worst-air-is-taking-6-hours-off-your-life.

[34] www.ted.com/speakers/kamal_meattle.

[35] www.ted.com/talks/kamal_meattle_on_how_to_grow_your_own_fresh_air?language=en.

[36] www.who.int/mediacentre/news/releases/2014/air-quality/en/.

[37] www.dharavibiennale.com/growing-fresh-air/.

328 **Space architectures**

Figure 12.23 "Living Screen" in action: a low-tech local intervention to make the air breathable. Credit: Nicola Antaki and Adrienne Thadani

GROWING FRESH AIR IN DHARAVI
Living Screen

The Living Screen is a simple intervention to use creeping Money Plant as a living air filter and space divider. In this case, the project focuses on altering the common terracotta pot to better suit the needs of the plants and save space.

Money Plant produces Oxygen and is one of the most effective plants at removing toxic agents such as benzene, formaldehyde, xylene and toluene from the air. Additionally, we selected money plant because it is very low maintenance, grows quickly and trellises well.

The terracotta pots were altered so they fit together more efficiently, making the addition of a cane and rope trellis simple and easy, and draining water effectively during the monsoon.

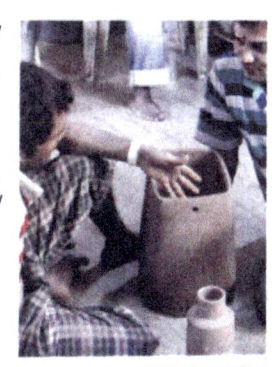

Square terracotta pots
The square pots fit together to create a linear screen that can vary in length.

Cane poles support a rope trellis that is tied to holes in the pot edge, creating a living screen with Money Plant leaves.

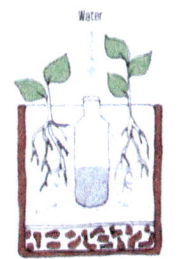

Section through a pot
Broken pieces of pot line the planting base to act as a water filter

An olla is inserted into the soil with a lid. This technique means water is absorbed by roots slowly without over-watering or providing a breeding ground for mosquitos

Figure 12.24 The "Living Screen" concept description. Credit: Nicola Antaki and Adrienne Thadani

330 Space architectures

While urban pollution is a daily concern in many megacities, there are seasonal climatic disasters around inhabited areas such as volcanic eruptions and forest fires that make the air unfit for human habitation. The recent wildfires across parts of British Columbia are a perfect example. These wildfires spewed smoke and ash that spread over several communities, prompting residents to share jaw-dropping photos of the eerily dark skies.[38] One does not have to quite make that long trek to Mars to see "red skies" after all. They are all here. City As A Spaceship (CAAS) Collective member Sue Fairburn was on her way from Aberdeen to Vancouver while these fires were raging and she tweeted: "We're flying to 'Mars' this week and hoping they see rain soon to relieve the firefighters and help rebalance the atmosphere."

12.3.3 Earth as a living laboratory

Earlier this year, KHOJ,[39] an autonomous, artist-led, interdisciplinary platform, invited the CAAS Collective[40] comprising Susmita Mohanty, Barbara Imhof, and Sue Fairburn to Delhi for an art-science residency called "The Undivided Mind."[41] CAAS is a metaphor

Figure 12.25 CAAS mash-up montage of low Earth orbit (LEO) farming on board the International Space Station (ISS). Credit: Image from the ISS

---

[38] www.cbc.ca/news/trending/smoky-skies-from-wildfires-snapped-by-residents-in-vancouver-1.3138949.

[39] http://khojworkshop.org/.

[40] www.facebook.com/pages/City-As-A-Spaceship-CAAS/227716587399312?fref=ts.

[41] http://khojworkshop.org/programme/the-undivided-mind-art-science-residency/.

for rethinking our relationship with our habitats, transporters, and the environment, through which the CAAS Collective explores the reciprocities between terrestrial and extraterrestrial architecture and design. CAAS Manifesto excerpts (Fig. 12.25):

> "We think of a wonderful, and yet obvious symbiosis—tomorrow's space ideas shape today's cities, and investment in today's cities serves as the vehicle and test bed to both subsidize and implement tomorrow's space endeavors. 'The earth as a spaceship' is not merely a metaphor—it is a tangible, viable way for the future survival of mankind.
>
> "We see the spaceship, and a space habitat as completely analogous to the modern, densely packed, technology driven hyper-metros of tomorrow and ideas and technologies for space that can immediately impact the development of these cities. In return, we see these living, thriving, survival-challenging uber-cities as collections of self-contained, super-redundant microcosms that prove themselves to be reliable, and hardy over time to be directly translatable to the space colonies of the future."

The Delhi residency culminated in a show open to the public, in which the CAAS Collective re-imagined the gallery space at KHOJ as a spaceship, with individual segments of the show conveyed as different "modules" of the ship. This intervention drew parallels between the challenges of living aboard a spaceship, with no gravity and not much by way of (recognizable) food, entertainment, or other creature comforts; and the everyday struggle of living in a city where resources are dwindling at the speed of light and sustainable alternatives are hard to implement. In one of the print materials that formed part of the show, the creators put forth the CAAS vision thus:

> "By grounding space innovations and uplifting earth innovations, CAAS can challenge and shape ideas and serve as curator and broker to the planning, designing, developing and inhabiting of near future cities. We are, by no means, propagating that the way we live in outer space is more eco-efficient than how we live on our Earth, or the other way round. There are parallels, there are differences and there are reciprocities."[42]

While Musk and fellow entrepreneurs are busy dreaming up Mars arks to colonize and conquer, scientists, architects, designers, engineers, and urban planners are busy using Earth as a living laboratory to experiment and develop robust man-made ecosystems that can then be transferred over to their extraterrestrial geographies. If one looks closely at, say, hydroponics or greenhouse farming on Earth, one finds that many countries including Israel, Finland, the Netherlands, Japan, Belgium, and others have made remarkable advances and developed sophisticated and efficient methods to grow everything from tomatoes, lettuce, basil, and other greens. The Internet is awash with impressive videos of automated hydroponic farms and greenhouses—the kind we need if we are to set up extraterrestrial human colonies (Fig. 12.26).

---

[42] http://www.sunday-guardian.com/artbeat/a-bleeding-city-a-spaceship-shelter.

Figure 12.26 Hydroponics farm on Earth (www.policyforum.net/feeding-our-fears/). Credit: Horticulture Group on Flickr (www.flickr.com/photos/mmwhortgroup/8949143931)

Compared with mainstream advances in food-growth systems, or even water-recycling systems, the millions of dollars that have been spent by space agencies over the past decades in creating advanced bioregenerative life-support systems for human missions has yielded rather modest results after so many decades of research and iterative design, likely because life-support system research at most space agencies is constrained by bureaucratic hurdles and funding uncertainties. This again reinforces the idea of why we should be looking outside of the aerospace realm to more mainstream applications of man-made ecosystem creation—the additional dimension being how to make the most from scarce resources, be they water or oxygen or growing mediums like soil.

It is perhaps time for a new biosphere experiment in which one can put to use the advances in greenhouse and hydroponic technologies in the context of confined living, as would be the case on future extraterrestrial space missions. This reconceived and redesigned approach could serve as an interdisciplinary and interplanetary experience-based test bed for future human settlements off the planet. The previous two Biosphere 2 experiments (the first from 1991 to 1993; the second from March to September 1994) ran into serious problems, including low amounts of food and oxygen, die-offs of many animal and plant species, not to mention management issues and squabbling among resident scientists.[43]

---

[43] https://en.wikipedia.org/wiki/Biosphere_2.

12.3.4 Case studies

The case studies that follow are an attempt to highlight innovative terrestrial technologies being developed that can, once mature, be applied to outer-space architecture and habitation systems.

12.3.4.1 Case Study 1: Urban algaetecture

The Milan Design Week 2014 featured design prototypes of the "Urban Algae Canopy" and "Urban Algae Façade" under the project name "Algaetecture."[44]

"The **Urban Algae Canopy**,[45] based on ecoLogicStudio's 'HORTUS' system, was presented at Milan Design Week 2014 with a 1:1 scale prototype of the world's first bio-digital canopy integrating micro-algal cultures and real time digital cultivation protocols on a unique architectural system. The potential of micro-algae have been integrated within a custom designed four-layered ETFE cladding system, whilst the flows of energy, water and CO_2 are controlled and regulated in real-time and made to respond and adjust to weather patterns and visitors' movements. Once completed, the Urban Algae Canopy will produce the equivalent amount of oxygen as four hectares of woodland, and up to 150 kg of biomass per day—60% of which are natural vegetal proteins."	"The **Urban Algae Façade**[46] is based on Cesare Griffa's 'WaterLilly 2.0' system, a project for a micro-algae vertical farm to be implemented as an architectural skin. The intention here is that, integrated into the green system of the cities, micro-algae can help in absorbing carbon dioxide and producing oxygen, while acting as a second skin of buildings, boosting passive cooling and increasing shading of the façade."

According to Carlo Ratti, director of the SENSEable City Lab[47] at the Massachusetts Institute of Technology (MIT):

"The functioning principle of the prototypes is based on the exceptional properties of micro-algae organisms, which are 10 times more efficient photosynthetic machines compared to large trees and grasses. The Algaetecture project aims to develop a natural man-made ecology and explore the use of algae as an integrated architectural cladding and urban agriculture system."

---

[44] www.carloratti.com/project/algaetecture/.

[45] Credits: Urban Algae Canopy prototype by ecoLogicStudio (Marco Poletto and Claudia Pasquero) and Carlo Ratti Associati; prototyping team—Taiyo Europe GmbH, Sullalbero Srl; consulting team—Nick Puckett, Paolo Scoglio, Catherine Legrand, Mario Tredici; lighting by: iGuzzini.

[46] Urban Algae Façade prototype by Cesare Griffa and Carlo Ratti Associati; prototyping team—Matteo Amela, Federico Borello, Marco Caprani; technical support by Environment Park Spa, Fotosintetica & Microbiologica Srl; lighting by iGuzzini.

[47] http://senseable.mit.edu/.

Figure 12.27 1:1 scale prototype of the "Urban Algae Canopy" presented at Milan Design Week 2014. Credit: Photo by Filippo Ferraris, courtesy: ALGAETECTURE|Carlo Ratti Associati, www.carloratti.com

Figure 12.28 Close-up of the "Urban Algae Façade" that can be implemented as a building's architectural skin. Credit: Photo by Filippo Ferraris, courtesy: ALGAETECTURE|Carlo Ratti Associati, www.carloratti.com

Microalgae perform an important photosynthetic activity, absorbing considerable amounts of carbon dioxide and producing oxygen, and growing into a biomass, which can be processed for energy, cosmetic, pharmaceutical, and nutrition markets. Once we master the use of microalgae through large-scale application here on Earth, we can also transfer this knowledge to extraterrestrial human habitats, both orbital and planetary. Thus, algae can be used as an innovative energy and food production system within our cities, and eventually in our space colonies of the future (Figs. 12.27 and 12.28).

12.3.4.2 Case Study 2: Growing As Building (GrAB)[48]

"Growing As Building" (GrAB) is an ongoing Vienna-based interdisciplinary project involving architects, scientists, and artists. The project commenced in spring 2013 and concludes in autumn 2015. GrAB takes growth patterns and dynamics from Nature and applies them to architecture with the goal of creating a new living architecture (Fig. 12.29).

The research goal of GrAB is to develop architectural ideas for growing structures. The GrAB team created visionary descriptions of what they thought would be achievable in the future, such as self-designing-growing buildings, changing wall openings according to environmental needs, and self-repairing buildings that allow upkeep of all vital functions of the building system. Seeking novel architectural approaches, the team researched methodologies that could connect these visions to biological growth principles and role models. Renowned biologist and biomimetic pioneer Julian Vincent, who collaborated with the GrAB team, proposed the application of the established methodology of quality function deployment[49] to

Figure 12.29 Mycelium facade panels integrated into high-rise structure. Credit: Growing As Building (GrAB) 2015, visualization: Rafael Sanchez

---

[48] www.growingasbuilding.org/.

[49] https://en.wikipedia.org/wiki/Quality_function_deployment.

336 Space architectures

Figure 12.30 Red Sands Maunsell forts. Credit: CC BY-SA 3.0

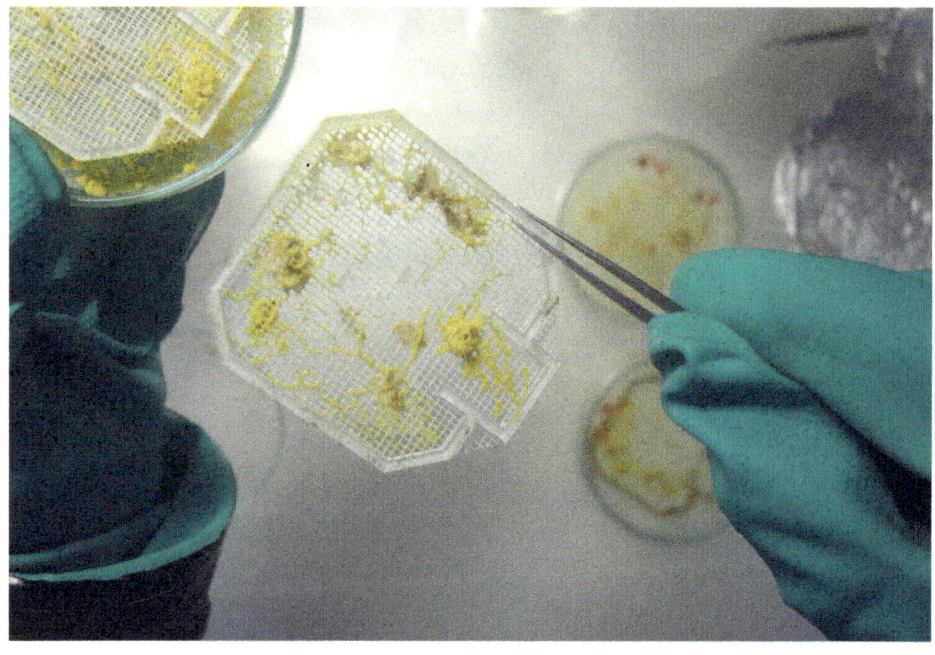

Figure 12.31 Slime mold (*mustard yellow*) finding its path through the top-view plan of the Maunsell forts. Credit: Growing As Building (GrAB) 2015, *photo*: Ceren Yönetim

achieve this. By transforming qualitative into quantitative relationships, including ranking the outcome, the architectural visions were correlated to important growth principles, which could then be matched to biological role models. This enabled the team to start laboratory-based investigations of selected role models related to their architectural visions. For example, the team ran a variety of experiments with "slime molds" (organisms that use spores to reproduce) in a glove box and used their exploring and trail-finding capabilities as a role model for creating architectural paths and circulation areas.

The two images in Figs. 12.30 and 12.31 represent one of the slime mold experiments inspired by the Maunsell forts. The image to the left shows the Maunsell forts cluster—they were small fortified army and navy towers built in the Thames and Medway estuaries during World War II to help defend the UK. The image to the right shows one of three Petri dishes (shaped like the floor plan of the Maunsell fort) with slime mold growth patterns in mustard yellow. When three of these Petri dishes are stacked one on top of the other, they resemble the bottom two floors and ceiling of each of the fort units. The slime mold growth is photographed, transformed into digital format, and overlaid with a digital model of the Maunsell fort unit. This digital model is then studied closely and used to interpret the connections and circulation patterns of the mold spores to design a new functional architectural space.

Figure 12.32 captures the relationship of the slime mold characteristics and their implications for GrAB-inspired architecture. GrAB designers identified slime mold traits as manifested by its multi-nucleated cell, protoplasmic tubes, spores, and fruiting body growing and spreading over the Petri dishes. These traits included network optimization, pattern formation, external memory, replication, random growth, convergence, chemical oscillations, layering, nutrition as a trigger, and resource quantity. These traits were then reinterpreted in architectural terms and a list of architectural qualities was drawn up: hierarchical design, optimized circulation pattern, multi-directional circulation, volume, optimized volumetric arrangement, building program, aesthetics, and circulation density.

GrAB thus explores living architecture in fascinating ways. GrAB's inherent goal is to work toward more resilient systems that can respond to disturbances and be damage-resistant. This approach can be useful to deal with the stresses encountered in designing and retrofitting cities and habitation systems from overcrowding resulting from the fact that, by 2030, 60 % of the world's population will live in urban areas.[50]

Further, GrAB is developing a local 3D printer that takes on current rapid prototyping technology, liberating it from its limiting frame, allowing free locomotion of the print head, and integrating calcium carbonate as print material. During the mixing of the print material comprising calcium carbonate, ethanol, and acetic acid, carbon dioxide is released, which is transformed by the algae bioreactor.

12.3.5 The future city

In the new millennium, the trend is to innovate using interdisciplinary collaboration, bringing together artists, biologists, architects, industrial designers, urban planners, anthropologists, programmers, robotic experts, game designers, interventionists, policy makers, and others to collaborate and converge their expertise, ideas, and research to create urban,

---

[50] www.un.org/ga/Istanbul+5/bg10.htm.

338 **Space architectures**

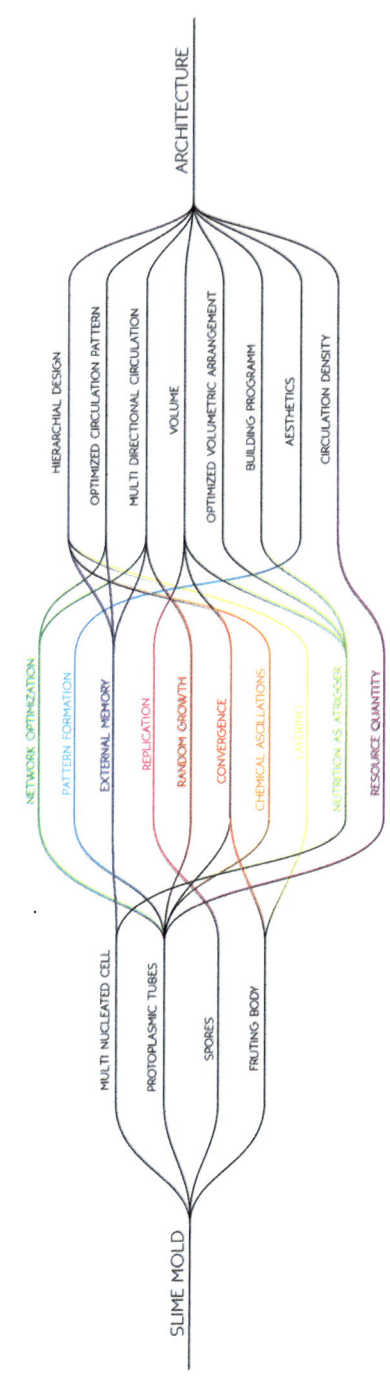

Figure 12.32 Relationship of the slime mold characteristics and their implications for architecture. Credit: Growing As Building (GrAB) 2015, visualization: Ceren Yönetim

Figure 12.33 Panorama of Vienna in the future. Credit: Concept by LIQUIFER Systems Group (Waltraut Hoheneder, Barbara Imhof, Susmita Mohanty, Damjan Minovski); *rendering*: Damjan Minovski; *scientific advice*: Bernhard Weingartner; *editor*: Bernhard Weingartner, Norbert Regitnig–Tillian

suburban, and rural systems for sustainable living. These collective ideas, products, and product systems can be precursors for future systems for living off the (home) planet.

Our cities, farms, factories, homes, studios, offices, and transporters make the best test beds for innovative new technologies that can be put to practical use over prolonged periods of time at any scale necessary. These can then be leveraged to create reliable, efficient, environment-friendly habitation, transportation, interaction, and exploration systems for living and working on neighboring planets and distant celestial bodies in orbital space and eventually in deep space.

Figure 12.33 shows a panorama of the city of Vienna in the future. This re-imagined cityscape shows feasible future utopias such as the space elevator and hyperloop high-speed transportation tubes, the harvesting of energy through deep-space solar farms and beaming it back to Earth, and microalgae building facades and roofs, creating photosynthetic surfaces to counter climate warming.

The city of the future will connect Earth and space in a seamless way. After all, as Buckminster Fuller would have pointed out, we are already in space. We are living on an orbiting spaceship called Earth.

COMBINED REFERENCE LIST FOR CHAPTER 12

W. J. Broad, The environment: oxygen loss causing concern in biosphere 2 (1993), January 5, www.nytimes.com/1993/01/05/science/the-environment-oxygen-loss-causing-concern-in-biosphere-2.html

M. Cohen, Water walls architecture: massively redundant and highly reliable life support for long duration exploration missions. GLEX-2012.10.1.9x12503, Global Space Exploration Conference, Washington, DC, USA, Copyright © 2012 by Marc M. Cohen, published by license by the International Astronautical Federation (2012)

J. Fisher Smith, Life under the bubble. Discover: Science for the Curious, December 20 (2010)

ISECG (International Space Exploration Coordination Group), Global exploration roadmap (2013), www.globalspaceexploration.org/wordpress/wp-content/uploads/2013/10/GER_2013.pdf. Accessed 23 Oct 2014

E. M. Reingold, Environment: Noah's Ark—the sequel, to test ideas for outposts on other planets, scientists have built a replica of earth in the Arizona desert. *Time* September 24 (1990)

13

Space bodies

Kevin Warwick, Arne Hendriks, Rachel Armstrong and Sarah Jane Pell

Four individual authors Warwick, Hendriks, R. Armstrong, and Pell propose different views of embodiment for inhabiting extraterrestrial environments, landscapes, and habitats.

13.1 Cyborgs: upgrading humans for a future in space

Kevin Warwick, Deputy Vice Chancellor (Research) at Coventry University.

In terms of space travel, many of the limitations faced by humans, in stand-alone form, are removed simply by the adoption of a cyborg persona, particularly in terms of neural upgrading. In this article, a look is taken at different types of brain–computer interface that can be employed to realize cyborgs as biology–technology hybrids. The approach taken is very practical, with actual space application in mind, although some wider implications

---

Kevin Warwick (✉)
Deputy Vice Chancellor (Research), Coventry University, Priory Street,
Coventry, United Kingdom, CV1 5FB
e-mail: aa9839@coventry.ac.uk

Arne Hendriks (✉)
Artist, Next Nature Department, Design Academy Eindhoven,
Emmasingel 14, 5611 Eindhoven, Netherlands
e-mail: arnehendriks@yahoo.com

Rachel Armstrong (✉)
Professor of Experimental Architecture, The Quadrangle, Newcastle University,
Newcastle-upon-Tyne, NE17RU, United Kingdom
e-mail: rachel.armstrong3@ncl.ac.uk

Sarah Jane Pell (✉)
Artist-astronaut, Australia Council Fellow, GPO Box 2476, Melbourne, Victoria 3001, Australia
e-mail: research@sarahjanepell.com

are also considered. Results from experiments are discussed in terms of their meaning and application possibilities. The article takes a scientific experimental approach, opening up realistic possibilities in the future of space travel, rather than providing conclusive comments on the technologies employed. Human implantation and the merger of biology and technology are important elements to the overall scheme.

13.1.1 Introduction

In science fiction, for many years, the view has been of a future in which robots are intelligent and cyborgs (a human–machine merger) are commonplace—*The Terminator* (1984), *The Matrix* (1999), *Blade Runner* (1982), and *I, Robot* (2004) are all good examples. However, until recently, any serious consideration of what this might actually mean in the future real world was not necessary, because it was only science fiction and not in any way scientific reality. Science has, though, not only done a catching-up exercise but, in bringing about some of the ideas initially thrown up by science fiction, has introduced practicalities that the original story lines did not extend to (and in some cases still have not extended to).

Discussed in this article are different experiments in linking biology and technology together in a cybernetic fashion. One must realize, though, that it is the overall final system that is important. Where a brain is involved, which in one form or another it is in each case, the brain should not be seen as an entity operating in isolation, but rather as part of an overall cyborg system—adapting to the system's needs as appropriate. In particular, we take a look here at what such hybrid systems could possibly contribute within the field of space exploration, travel, and living.

It has to be said that there is clear overlap between the experiments described; however, they also throw up individual considerations. In each case, firstly, suitable background on the subject in terms of a description of practical investigations is given, and then pertinent issues on the topic are discussed. Specifically, issues have been raised with a view to near-term future technical advances and what these might mean in a practical space scenario. By no means has it been the case of an attempt to present a fully packaged conclusive document; rather, the aim here has been to open up the range of practical cyborg research actually carried out, with a look being taken at some of its implications.

13.1.2 Robots with biological brains

When one thinks of brain–computer interaction, then it is usually in terms of a brain already functioning and settled within a body—often a human body. Here, however, we consider the possibility of a fresh merger, where a brain is grown from scratch and is subsequently given an engineered body, of our design, in which to operate.

In conceiving of a robot, at first it may be a little wheeled device that springs to mind (Bekey 2005) or perhaps a metallic head that looks roughly human-like (Brooks 2002). Whatever the physical appearance, our concept tends to be that the robot may be operated remotely by a human, as in the case of a bomb-disposal robot, is being controlled by a simple computer program, or even may be able to learn with a microprocessor/computer as its brain. In all these cases, we regard the robot simply as a machine.

But it is quite possible for the robot to have a biological brain made up of brain cells (neurons), possibly human neurons. Neurons cultured under laboratory conditions on an array of non-invasive electrodes provide an attractive method with which to realize a form

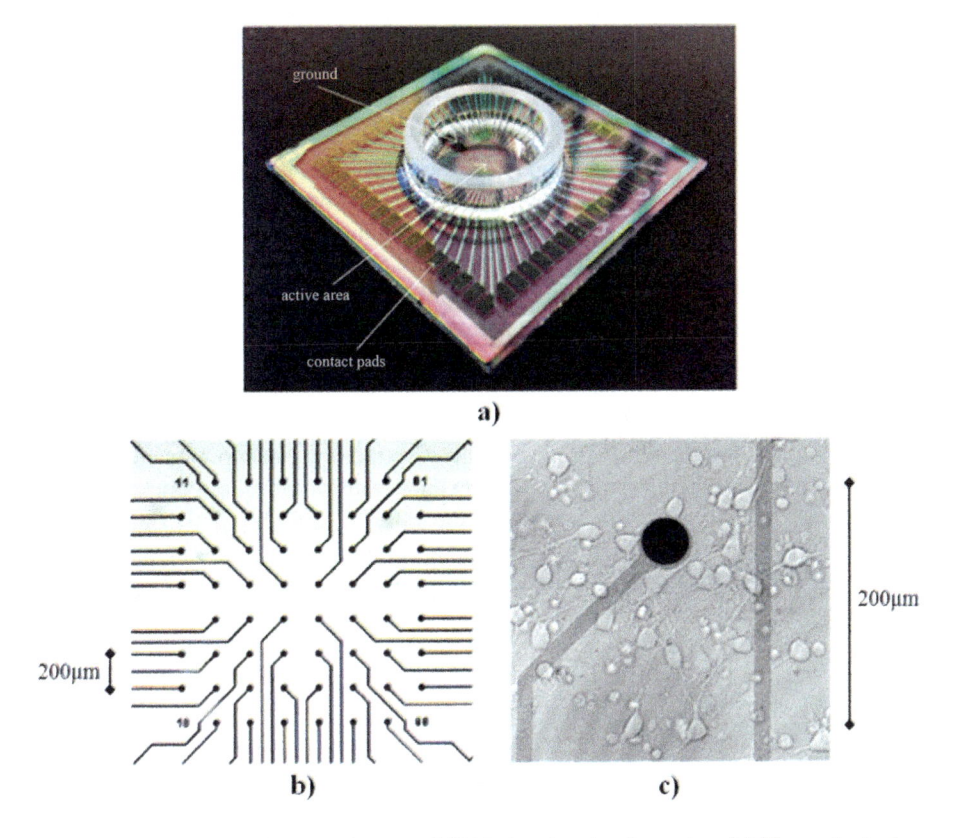

Cyborgs: upgrading humans for a future in space   343

Figure 13.1 (a) A multi-electrode array (MEA) showing the electrodes. (b) Electrodes in the center of the MEA seen under an optical microscope. (c) An MEA at ×40 magnification, showing neuronal cells in close proximity to an electrode. *Credit*: Kevin Warwick

of robot controller. An experimental robot body can move around in an area purely under the control of such a network/brain, and the effects of the brain, in controlling the body, can be witnessed. This is not only extremely interesting from a robotics perspective, but it also opens up a new approach to the study of the development of brains themselves, because of the sensory-motor embodiment.

Investigations are therefore being carried out on such brains into memory formation and reward/punishment scenarios. Typically, culturing networks of brain cells (around 100,000–150,000 at present) *in vitro* commences by separating neurons obtained from fetal rodent cortical tissue using enzymes. The neurons are grown (cultured) in a specialized chamber, in which they are provided with suitable environmental conditions (e.g. appropriate temperature) and fed with minerals and nutrients. An array of electrodes embedded in the base of the chamber (a multi-electrode array (MEA)) acts as a bidirectional electrical interface to/from the culture. The neurons in such cultures spontaneously connect, communicate, and develop within a few weeks giving useful responses.

Such a culture is grown in a glass specimen chamber lined with a planar "8×8" MEA which can be used for real-time recordings (see Fig. 13.1). The firings of small groups of neurons can be monitored via the output signal on the electrodes. So, a picture of the global activity of the entire network can be formed. The culture is electrically stimulated by means of biphasic electrical pulses via the electrodes to induce neural activity. The MEA, therefore, is a bidirectional interface to the cultured neurons (Chiappalone et al. 2007; DeMarse et al. 2001).

Initial growth and brain development last around 10 days, following which the culture can be coupled to its physical robot body (Warwick et al. 2010). Sensory data from the robot are subsequently fed back to the culture, thereby closing the robot–culture loop. This can be broken down into two discrete sections: (i) "culture to robot," in which live electro-chemical neuronal activity is used as the decision-making mechanism for robot control, and (ii) "robot to culture," which involves an input from the robot sensors (typically ultrasonic) to stimulate the culture. The actual number of neurons in a culture depends on natural density variations in seeding.

An existing neuronal pathway is identified by searching for strong relationships between pairs of electrodes. A rough input–output response map of the culture can then be created by cycling through all electrodes. In this way, the best input–output electrode pair can be chosen to provide an initial decision-making pathway for the robot. This is then used to control the robot body; for example, if the ultrasonic sensor is active, this indicates that an object is nearby and, possibly, we wish the culture's response to cause the robot to turn away from the object being located ultrasonically (possibly a wall) to keep moving.

In experiments, the robot follows a forward path until it reaches a wall, at which point the front sonar value triggers a stimulating pulse to the culture. As a result, if the output electrode registers activity, this drives the wheel motors and the robot turns to avoid the wall. The relevant result is the chain of events: wall detection—stimulation—response. From a neurological perspective, it is, of course, also interesting to speculate why there is activity in experiments on the response electrode when no stimulating pulse has been applied.

As a control element for direction and wall avoidance, the cultured network acts as the sole decision-making entity within the feedback loop. One important consideration then involves neural pathway changes in the culture with respect to time. Usually, the robot improves its performance over time in terms of its wall avoidance ability, in the sense that neuronal pathways that bring about a satisfactory action tend to strengthen purely through the process of being habitually performed. On many occasions, the culture responds as expected; on other occasions, it does not, and in some cases it provides a motor signal when it is not expected to do so. But does it "intentionally" make a different decision to the one we would have expected? We cannot tell.

It has been shown by this research that a robot can successfully have a biological brain to make all its "decisions." The 150,000-neuron size is merely due to the present-day limitations of the 2D experimentation described. Three-dimensional structures are also being investigated. Increasing from two to three dimensions realizes a figure of approximately 30 million neurons—not yet reaching the 100 billion neurons of a "perfect" human brain, but well in tune with the brain size of many other animals.

The range of sensory inputs applied is also being expanded to include audio, infrared, and even visual. Such stimulation richness will have a dramatic effect on culture development. The potential of such a system, including the range of tasks it can deal with, also means that its physical body can take on different forms. The body could be a two-legged walking robot, with a rotating head.

It seems quite realistic, looking to the future, to assume that such cultures will become larger, potentially growing into sizes of billions of neurons. On top of this, the nature of the neurons may be diversified. At present, rat neurons are usually employed in studies.

However, human neurons are also being cultured, enabling robots each with a sort of human-neuron brain. Clearly, when this brain consists of billions of human neurons, many social and ethical questions will need to be asked (Warwick 2010).

For example, if the robot brain has roughly the same number of human neurons as a typical human brain, then could/should it have similar rights to humans? Also, what if such a creature has far more human neurons than in a typical human brain—for example, a million times more—would such cyborgs make all future decisions rather than ordinary humans?

The technology opens up significant opportunities when we look at space travel. Sending living humans through the considerable distances required for space travel, especially if we wish to explore outside our own Solar System, is extremely problematic due to: (i) the travel time taken potentially extending well over a lifetime or two; (ii) the requirements to keep a human alive for such a period in a remote environment; (iii) the unknown hazards that could be faced on arrival; and (iv) the rigors on the human body during the trip, such as the effects of gravity loss on the body and brain.

Here, however, there is a possibility to freeze human neurons for the period of travel and to defrost them when within the gravitational pull of the distant planet. Robot technology could be employed to culture the neurons and embody them on arrival and not before. All that would be required would be a method to retain their feedstock in a reasonable state over the necessary time. Educational aspects could be provided to cause the newly embodied brain to investigate the planet as desired and to communicate any results in a suitable fashion.

Advantages of such space travel are considerable. Costs to send such a creature to a distant planet would differ very little from sending a mere technological robot. Such costs would be far lower than sending a human expedition. Further, if anything was to go wrong either on arrival or during the journey, then no life, in the normal (ordinary) sense of the word, would be lost; hence, there would be little or no negative political outcry.

13.1.3 General-purpose brain implants

Here, we consider, as a starting point, a regular human body and brain. It is certainly possible nowadays to employ implants within the human brain to attempt to counteract the effects of neurological problems such as Parkinson's disease (Pinter et al. 1999; Wu et al. 2010a, b); that is to say, the use of implants for therapeutic purposes. Even in such cases, it is in fact quite possible to consider employing such technology to give individuals abilities not normally possessed by humans. Here, however, we look at the possibility of neural implants being employed directly to extend human capabilities.

In some cases, it is possible for those who have suffered an amputation or have received a spinal injury due to an accident to regain control of devices via their (still functioning) neural signals (Donoghue et al. 2004). Meanwhile, stroke patients can be given limited control of their surroundings, as indeed can those who have motor neurone disease. However, in these cases, the situation is not simple, as each individual is given abilities that no ordinary human has, such as the ability to move a cursor around on a computer screen from neural signals alone (Kennedy et al. 2004). The same situation exists for blind individuals who are allowed extrasensory input, such as sonar (a bat-like sense)—it does not repair their blindness but, rather, allows them to make use of an alternative sense.

Some of the most important practical human research to date has been carried out using the MEA, shown in Fig. 13.2 (Nordhausen et al. 1996; Warwick and Gasson 2004; Gasson et al. 2005). The individual electrodes are 1.5 mm long and taper to a tip diameter of less than 90 μm.

Figure 13.2 A 100-electrode, 4×4 mm microelectrode array, shown on a UK one-penny piece for scale. *Credit*: Kevin Warwick

A number of trials not using humans as test subjects have occurred; however, human tests are at present limited to two groups of studies. In the second of these, the array has been employed in a recording-only role, most notably as part of (what was called) the "braingate" system.

Essentially, electrical activity from a few neurons monitored by the array electrodes was decoded into a signal to direct cursor movement. This enabled an individual to position a cursor on a computer screen, using neural signals for control combined with visual feedback.

The same technique was later employed to allow the individual recipient, who was paralyzed, to operate a robot arm (Hochberg et al. 2006).

The first use of the microelectrode array (shown in Fig. 13.2) has considerably broader implications, however, which extend the capabilities of the human recipient. As a step toward a broader concept of brain–computer interaction, in the first study of its kind, the microelectrode array was implanted into the median nerve fibers of a healthy human individual (the author), during two hours of neurosurgery, to test bidirectional functionality in a series of experiments. A stimulation current directly into the nervous system allowed information to be sent to the user, while control signals were decoded from neural activity in the region of the electrodes (Warwick et al. 2003). In this way, a number of experimental trials were successfully concluded (Warwick et al. 2004), in particular:

1. Extrasensory (ultrasonic) input was successfully implemented.
2. Extended control of a robotic hand across the Internet was achieved, with feedback from the robotic fingertips being sent back as neural stimulation to give a sense of force being applied to an object (this was achieved between Columbia University, New York, and Reading University, UK).

3. A primitive form of telegraphic communication directly between the nervous systems of two humans (the author's wife assisted) was performed (Warwick et al. 2004).
4. A wheelchair was successfully driven around by means of neural signals.
5. The color of jewelry was changed as a result of neural signals, as was the behavior of a collection of small robots.

In all of the above cases, it could be regarded that the trial proved useful for purely therapeutic reasons; for example, the ultrasonic sense could be useful for an individual who is blind, or the telegraphic communication could be very useful for those with forms of motor neurone disease. However, each trial can also be seen as enhancement beyond the human norm for an individual. The author did not need to have the implant for medical purposes to overcome a problem, but rather for scientific exploration. But, how far should things be taken? Enhancement by means of brain–computer interfaces opens up technological and intellectual opportunities, but it also throws up ethical considerations that need to be addressed directly.

Experiments of the type just described involve healthy individuals, and there is no reparative element in the use of a brain–computer interface but, rather, the main purpose of the implant is to enhance an individual's abilities; it is difficult to regard the operation as being for therapeutic purposes. The author, in carrying out the experimentation, wished to investigate actual, practical enhancement possibilities (Warwick et al. 2003; 2004). From the trials, it is clear that extrasensory input is one practical possibility that has been successfully achieved; however, improving memory, thinking in many dimensions, and communication by thought are other distinct potential, yet realistic, benefits, the latter of these also having been investigated to an extent. There is no reason why all these things cannot be possible for all humans who upgrade to become cyborgs.

If we consider the possibilities with this type of implant when it comes to space applications, then these are quite different in comparison with those considered earlier for the brain grown within a robot body. In this case, any technology is regarded by the recipient as merely being a new extension to their body, rather akin to an extra leg or arm. In reality, however, the extension can take on any desired technical form. It does not have to be an actual leg or arm; rather, it can be a wheeled device or a building. Whatever form it takes, the individual whose brain is connected to it as the sole controlling element regards it as being themselves.

When looking at space travel, which of course is a big problem for ordinary humans, the opportunity exists here for the individual cyborg to remain on planet Earth—they do not need to travel—however, their new body parts can travel to distant solar systems. By means of an implant, as discussed here, once the required items of technology have safely landed on a distant planet, then the connection can be made with the individual who has remained on Earth, and the distant parts are part of the cyborg's body.

While the individual is safe on Earth, their new body parts can investigate the planet of choice as though the individual were there themselves. The one negative in this plan is the time lag between a signal being transmitted from the brain of the individual on Earth, bringing about an action in the distant body part and receiving a response from any sensors on the body part. Practical experiments to this end have thus far only involved such a loop from Columbia University, New York, to/from Reading University, UK (Warwick et al. 2004).

From such experience, it can be reported that the human brain can cope with the different parameters that arise. It has to be acknowledged, though, that the time lag between the actuation and sensing in terms of controlling devices in space by means of an implant may be a significant problem. Unless our present understanding of physics changes, then there appears to be no way of avoiding this. So, either brain-coupled control of devices in space from Earth will be limited to low to medium space orbits or, for control in distant solar systems, then the time delays involved will be considerable.

However, there is an enormous cost saving for this type of space travel when compared with manned missions. While a neural implantation is indeed required, the costs are as almost nothing when compared with those of space travel. Time is also important. The individual involved can lead a perfectly normal life until their distant body parts are switched on. Meanwhile, space travel would have tied up the individual for many years and presented a large number of dangers. Although a neural implant is required, as of yet, there have been no reported problems with this type of implant. Indeed, it is several years since the author experienced the implant and there are no bad effects whatsoever to report. Space travel, meanwhile, after considerable expense, has many associated hazards. These must also be coupled with the potential extra hazards of traveling farther, for longer, than ever before and visiting, for the first time, new, relatively unknown, planets.

One final question that might be raised with regard to an implant of this type is the potential durability of the connection between the human nervous system and technology. It must be admitted that, in terms of experimentation involving an able-bodied individual, then the length of functionality reported is just over 3 months, this limitation being due to the length of the experiment rather than any problem with the implant (Warwick 2004). However, it has been found that the MEA can provide a reliable computer interface to an individual with tetraplegia 1000 days after implantation (Simeral et al. 2011), so there appears to be no reason why long-term use of the implant cannot be possible.

13.1.4 Non-invasive brain–computer interfaces

The most studied brain–computer interface is that involving electroencephalography (EEG), which is due to several factors. Firstly, it is non-invasive; hence, there is no need for surgery with potential infection or other side effects. As a result, ethical approval requirements are significantly lower and, due to the ease of electrode availability, costs are significantly lower than other methods. It is also a portable procedure, involving electrodes which are placed on to the outside of a person's head and can be set up in a laboratory with relatively little training and little background knowledge, taking little time—it can be done then and there, on the spot.

The number of electrodes employed for experimental purposes can vary from a small number—for example, four to six—to the most commonly encountered 26–30, to well over 100 for those attempting to achieve higher resolutions. As a result, it may be that individual electrodes are attached at specific locations, or a cap is worn in which the electrodes are pre-positioned. The care and management of the electrodes also vary considerably between experiments from those in which the electrodes are positioned dry and external to hair to those in which hair is shaved off and gels are used to improve the contact made.

This method is often focused in the medical domain, such as to study the onset of epileptic seizures in patients. The range of applications is widespread. A few of the most typical and interesting ones are included here, as much to give an idea of possibilities and ongoing work, rather than for a complete overview of the present state of play. Typical are those in which subjects learn to operate a computer cursor in this fashion (Trejo et al.

2006). It must be pointed out here, however, that, even after significant periods of training, the process is slow and usually requires several attempts before success is achieved.

Several research groups have used EEG recordings to switch on lights, control a small robotic vehicle, and control other analog signals (Millan et al. 2004; Tanaka et al. 2005). A similar method was, however, employed, with a 64-electrode skull cap, to enable a quadriplegic to learn to carry out simple hand-movement tasks by means of stimulation through embedded nerve controllers (Kumar 2008).

The uniqueness of specific EEG signals can be considered in response to associated stimuli, potentially as an identification tool (Palaniappan 2008). Meanwhile, interesting results have been achieved using EEG for the identification of intended finger taps, whether the taps occurred or not, with high accuracy. This is useful as a fast interface method, as well as a possible prosthetic method (Daly et al. 2011).

Some alternative non-invasive methods are quite different. Both functional magnetic resonance imaging (fMRI) and magnetoencephalography (MEG) have been successfully employed. fMRI brain scans use a strong, permanent magnetic field to align nuclei in the brain region being studied to ascertain blood flow at specific times in response to specific stimuli. They can, therefore, be used as a marker to figure out where there is activity in the brain when an individual thinks about something specific, such as moving their hand.

The equipment for such techniques, though, is cumbersome and expensive. Hence, experimentation in this area is by no means as widespread as that for EEG. Results have nevertheless been obtained in reconstructing images from such scans (Rainer et al. 2001) and matching visual patterns from watching videos with those obtained in a time-stamped fashion from the fMRI scans being recorded (Beauchamp et al. 2003).

It is not so convenient to assign advantages in space travel to this type of brain–computer interaction. On the one hand, if we can learn to recognize more easily and accurately intent from neural signals, then potentially the technique could be of some use remotely. But, without the concept of feedback and hence feelings, it is difficult to see how traveling astronauts could immediately be replaced in this way. Nevertheless, the potential for monitoring the brain activity of astronauts by means of this technology is clear. This is pertinent to research the effects of the long-term lack of Earth gravity on brain functioning. A good review of the overall potential of such interfaces in terms of space travel can be found in Rossini et al. (2009).

13.1.5 Subdermal magnetic implants

One final area to be considered is that of subdermal magnetic implants (Hameed et al. 2010). This involves the stimulation of mechanoreceptors by an implant affected by an external electromagnet. Implantation is an invasive procedure and, hence, implant durability is an important requirement. Only permanent magnets retain their magnetic strength over a very long period of time and are robust to different conditions. Typically, such magnets are implanted into an individual's fingers, for reasons which will be described.

Hard ferrite, neodymium, and alnico are all easily available, low-cost permanent magnets suitable for this purpose. The magnetic strength of the implant magnet contributes to the amount of agitation the implant magnet undergoes in response to an external magnetic field, and also determines the strength of the field that is present around the implant location.

Skin on the human hand contains a large number of low-threshold mechanoreceptors that allow humans to experience, in great detail, shape, size, and texture of objects in the physical world through touch. The highest density of mechanoreceptors is found in the fingertips, especially of the index and middle fingers. They are responsive to relatively high frequencies and are most sensitive to frequencies in the range of 200–300 Hz.

For reported experiments (Hameed et al. 2010), the pads of the middle and ring fingers were the preferred sites for magnet implantation. A simple interface, containing a coil mounted on a wire frame and wrapped around each finger, generates the magnetic fields to stimulate movement in the magnet within the finger. The idea is that the output from an external sensor is used to control the current in the wrapped coil. So, as the signals detected by the external sensor change, these affect the vibration experienced by the individual through the implanted magnet.

Several application areas have already been experimented on (Hameed et al. 2010). The first was ultrasonic range information. In this scenario, the magnetic interface was connected to an ultrasonic ranger for navigation assistance. Distance information from the ranger was encoded via the ultrasonic sensor as variations in frequency of current pulses. These were then transmitted to the electromagnetic interface. This mechanism allowed a means of providing reasonably accurate information about the individual's surrounding for navigational assistance. The distances were intuitively understood within a few minutes of use and were enhanced by distance "calibration" through touch and sight.

A further application involved reading Morse signals. In this application scenario, the magnetic interface was used to communicate text messages. Morse code was chosen for encoding, due to its relative simplicity and ease of implementation. In this way, text input was encoded as Morse code and the dots and dashes, which could be represented as either frequency or magnetic-field strength variations, transmitted to the interface.

The invasiveness of such implants is relatively trivial. Many individuals who have piercings see the operation as of no concern whatsoever. We are not looking here at a new form of motor control, but rather a way to sense other signals not normally sensed by humans—infrared being a good, immediate example. In this case, therefore, it may be simply an extra tool for an astronaut. Rather than using technology to take measurements of different signals, they could, potentially, with an implant or two of this type in place, experience sensations themselves of the different signals measureable on another, possibly quite remote, planet. Also, if it was felt likely that certain signals, above a threshold, could spell danger for that person, the implant could be very useful as an early-warning indicator of danger.

13.1.6 Conclusions

A look has been taken here at several different types of brain–computer interface. In particular, experimental cases have been reported on to indicate how humans can merge with technology; in doing so, this has thrown up a plethora of social considerations as well as technical issues. It can be seen, though, that, in each case, reports on actual practical experimentation results have been given, rather than just theoretical concepts.

Firstly, when considering robots with biological brains, this could at some point mean human brains, or at least human-like brains, operating in a robot body. Therefore, is it appropriate to consider whether such a robot should be given rights of some kind? If a robot of this type was switched off, would this be deemed as cruelty to robots? Should there be some sort of law dependent on the number of human brain cells and, if so, should this also apply to ordinary humans? But, at this time, should such research forge ahead regardless? Before too long, we may well have robots with brains made up of human neurons that have the same sort of capabilities as those of the human brain, and the ethical aspects of such an eventuality really need to be discussed now.

In the section considering a more general-purpose invasive brain implant, as well as implant employment for therapy, a look was taken at the potential of using these implants for human enhancement. By this technique, extrasensory input has already been scientifically achieved, along with extending the nervous system over the Internet and a basic form of thought communication. But, if large numbers of humans upgrade and become cyborgs, then that could have a significant impact also on those who do not. Indeed, if ordinary humans are left behind as a result, then this could bring about the digital divide. It will be interesting for each person to consider whether, if they as an individual could be enhanced, they would even question it (Warwick 2007).

Then followed a section on more standard EEG electrodes, which are positioned externally and which are encountered far more frequently. However, the resolution of such electrodes is relatively poor and they are, in practice, only useful for monitoring brain activity, and not for other uses such as stimulation. Bidirectional communication with a brain in this way is not a possibility. Hence, issues surrounding them are somewhat limited. We may well use them to learn a little more about how the brain operates and to monitor which part of a brain is active at a particular time, but it is difficult to see them ever being used for highly sensitive control operations when several million electrodes feeding into the information transmitted by each electrode.

Finally, a look was taken at subdermal magnetic implants. This method has, until very recently, been investigated more by body modification artists rather than by scientists, and hence scientific application areas are still relatively few. In its favor, while involving an invasive procedure, it is still relatively straightforward in comparison with such techniques as deep brain stimulation or MEAs fired into the nervous system. It is expected, therefore, that this will become an area of considerable interest over the next few years, with many more potential application areas being revealed as practical experimentation is carried out.

Some technological issues have been pondered on here to open a window on the direction in which developments are heading, particularly with regard to the possible use of such methods within space travel. In each case, however, a firm footing has been planted on actual practical technology obtained thus far, rather than on speculative ideas.

In each case, the possibilities of how this interface technology could play a part in space travel have been considered. In this respect, the first two examples appear to be potentially most useful and certainly disruptive. It is felt that both culturing brains and embodying them within a robot body and the use of neural implants offer significant advantages for the cyborgs which result, in reality partly for cost, safety, and realizability reasons, in comparison with the present (ordinary human) manned space travel programs.

Acknowledgment

An earlier version of this article appeared in an online form only in *Acta Futura*, **6**, 25–35 (2013).

13.2 Shrinking into the universe

Arne Hendriks, artist, exhibition maker, researcher, and historian, Next Nature department of the Technical University in Eindhoven

> "I was continuing to shrink, to become ... What? The infinitesimal? What was I? Still a human being, or was I the man of the future?"[1]

13.2.1 Introduction

After being exposed to a radioactive cloud and pesticides, Scott Carey, the protagonist in the 1957 film classic, *The Incredible Shrinking Man*, starts shrinking. Just before he disappears from view, he speaks his dramatic but prescient words. The fantasy of shrinking is deeply embedded within our culture. The *pygmaloi* in Greek mythology, the Lilliputians in *Gulliver's Travels* (by Jonathan Swift, 1726, amended 1735), the *Alice in Wonderland* (by Lewis Caroll, 1865) and *The Incredible Shrinking Man* (1957) books, and even a more recent translation like the film *Honey I Shrunk the Kids* (1989) all manifest the awareness that present human size is just one outcome of infinite possibilities. Still, the film is a typical product of the 1950s, when anxieties about technological developments and the Cold War were expressed in dystopian scenarios predicting disaster for humanity. Shrinking people became a popular metaphor for power games, Man's vulnerability, and a warning that technological progress has a price. Yet, looking at what technology has meant for the development of the human body, the enormous giantess in the film *Attack of the 50 Foot Woman* (1958) represents a more appropriate warning. In the last two centuries, innovations in the realms of agriculture, housing, medicine, and hygiene have caused a dramatic increase in average global human height. As a result, we need more food, more space, more energy, more everything. The height of the Dutch, the tallest people in the world at an average 184 cm, increased by almost 20 cm in less than two centuries. Unfortunately, because of proportional growth, this 9% increase represents a 30% increase in size and weight. In a world with limited resources, such an increase is a bad idea.

In 2004, Dr. Andrew Danneberg calculated that, if the weight of every US citizen increased by 10 lb, as it did during the 1990s, domestic air travel in the US alone would expend 1.3 billion extra liters of kerosene to transport the extra weight at a cost of US$300 million.

---

[1] These visionary words of hope were not in the original book by Richard Matheson but added by film director Jack Arnold. *"I was continuing to shrink, to become ... What? The infinitesimal? What was I? Still a human being, or was I the man of the future? If there were other bursts of radiation, other clouds drifting across seas and continents, would other beings follow me into this vast new world? So close, the infinitesimal and the infinite. But suddenly I knew they were really the two ends of the same concept. The unbelievably small and the unbelievably vast eventually meet, like the closing of a gigantic circle. I looked up, as if somehow I would grasp the heavens, the universe, and worlds beyond number. God's silver tapestry spread across the night. And in that moment I knew the answer to the riddle of the infinite. I had thought in terms of Man's own limited dimension. I had presumed upon Nature. That existence begins and ends is Man's conception, not Nature's. And I felt my body dwindling, melting, and becoming nothing. My fears melted away and in their place came acceptance. All this vast majesty of creation, it had to mean something. And then I meant something too. Yes, smaller than the smallest, I meant something too. To God, there is no zero. I STILL EXIST."*

What complicates matters is that auxologists like Nobel Prize winner Robert Fogel continue to regard tall stature as a sign of well-being. While it is true that our increased height is the result of greater food security, living conditions, hygiene, and medical care, Fogel fails to address that there is a difference between the conditions influencing height and the actual consequences of tall stature itself. Plants growing in a greenhouse might soar high but are not necessarily stronger than other plants. On the contrary, we have far outgrown the height optimum for our present body frame. It is a statistical fact that every extra centimeter above 150 cm lowers our life expectancy by 6 months. Professional basketball players rarely live to the age of 60. The average age of the 10 tallest people that ever lived is only 36 years old. Robert Wadlow, the tallest person in history at 272 cm, died at 22. Geneticist and evolutionary biologist J.B.S. Haldane pointed out that bigger size leads to complexity and that complex systems are intrinsically hazardous. This paper would like to answer Scott Carey's question about his being the man of the future with a resounding yes! Yes, you are the man of the future! If the human species is to live in balance with Earth, it must shrink.

Shrinking could also make affordable space travel a lot more realistic. Smaller astronauts offer an economic and ecological advantage in interstellar space exploration, since they need considerably less food, water, oxygen, and energy. Moreover, miniature space travelers would possess greater agility, relative strength, would be better equipped to deal with radiation, and suffer fewer problems arising from alterations in gravity. Macroevolution has shown that, for most species, shrinking is a lot easier than growing large. Maximum rates of the decrease in body mass are up to 30 times greater than the rate of mass increase. This may be due to the concurrent anatomical, physiological, environmental, genetic, and other constraints that must be overcome by evolutionary innovations before further increases in size are possible, whereas shrinking species are able to return to formats and solutions that proved successful in the past. Challenging environmental conditions may also favor small species that mature and reproduce faster. The question is whether the human species can return to these smaller prototypologies if so desired or when circumstances demand it. Perhaps the small astronaut of the future exists already as a homunculus within our anatomical programs. The height of any particular individual depends on a great number of environmental and cultural factors: where and how we live, what we consume or desire, which are entangled with a complex cocktail of height-related genes. What genes do we bring on board and how will their environment influence gene expression?

Extreme differences in size in the human species are relatively common. In 2013, a small group of men and women took part in an ongoing NASA-funded research project called HI-SEAS (Hawaii Space Exploration Analog and Simulation). For 4 months, they were cooped up in a geodesic dome on the side of the very Mars-like Mauna Loa volcano in Hawaii, simulating living conditions on the Red Planet. One of the outcomes was that female crewmembers expended only half the calories of male crewmembers. During one week, the most metabolically active male burned an average of 3450 cal per day, while the least metabolically active female expended just 1475 cal. In the early 2000s, Alan Drysdale, a former systems analyst in advanced life support with NASA, while thinking about the problem of astronaut bodies, turned to a NASA document on physiological metrics called STD-3001, "Man-Systems Integration Standards," which details needs and effluents for a range of body types. Drysdale calculated that, all other things being equal, a female crew would launch for about half the payload cost. Ultimately, however, it is not about sex, but about size. Small men yield similar results to women, and a permanent departure from Earth would probably include members of both sexes. Because of the highly varied ways in which small stature may be attained, a variety of case studies in size reduction are discussed here to develop a feel for what kinds of adaptations those who seek to become smaller may need to master, or be subjected to. The cases include environmental studies, calorie restriction, cultural influences, and a series of genetic factors.

A relatively speedy decrease in average human size could be accomplished if women would be attracted to short men. Perhaps pygmy squid could inspire a directional change in desire. Their mating rituals include neither pleasant courtship nor aggressive behavior. Males copulate freely with females and several males attach a capsule containing sperm to the base of the female's arms. The female then selects which donated capsule is to be the lucky one and gets to spread its genes. The female chooses. In the case of the pygmy squid, it is most often the capsule donated by the smallest males. Although existing formats such as this one cannot be translated directly into human practice, human mating behavior is not a fixed given. The introductions of religious, political, and economic systems, and their subsequent social consequences, have considerably changed our choice of partner. How our mating rituals and attraction values will develop depends at least as much on societal change as it does on the laws of Nature. It is not entirely impossible that, under certain environmental circumstances, frogs and squid will show us a way out of our growth-obsessed mating behavior. In the future, women may very well feel more attracted to shorter men. In fact, women already prefer shorter (more dependable) men in times of crisis.

Since 2007, researchers of the GIANT consortium (Genetic Investigation of Anthropometric Traits) have been uncovering the polygenic traits that influence human height. After analyzing data from the genomes of 253,288 subjects, they identified 697 gene variants as related to height. The variants were enriched for genes, pathways, and tissue types known to be involved in growth. The results indicate a genetic architecture for human height that is characterized by a very large but finite number (several thousands) of causal variants. This suggests that a complete map of all genes related to height can be compiled. At this point, we may begin to design and engineer bodies to become smaller to suit the resource and economic challenges implicit in interstellar travel. But, let us not overlook the fact that spontaneous genetic experimentation has been going on for as long as Man can remember.

13.2.2 Genetics

13.2.2.1 Primordial dwarfism

There are over 200 known forms of dwarfism, each allowing for the exploration of other and shorter ways of being human. At the far end of height, we find primordial dwarfism. Primordial dwarfism is a diagnostic category including specific types of profoundly proportionate dwarfism, in which individuals are extremely small for their age, among them Chandra Bahadur Dangi, at 54.6 cm, and Junrey Balawing, the current shortest living human at a height of 59.8 cm. Despite the considerable challenges with which most primordial dwarfs have to deal, it would be shortsighted to dismiss them as possible human models. One could, in fact, regard them as the man of the future, today. Small size should not primarily be read as a defect. With directed and positive research, perhaps it is possible to embed the genes that lead to shortness into a framework that repairs its negative consequences, and make a big step toward a smaller astronaut.

During World War II, little people were actively recruited to work at Henry Ford's Ypsilanti plant, bucking rivets in the more inaccessible places inside the wing of the B24 bombers that were built there. Only adult little people had the strength for this job. Relative strength greatly increases as the body gets smaller. A 10 % decrease in height results in a 19 % decrease in average strength, yet it creates a surplus in power since, at the same time, weight decreases by 27 %. Compared with their height, small people are a lot stronger.

13.2.2.2 Laron syndrome

The May 2013 cover of *National Geographic Magazine* (*NGM*) showed a portrait of a young Ecuadorian child with a very rare growth condition called Laron syndrome. People with this syndrome, or should we say genetic ability, are able to resist the growth hormone (GH) because of a mutation in the growth hormone receptor (GHR) gene that transcripts the protein that receives the body's growth signals. As a result, they also display very low levels of insulin growth factor-1, a hormone that promotes cell proliferation, inhibits programmed cell death, and is a key agonist in the specific hormonal pathways involved in growth. Laron syndrome people grow, proportionally, to a height of less than 130 cm, and none of them develops malignant cancer, or diabetes. The baby on the *NGM* cover is expected to reach the ripe old age of 120 years.

In the late 1990s, scientists managed to create model mice bearing a disrupted GHR/binding protein similar to the one in people with Laron syndrome, greatly facilitating a full investigation of GHR dysfunction. The Laron mice were 40 % smaller and had an extended lifespan of up to 40 %.

13.2.2.3 Supercentenarians

Supercentenarians, those aged 110 years and above, often have an altered genetic profile that, as with the Laron syndrome people, affects the insulin growth factor-1 signaling pathway, and interferes with its growth-promoting properties. If we take a look at the oldest people in history, we see that almost all of them are shorter than 160 cm. Jean Calment, the French lady that holds the official old-age record, measured just under 150 cm and lived to be 122 years and 164 days old, adding well over 40 % to contemporary French average life expectancy (85 years for women). This strong relationship between body size and longevity could convince humanity, or a group of aspiring astronauts, to shrink. Many of us would prefer to live longer while enjoying good health and would not mind giving up a few inches as a trade-off. Another thing not to be wasted in space is experience. Longer lifespan is an immediate advantage to astronauts, allowing them a longer active career in a field of highly educated and specialized professionals. Perhaps future space travelers would be wise to study bonsai trees. The Japanese believe that to develop a bonsai tree is an act of self-cultivation. The bonsai represents Man. Just like an astronaut, a bonsai tree exists under extreme conditions. It is confined to a small space with limited supplies of water and nutrients. Because of the tree's high level of phonotypical plasticity, the ability to change phenotype according to external circumstances, it can survive. The bonsai expresses phytohormones that scale down growth to within the possibilities of its specific environment. If it were not for this endocrinological self-therapy, the tree would not be able to survive such scarcity. In general, animals, including people, are less able to respond quickly to environmental challenges because they are more able to manipulate their environment by moving somewhere else. Still, people also display several plant-like abilities. In Nature, a tree that grows under perfect conditions will grow until it reaches the predetermined height and width for that species. On reaching and slightly surpassing its potential, the now massive amount of foliage at the incalculable number of branch tips is just too vast. The tree starts to weaken and eventually dies, because the foliage has grown too far away from the active roots. Conversely, a bonsai tree, which is prevented from ever reaching its maximum dimensions through regular pruning of the roots and branches, could, theoretically, live forever. The bonsai, just like its full-size cousin, is genetically programmed to achieve maturity. By preventing the bonsai from reaching maturity, you prevent it from reaching old age and, inevitably, death. Could this cultivation of a smaller tree-self relate to the idea of shrinking the human body, as well as nurture the desire to do so? Should we remain in a permanent state of physical immaturity, never quite reaching our full potential and at the same time, paradoxically, allowing ourselves to continue to grow and live forever?

13.2.3 Environment

13.2.3.1 *Homo floresiensis*

Although not a tree, and extinct, there is much we can learn from studying that most peculiar branch of Man's evolutionary tree: *Homo floresiensis*. Although its origins are still much debated, because its remains were only discovered in 2003, *H. floresiensis* probably was a hominin, descending from *H. erectus*. Members of the species lived between 100,000 and 12,000 years ago on the Indonesian island of Flores and were just a little over a 100 cm tall. Where the "Hobbit of Flores" came from and how it developed once it became isolated on Flores is still not entirely clear. Some think *H. floresiensis* might have suffered from a form of dwarfism, perhaps even Laron syndrome. But, this does not seem to correspond to the findings of partial skeletons of nine individuals within the same size range. The most interesting hypothesis on the reasons for its diminutive size is that *H. floresiensis* was subjected to the process of insular dwarfism. Insular dwarfism is the reduction in size of large animals, and in this case humans, over a number of generations, when their population's range is limited to a small environment. Because of caloric scarcity on islands, smaller individuals with lower energy requirements have an advantage. There are numerous examples of this dwarfing process throughout evolutionary history, including dinosaurs, elephants, and rhinoceros. In one case, a population of red deer, isolated on the island of Jersey, shrank to just one-sixth of their original weight over the relatively short time span of 6000 years. Evolutionary survival strategies are the driving force behind insular dwarfism. But, in some cases, it is the relaxation of harsh conditions that leads to size change. Foster's rule states that, when an animal gets trapped on an island, small species get taller and tall species become smaller. The greater competition on the continent may actually force species beyond the limits of their optimum size. They grow larger as a defense against predators, or shrink to hide. On islands, their natural competitors are absent. Without this pressure, environmental conditions allow them to return to the size of their genetic preference. Perhaps for *H. sapiens* the optimum size would resemble the size of the insularly dwarfed *H. floresiensis*, evolution's most relaxed version of ourselves. Perhaps there already exists a 100-cm-tall astronaut within all of us.

For vertebrates, it is a truism that change in body length is considered to be unidirectional. But, there is an exception. Studies of two island populations of the Galápagos marine iguana found that individuals were able to shrink all of their body, including the skeleton, by as much as 20 % within 2 years. The smaller lizards had higher survival rates during harsh periods with low food availability, because their foraging efficiency increased while energy expenditure decreased. The mechanisms that determine whether and to what extent they shrink remains largely unclear but it could well involve re-absorption of bone and the quantum biological principles involved in insect metamorphosis and tadpoles becoming frogs. The research in this field continues.

13.2.3.2 Mbuti

Although *H. floresiensis* disappeared 12,000 years ago, there is a hypothetical possibility that they merged with the Rampasasa, a pygmy tribe still in existence today. Pygmies are recognized for their short height below 155 cm. Among the very shortest are the Mbuti of northeast Congo. At an adult height of only 140 cm, the Mbuti are about 25–30 % shorter than an average Western European male. Their average weight of 40 kg constitutes a much more intelligent and efficient body design. The extreme conditions of their natural environment, the warm, moist, and dark Ituri rainforest, turned out to be the perfect platform for the origin of small people. The exact driving forces are the subjects of scientific debate. Mbuti may have adapted to a nutrient-low environment, much like some insular dwarfs. It could also be that smaller bodies are better suited to move through dense forest, dissipate heat, or deal with low ultraviolet light conditions. Possibly, it is because Mbuti are genetically programmed to stop growing right after puberty, so that the body's resources can be channeled into bearing offspring. One study raises the intriguing possibility that short stature makes for better disease resistance. The research suggests that genes related to height may, at the same time, boost immunity for malaria and tuberculosis, giving its bearers an evolutionary edge for survival. Ultimately, perhaps it is the combined advantages of small stature within this very specific environment. The embrace of the Ituri rainforest as a functional shrink-environment allow for a vision of a spaceship environment that has a proven positive effect on size reduction.

13.2.4 Culture

13.2.4.1 Oskar matzerath

The actual functional steps toward creating an environment that shrinks its inhabitants shows that it will take a monumental leap of the imagination and of courage, much like that brave leap of the stairs Oskar Matzerath takes in the famous Günter Grass novel *The Tin Drum* (1959). The human species is biologically and culturally programmed to think, feel, and act in terms of growth. We find the increase in height over the last century easy to accept, while shrinking is still against what we wrongfully feel is right. We are looking for those rare individuals willing to let go of personal certainties and comfort to perhaps find a greater good for all. Our desire for height, with all its obvious advantages, needs to be matched by an even stronger desire for small stature. Although Grass passes the thin line between the real and the imagined, Oskar does represent how circumstance, time and again, pushes humanity to reinvent itself. The vision of embarking on a space odyssey, or the depletion of Earth's resources, may just be the necessary enzyme to activate this ability and give the strength to face fear, ridicule, perhaps even aggression. The initial public enthusiasm to sign up for the Mars One mission gave an insight in how far some people are willing to go to be part of a deep-space mission. Almost 3000 people agreed to leave for Mars, although they would never return to Earth. Could such dedication and sacrifice mean they would also be willing to change physically if it would allow them a place on board an interstellar mission? And are we able to look beyond our own generation? Do we allow our children a shorter-sized, more abundant future? One that perhaps gives them the possibility of interstellar space travel, or a life in abundance on the planet we inhabit presently? Can we make this choice for our children? Are we allowed?

When Francis Kenter of Amsterdam decided to raise her son, Tom, on raw vegan food only, it created quite a stir in the Dutch media. Mostly, it seemed to be because doctors projected that the diet meant that Tom would grow 6–12 cm shorter than his peers brought up on regular diets. People were outraged and Francis was all but crucified. Both she and Tom have never capitulated to popular opinion—an act that, in the context of the tall-bodied Dutch, shows precisely the sort of activism needed to redirect the path away from becoming taller. In fact, in the nineteenth century, the Dutch were among the shortest people in Europe, until they started producing industrial quantities of cheese, which is known to stimulate the production of GH. Since casein, the main active ingredient in cheese, turns into the mildly addictive caseomorphine in the stomach, one might say that the Dutch are literally addicted to growth.

13.2.4.2 *The Incredible Shrinking Man*

What *The Incredible Shrinking Man* shows us, perhaps more than anything else, is how challenging it is to fundamentally change paradigms when the body is involved. Yet, the examples above give direction to the real possibility of becoming smaller. *The Incredible Shrinking Man* exists already. He is among us, trapped in isolated pieces of a complex global puzzle of human potential. It is up to humanity to find those pieces, of which we have presented only a few, and to put them together. What we already know is that even just shrinking the human species to as much as 150 cm would be an enormous reduction in our need for resources. If relatively simple animals like the marine iguana, the pygmy squid, and several species of frogs are able to deal with scarce conditions, then nothing should be able to stop Man from acquiring this ability, except Man himself. Paradoxically, the proof that we can shrink is the enormous increase in height over the past century or so. We need to tap into latent desires already embedded within society. We fantasize about shrink beams and tiny pets; we play with model trains and grow giant vegetables. Some even literally desire to shrink in fantasies of shrink beams, magic potions, or more obscure sexual fantasies of giantesses or Tinkerbell-sized lovers. Who has never fantasized about the incredible abundance one ripe strawberry provides us with if we would be the size of a fairy? We need to return to the realm of fantasy to light a fire in reality. Perhaps the latent societal desires for smaller, reinforced by its advantages in regard to health issues, abundance, and sustainability, allows for an unprecedented leap into the future. Once we leave Earth, what we are is up for grabs. Our departure body will only be the first sketch in a long line of genetic and environmental experiments, both wanted and unwanted, toward a species specifically equipped to deal with space travel. The worldship will function as an inverted planet, initiating evolutionary processes of interstellar dwarfism necessary for successful existence in space. The interstellar island must reach a threshold of complexity that allows the organism that inhabits it a dynamic relationship with its surroundings— much like the bonsai in its tray, the fish in its aquarium, *H. floresiensis* on Flores, or the bantam astronauts on the 100YSS.

13.2.5 View on the future

So, this is what might happen. A hundred years from now, after several generations of careful selective breeding, a crew of Laron Mbuti astronauts ascends the 100YSS. Their average height is 107 cm—not quite as small as they projected to be at the onset of the program, 75 years ago, to downsize the human species, but an incredible improvement from the previous 130 cm. In any case, the process of becoming smaller will continue on board the 100YSS. Its climate, its functional food crops, its endocrinological acclimatization, as well as behavioral protocols have all been designed in such a fashion that the inhabitants are projected to shrink to an average height of only 50 cm over the course of the next 1400 years. Because of this steady decrease in size, the spaceship becomes a little larger with each generation, allowing for a gradual increase in its population. When finally they approach a habitable planet, the notion of living on a planet has become so foreign, so strange, so futuristic, the stuff of mythology rather than reality, that most decide to remain on the ship, their natural habitat. Yet a small group of pioneers gets ready to make the almost unimaginable leap to go and live on a planet, never to return. "Perhaps," one of the pioneers offers as an idea, "we need to become larger so we can better deal with whatever challenges our new habitat will throw at us." The others just laugh at him. Now there's a weird thought.

13.3 Human reproduction in space

Rachel Armstrong, Professor of Experimental Architecture, Newcastle University.

13.3.1 The imperative for human space exploration

Human civilization as we recognize it today is established on planet Earth, in a solar system that is 4.55 billion years old and already halfway through its life.

At 10 billion years, the Sun, which is a "yellow dwarf star," will use up all its hydrogen and begin to cool and gradually collapse due to the force of gravity. The energy created by the collapse of the Sun will raise its temperature to hundreds of millions of degrees, which is hot enough to start burning the helium in its core. These higher temperatures will cause the Sun to expand and change color, becoming a "red giant star."

As the Sun expands, it will destroy Mercury, Venus, and possibly Earth.

Even if our planet is not completely destroyed by the aging Sun, its surface will become dry and inhospitably hot. The familiar and unique aqueous conditions that have so rarely supported life and our own evolution will no longer exist. Instead of a lush green, water-soaked world, planet Earth will be a hostile, radiation-baked, lifeless planet, and the human race will face its greatest challenge of all time: its continued survival in an alien universe.

Over the course of 150,000 years, our species has developed technologies that have changed the world. Despite our extraordinary abilities and collective wisdom, we will not be able to prevent the eventual destruction of our planet from cosmic forces.

Space habitation is inevitable for the survival of the species. Our ongoingness depends on our ability to foray into space to resettle on Earth-like planets elsewhere in the universe, or terraform those of suitable size.

Scientists already know that only a tiny fraction of the 400 billion stars in our galaxy seem to have what it takes to support life on orbiting planets. They have spent years identifying and studying the basic characteristics of stars in our galaxy, the Milky Way, and have found that stars vary in their characteristics and that some probably contain a habitable zone, the region around a star where liquid water can exist on a planet's surface. The nearest Earth-like planet, so far unnamed, is 20.5 light years away and orbits a red dwarf star called Gliese 581. Nearer Earth-like planets may exist in the solar system of Alpha Centuari, our nearest neighbors, at only 4.3 light years away.

13.3.2 Human space migration

Even if we started traveling to the Alpha Centauri system today, we would reach it in 4.3 years if we traveled at the speed of light, which is around 300,000 km per second, and impossible to accomplish with today's technology. The actual time that humans may expect to take to reach the Alpha Centauri system is dependent on the type of propulsion used, as this will determine the speed of the spacecraft. The New Horizons probe, launched in January 2006, which is now heading toward Pluto, reached the highest speed for a modern spacecraft at around 60,000 km/h. Even at these speeds, using the latest technology such as ion-drive propulsion, the travelers would be looking at around 80,000 years. It is estimated that with the still purely theoretical nuclear pulse propulsion system, it may be possible to reduce this time to around 80 years.

Science-fiction authors have suggested many creative ways in which humans may accomplish and survive migration away from Earth. Solutions range from solving time travel, as in Richard Matheson's "Death Ship," the creation of space arks such as in Arthur C. Clarke's *Rendezvous with Rama* (1973), the construction of artificial habitats like Iain M. Banks's "Culture," holding human biology in a dormant state using cryosuspension such as in Dan O'Bannon's *Alien* (1979) movie, bodily modification to create an altered physiology as in Rachel Armstrong's "plant people" in *The Grays Anatomy* (2001), or the complete abandonment of human biology altogether and replacement by much hardier machines, which are typical of Isaac Asimov's 2010 stories about intelligent machines such as *I, Robot*.

13.3.3 The survival imperative

Arthur Schopenhauer, like most scientists and philosophers of his day, attributed the "*will to live*" as the highest motivational life force in Nature. Frederich Nietzsche observed that the "*will to live*" was not life affirming enough, and that humans had a higher need, which he called the "*will to power*" that caused people to dominate and impose their resolve on others. In fact, Nietzsche believed that one would apply his concept of the "*will to power*" further than its relevance to individuals, and use it to explain the motivation of whole societies and nation-states. In other words, Nietzsche believed that humankind is driven to take control of its own destiny and to impose its will on the immediate environment, wherever that may be.

Such plans are implemented through the use of technology.

In *The Production of Space* (1974), Henri Lefebvre observes that "the space of nature remains open on every side, and thanks to technology we can 'construct' whatever and wherever we wish, at the bottom of the ocean, in deserts, or on mountaintops—even, if need be, in interplanetary space."

William E. Burrows suggests a catalog of catastrophes that may destroy civilization as we know it and that the impact of potential catastrophic events ranging from the devastating impacts of asteroids, comets, nuclear holocaust, famine, earthquakes, or hurricanes might be mitigated if we intelligently develop a program for humans to colonize space. Burrows claims that NASA has lacked a well-defined mission since the Moon landings and should dedicate its resources to such an endeavor. He poignantly observes that our extinction will be inevitable if we remain on a localized planet. Stephen Hawking agrees that:

> "Life on Earth is at the ever-increasing risk of being wiped out by a disaster such as sudden global warming, nuclear war, a genetically engineered virus or other dangers … I think the human race has no future if it doesn't go into space …. There are too many accidents that can befall life on a single planet." (Sato 2007)

In *Beyond Earth: The Future of Humans in Space* (2006), Bob Krone provides the methodology for human settlement in the Solar System in the near future—a perspective supported by Carl Sagan in *Pale Blue Dot* (2007) near the end of his life, who had been a long-time opponent of manned spacecraft.

13.3.4 Spacefaring humans

"Despite our innate instinct to survive beyond the destruction of Earth, we have only really just taken our first steps into space. Already, the experiences of astronauts and experiments performed in weightlessness demonstrate that our terrestrial anatomy is poorly designed for and profoundly affected by space travel.

   The early space program focused obsessively on keeping the astronaut-trainees in perfect physical shape, but it soon became clear that a body like a Greek demigod's was no defense against that horrible feeling that your stomach was falling out from under you and you were never going to catch up. Our inner ear, which normally tells us which way is down, tortures us when down is nowhere to be found. There is contradictory information about whether anyone ever gets over it; the 'right stuff' culture creates a strong incentive for astronauts to deny that they are sick." (Crowell 2003)

Weightlessness is not benign. It has a multi-system impact. Kevin Fong, co-director of the Centre for Aviation, Space and Extreme Environment Medicine (CASE), observes that "Space really screws you up" (Sladden 2004).

Reduced gravity, lack of atmosphere, and cosmic radiation collectively predispose astronauts to a wide range of conditions, including osteoporosis, motion sickness, and decompression illness. Under these conditions, the human body redistributes its nutrients and fluids, causing general physical atrophy and psychological disturbance—a syndrome called space sickness. Although some initial adaptation takes place within the first day or so to compensate for the initial disorientation in weightlessness, the overall readjustment to the alien environment becomes more profound as the duration of the stay in space increases. These macroscopic changes are caused by the disruption of the tissues at a microscopic level, but microgravity also has dramatic effects at the cellular level. It affects cells and tissues either through direct response elements or by influencing the environment in which the cell lives. Consequently, cells respond to decreased gravity and the reordering of forces by changes in gene expression and cellular function. The extent of the changes that microgravity has on our biology is still very much under investigation.

Yet, despite these challenges, astronauts are already staying in space for increasingly longer episodes. Valeri Polyakov currently holds the record for the longest single spaceflight in human history, staying aboard the Mir space station for 14 months (437 days 18 h). After their marathon endurance tests, returning astronauts suffer from symptoms of weakness and disorientation as a result of re-adaptation to the terrestrial environment. Fortunately, astronauts, who are selected and trained to be extremely fit individuals, are known to recover with remedial treatment yet, despite their exceptional basic level of health, the readjustment to gravity may still be prolonged and difficult and the full extent of the long-term effects of microgravity are still being evaluated.

13.3.5 Space colonization

In 2005, NASA administrator Michael Griffin identified space colonization as the ultimate goal of current spaceflight programs:

"… the goal isn't just scientific exploration … it's also about extending the range of human habitat out from Earth into the solar system as we go forward in time …. In the long run a single-planet species will not survive …. If we humans want to survive for hundreds of thousands or millions of years, we must ultimately populate other planets. Now, today the technology is such that this is barely conceivable. We're in the infancy of it. … I'm talking about that one day, I don't know when that day is, but there will be more human beings who live off the Earth than on it. We may well have people living on the Moon. We may have people living on the moons of Jupiter and other planets. We may have people making habitats on asteroids … I know that humans will colonize the solar system and one day go beyond." (*The Washington Post* 2005)

A manned mission to Mars is the next goal for human space exploration and it will take a crew around 3 years for a round trip to the Red Planet. Space agencies still face huge technological and physiological challenges if there is to be the prospect of a return mission in the next few decades. The European Space Agency (ESA) is planning to address some of these challenges by running simulations of the experience (ESA 2007). Also imminent is the prospect of mixed crews (Schneider 2007) and, with the proximity of the crew in the context of such a difficult and intense mission, the specters of conception, pregnancy, and even birth become apparent.

13.3.6 Sex in space

Lawrence Palinkas, a medical anthropologist at the University of Southern California in Los Angeles and an author of the report published by the US National Academy of Sciences (NAS), observes that, with the prospect of a very long-term mission, it is hard to ignore the question of sexuality (Bergin 2005). Despite the ultimate goal of space exploration being human colonization, as asserted by Griffiths, national space agencies are vague about the interpersonal relationships of their crewmembers. Human biology is bound by sexual reproduction as the unique way of creating new humans and is intimately linked to sexual relations. These extend beyond the social and psychological benefits of selecting mixed crews and influence crew dynamics.

NASA spokesman Bill Jeffs asserts that "We don't study sexuality in space, and we don't have any studies ongoing with that" (Fox News 2008).

This guarded position is understandable, as national space agencies avoid the risk of raising criticism, not only aimed at their professionalism, but also at the significant public investment made on their behalf. There is nothing formally written down about sex in space. Indeed, space agencies depend on the professionalism and discretion of their astronauts. Yet, the human interest in this matter leaves much open to speculation. Many experiments in space biology have been conducted, and dozens of papers have been published on the subject, but many of the key questions have not been answered.

Dr. Lyubov Serova from the Institute of Biomedical Problems (IBMP), a leading Russian research institute in the field of space medicine and biology, has been involved for decades in the sex-related studies of living species in space. She speculates that cosmonauts do not really need a psycho-emotional relief like sex during long-lasting spaceflights, and that abstinence does not have any negative effect on the human mind or body. Polyakov, one of Serova's colleagues, who holds the record for the longest single spaceflight, observes that sexual satisfaction was never a reason for the premature interruption of any spaceflight or at the root of any considerable problem on board a spacecraft (Peakin 2000).

Yet, something may be happening. Pregnancy tests are included in International Space Station (ISS) medical packs, which appear to be there in a diagnostic capacity to rule out the differential diagnosis of ectopic pregnancy in a female crewmember presenting with abdominal pain (Marshburn 2008).

Sensationally, French astronomer Pierre Kohler claimed in *The Last Mission: The Human Adventure of Mir* (2000) that NASA had studied the feasibility of 10 sex positions in space during a Space Shuttle mission in 1996—a claim that NASA vigorously denies. Kohler also speculated that the most likely case of sexual relations in space would have been aboard the US Space Shuttle when the married couple Jan Davis and Mark Lee flew on the same flight in 1989. Yet, none of these rumors is substantiated (Beam n.d.).

Other cases of sexual conduct have been reported during an 8-month space-station simulation on Earth in 2000, where a Russian man allegedly twice tried to kiss a female Canadian researcher, prompting the installation of locks between the Russian and international crews' compartments (*New Scientist* 2005).

Indeed, the novelty of human anatomy in space appears to become fascinating enough to invite sexual comment even from the most seasoned science-fiction authors:

> "Some women, Commander Norton had decided long ago, should not be allowed aboard ship; weightlessness did things to their breasts that were too damn distracting. It was bad enough when they were motionless; but when they started to move, and sympathetic vibrations set in, it was more than any warm-blooded male should be asked to take. He was quite sure that at least one serious space accident had been caused by acute crew distraction, after the transit of an unholstered lady officer through the control cabin." (Clarke 1973)

Yet, reports suggest that, while weightlessness may arouse sexual curiosity, it does not follow that the body is able to respond normally to the otherwise excitatory cues, as reduction in gravity is associated with disturbances in sex hormones (Xia et al. 2014).

Even more disabling is that the strange biophysical adaptations that take place as an astronaut adjusts to weightlessness are indiscriminate with regard to which organs they affect:

> "… no matter how stressed anyone gets, they can't even enjoy a little release by manipulating their own joystick: One of the effects of weightlessness is reduced blood flow to the lower half of your body. The rumour in Star City is that many have tried in vain to get it up out there … Viagra will not help." (Kushner 2008)

Whether other appliances have been tried to remedy the reduced turgor of weightlessness, it is apparent that normal sexual activity in space requires a huge investment of creativity, stamina, and commitment. This is often difficult enough to achieve in a normal situation, and possibly quite out of the question under the stressful conditions of weightlessness where other people will be very close by. It seems that the idea of space sex works a whole lot better in theory than in reality. No confirmed reports of a sexual experience in space exist, and the physiological evidence suggests that, in microgravity, the physics and biology of sex could make the act difficult, if not completely awkward (Boyle n.d.).

Perhaps the main problem with sexual relations in space is that it requires more exertion than its equivalent on Earth. With all the water-logging in of extracellular spaces, it is likely that the pleasurable sensations in critical neurones will be damped down to such an extent that the positive reinforcement for making the effort will be lost. Humans in weightlessness may decide that they cannot be "bothered" with making the necessary effort at all! There are reports that Vanna Bonta found it challenging even to kiss her husband during a zero-G simulation flight and that they had to "struggle" to connect and stay connected.

Even if it is not possible for babies to be made in the "natural" way in weightlessness, the notion of assisted conception also poses a public-relations minefield for rumor-sensitive space agencies. Although human reproductive technologies have advanced rapidly since the 1980s, with the advent of genetic engineering, cloning, and the sequencing of the human genome, all these methods are controversial and introduce impossible ethical dimensions to the issue.

13.3.7 Human reproduction in space

While national space agencies assert that the public fascination with sex is space is mere titillation, there is also genuine human interest in the future of the species beyond our home planet.

Although the likelihood that humans will spend a lifetime in weightlessness remains a long way off, space agencies are approaching a time at which they will need to issue information regarding human fertility, reproduction, and sexual relations in space. The notion of space travelers as being exclusively military professionals is changing, as the first wave of private citizens make their way into the extraterrestrial environment on commercial flights:

> "… [Richard] Garriott will become the sixth private citizen to join the most exclusive, most high-octane clique on the planet: Call it the 240-mile-high club. Membership includes Greg Olsen, who made his fortune developing infrared cameras; Mark Shuttleworth, the software engineer who spearheaded Ubuntu; and Charles Simonyi, former chief architect of Microsoft. What they have in common, other than tremendous success in the tech industry, is a willingness to pay tens of millions of dollars for a week and a half in space." (Kushner 2008)

With the foray of the public into space, the trappings of holidays are likely to make their way into the expectations of these environments too. In the 1960s, Hilton proposed an orbiting hotel. Architects Wimberly Allison Tong and Goo (WAT&G) and Galactic Suite are also anticipating a booming space-tourism industry by designing space hotels (Gray 2007). More recently, the American company Bigalow Aerospace has been building the prototype of an inflatable space hotel. They even plan to host research into animal propagation on their commercial space modules (Boyle n.d.).

While sex between humans does not appear to have been successful in space, the first space honeymooners will take up residence in their orbital suites with a view and will watch Earth rise. They will become curious about which sexual positions are feasible in weightlessness and wonder whether contraception is effective. Perhaps they will wonder whether they can conceive a healthy child in space. Yet, we still do not know exactly what kind of impact spaceflight conditions, like weightlessness or radiation, have on fertilization or embryogenesis. NASA's advice on human conception in weightlessness does not go as far as to "ban" human conception in antigravity. Rather, pregnancy is contraindicated on the basis of animal studies that demonstrate that gametogenesis and early cell development are deleteriously affected.

13.3.8 Data from animal experiments

Research on other mammals suggests that implantation is normal. However, embryos are more likely to have defects at an early stage. In experiments with mice, Japanese researchers found that embryos created in low-gravity conditions that simulated space travel went on to implant and develop normally (Wakayama et al. 2009). However, although reproductive issues have been contemplated by space agencies for decades, the sum effects of weightlessness plus the physical trauma during re-entry remain unknown. Susan Crawford-Young analyzed NASA experimental findings in a wide range of animal species, which offered a fascinating insight into reproductive capacities or organisms in microgravity (Crawford-Young 2006).

13.3.8.1 Insects

Varied results were obtained in the fruit fly *Drosophila* during embryogenesis. Some experiments resulted in widespread embryo death, while others demonstrated increased mutation rates. Control experiments were used to distinguish between the effects of background radiation and microgravity, which had a significant negative effect on embryonic development, even when radiation was taken into account.

13.3.8.2 Axolotl

This strange animal is an example of the earliest kinds of vertebrate. Neural-tube defects are seen in embryos reared in microgravity.

13.3.8.3 Amphibians

Salamanders were shown to have problems with neural-tube closure in the head or cephalic region in microgravity. Four out of five embryos had this neural-tube defect, and two out of five had another deformity at the early tail bud stage called "microcephaly," in which the brain was poorly developed.

13.3.8.4 Clawed toad

Xenoupus embryos showed widespread changes in the tissues that developed from the primitive streak, which forms the nervous tissue and resulted in a range of neural-tube defects.

13.3.8.5 Mammals

No mammals have been conceived and born in space. Rats that have been impregnated in the terrestrial environment, then brought into microgravity, have given birth to normal offspring. These results suggest that something fundamental is happening at an early stage of mammalian development that completely disrupts cell organization.

13.3.8.6 Human tissue culture

Although it is unethical to conduct human studies in this context, researchers have used immortal cell lines from just about every organ in the human body, including kidney, hemopoietic tissue, bone, skin, brain, and thyroid. The results unanimously show fundamental disruption of cell division and structure.

Crawford-Young's compelling review suggests that early stages in cell development and differentiation in all experimental animals seem to be affected. Embryos that suffer from disruption at early stages of their development are known not to be viable. She speculates that the deformities observed in early embryogenesis may be attributed to factors such as disruption of the cytoskeleton, low calcium, disturbance of differentiation waves that distribute organizing molecules throughout the embryo, and the general dysfunction of ion channels. Interestingly, gene expression was also noted to have altered regions of the DNA that coded for cytoplasmic structures. The organ system that appeared to be most affected by the disruption of cytoskeletal organization was the nervous system.

13.3.9 Implications of animal studies for humans

Crawford-Young notes that it is difficult to make sweeping statements about embryonic development in microgravity, as the animal studies are ad hoc and non-standardized. As different types of animals appear to be affected differently by alterations in gravity, it is difficult to make generalizations. However, it is likely that critical, early developmental stages are affected by microgravity, which makes human conception in space by normal methods impossible. Should any human fetuses go to term, their most likely abnormalities would be in the nervous system. She recommends that more research is needed in the field of developmental pathology to more fully characterize the fundamental cell processes that are disrupted in the extraterrestrial environment.

Interestingly, animals with aquatic biology seem to be best suited to microgravity. Medaka fish are used as an experimental model as they do not exhibit "looping" — a pathological behavior — and are the only vertebrate to successfully breed in space (Ijiri 1995). Results in breeding higher animals, such as mammals, have been disastrous.

13.3.10 The future of human reproduction in space

Human biology is so poorly designed for the extraterrestrial environment that a journey into space is actually a form of species contraception. Microgravity so radically affects our anatomy that it seems impossible for us to survive in space without gravity. Crawford-Young's review also highlights that there is much more research to be done in terrestrial laboratories to understand the role of the cytoskeleton more fully, the structural systems through which gravitational forces are transduced throughout the cell. Moreover, the long-term effects on "wobbles" in artificially induced gravitational fields are unknown. High-altitude terrestrial cultures demonstrate physiological traits that are different from those that live at sea level. Populations living in mountains have much denser red blood cell counts, which have been attributed to the production of the hormone erythropoietin (EPO), which is stimulated by low oxygen concentrations. However, the gravitational effects of EPO production are not controlled for in these investigations, as gravity is assumed as being constant. NASA physician Jim Logan believes that we need further research into the link between gravity and biology to better understand the relationship between the two:

> "We still do not have an inkling of what the 'gravity prescription' is ... think of gravity as a medication. We don't know the dose, we don't know the frequency, and we don't know the side effects." (Boyle n.d.)

With increasing knowledge about the fundamental effects of variations of gravitational fields, it may be possible to reengineer our cells at the most fundamental level so that we may survive in the longer term. Of course, the simplest move would be to produce a spaceship capable of producing a constant gravitational field that matched Earth's. However, even our own planet's gravitational field is not entirely homogenous. Earth's gravitational model, nicknamed the "Potsdam Potato," is based on data from the LAGEOS, GRACE, and GOCE satellites and surface data. It demonstrates that our planet's gravitational field is subject to variations that occur over time. This is due to a range of overlapping factors, such as the uneven distributions of mass in the oceans, continents, and deep interior. Climate-related variables also contribute to variations in our gravitational fields, like the water balance of continents and changes in glaciers. Imaging the world through gravitational forces produces a topology that is different from, but still complimentary to, approaches based on light, magnetism, and seismic waves. Through this lens of observation, it is possible to view everything anew, not only within the context of a worldship, but also on Earth, including the speed of ocean currents, rising sea levels, melting ice sheets, hidden geographical features, and even convection force driving plate tectonics (Williams 2014).

Variable gravity fields are not the only challenges for long-term forays into space. Background radiation may prove disabling, if not lethal. Radiation exposure poses acute and long-term risks to space travelers by producing free radicals from the water in cells, which go on to break DNA molecules. Acute effects are determined by the type and amount of exposure to radiation fields. They range from mild and recoverable effects, such as nausea and vomiting, to central nervous system damage and even death. In the longer term, radiation exposure can cause cataracts and cancer. Central nervous system damage can also be caused by heavy ions, the effects of which are similar to aging. These risks are ongoing, so that even astronauts that have returned to Earth are at continued risk of cancer, as prolonged radiation exposure can damage the ability of cells to repair themselves properly. Astronauts have been classified as radiation workers since Project Mercury, as terrestrial radiation guidelines are considered too restrictive for space activities. Therefore, NASA has adopted the recommendations of the National Council on Radiation Protection (NCRP) for spaceflight activities, and the long-term monitoring their radiation exposure is a key requirement for spaceflight (NASA n.d.).

To combat such deleterious effects, spacefarers may be genetically modified using sequences from radiation-resistant extremophile bacteria such as *Deinococcus radiodurans*. The bacteria possess a biological toolset that prevents them being harmed by radiation on longer missions and it is possible that these biological survival systems could be incorporated into our cells. An alternative would be to engineer our bacterial microbiomes to include the *Deinoccocus* species.

Figure 13.3 Hair machine. *Credit*: Lucy McCrae

But perhaps biological systems alone will simply not be sufficient to cope with the extreme challenges of living away from our home planet. The idea of hybrid organisms that are forged by unions of organic matter and machines was first conceived by Manfred E. Clynes and Nathan S. Kline, as self-regulating human–machine systems to survive in outer space (Clynes and Kline 1960) (Fig. 13.3). Today, advances in biotechnologies and information technologies offer new methods (Armstrong 1997) through which human anatomy may be modified to cope better with the extraterrestrial environment:

> "Humans embarked on an extraterrestrial odyssey would find the body's 'complexity, softness, and wetness … hard to sustain,' predicts Stelarc. The body must be hollowed, hardened, and dehydrated, its inessential innards scooped out so that it may be 'a better host for technology,' its skin peeled off and replaced by a synthetic dermis capable of converting light into chemical nutrients and absorbing all oxygen necessary to sustain life through its pores. An internal early warning system would monitor what few organs remain, and 'microminiaturized robots' or nanomachines could 'colonize the surface and internal tracts to augment the bacterial populations—to probe, monitor and protect the body'." (Dery 1996)

13.3.11 Self-preservation

Perhaps the strategy of reproduction to ensure the survival of the species is not effective for space travel. Alternative ways of preserving the organism are used by plants such as fungi and bacteria that retreat into sporulation during stressful times, which is a biological form of cryosuspension. Humans may need to develop effective cryonics (Cryonics Institute n.d.) to preserve gametes or other matter for reanimation at a later date, until an Earth-like planet is reached, since the technology is currently problematic. However, although technology continues to advance, the structure of the body and its complexity are extremely challenging for near-term rejuvenation of human corpses. Advocates propose that adjunct technologies such as nanotechnology that can deal with cellular damage will need to be invented to turn the research into reality.

Immortality is another survival strategy, and Aubrey de Grey argues that, since aging is the consequence of a series of errors, then their repair could greatly preserve biological function and render us effectively immortal. He proposes that humans can reasonably look forward to a 1000-year lifespan (TED.com 2005).

> "If Bezos wants to see colonization happen," Friedman says, "a concurrent problem he will have to solve is eternal life." (Grainger 2004)

13.3.12 Augmented reproduction

With advances with in-vitro techniques of stem cell differentiation, cloning, and the orchestration of cell growth in specially engineered environments, innovative survival strategies may be developed to produce new humans that are better adapted to extreme environments than their ancestors. The implications for using these technologies are that, potentially, fertilization and embryogenesis may be externalized from the human body using reproductive technologies to control and monitor the process in an artificial uterus using simulated gravity.

Developments in a few research laboratories around the world promise the advent of an artificial uterus, such as pioneering work done by Hung-Ching Liu at Cornell University (Reynolds 2005) and Dr. Yoshinori Kuwabara at Juntendo University in Tokyo (McKie 2002); yet, these developments have not been without significant issues such as the considerable mortality of implanted embryos as well as the associated ethical issues of the technique.

13.3.13 Symbiotic environments

The architecture of the spacecraft and immediate environment would ideally need to simulate Earth-like conditions and find energy sources that will enable its inhabitants to survive under minimum stress. Architectures that are regarded as prostheses for human survival may allow users to move freely outside the spacecraft and wearable architectures that may eventually become symbiotic with our changing, extraterrestrial biology (Armstrong 2002) (Fig. 13.4).

Figure 13.4 Human space prosthetics may enhance our function in extreme environments.
Credit: Lucy McCrae

13.3.14 Succession of humans

Many lifetimes will be spent in pursuit of the further exploration of the universe, long before our greatest thinkers hope to conquer the issue of time travel. Those who spend the longest time in the extraterrestrial environment will inevitably adapt physically and mentally to the various extreme conditions that will face them. In fact, Louis Bec postulates that humans are extremophiles by nature, seeking and colonizing the most difficult terrestrial habitats (Bec 2007). He argues that, by using technologies that promote rapid evolution, we are capable of surviving extraterrestrial environments. However, when we leave our planet in the search of new habitation, we may no longer be *Homo sapiens*, but *Homo galactica*.

Sadly, it seems that, at least for the foreseeable future, the human race is faced with an almost insurmountable task if it is to survive for any extended periods in the extreme environment of space. Hans Moravec and Ray Kurzweil ominously predict that our "wet" biology is nothing more than a transitional evolutionary stage that moves toward the evolution of spacefaring robots, or that transhuman species, which no longer identify as "human," will replace us:

> "Exploration and colonization of the universe awaits, but earth-adapted biological humans are ill-equipped to respond to the challenge. Machines have gone farther and seen more, limited though they presently are by insect-like behavioral inflexibility. As they become smarter over the coming decades, space will be theirs. Organizations of robots of ever increasing intelligence and sensory and motor ability will expand and transform what they occupy, working with matter, space and time. As they grow, a smaller and smaller fraction of their territory will be undeveloped frontier." (Moravec n.d.)

Or maybe Douglas Adams had it right when he said "*So Long, and Thanks for All the Fish*," which is the message left by the dolphins when they departed planet Earth just before it was demolished to make way for a hyperspatial express route (Adams 1979). However, in light of ongoing developments, perhaps the "fish" are no longer simply food for the dolphins, but our successors—the medaka fish that are the "highest" living order of species to date that can survive and persist within the hostile extraterrestrial environment.

This essay was originally presented at the Less Remote, satellite conference of the International Astronautical Congress, as part of the Arts Catalyst event, Glasgow, September 2008.

13.4 Bodies *in extremis*

Sarah Jane Pell, artist-astronaut, Australia Council Fellow, Australia.

13.4.1 Fluid architectures for human survival in space

I conceived my daughter Amulet during an unscheduled activity on the International Space Station: the marriage of my body and another within the body of space. Mission Control discussed an attempt to create artificial gravity by centrifuge to isolate exactly what gravity does to fetus development. Others suggested genetically engineering a substitute for the lack of gravity. It was all too late. Gravity played a vital role in the body's circulatory feedback responsible for the development and maintenance of healthy bones and blood vessels. Without it, physicians predicted that my child's skeleton would be more cartridge-like than bone. Born "*en caul*," Amulet was splendidly shielded from cosmic radiation in her thermally regulated cocoon and floated out from me like a rare jewel. Initially, an infinite loop of biology, technology, vacuums, and pulses supported and confined our precarious lifeline. I delivered oxygen through the placenta, and her oral–nasal cavities meshed with the nanoscaffold stem-cell cannulae feeding ancillary O_3 to maintain circulation and keep her chest cavity supported around her heart.

Those 3 months waiting for an inflatable quarantine in low Earth orbit taught us a great deal about caring for bodies *in extremis*.

How we move or transport our bodies has been the cornerstone of human adaptation, he said, taking my hand. By expanding our biological capabilities with the support of animals and tools, from two feet to the wheel and the Space Shuttle, we directly impact our reach and capacity for genomic diversity, trade, culture, knowledge, resilience, and growth. If we mapped the human species according to the era of the aquatic, climber/swinger, crawler/walker, rider, sailor, driver, diver, pilot, astronaut, and avatar, each transportation epoch demonstrates our cognitive evolution of agency, real and imagined. Humans now constitute part of the system of technics.

Nonetheless, I replied, our cultural imagination of human movement toward the stellar epoch has remained locked to a vision of recognizable humanoid forms in suits decorated with tailored instrumentation. We view cyborgs in vehicles traveling from one celestial body to another, exploring a road map of fixed, linear waypoints. This vision pervades our approach and drives our technology, politics, and poetry. Why? Because, historically, bodies *in extremis* from deep earth, deep sea, to the heights of terrestrial summits and, more obviously, in the vacuum of space depend on tools such as advanced life-support systems and return vehicles. These are not simply exoskeletons or orthotics, but devices enabling a biotech fission and protection from the harsh atmosphere in a respiratory loop (metaphoric and real). In this narcissistic inscription, we maintain a reassuring Earth-bound body (Latour 2014), and make sure the human body is a temporal alien to space, by identifying future-oriented concepts dependent on returning home to planet Earth.

We mused over Bernard Steigler's volumes that described technics as organized (performing) matter and memory (of the human). I wondered whether the inverse of inscribing intent and experience on the body *in extremis* would be the *logos* of the unbounded body. In principle, such fluid architectures would support space-specific morphological and ethological differences and, by recognizing that space art is breaching familiar genomic architectures, microbial behaviors, and dynamics within technical objects (for space-normal life), they may also ensure human survival in space. So inspired, we began to explore the embodiment, metaphor, and system of an unbound body and fluid space exchange …

13.4.2 Wet dreams and politicized bodies *in extremis*

I had an idea that my expertise in the aesthetics and embodied knowledge of human-aquatic performance would be such an asset for space adaptation. It was my first time stationed at the International Space Station (ISS), conducting choreographic microgravity experiments, testing saline-filled porous flight suits for long-duration spaceflight. In microgravity, redistributed blood flows toward the thorax and cranium, producing change in all systems of the body from the endocrine system to the cardiovascular system, as if underwater. Women's breasts engorge, their stomachs flattens, cheeks blush, and lips swell. All fluid membranes expand, heat up, and excrete hormones. As Oscar Wilde noted, the body *in extremis* is simultaneously the body *in ecstasy*. Indeed.

The radiation shielding of the aquabatic suit protected our intermingled cells long enough for fertilization and embryonic maturation. Medical teams noted hormonal anomalies but ruled out conception due to context and exposure. They routinely prescribed countermeasures for rapid onset signs of "space motion sickness," including an exercise régime (sensorimotor), mindfulness (psychological), and pharmacology (oxygen), until vestibular disturbance subsided. The demands of the ISS workload facilitated an unnoticed near-natural pregnancy until 22 weeks, when I reported a distinctly metallic metal taste leading to many tests and a contamination review for any non-standard biological pathogens. Our commander performed a private wedding before we told the capcom, our family, and friends.

All of Earth hotly debated our legal status, our rights to life, our rights to support, our humanity, and the implications of rebirth through any attempt at re-entry to an atmosphere. In space biomedical terms, I was alive, stable, and capable of "high-value" performance within the routine limits of "space normal" (Olson 2010). What defines an astrobiological entity incapable of human life beyond extreme or artificial environments is more challenging: Amulet was categorized as a normal juvenile astronaut, and simultaneously she was an at-risk living system, incapable of seamless integration with the mechanical and environmental ISS systems. Amulet was not "mission critical," nor did she pose any higher chance of cross-contamination than every organism/environment interaction existing from the microbial level to the cosmic; yet, her vitality created an unprecedented distance from terrestrial systems of care, and her body inscribed the future generation.

We were therefore held in a static system, under non-consensual pharmacological control, and closed off from direct exchange with other living matter on board the ISS. I had fully completed the flight assignment and program of observation during all extreme environmental transitions, exposures, integrations, and adaptations. A remote team of physicians performed the cesarean section following strict asepsis and telepresent surgical protocols. As my uterus opened, Amulet ascended unaided within a full amniotic sac, floating, pulsing, and falling in microgravity. Total communication silence followed.

I breathed, cried, prayed love.

Soon, I would be quarantined for extraterrestrial microbes, and Amulet incubated: immersed in a liquid-oxygen pod for life, to be denied Earthly matter. What the term astronaut implies (Langston and Pell 2015) might include other agents on our biological systems, but it did not yet include our offspring.

We reconcile politicized bodies *in extremis* to interdependency with advanced systems of confinement, control, and order, which expose and support the body at the limits of survival and bare existence. The mechanisms of war, imprisonment, camps, and political theater of the masses subject people to abuses (Balfour 2001). They do not fall within the range of any normal human experience, but those who become "institutionalized" to adverse conditions show that physical and emotional survival often depends on the system for continued learning. As physical workers *in extremis*, miners, divers, soldiers, tactical units, slaves, and astronauts adapt to conditions of *extremis*; they develop leadership skills including the consequential reasoning in the case of danger, threat to life or being, which does not deny dehumanization, but rather attempts to affirm human values to save a life or their own life, even if it brutalizes ethics, morals, and laws (Ripstein 2004).

For this reason, space missions needed artists working on discovery-driven-research within the technoscientific domains to make sense of omic information at the human-environment-scale and within the context of the lived experience (political, societal, and cultural) relating natural, constructed, and virtual worlds. Artists and scientists have examined human performance in microgravity from aesthetic and biometric angles. For example, the MIR network (Triscott and la Frenais 2005) explored the choreographic potentials afforded by the body in short-duration microgravity parabolic flights (Dubois 1994); expressive painting in microgravity (Pietronigro 2000); the transmission of "embodied knowledge" within dance as a technology of proprioception (de Lima 2013); and inhabiting fluid forms in architecture (Oksiuta 2007b).

Co-opting the lens of art and science, I too created studio/laboratory/field studies to examine my bodily performance in a state of *extremis*, usually under aquatic and hyperbaric conditions. The aim was to map the status, behavior, and limits of human-aquatic performance in occupational and extreme environments and related interactions with life support, ranging from physicochemical systems to bioregenerative life support. Aesthetic, metric, designed, and philosophical outcomes resulted that were speculative or provided new data with tangible meaning in space and related analog contexts. Over a series of investigations, I had articulated an understanding of extreme performance between the body and its breathing technology and body of water as a singular action perception (Pell 2014). I called this aquabatic activity, identifying extreme performance between the body and body of water as singular action perception in and of itself within the neutral buoyancy environment. The many dimensions of art, such as aquabatics, provided the methods and interdisciplinary expertise to transfer capabilities to overcome innovation deficits in analog spaces, while affirming human values without overt disruption to existing fields of practice, laws, timelines, and budgets. Immersive epistemologies arising further interrelated the hydrodynamic behavior of all matter, flow, and human expression.

While theoretical and methodological understandings of human performance in safety-critical environments, from sea to space, are well supported by the field of human factors (Reason 1995; Redmill and Rajan 1996), rapidly evolving technologies, private exploration, and public–private partnerships influence priorities and capabilities, mission-related engineering, and imagination needed to adapt to support crew within these extreme performance systems (Moore and Gast 2010).

Aquabatics highlighted that the short-to-midterm understanding required for enhancing human performance in complex critical systems depended on a new interdisciplinary user-centered approach for the evolutionary redesign of systems and their values. Like other astronauts, I trained for extravehicular activity (EVA) or space walking through immersion in simulated working environments. A neutral buoyancy environment provided a reasonable analog for the cognitive preparation and training for the interactions we would encounter in the microgravity environment. By leading underwater training workshops for EVA training (Pell et al. 2013), I noted an opportunity existed to relate aquabatic experience to cognitive and biological performance to extend knowledge and technology transfer from underwater to space (Pell 2014) and classifying immersion by technology, perception of the user, or experience.

Historically, practical EVA or space walk training involved separate elements: complex, remote, or digital simulation exercises (dry runs); neutral buoyancy training (underwater) and activities performed in short exposure to microgravity (parabolic flight). Notwithstanding, astronauts reported that no digital format could convey the microgravity embodiment that framed the experience (Garan and Hoffman 2013; Leonov and Lebedev 2001), and recommended redeveloping the infrastructure for EVA training (Ney et al. 2006). I replied with the view that the aquatic environment was a fluid space, and ripe for crafting a cinematic architecture and sonic resonance around the body. I worked out how to "waterproof" the DomeLab (Bourke 2012) and sink it in a local diving pool. I designed a full-bodied 360° HD immersive and responsive EVA training protocol for classes underwater. By developing specialist somatic, kinesthetic, and embodied immersion methods for utilization in the underwater domain, my aquabatics research had built leadership in the embracive approach to space adaptation. It was radical to suggest that we "go with the flow" and accept the way in which the body becomes more languid and the head becomes more bulbous, trusting that the skeletal and muscular change, for instance, reflected the demands of the bodily workload in space, and perceived disfigurement did not deny our humanness.

13.4.3 Human relationships are intricate

Amulet looked like a baby curled inside a translucent stingray suit.

Her spine had a splendid single curvature, which rounded her core when she moved. Her heart visibly fluttered and fibrillated her fluid encasing. Hands and feet webbed and flattened moved in tandem. Her head was the same size as her body and her eyes transparent. Her optical nerve and cerebral cortex appeared well formed, but her balance, vision, and motion awareness may develop differently. Frequency-activated interactions may be more meaningful than expecting Amulet to develop the manual dexterity of her Earthling cousins, and so I hummed to her.

The crew guided her close to my chest. As we connected hearts, skins sticking together again as one, I released a flood of hormones and transferred microbes that would bond and strengthen us forever. The unstable happenstance of phytochemical exchange between our porous bodies created the conditions for growth.

We were left to perform her second birth in private as a family ...

Immersive epistemology seeds the possibility for evolutionary expansion from star sailors to fluid actors. The extreme performance of the post-turbulent body—underwater or interstellar—disrupts the dominant vision of future spaceships, starships and human-rated vehicles in favor of a complex environmental and transcendental human experience of extreme new worlds ...

It took us several hours to build the courage to knowingly expose her to the intense radiation and the volatile organics, allergens, and toxins in the closed environment. Breaking the sac and chewing the placenta felt like an ancient rite of passage: appropriately animal and fluid ...

When humans migrate to distant stars, the concept of human embodiment—in a performative and conceptual way—will become indistinguishable from these interstellar interactions.

As the primordial soup of time space expands, we grow.

Old.

Part of the space-travel allure is the promise of defying aging and bending time. Living forever through our progeny, our fluid clusters of living organisms and coded DNA and RNA become immortal. Fluid phenomena create conditions that give rise to the autonomous self-organizational processes and the creation of new biological forms (Oksiuta 2007a; Pell 2014). As we develop a heightened awareness to natural systems in complex working environments, there emerge new modes and strategies for adaptation to rapidly changing technological and natural environments. The maelstrom of cultural, political, scientific, and technological modalities leading to a socially coded model of extreme performance—epistemological and physical—represented by this event provoked new understandings of the idiosyncratic human experience to support the principle of sustaining life in post-turbulent ecologies.

When Amulet opened her big blue eyes to a liminal world, a flood of intuitive, remembered, imaginative, spontaneous, and affected factors resulted, and she gave hope for our collective survival. In a fluid, orbital, international, cooperative environment, our species moved, in principle, from microgravity adjustment and space-normal-filtered adoption toward an authentic existence and adaptation to high levels of environmental stimulation in space (Wohlwill 1974). Over time, classification and declassification of records and contexts will eventually shift our perception of the mission, not as an illusion or folly of early human spaceflight, but as the miracle determination of life. An evolutionary long view may support a novel understanding of our species as transient bodies with masses of microbial ecologies as part of a system of technics having essential and distinct dynamics, infinitely connected with the primordial flow of the expanding universe, and requiring critical care (Figs. 13.5 and 13.6).

Figure 13.5 We are all Explorer Fish' artist Sarah Jane Pell, cinematographer Shaun Wilson, film still, 2016

Figure 13.6 We are all Explorer Fish' artist Sarah Jane Pell, cinematographer Shaun Wilson, film still, 2016

COMBINED REFERENCE LIST FOR CHAPTER 13

D. Adams, *The Hitchhiker's Guide to the Galaxy: A Trilogy in Five Parts* (William Heinemann, London, 1979)

R. Armstrong, What is Sci-Fi aesthetics? in *Sci-Fi Aesthetics (Art & Design Profile 56)*, ed. by R. Armstrong (Academy Group, London, 1997), pp. 2–5

R. Armstrong, *The Gray's Anatomy* (Serpents Tail, London, 2001)

R. Armstrong, *Space Architecture* (Wiley, New York, 2002)

I.I. Asimov, *Robot* (Harper Voyager for The Times, London, 2010)

M. Balfour, *Theatre and War, 1933–1945: Performance in Extremis* (Berghahn Books, New York, 2001)

C. Beam, Have astronauts ever had sex in space? *Slate* (n.d.), www.slate.com/articles/news_and_politics/explainer/2007/02/do_astronauts_have_sex.html. Accessed 15 Aug 2015

M.S. Beauchamp, K.E. Lee, J.V. Haxby, A. Martin, fMRI responses to video and point-light displays of moving humans and manipulable objects. J. Cogn. Neurosci. **15**(7), 991–1001 (2003)

L. Bec, We are extremophiles. Mutamorphosis conference (2007) www.mutamorphosis.org/index.php?lang=en&node=120&catid=108&id=92. Accessed 22 Oct 2015

G.A. Bekey, *Autonomous Robots* (MIT Press, Cambridge, MA, 2005)

C. Bergin, No sex please — we're astronauts. NASASpaceFlight.com (2005), October 23, www.nasaspaceflight.com/2005/10/no-sex-please-were-astronauts/. Accessed 15 Aug 2015

P. Bourke, Immersion: the challenge for commodity gaming. *Computer Games, Multimedia and Allied Technology CGAT1* (2012)

A. Boyle, Outer-space sex carries complications. NBC News (n.d.), www.nbcnews.com/id/14002908/print/1/displaymode/1098/. Accessed 15 Aug 2015

R.A. Brooks, *Robot* (Penguin, London, 2002)

M. Chiappalone, A. Vato, L. Berdondini, M. Koudelka-Hep, S. Martinoia, Network dynamics and synchronous activity in cultured cortical neurons. *Int. J. Neural Syst.* **17**(2), 87–103 (2007)

A.C. Clarke, *Rendezvous with Rama* (Bantam, New York, 1973)

M.E. Clynes, N.S. Kline, Cyborgs and space. *Astronautics* **14**(9), 26–76 (1960)

S.J. Crawford-Young, Effects of microgravity on cell cytoskeleton and embryogenesis. *Int. J. Develop. Biol.* **50**, 183–191 (2006)

B. Crowell, Biological effects of weightlessness. Lectures on physics in 2010–11 (2003), www.vias.org/physics/bk1_05_07.html. Accessed 20 Mar 2016

Cryonics Institute, Cryonics FAQ: answers to your questions (n.d.), www.cryonics.org/. Accessed 15 Aug 2015

I. Daly, S.J. Nasuto, K. Warwick, Single tap identification for fast BCI control. *Cogn. Neurodyn.* **5**(1), 21–30 (2011)

C. de Lima, Trans-meaning: dance as an embodied technology of perception. *J. Dance Somat. Pract.* **5**(1), 17–30 (2013)

T.B. DeMarse, D.A. Wagenaar, A.W. Blau, S.M. Potter, The neurally controlled Animat: biological brains acting with simulated bodies. *Autonomous Robots* **11**(3), 305–310 (2001)

M. Dery, *Escape Velocity: Cyberculture at the End of the Century* (Grove, New York, 1996)

J.P. Donoghue, A. Nurmikko, G. Friehs, M. Black, Development of neuromotor prostheses for humans. Suppl. Clin. Neurophysiol. **57**, 588–602 (2004)

K. Dubois, Dance and weightlessness: dancers' training and adaptation problems in microgravity. *Leonardo* **27**, 57–64 (1994)

ESA, The ESA seeks candidates for simulated 'Missions to Mars' in 2008/2009 (2007), June 19, www.esa.int/Our_Activities/Human_Spaceflight/ESA_seeks_candidates_for_simulated_Missions_to_Mars_in_2008_2009. Accessed 15 Aug 2015

Fox News, Sex in space: soon if not already. (2008), July 8, www.foxnews.com/story/2008/07/08/sex-in-space-soon-if-not-already.html. Accessed 15 Aug 2015

R.J. Garan Jr., J.A. Hoffman, The overview effect: Freethink@Harvard (2013), http://vimeo.com/55073825. Filmed 22 November

M. Gasson, B. Hutt, I. Goodhew, P. Kyberd, K. Warwick, Invasive neural prosthesis for neural signal detection and nerve stimulation. Int J Adapt Contr Signal Process **19**(5), 365–375 (2005)

D. Grainger, Space nerds face the final frontier billionaire blastoff (2004), November 1, http://money.cnn.com/magazines/fortune/fortune_archive/2004/11/01/8189586/index.htm. Accessed 15 Aug 2015

R. Gray, Space hotel to offer guests 18 sunrises a day. *The Telegraph* (2007) August 12, www.telegraph.co.uk/news/science/science-news/3303113/Space-hotel-to-offer-guests-18-sunrises-a-day.html. Accessed 18 Aug 2015

H. Moravec, Pigs in cyberspace. (n.d.), www.primitivism.com/pigs.htm. Accessed 15 Aug 2015

J. Hameed, I. Harrison, M.N. Gasson, K. Warwick, A novel human–machine interface using subdermal magnetic implants, in *Proc. IEEE International Conference on Cybernetic Intelligent Systems*, pp. 106–110 (2010).

L.R. Hochberg, M.D. Serruya, G.M. Friehs, J.A. Mukand, M. Saleh, A.H. Caplan, A. Branner, D. Chen, R.D. Penn, J.P. Donoghue, Neuronal ensemble control of prosthetic devices by a human with tetraplegia. *Nature* **442**, 164–171 (2006)

K. Ijiri, Fish mating experiment in space: what it aimed at and how it was prepared. Biol. Sci. Space **9**(1), 3–16 (1995)

P. Kennedy, D. Andreasen, P. Ehirim, B. King, T. Kirby, H. Mao, M. Moore, Using human extra-cortical local field potentials to control a switch. *J. Neural Eng.* **1**(2), 72–77 (2004)

P.L. Kohler, *Dernière Mission: Mir, l'aventure humaine [The Last Mission: The Human Adventure of Mir]* (Calmann-Lávy, Paris, 2000)

R. Krone, *Beyond Earth: The Future of Humans in Space* (Apogee Books, Burlington, 2006)

N. Kumar, Brain computer interface. Cochin University of Science & Technology Report, Kochi (2008)

D. Kushner, Going to space? First stop: eight months of gruelling training in Russia's star city. *Wired* (2008), August 18, http://archive.wired.com/techbiz/people/magazine/16-09/ff_starcity?currentPage=all. Accessed 15 Aug 2015

S. Langston, S.J. Pell, What is in a name? Perceived identity, classification, philosophy, and implied duty of the "Astronaut". *Acta Astronautica* **115** (October–November), 185–194 (2015).

B. Latour, Agency at the time of the Anthropocene. New Lit. His. **45**(1), 1–18 (2014)

H. Lefevbre, *The Production of Space* (Anthropos, Oxford, 1974)

A. Leonov, V. Lebedev, *Space and Time Perception by the Cosmonaut* (Minerva Group, Denmark, 2001)

T. Marshburn, Acute care, in *Principles of Clinical Medicine for Space Flight*, ed. by M. Barratt, S.L. Pool (Springer, New York, 2008), p. 112

R. McKie, Men redundant? Now we don't need women either. *The Observer* (2002), February 10, www.guardian.co.uk/world/2002/feb/10/medicalscience.research. Accessed 15 Aug 2015

J. Millan, F. Renkens, J. Mourino, W. Gerstner, Non-invasive brain-actuated control of a mobile robot by human EEG. IEEE Trans Biomed Eng **51**(6), 1026–1033 (2004)

S.K. Moore, M.A. Gast, 21st century extravehicular activities: synergizing past and present training methods for future spacewalking success. *Acta Astronaut.* **67**(7), 739–752 (2010)

NASA, Why is space radiation an important concern for human spaceflight? National Aeronautics and Space Administration (n.d.), http://srag-nt.jsc.nasa.gov/spaceradiation/Why/Why.cfm. Accessed 15 Aug 2015

New Scientist, Out of the world sex could jeopardise missions (2005), October 21, www.newscientist.com/article/dn8195-out-of-this-world-sex-could-jeopardise-missions/. Accessed 15 Aug 2015

Z. Ney, C. Looper, S. Parazynski, Developing the infrastructure for exploration EVA training, in *Space* (2006) p. 7451

C.T. Nordhausen, E.M. Maynard, R.A. Normann, Single unit recording capabilities of a 100 microelectrode array. Brain Res. **726**(1–2), 129–140 (1996)

Z. Oksiuta, *Formy, Procesy, Konsekwencje\Forms Processes, Consequences, Galarie Arsenat, Bialystok* (Centrum Sztuki Wspotczesnej Zamek Ujazdowski, Warszawa, 2007a)

Z. Oksiuta, New biological habitats in the biosphere and in space. *Leonardo* **40**(2), 122 (2007b)

V.A. Olson, *American Extreme: An Ethnography of Astronautical Visions and Ecologies* (Outside Reader, Anthropology, Rice University, Houston, 2010)

R. Palaniappan, Two-stage biometric authentication method using thought activity brain waves. Int. J. Neural Syst. **18**(1), 59–66 (2008)

W. Peakin, William Peakin on sex in space. *The Guardian* (2000), November 19, www.theguardian.com/uk/2000/nov/19/theobserver.uknews. Accessed 15 Aug 2015

S.J. Pell, Aquabatics: a post-turbulent performance in water. *Perform Res* **19**(5), 98–108 (2014)

S.J. Pell, P. Nespoli, M. Mackay, EVA simulation training underwater with a remote mission "control", in *Proceedings 64th International Astronautical Congress, Beijing, ROC (Republic of China)* (International Astronautical Federation, IAC, 2013)

F. Pietronigro, Project no. 33: investigating the creative process in a microgravity environment. *Leonardo* **33** (3), 169–177 (2000).

M.M. Pinter, M. Murg, F. Alesch, B. Freundl, R.J. Helscher, H. Binder, Does deep brain stimulation of the nucleus ventralis intermedius affect postural control and locomotion in Parkinson's disease? Mov. Disord. **14**(6), 958–963 (1999)

G. Rainer, M. Augath, T. Trinath, N.K. Logothetis, Nonmonotonic noise tuning of BOLD fMRI signal to natural images in the visual cortex of the anesthetized monkey. Curr. Biol. **11**(11), 846–854 (2001)

J. Reason, Understanding adverse events: human factors. Qual. Health Care **4**(2), 80–89 (1995)

F. Redmill, J. Rajan, *Human Factors in Safety-Critical Systems* (Butterworth-Heinemann, Oxford, 1996)

G. Reynolds, Artificial wombs. *Popular Science* (2005), August 1, www.popsci.com/scitech/article/2005-08/artificial-wombs. Accessed 15 Aug 2015

A. Ripstein, In Extremis. Ohio State J. Crim. Law **2**, 415 (2004)

L. Rossini, D. Izzo, L. Summerer, Brain machine interfaces for space applications, in *Proc. IEEE International Conference on Engineering in Medicine and Biology* (IEEE, 2009), pp. 520–523

C. Sagan, *Pale Blue Dot: A Vision of the Human Future in Space* (Ballantine Books, New York, 2007)

R. Sato, The "Hawking Solution": will saving humanity require leaving Earth behind? *Daily Galaxy* (2007), May 9, www.dailygalaxy.com/my_weblog/2007/05/the_hawking_sol.html. Accessed 14 Aug 2015

M. Schneider, Mission to Mars requires NASA to mull death. Sex (2007), February 5, http://usatoday30.usatoday.com/tech/science/space/2007-05-01-mars-mission-issues_N.htm. Accessed 15 Aug 2015

J.D. Simeral, S.P. Kim, M.J. Black, J.P. Donoghue, L.R. Hochberg, Neural control of cursor trajectory and click by a human with tetraplegia 1000 days after implant of an intracortical microelectrode array. J. Neural Eng. **8**(2), 025–027 (2011)

J. Sladden, Space medicine in the United Kingdom. Student BMJ **12**, 437–480 (2004)

K. Tanaka, K. Matsunaga, H.O. Wang, Electroencephalogram-based control of an electric wheelchair. IEEE Trans. Robot. **21**(4), 762–766 (2005)

TED.com, A roadmap to end ageing (2005), July, www.ted.com/talks/aubrey_de_grey_says_we_can_avoid_aging?language=en. Accessed 15 Aug 2015

The Washington Post, NASA's Griffin: "Humans Will Colonize the Solar System" (2005), September 25, www.washingtonpost.com/wp-dyn/content/article/2005/09/23/AR2005092301691.html. Accessed 14 Aug 2015

L.J. Trejo, R. Rosipal, B. Matthews, Brain computer interfaces for 1-d and 2-d cursor control: designs using volitional control of the EEG spectrum or steady-state visual evoked potentials. IEEE Trans. Neural Syst. Rehabil. Eng. **14**(2), 225–229 (2006)

N. Triscott, R. la Frenais, *Zero Gravity: A Cultural User's Guide* (The Arts Catalyst, London, 2005)

S. Wakayama, Y. Kawahara, C. Li, K. Yamagata, L. Yuge, T. Wakayama, Detrimental effects of microgravity on mouse preimplantation development in vitro. PLoS One **4**(8), e6753 (2009)

K.I. Warwick, *Cyborg* (University of Illinois Press, Champaign, 2004)

K. Warwick, The promise and threat of modern cybernetics. South. Med. J. **100**(1), 112–115 (2007)

K. Warwick, Implications and consequences of robots with biological brains. Ethics Inform Technol **12**(3), 223–234 (2010)

K. Warwick, M. Gasson, B. Hutt, I. Goodhew, P. Kyberd, B. Andrews, P. Teddy, A. Shad, The application of implant technology for cybernetic systems. Arch. Neurol. **60**(10), 1369–1373 (2003)

K. Warwick, M. Gasson, Practical interface experiments with implant technology, in Computer Vision in Human-Computer Interaction, Lecture Notes in Computer Science, vol. 3058, (2004), pp. 7–16

K. Warwick, M. Gasson, B. Hutt, I. Goodhew, P. Kyberd, H. Schulzrinne, X. Wu, Thought communication and control: a first step using radiotelegraphy. IEEE Proc Commun **151**(3), 185–189 (2004b)

K. Warwick, S. Nasuto, V. Becerra, B. Whalley, Experiments with an in-vitro robot brain, in *Instinctive Computing* ed. by Y. Cai. Lecture Notes in Artificial Intelligence, vol. 5987, (2010), pp. 1–15

M. Williams, The "Potsdam Gravity Potato" shows variations in Earth's gravity (2014), November 29, www.universetoday.com/116801/the-potsdam-gravity-potato-shows--earths-gravity-variations/. Accessed 15 Aug 2015

J.F. Wohlwill, Human adaptation to levels of environmental stimulation. Hum. Ecol. **2**(2), 127–147 (1974)

D. Wu, K. Warwick, Z. Ma, J.G. Burgess, S. Pan, T.Z. Aziz, Prediction of Parkinson's disease tremor onset using radial basis function neural networks. Expert Syst Appl **37**(4), 2923–2928 (2010a)

D. Wu, K. Warwick, Z. Ma, M.N. Gasson, J.G. Burgess, S. Pan, T.Z. Aziz, Prediction of Parkinson's disease tremor onset using a radial basis function neural network based on particle swarm optimization. Int. J. Neural Syst. **20**(2), 109–116 (2010b)

X. Xia, H. Zhu, Y. Hu, J. Guo, X. Guo, Effect of gravity on sex hormones. *Science*, Human Performance in Space: Advancing Astronautics Research in China, special issue sponsored by the National Key Laboratory of Human Factors Engineering, Produced by the *Science*/AAAS Custom Publishing Office, (2014), p. 9, www.sciencemag.org/site/products/collectionbooks/HFE_booklet_lowres_12sep14.pdf. Accessed 15 Aug 2015

14

Connecting with the divine and the sacred, and becoming cosmically conscious

Steve Fuller, Roberto Chiotti and Krists Ernstsons

Three individual authors Fuller, Chiotti, and Ernstsons each consider unique belief systems and modes of self-understanding that locate us as present and future beings within the cosmos.

14.1 Humanity's lift-off into space: prolegomena to a cosmic transhumanism

Steve Fuller, Auguste Comte Professor of Social Epistemology in the Department of Sociology at the University of Warwick.

14.1.1 What is transhumanism?

In the most general terms, "transhumanism" says that the indefinite projection of those qualities that most clearly distinguish humans from other natural beings is worth pursuing as a value in its own right—even if that means radically altering our material nature

---

Steve Fuller (✉)
Auguste Comte Chair in Social Epistemology, Department of Sociology,
University of Warwick, Coventry CV4 7AL, United Kingdom
e-mail: s.w.fuller@warwick.ac.uk

Roberto Chiotti (✉)
Assistant Professor, Ontario College of Art and Design University,
100 McCaul Street, Toronto, Ontario, Canada, M5T 1W1
e-mail: rchiotti@faculty.ocadu.ca

Krists Ernstsons (✉)
Architectural Assistant, Hawkins/Brown, 159 St John Street, EC1V 4QJ, United Kingdom
e-mail: k.ernstsons@gmail.com

(More and Vita-More 2013). This rather open definition of transhumanism makes it clear who might *oppose* such a movement, not least those—often of a "green" persuasion—who believe that humanity's current global crises stem from our attempts to minimize, if not deny, our commonality with the rest of nature. In this respect, *transhumanism* needs to be distinguished from *posthumanism*, which aims to de-center the human as the locus of value altogether, thereby rendering it friendlier to green concerns (Fuller 2012). Whereas posthumanism may be seen in the broad sweep of Western intellectual history as *Counter-Enlightenment*, transhumanism is better seen as *Ultra-Enlightenment*: The former sees the Enlightenment as having gone too far, the latter not far enough.

The word "transhumanism" was coined in the 1950s by Julian Huxley (1957), a founder of the dominant paradigm in biology today, the neo-Darwinian synthesis, which integrates Darwin's account of natural history with the experimental principles of modern laboratory-based genetics. Huxley, following the lead of his grandfather, Thomas Henry Huxley, accepted that Darwin fundamentally challenged anyone who wanted to uphold the superiority of *Homo sapiens* as a species, given that natural selection implies that all forms of life are limited by their largely innate capacities to adapt to a changing environment. In the end, any given species—including humans—should expect extinction, not immortality. From that standpoint, all the promises made by Christianity and Islam of an eternal "afterlife" looked empty. Nevertheless, the Huxleys believed that there was something fundamentally correct about these religious intuitions—something that Julian thought could be addressed by transhumanism. Whereas his grandfather held that advances in law, medicine, and engineering served to push back, if not reverse, the default tendencies of natural selection, Julian Huxley argued that *Homo sapiens* is the only species equipped to comprehend the entire evolutionary process, in which case we incur a unique moral obligation to administer and direct its future course (Fuller and Lipinska 2014, Chapter 3).

In terms of religious precedents for transhumanism, two of the oldest Christian heresies—both already opposed by St. Augustine in the fifth century AD—stand out for their persistent and countervailing visions of the transhumanist utopia: *Pelagianism* and *Arianism*, each named for their originators, Pelagius (a Celtic lawyer) and Arius (a Libyan bishop). Both Pelagians and Arians believe that it is within the power of humans to achieve godhood, understood as a recovery from humanity's fall from divine grace recounted in Genesis. This shared belief is based on the heterodox Christian idea that the death of Jesus effectively cancelled Adam's sin, putting humans back on course to become embodied deities, modeled on the person of Jesus (Fuller and Lipinska 2014, Chapter 2). Pelagians and Arians differed over what the end state of a fully deified humanity would look like. In a nutshell, Pelagians imagined a "heaven on Earth," whereas Arians imagined an "Earth in heaven." These alternative visions have resonated with the process of modernization. Thus, the technological transformation of the life-world to maximize human convenience is a Pelagian project, just as the scientific aspiration for a maximally comprehensive theory of reality (aka "entering the mind of God" or the "view from nowhere") is an Arian project. Whereas the Pelagian aims to reduce the time it takes to realize the human will, the Arian aims to expand indefinitely the scope of humanity's intellectual horizons. In the modern era, the two movements worked in tandem. The seventeenth-century scientific revolution in Europe marked the triumph of the Arian vision, on the basis of which the eighteenth-century Industrial Revolution began to make the Pelagian vision a reality (cf. Passmore 1970, Chapters 5–7).

Latter-day descendants in the transhumanist movement who bring this contrast into high relief are, on the Pelagian side, Aubrey de Grey's vision of indefinite longevity for *Homo sapiens* through various biologically based enhancements and, on the Arian side, Ray Kurzweil's vision of *Homo sapiens* evacuated from its carbon-based platform to a supercomputer capable of colonizing the universe with its ever-expanding consciousness. In what follows, the Pelagian vision is captured by the ecomodernist project, while the Arian vision is captured by the star ark project of Icarus Interstellar.

One reason why Arianism and Pelagianism are not normally factored into the prehistory of transhumanism is that their twentieth-century secular heirs morphed into cultural projects that sympathetically tracked that period's two main totalitarian ideologies, Communism and Fascism. These projects were, respectively, Russian *Cosmism* and Italian *Futurism*. Both movements attracted those who were frustrated by the short-termist and indecisive character of liberal democracy—modernity's original political innovation—in a rapidly changing world which lacked the normative focus that in the past had been provided by religious authority. Nevertheless, both Cosmism and Futurism have left an indelible impression in contemporary transhumanism.

Although the Soviet Union advertised itself as a staunchly atheistic regime which removed the Russian Orthodox Church of all official standing, it managed to nurture a materialistic version of an ideology that had been promoted in the nineteenth century by the Orthodox philosopher Nikolai Fedorov, who had argued on theological grounds that science should aim to make all humans—both living and dead—immortal by acquiring control of the mechanisms of heredity (Young 2012, Chapter 6). This idea, which was presented as a solution to our fallen state (courtesy of Adam's sin), was advanced in Federov's posthumous work, *The Philosophy of the Common Task* (1904). Far from idiosyncratic, this idea was presaged in the Orthodox doctrine of *theosis*, which interprets the "transfiguration" of Jesus (i.e. the moment he realizes his divine nature) as something available to all humans. It was picked up by one of Julian Huxley's fellow-travelers, Theodosius Dobzhansky (1967), a Ukrainian Orthodox geneticist who spent most of his career at Columbia University. The two were instrumental in supporting English translations of the work of the heretical Jesuit paleontologist, Pierre Teilhard de Chardin (1961), who proposed the idea of the "noosphere," a kind of cosmic consciousness that was an emergent effect of mass communications and the sheer physical dominance of humanity on Earth.

Teilhard, in turn, had been influenced by another "Cosmist," the geochemist Vladimir Vernadsky, the main early Soviet booster of nuclear energy, who is now seen as an anticipator of the current idea of the "Anthropocene" (Guillaume 2014). But the Cosmist who left the strongest imprint on Soviet thinking was the astrophysicist Konstantin Tsiolkovsky, an inspiration for Sputnik, who presented interstellar travel as breaching the final frontier of humanity's quest for cosmic consciousness (Young 2012, Chapter 9). Today, such ideas are most explicitly pursued by trans-humanists who believe that an expansion of our cosmic horizons is not merely desirable, but required for humanity's survival, should it turn out that one or more global catastrophes prove that our stewardship of Earth has been a failed experiment. The star ark is clearly a latter-day development from this line of thought—and its relationship to the Anthropocene will be discussed below.

In contrast to the cosmists, who were mostly scientists, the Futurists were artists practicing in different, often multiple, media who believed that rapid advances in science and technology—especially in terms of communication and transportation—were hastening a revolution in human consciousness that would empower humanity to unprecedented levels. They were fixated on the speed of production enabled by advanced industrial technology, including such emerging consumer-oriented products as automobiles, airplanes, telephones, and radios, all of which allowed more to be done in less time. Indeed, the Futurists were among the first to recognize the distinctive role that science plays in ensuring the continual improvement of existing technologies. In effect, science permits a speedier understanding of how increases in speed are made possible. Although the reputation of the original Futurist theorist, Filippo Marinetti, has been tarnished by his association with Mussolini, his ideas have been updated and extended in the name of "transhumanism" by a Milanese lawyer who writes under the pen name "Stefano Vaj" (2005). Vaj argues that "smart regulation" of rapidly advancing technologies requires that they be allowed to advance more rapidly via a two-pronged strategy: on the one hand, by removing legal obstacles to such innovations being brought to market; on the other hand, by incentivizing public- and private-sector uptake of those innovations.

As just suggested, a commitment to "accelerationism" distinguishes Futurism from the plethora of early-twentieth-century "modernist" ideologies. Moreover, its legacy can be felt in both contemporary posthumanism and transhumanism. The most extreme posthumanist version is what Shanghai-based British philosopher Nick Land (2013) has called the "Dark Enlightenment," which involves a desire to hasten what he takes to be inevitable, namely the self-destruction of "humanity," understood as a hegemonic species life-project imposed on the planet, which, over the centuries, has repressed the natural order of things. However, the natural order will return—with humanity in a clearly subordinate position—after what Land envisages as an ecological apocalypse. Moreover, Land's "reactionary modernist" vision of the future—one involving a sense of "natural law" based on E.O. Wilson rather than Thomas Aquinas—is shared even by those who would wish to prevent it (e.g. Oreskes and Conway 2014). In contrast, accelerationism is most conspicuous in contemporary transhumanism via Ray Kurzweil's (2005) invocation of "Moore's law," which is a historically based principle of exponential improvement in computational power. Like the Futurists, Kurzweil argues that the main reason people think the convergence of human and machine consciousness (aka the "singularity") lies in the distant future is that we fail to realize that progress is happening at an ever more rapid pace, which means that the speed with which change happened in the past is a poor indicator of how it will happen in the future. This provides a quantitative basis for thinking that a qualitative change in the human condition is just around the corner—one that Kurzweil and other transhumanists see in largely positive terms.

14.1.2 Which way is up for the human condition?

I have argued for a 90-degree rotation of the ideological axis, from *left–right* to *up–down*, implying that the axis defined by the latter polarity bisects the axis defined by the former one (Fuller and Lipinska 2014, Chapter 1). To be sure, left and right have been always unstable ideological positions. The left has consisted of top-down technocrats and bottom-up community activists, while the right has consisted of radical libertarians and religious traditionalists. In retrospect, it is amazing that the two halves of the "left" and the "right" have cohered as well as they have over the past two centuries. However, an exemplar of the new ideological order suggested by the 90-degree rotation is the Roman Catholic Pope Francis I, a "down-winger" in my sense, who combines a communitarian concern for the poor with a traditionalist understanding of humanity's theological rootedness in the natural world. The "radical Orthodox" Anglican theologian John Milbank (1990) would count as

Humanity's lift-off into space: prolegomena to a cosmic transhumanism

another such down-winger. A third would be the Australian "public ethicist" Clive Hamilton (2010), who has acquired international notice for his openly moralistic critiques of humanity's role in global climate change—and to whom I will refer in the rest of this piece.

Hamilton has recently made a frontal assault on a signature "up-winger" project, popularly known as the "star ark," the brainchild of Icarus Interstellar, an Anglo-American non-profit organization dedicated to starship research and development (Hamilton 2015b). Due to be built by 2100, the ark would carry a few thousand humans, equipped with other living species, the DNA of many of the rest, and a renewable energy source which would permit the ark, at least in principle, to sail indefinitely into space without ever having to return to Earth. It would literally be a floating ecosystem, which would testify to humanity's claim to the entire cosmos as its home. As Hamilton rightly observes, the star ark project has received support from the two biggest American public funders of cutting-edge research—NASA and DARPA (Defense Advanced Research Projects Agency)—very much in the spirit of "necessity is the mother of invention." In other words, if the worst predictions concerning global climate change turn out to be true, humanity might just need such a space-bound vehicle to preserve its civilization. But, even if those predictions turn out to be false, the project will have still made a deep metaphysical point about humanity's capacities to survive and flourish.

Hamilton's skepticism about human resilience runs quite deep. I encountered him as the most articulate and fearsome opponent of "ecomodernism" at a conference where the ecomodernists launched their manifesto (Breakthrough Institute 2015a). Ecomodernism is the brainchild of the Breakthrough Institute, a San Francisco Bay Area think tank which is unashamedly "progressive" in its attitude toward the natural environment—which is to say, "anthropocentric." Thus, even the value attached to "wild" nature or the "rewilding" of nature—including perhaps by resurrecting extinct species—is mainly to do with the spiritual and intellectual benefits that nature affords to humans. To be sure, this is the attitude that had inspired the original "conservation" movement, which was as an outgrowth of the US Progressive movement in the early twentieth century. In this spirit, the conference where Hamilton and I spoke was entitled "The Good Anthropocene," a challenge to the presumptive meaning of the neologism "Anthropocene"—which refers to humanity's overriding footprint on the biosphere since the Industrial Revolution, normally for the *worse* (Breakthrough Institute 2015b).

Hamilton would have none of this optimism concerning humanity's ability to get itself out of problems that may well be largely of its own creation. In particular, he believes that we underestimate the amount of suffering that we have already caused to the planet—and are likely to cause by more technologically extravagant proposals such as rewilding and geoengineering—as well as the star ark. What is puzzling about Hamilton's position is not its moral clarity, its deep concern for nature as an end in itself, or even its sympathy for the fallen character of the human condition. Rather, it is its provenance—where does it come from, philosophically or theologically speaking (cf. Hamilton 2015a)? I shall address this question more explicitly in the next section. But, this much can be admitted at the outset. While a Hamilton-like view might be upheld by a pure Darwinist who evaluates even our most compelling visions in terms of their adaptive value to our Earthly existence, Darwinists are not given to Hamilton's style of moralizing. More to the point, "humanity" is not a Darwinian concept. It is both *pre-* and *post*-Darwinian: it refers both to a time at which we humans were blissfully ignorant of our biological nature and to a time at which we are intimately knowledgeable and can thus take our biology into our own hands, for better or worse. The phrase "for better or worse" is crucial, as it points to the element of *faith* involved in sticking to a meaningful conception of humanity, despite the radical transformations that the "Anthropocene" has wrought to both humanity's material conditions and our self-understanding as human beings.

Faith may be understood as a creative response to radical uncertainty. One side of faith is humility, the recognition of likelihood of error, which, in today's terms, is interpreted as "precautionary." This is a common trope in many of the world's religions. It is also very much in the forefront of Hamilton's thinking, typically as a charge of hubris lodged against those who believe that humans can "simply" blast off from Earth if the going gets too tough. However, crucial to distinctly Christian conceptions of faith has been the idea of self-empowerment. This requires taking our deep ignorance as an opportunity for exploration and taking risks — but why?

Here, the idea of our having been created "in the image and likeness of God" plays an overriding role (Fuller 2011, Chapter 5). However much day-to-day empirical realities remind us of our Earth-bound nature, we are, nevertheless, more than just that. The question, then, becomes how to give that "transcendental" aspect of our being its proper due: is it just something that we release on special occasions, such as a church service, or is it integral to our ordinary being in the world, propelling us to realize our godlike potential? Initially, Catholics and Protestants split over this matter. In particular, the Protestant mentality, which fueled the scientific revolution, was about humanity's fallible exercise of its godlike potential, not least through the experimental method — which extended beyond the realm of science to that of politics. In other words, the same spirit that informed Francis Bacon's call for national laboratories in the early seventeenth century also motivated contemporaneous efforts to find new lands — notably in the Americas — to try out new schemes of governance. In this context, "nature," understood as a safe haven for animal beings, inhibited humanity's realization of its godlike potential (Noble 1997).

A now much ridiculed saying, normally attributed to Benjamin Franklin, captures this sensibility: "God helps those who help themselves." No, this does not mean that God favors only the winners in life. After all, "help" does not mean "save." Rather, those who are saved will have demonstrated a godlike capacity in their being: they will have taken the major existential risks, the greatest of which is to envisage that our humanity might survive and even flourish in a radically different material form. This may involve a different country or a different planet — or even a different mode of embodiment. The challenge is to remain "human" in these various transcendences of our animal nature. But, there is no guarantee of success under what transhumanists call conditions of "morphological freedom" (Bostrom 2005).

For Christians, humanity's fallen nature is twofold: we are much less powerful and much less knowledgeable than God. The only chance we have to recover from that dual loss is by trying to approximate God in our own way, fully realizing that both our efforts and our feedback from those efforts are bound to be imperfect. In this context, faith is expressed as a belief in Providence, namely that God will always provide what we need to know to improve our position — but the trick is for us to figure what that is (Passmore 1970, Chapter 9). A secular display of this attitude routinely occurs in philosophy of science discussions of how to move forward from the falsification of a hypothesis. Interestingly, stopping the line of inquiry altogether is not an option. Indeed, those who continue to believe in scientific progress — in the face of the massive error and harm committed in the name of science in the modern era — are the truest of true believers in Providence. As one of their number, I find it unfortunate that these people of faith refuse to recognize the theological dimension of their thought.

Humanity's lift-off into space: prolegomena to a cosmic transhumanism

At the ecomodernist summit, Clive Hamilton rightly portrayed his opponents as believers in Providence, but mischaracterized it as a complacent faith, an effective offloading of human responsibility to a deity who is little more than a projection of wishful thinking. While this may have been Voltaire's caricature of the idea in *Candide* (1759), it does not do justice to the spirit of those who have staked their lives on the idea — especially given how many have succeeded, despite the odds stacked against them. The star ark is in that tradition of metaphysical optimism that extends from the original dissenting Christians who sparked the scientific revolution and settled the Americas — only now taken to a much higher level. There is no stronger test of our faith as beings created in the image and likeness of God.

14.1.3 Who's afraid of the human? a theological defense of ecomodernism

Whether one expresses things in Greek or Latin seems to make an enormous difference — at least to English-speakers — when it comes to discussing who we are. Only brave souls distance themselves from "humanity" or even "humanism." These Latin coinages are normally seen in a positive light. This is why postmodern thinkers such as Michel Foucault and Jacques Derrida were at first treated to such fear and loathing when they were seen as "anti-humanist." On the other hand, people seem more than happy to distance themselves from "anthropocentrism" and "anthropomorphism," Greek coinages. The same applies to the more recent coinage, "Anthropocene" — and certainly to Clive Hamilton's pessimistic reading of the idea. But what hangs in the balance is how the two great ancient Mediterranean languages refer to who we are.

The implied difference is metaphysical. The Latin coinages refer to features of a being that qualify it as human, whereas the Greek coinages refer to the projection of generically human features on beings that might not be otherwise seen as human. Thus, the two languages address, one might say, the *depth* and the *breadth* of "being human," respectively. This is not just an interesting philological point, but one of profound political significance.

On the one hand, we might say that, until we figure out what it means to be "human" (especially whether it is a good thing), we should restrict our usage of "human." On the other hand, we might say that only by extending our implicit sense of humanity to others can we make explicit to ourselves what it really means to be human. In policy terms, the former adopts a *precautionary* and the latter a *proactionary* attitude toward risk — specifically, the risk involved in a pro-human agenda (Fuller and Lipinska 2014). So, while I do not dispute the evidence for anthropogenic climate change, Hamilton and I disagree over the spirit in which those facts should be taken. What he sees as a threat, if not an outright cosmic condemnation, I see as a challenge or perhaps even an opportunity.

Much depends on whether humanity is seen as partners with nature in some larger divine "project." (I put "project" in scare quotes, because what God has always already done from the standpoint of eternity is something that we as fallen creatures might realize only gradually in the fullness of time — which is to say, as a "project.") If so, then the slide from "humanist" to "anthropic" is justified. I shall return to this point below but, for now, the relevant point is that Hamilton denies any such slide. He sees us clearly as a subordinate, albeit morally interesting part of nature. In him travels the spirit of the more enlightened Hellenistic thinkers whom St. Paul *failed* to convert to Christianity. More to the point, Hamilton's position has no basis in Christian theology, where humanity's creation *imago dei* is non-negotiable. To be sure, humans find themselves in a nature that is equally created by God—but *not* in his image. Moreover, our fallen state continues to be demonstrated by our failure to distance ourselves from nature in how we conduct our lives.

The current environmental crisis may be seen as the latest version of humanity's fallen state. But, its resolution need not mean a return to nature in the sense of a scaled-back "reabsorption" or "restoration" to some imagined state of cosmic equilibrium. Rather, it means pushing nature to a level closer to what we hypothesize as a divine state of being. Again, I do not wish to underestimate the risk involved, but risk is precisely what we are about as human beings. "Ye shall know them by their bets" might be a good contemporary way of putting the matter, which assumes that people know enough about the bets they are making on the future to take responsibility for them, even if they turn out to have been in error. Put another way, it is exactly by taking bold calculated risks that we acknowledge that we are not fully in control of our fates, and hence pay homage to God. The early modern Christian idea of Divine Providence, previously discussed, captures this sensibility. In the past three or four centuries, it has been popularized in various "self-help" ways as "you can't succeed if you don't try" and, more brutally, "no pain, no gain." In contrast, creatures who fail to embrace risk as a self-defining concept are never in a position to recognize—let alone take the full measure of—a force beyond their control; hence, we speak of their responses as "programmed": they stay within their biologically prescribed limits. They have no sense of the transcendent.

As a metaphysical backdrop to this pro-human discussion, one needs to imagine that God is the source of *logos*: that is, "logic" in an extended sense that includes both computer and genetic codes, as well as physical laws. It follows that one can divide God's creatures into those capable of willing the *logos* for themselves (aka "human") and those capable of no more than conforming to the *logos* (aka "nature"). In that case, the "capable" must incorporate the "incapable" for humanity's divine entitlement to be fully redeemed. From this standpoint, the radical "otherness" of nature—whereby it is seen as a threat to the human way of being—is no more than a dramatic reminder of our fallen yet corrigible nature.

14.1.4 Ecomodernism as the politics of theological energeticism

Ecomodernism has a distinctive way of gagging our *alienation* from nature. *Contra* Hamilton, who associates alienation with excessive *detachment*, ecomodernists associate alienation with excessive *dependency*—specifically, humanity's reliance on biomass as an energy source. As long as we need to fell trees to make and heat our homes or enslave and kill animals for transport, food, and clothing, we have been unable to accord nature the respect it deserves—and, in the process, we display our own fallen nature. Such an "us versus them" attitude toward nature speaks to humanity's inability to rise above nature's own default mode of being. Thus, we engage in zero-sum struggles for resources. The sort of scarcity-based economics first advanced by Reverend Thomas Malthus and later generalized by Charles Darwin into a theory of natural selection has been based on a Hamilton-like pessimism vis-à-vis our prospects for overcoming this predicament. At most, we will be the supreme predatory animal for a brief moment in cosmic history. Ecomodernism speaks unequivocally against this entire line of thought, and in so doing aims to reclaim our divine entitlement.

At this point, it is worth considering a precedent for this line of thinking, which goes much of the way—but not entirely—to contemporary ecomodernism. The mid-twentieth-century

US anthropologist Leslie White (1959) proposed an energy-driven view of cultural evolution, which is encapsulated in what is dubbed "White's law"—namely "$C=ET$," which measures cultural development (C) as a function of the amount of energy consumed per capita (E) and the technical efficiency of the energy so consumed (T). Like contemporary ecomodernists, White envisaged a long-term decline in our dependency on biomass-based energy sources, which, in turn, reflected an increasingly abstract understanding of energy as "fuel"—rather than as the concrete work performed by particular humans or animals. Also like the ecomodernists, White concluded his world-historic energy narrative with the advent of nuclear energy. However, from the ecomodernist standpoint, the missing piece of White's worldview was the idea that increasingly energy-dense fuels create the opportunity for us not to be reliant on nature for energy altogether, which, in turn, allows us to think more deeply about what it might mean to grant nature a sense of autonomy previously reserved for humans.

If history teaches one overarching lesson, it is that humans have gradually but clearly "decoupled" their own survival needs from those of the rest of nature. In other words, we have discovered—and, increasingly, invented—"energy-dense" resources. Such resources typically involve significant human intervention, if not outright creation, yet the overall result is to leave more of nature unexploited. In the twentieth century, artificial fertilizers and nuclear energy stand out in this regard. However, these achievements are often not fully appreciated, because they have allowed very many more people to flourish, even though the burden that each person places on nature is less than before these innovations, when many fewer people inhabited the planet.

The long-term tendency to "energy density" can be understood theologically as humanity's attempt to approximate the Abrahamic deity's creation of everything out of nothing. Put another way: in our most godlike state, our survival would be based on resources entirely of our own creation—"pure chemistry," as might have been said a century ago. This divine benchmark has never been far from the minds of scientists or the scientific public. Indeed, the first advertising campaigns for artificial fertilizers, which involved the chemical harnessing of atmospheric nitrogen, spoke of "food from the skies," a secularized allusion to "manna from heaven." Around the same time (the 1920s), divine creation itself was assimilated into secular science as the "big bang," courtesy of the Belgian Jesuit cosmologist, Georges Lemaître.

At one level, comparing even a world fed by artificial fertilizers and powered by nuclear energy with *creatio ex nihilo* looks crass and sacrilegious. But, at another level, the sort of reasoning and processes associated with the pursuit of energy density have provided the material conditions for human beings first to treat each other, then animals, and finally nature as a whole with dignity—as less biomass-based energy is required to achieve comparable results. In this respect, efficiency can be a great emancipator, not only of those who now can do more with less, but also of those whose lives no longer need to be exploited and, hence, may join those allowed to do more with less.

Implied here is a four-stage moral journey to a more comprehensive sense of "humanity" that parallels that of increasing energy density: (i) our energy needs cause the death and suffering of other natural creatures (e.g. biomass-based energy); (ii) our energy needs take advantage of the death and suffering of other natural creatures (e.g. fossil fuels); (iii) our energy needs track and imitate the life patterns of other natural creatures without harming them; and (iv) our energy needs enable us to give something back to enhance the existence of other natural creatures.

Nevertheless, even someone like myself, who endorses ecomodernism, realizes that it faces two interesting problems that are largely of its own creation. The first is "easier" in the sense that it is more readily recognized and dealt with, albeit perhaps too superficially for critics. The second is "harder" in the sense of questioning whether the "anthropic" agenda can be pursued indefinitely.

The first problem is that humanity might become victim to the success of an ecomodernist agenda, as too many of us inhabit the planet for too long. Against this gloomy forecast is that lower infant mortality rates typically mean fewer offspring. Indeed, thanks to improvements in healthcare, education, nutrition, and hygiene, global birth rates have declined over the past 40 years. But that only takes care of the material problem, not the spiritual one. Ruptures in the history of human thought and action have often required new generations unburdened by the experience of their elders. Will such independently minded cohorts exist in the future, especially as we develop biomedical resources to enable, say, today's adults to pursue healthy lives indefinitely? The sorts of aspirations that in the past might have been invested in children—due to one's own mortality—might eventually disappear.

Of course, there are ways around this prospect of radical zero population growth. One is simply to create a social and legal environment friendlier to rational suicide. The other is to provide incentives for healthy older people to leave Earth to their young successors, while they navigate the cosmos, spreading their wisdom to all and sundry.

At this point, we arrive at an important motivation for the star ark, which is addressed by the second problem—namely how it might be organized and governed. As in Noah's Biblical ark, the star ark promises to carry some representative sample of humans and other creatures—albeit not for 40 days on Earth, but for an indefinite extent of time and space. Is the ark, then, supposed to preserve as much of our current ecosystem as possible, or be a literal floating laboratory of organic transformation? The latter seems scientifically more plausible—and existentially more interesting, as the space travelers are continually challenged to redefine the distinctly "human" in what would become an ever more artificial version of the "human condition."

14.1.5 Expanding the scope of the human in search of the divine

The task ahead may be formidable, but hardly insurmountable. Indeed, arguably we have always been engaged in it, even on Earth. After all, "humanity" is a normative category that has been extended only with great difficulty to all members of *Homo sapiens*—and, arguably, that task has yet to reach completion. And each time a new class, gender, race, or religion has been incorporated, the meaning of "human" has changed, typically by shifting the default position of what counts as "normal." (And along the way there were those who regarded these extensions as perversions.) Education and legislation were the original "enhancement" strategies for upright apes but, over time, we have developed more physically invasive and perhaps more direct ways of achieving many of the same effects—along with some new and unexpected ones, to be sure.

To appreciate the challenge ahead, let us return to the significance of the Greek "anthropic," which implies an extension of the human to the non-human. Consider two English words derived from this root, whose meanings are often conflated: *anthropocentric* and *anthropomorphic*. The former concerns the human as the conceptual framework in which knowledge is created, the latter the character of the knowledge that is so created.

It follows that science is an anthropocentric enterprise that aspires not to be anthropomorphic. However, politics (especially in the cosmic terms entertained here) is an anthropomorphic enterprise that aspires not to be anthropocentric.

Most of what follows concerns the latter, more difficult point. The former point has been captured well in the history of science in terms of nature's "intelligibility": the fact that nature is tractable to human intelligence—not only in the parts that we normally experience, but also in the vast majority of it that we do not experience. Our success in launching spacecraft based on imagined physical laws underwritten by exclusively Earth-bound evidence proves the point. But, of course, whatever is learned about the cosmos by such divinely inspired human intelligence need not itself be human in character, and hence not "anthropomorphic." This then opens the difficult political question of how to treat any creatures we discover that have an existence independent of our own. While this question is especially relevant to animals, it potentially extends throughout cosmic nature—not least to whatever hybrid life forms might be bred on the star ark.

The bottom-line question is this: what is the appropriate relationship between humanity and nature? Clearly, it should *not* be absolute submission, either of the natural to the human or, for that matter, of the human to the natural. Humans are not merely upright apes and nature is not just raw material. The right relation is a kind of integration. But, unlike certain fashionable ecofriendly versions of "theistic evolution," the human and the natural are not *equal* partners in the work of divine creation. Rather, the human is the superior partner by virtue of being the place where the divine is manifest in nature as a whole, as epitomized in Christianity by the person of Jesus. After all, God could have communicated via other aspects of our material being—not necessarily the specifically human. Moreover, this is no mere theological nicety, but a point that bears on how humans disclose their identity to potential intelligent beings elsewhere in the cosmos: what is the best medium through which to convey who we are?

It is here that ecomodernism reveals its theological orientation to be closer to intelligent design theory than theistic evolution. *Creatio ex nihilo*—or "big bang"—is not simply a break point between a mysterious deity who happens to create everything out of nothing and a nature that evolves in a way that points to humanity's ultimate extinction—perhaps with some promise of an equally mysterious afterlife. Rather, the deity's creative capacity is a model for human emulation, a central feature of human redemption from sin, hence the theological significance of efficiency. But, this suggests a further challenge, one that the star ark brings into sharp relief. Put in ecomodernist terms, even once humanity is fully "decoupled" from nature in terms of energy needs, there will remain the challenge of whether we are learning sufficiently from nature to enhance the human condition.

Humanity is already going down this trajectory. The first stage is, broadly speaking, *biomimicry*—that is, the technological modeling, if not outright reproduction, of functional properties of natural things. Thus, we have been fascinated by how birds fly, fruits hold their contents, and ants carry vast loads. To be sure, our inventions have strayed significantly from nature's own designs. Yet, arguably, these uses remain intellectually exploitative of their natural models: after all, what do we give back to the birds, fruits, and ants that inspire our inventions—other than simply permitting them to live?

I can only sketch a responsive strategy here. It goes by the general name of "uplifting," after a series of science-fiction novels by David Brin. Brin (1980) envisages a cosmos in which a wide variety of species are cognitively "uplifted" so as to communicate not only with humans, but also—at least in principle—with each other. This opens up new realms of social, political, and economic relations, as well as attendant risks. The strategy is clearly anthropic in inspiration, yet, equally clearly, it goes against the usual imperial image of anthropomorphism. Indeed, one possible outcome is that humans might themselves become more adept in ways of knowing and being that were formerly limited to particular non-human species. In this vein, the Canadian political philosopher Will Kymlicka, long known for his work on multicultural justice, has even begun to entertain the concept of "zoopolis" to capture a regime in which a set of mutually binding laws could be enforced to govern a biodiverse ecosystem modeled on a city (Donaldson and Kymlicka 2011). Such a polity would be much more integrated than the species-segregated sanctuaries typically favored by today's animal rights activists (cf. Fuller 2012, Chapter 2). At the extreme end of this line of thought lies what the visionary architect behind the star ark project, Rachel Armstrong (2015), calls "protocell architecture," whereby sustained study of the lifelike properties of complex molecules, such as Bütschli oil droplets, could lead to the design of ultra-smart environments which would condition us to explore creatively the possibilities that a dynamic ecosystem opens up not least different physical forms in which "humanity" itself might be expressed (cf. Church and Regis 2012).

It is common nowadays both among the secular and the religious—even among Christians—to reduce the value of being "human" to the value of life itself. Clive Hamilton belongs to this ancient tradition that is currently *en vogue*. However, for those who still appreciate the *human* as the form of divine self-disclosure (Jesus, in the case of Christianity), there is everything to play for, as we stretch our imaginations and our capacities in search of what it means to be "human," not only on Earth, but across the cosmos.

14.2 The dream drives the action: toward a functional cosmology for interstellar travel

Roberto Chiotti, Principal at Larkin Architect Limited and Assistant Professor at the Faculty of Design, OCAD University, Toronto.

> "There is eventually only one story, the story of the universe. Every form of being is integral with this comprehensive story. Nothing is itself without everything else."
> (Swimme and Berry 1992, p. 268)

Over the past 100 years or so, with the help of quantum physics and highly sensitive instrumentation, the scientific community has been able to provide us with an immense amount of data concerning the evolution of the universe. However, for physicist Brian Swimme and cultural historian Thomas Berry, the meaning of this epic narrative in its unfolding from an explosive origin roughly 14 billion years ago to the emergence of life and human consciousness here on planet Earth has yet to be fully comprehended or integrated in a way that can help us achieve a functional mode of human presence (Swimme and Berry 1992, pp. 1–2).

The contribution of this essay will be to examine the implications of the universe story as we seriously direct our energies and creativity towards achieving interstellar travel in an "ecological age." It will then consider how the emerging cosmological perspective based upon this story, as articulated by Swimme and Berry, can illuminate intrinsic values that have the power to inspire every question, every investigation, and every decision we make towards realizing this visionary quest.

As Rachel Armstrong suggests in her introductory essay, our imaginations have been attracted to the notion of star travel for a very long time:

> "It spans back to the dawn of human history where we formed a deep relationship with the stars. We imagined these bodies as gods and heavenly places, but so far we have not been able to reach and experience them directly."

Primal peoples first understood this sense of intimacy between the human and a numinous universe intuitively. Their cosmology, or mythical stories of creation, and their relationship with the phenomenal world were informed by this deep sense of connectedness. In Berry's view:

> "For peoples, generally, their story of the universe and the human role within the universe is their primary source of intelligibility and value. Only through this story of how the universe came to be as it is, does a person come to appreciate the meaning of life or to derive the psychic energy to deal effectively with those crisis moments that occur in the life of the individual and in the life of the society. Such a story … communicates the most sacred of mysteries … and not only interprets the past; it also guides and inspires our shaping of the future." (Berry 1988, p. xi)

14.2.1 The big picture: how did we get here and where are we going?

In these early decades of the twenty-first century, we are coming to the realization that we may have irreversibly made our current home inhospitable to human survival (Berry 1988, p. 2). Consequently, there is a sense of increasing urgency to transcend the limits of this world in order to find a new one. Like our early ancestors gazing at the heavens above, we must first consider our point of orientation before we begin to imagine ways in which interstellar travel can be achieved within an "ecological age." Before we even attempt to design our way into an extraterrestrial future, it may be of value to re-examine the "big picture" context for understanding meaning in our own world and, ultimately, the purpose behind this ambitious goal. For me, the challenge remains one of cosmology. What cosmological framework will be adequate to inspire our psyches, direct our enquiries, and sustain our efforts? What dream will drive our actions and ensure the desired outcomes?

Since we are Earthlings, perhaps we might first reflect upon the cosmology that has for a long time defined the way in which we relate to our own planet before we embark upon our search for new ones. Our dominant, Western worldview remains rooted in an old cosmology based upon the Biblical Creation narratives of Genesis that established human exceptionalism and our dominion over the rest of Creation as its sacred "a priori" values (Genesis 1:26–28). Dominion gave us the freedom to probe and explore Nature, to discover how things were created. Because we saw ourselves as separate from, and superior to, the rest of the natural world, we had the detachment necessary for Biblical cosmology to be eclipsed by the search for empirical knowledge in an attempt to exploit and manipulate Nature for the protection and benefit of humankind. The ability to transcend the human condition became of paramount concern to the scientific community. With the development of a wide variety of technologies, such as the microscope and telescope, scientists could learn how the universe worked (Dowd 1991, pp. 9–10). Under objective, reductionist scrutiny, the universe was no longer perceived to be alive, but was now understood in mathematical and mechanistic terms. Traditional cosmology became "mathematical cosmology." The disciplines of philosophy and theology were left to deal with the questions concerning the role and meaning of humans.

With the coming of the Industrial Revolution, and the widespread acceptance of Darwin's Theory of Evolution that placed the human as the crowning glory of the evolutionary process, the Modern Era emerged. The understanding of natural selection based upon a cooperative relationship between a particular species and its environment became distorted. "The survival of the fittest" soon became the justification for an insatiable quest for economic wealth and power under the auspices of the industrial, commercial, and financial corporations (Berry 1999, pp. 117–128). Earth was seen solely as a resource to be exploited for its raw materials to fuel the engines of the economic enterprise. It also became the dumping ground for the waste by products of these same industrial processes. Earth's ability to provide and absorb was assumed to be limitless.

The modern human adventure, particularly here in the Western world, remains articulated and driven by a powerful myth deeply rooted in our collective consciousness, which is defined in great part by our dysfunctional cosmology. Swimme and Berry have identified this as "the myth of Wonderland, the Wonderland that is coming into existence by some inevitability if only we continue on the path of Progress, meaning by Progress the ever increasing exploitation of the Earth through our amazing technologies" (Swimme and Berry 1992, p. 218). Our pervasive human presence on the planet is quickly bringing to an end the Cenozoic age of geological history (Tucker 2006, p. 43). The complexity and diversity that have evolved over the past 65 million years of planetary creativity are being undermined at an unprecedented rate through human intervention. In our race to control and exploit Earth's natural resources for the benefit of humankind, we have been blind to the fact that we are shutting down the very life-supporting systems of the planet that we depend upon for our survival. Like similar addictive behaviors, our thirst for unmitigated production and consumption is accelerating the demise of the Cenozoic. We continue to believe in and trust that technology will find solutions to our problems. This is the denial of our addiction (Berry 1988; p. 32).

Even if our technologies turn out to be inadequate to preserve life here on Earth, as we know it, do we still believe and trust that our technologies will help us to find a new home beyond our Solar System? If and when that possibility does exist, will we catapult ourselves into space with the same worldview that privileges our own needs over all others, treating the universe as nothing more than a collection of objects to be exploited ... or will we proceed based upon a cosmology that seeks a universe which is understood as a community of subjects ... our kin? (Swimme and Berry 1992, p. 243) Are we going to be just another invasive species that transfers to someplace else our dysfunctional attitudes and relationships towards our Earthly environment or do we go as primal peoples, if they could have, with a sense of awe and wonder, seeking intimacy with a numinous universe whose chemistry and story we share? How will our cosmology affect the way in which we travel to our destination, how we design our star ships, what we bring with us, how we settle upon arriving, how we respond to our new context, and what we bring back if indeed we can come back? If we address Armstrong's challenge identified in her opening essay—"to ensure that life can be established on non-terrestrial habitats in a considered, responsible and humane manner"—then we must first relinquish our current dysfunctional cosmology in favor of one that will provide the meaning and values necessary to move towards our goal accordingly.

As Thomas Berry reminds us:

> "It is all a question of story. We are in trouble just now because we are in between stories. The Old Story, the account of how the world came to be and how we fit into it, sustained us for a long time. It shaped our emotional attitudes, provided us with life purpose, and energized action …. Today however, our traditional story is no longer functioning properly, and we have not yet learned how to integrate the New Story." (Berry 1988, p. 123)

14.2.2 The universe story as our new story, our new cosmology

Western science, which itself emerged out of this old cosmology, is now giving birth to a new empirically observed, 14-billion-year creation story—a new cosmology that defines the universe story. More specifically, modern science is providing an increasingly detailed account of the physical and biological evolution of the universe story that compels us to view reality in a single unfolding process … what Swimme and Berry have characterized as "cosmogenesis" (Swimme and Berry 1992, p. 2).

This is good news as we consider the "big picture" context for comprehending the universe as destination and the reality of interstellar travel as an essential part of our future. The story of the universe being articulated by the scientific community then becomes the single, most powerful text of reference for our investigations. As Swimme and Berry suggest:

> "The most significant change in the twentieth century, it seems, is our passage from a sense of cosmos to a sense of cosmogenesis … we have moved from that dominant spatial mode of consciousness, where time is experienced in ever-renewing seasonal cycles, to a dominant time-developmental mode of consciousness, where time is experienced as an evolutionary sequence of irreversible transformations. Within this time-developmental consciousness we begin to understand the story of the universe in its comprehensive dimensions and in the full richness of its meaning …. This story is the only way of providing, in our times, what the mythic stories of the universe provided for tribal peoples and for the earlier classical civilizations in their times." (Swimme and Berry 1992, pp. 2–3)

Berry's descriptive prowess infuses the technical realities associated with the evolution of the universe into a poetic, epic narrative that is both alluring and transformative:

> "The story of the universe is the story of the emergence of a galactic system in which each new expression emerges through the urgency of self-transcendence. Hydrogen in the presence of some millions of degrees of heat emerges into helium. After the stars take shape as oceans of fire in the heavens, they go through a sequence of transformations. Some eventually explode into the stardust out of which the solar system and the earth take shape. Earth gives unique expression of itself in its rock and crystalline structures and in the variety and splendor of living forms, until humans appear as the moment in which the unfolding universe becomes conscious of itself. The human emerges not only as an earthling, but also as a worldling. We bear the universe in our beings and the universe bears us in its being. The two have

a total presence to each other and to that deeper mystery out of which both the universe and ourselves have emerged." (Berry 1988, p. 132)

Berry also compares this identity to the "anthropic principle" as expressed by physicists:

"In this perception the human is seen as a mode of being of the universe as well as a distinctive being in the universe. Stated somewhat differently, the human is that being in whom the universe comes to itself in a special mode of conscious reflection." (Berry 1988, p. 16)

This contradicts the Biblical cosmology that set humans apart from everything else and enabled us to imagine having dominion over Earth. By contrast, the new cosmology suggests that what we do to Earth, we do to ourselves and, by extrapolation, what we do to the universe, we do to ourselves.

For ecologist Michael Dowd:

"The scientist looking through a telescope is literally the universe looking at itself. The child entranced by the immensity of the ocean is Earth enraptured by itself. The student learning biology is the planet consciously learning about how it has functioned unconsciously for billions of years." (Dowd 1991, p. 17)

This sense of humans as having emerged as the self-reflective consciousness of the universe begs us to consider the question: if we embark upon interstellar travel, is it in fact the universe attempting to consciously seek out itself in order to greater understand its own complexity?

According to Berry:

"The scientist in the depths of the unconscious is drawn by the mystical attraction of communion with the emerging creative process. This would not be possible unless it were a call of subject to subject, if it were not an effort at total self-realization on the part of the scientists."(Berry 1988, p. 133)

To invoke a more visceral response, Berry recites the poetic words of his colleague and friend, Brian Swimme: "The universe shivers with wonder in the depths of the human" (Berry 1988, p. 16).

The concept of cosmogenesis suggests that certain form-producing dynamics have guided the universe in its 14 billion years of time-developmental evolution. Swimme and Berry have characterized these creative dynamics as inner spontaneities governed by the primordial orderings of differentiation, subjectivity, and communion:

"These orderings are real in that they are efficacious in shaping the occurrences of events and thereby establishing the overriding meaning of the universe. Indeed the very existence of the universe rests on the power of this ordering. Were there no differentiation, the universe would collapse into a homogeneous sludge; were there no subjectivity, the universe would collapse into inert, dead extension; were there no communion, the universe would collapse into isolated singularities of being." (Swimme and Berry 1992, pp. 72–73)

According to Berry, these principles, which were understood by our primal ancestors as intuitive processes, are now understood by scientific reasoning. However, the significance of their implications for how we relate to the universe and more poignantly Earth has yet to be fully understood or acted upon in any effective manner (Berry 1988, p. 44).

"Differentiation" or "Diversity" is the primordial expression of the universe. Out of the fiery violence of the "Big Bang" came radiation and differentiated particles that, through a certain sequence of events, found expression in an overwhelming variety of manifestations (Berry 1988, p. 45). The universe is coded for an ever-increasing, non-repeatable biodiversity as exemplified by the incredible variety of life that has evolved on Earth. For Swimme and Berry, "But in the universe, to be is to be different …. At the heart of the universe is an outrageous bias for the novel, for the unfurling of surprise in prodigious dimensions throughout the vast range of existence" (Swimme and Berry 1992, p. 74).

From its rich and abundant tropical rainforests to the stark beauty of its polar regions, the evidence of this tendency on our own planet towards biodiversity is obvious. Humankind would not have appeared as a species if somehow this process towards increasing biodiversity had been allowed to shut down. As an architect, my imagination and creativity are directly born of and related to the creativity of the universe and continue to be sustained by the awesome magnificence of diversity that declares itself around me within the natural world here on Earth. Berry reminds us that, if we were born on the Moon, our imaginations would reflect the desolation of the lunar landscape (Berry 1988, p. 11). If we lived our entire lives there, we would not develop the necessary inspiration or the imaginative capacity to consider interstellar travel let alone consider what it means to be conscious human beings within a universe characterized by differentiation.

The second primary creative principle of the universe is that of increased "subjectivity" or "interiority." Together, every reality that makes up a part of the universe is not just a collection of objects, but is a community of subjects (Swimme and Berry 1992, p. 243). As subjects, all realities that make up the universe, from individual atoms, to individual persons, to individual solar systems, to individual galaxies have an inner dimension—an interior depth of reality that not only reflects the diversity that surrounds us, but reflects the original bursting-forth of energy at the beginning of time. By way of example, the essay contributions in this book emerge from each individual author's "interiority" and, as such, represent a unique "universe" voice towards the conversation surrounding interstellar travel. Therefore, every expression is deserving of our attention and respect, however similar, disparate, or even contradictory.

"Each being in its subjective depths carries that numinous mystery whence the universe emerges into being" (Berry 1999, p. 163). For Berry, "the reality and value of the interior subjective numinous aspect of the entire cosmic order is being appreciated as the basic condition in which the story makes any sense at all" (Berry 1960, p. 135). He often refers to the influence of paleontologist Teilhard de Chardin, whom he credits as being "the first person to describe the universe as having, from the beginning, a psychic-spiritual dimension as well as a physical-material dimension" (Dunn and Lonergan 1991, p. 24). Perhaps this is why the idea of interstellar travel has been and continues to be so enchanting and alluring. Like all good stories, we are drawn by the mystery.

Furthermore, as Swimme and Berry suggest:

> "The sentience of today's world is an ontological creation of the evolving universe. In former times it existed as a latent possibility; now it exists in its activated or historical realization. Because creatures in the universe do not come from some place outside it,

we can only think of the universe as a place where qualities that will one day bloom are for the present hidden as dimensions of emptiness." (Swimme and Berry 1992, p. 76)

This has profound implications for us as we consider "whom" and not "what" we might encounter in space when we set out into its depths. In support of Armstrong's speculations, this may mean "that intrinsically, 'we' are not alone—as 'we' were never singular in the first place ... there are other bodies, other agencies, other stories that are writing futures ... and events ... and meaning" (Rachel Armstrong, e-mail, August 26, 2015). If she is correct, then, as we ourselves contemplate the notion of travel to other galaxies, is it likely that we will indeed discover as yet unseen "others," out there, who are already considering the very same journey?

The third creative principle of the universe is "communion" or "interconnectedness." As mentioned before, we are an inextricably related community of subjects. According to Berry, this genetic interrelatedness of everything in the universe to everything else means that the universe is in dialogue with itself as a single community, "for every reality of the universe is intimately present to every other reality of the universe and finds its fulfillment in this mutual presence" (Berry 1988, p. 106). All of this seems to affirm the idea that interstellar travel is part of our human destiny, that, indeed, it is perhaps even an ontological imperative. If the universe is our "hood," are we then compelled to get to know our "neighbors"? If we respond with a resounding "yes," then we may want to consider the following words of encouragement for our journey:

"To be is to be related, for relationship is the essence of existence. In the very first instant when the primitive particles rushed forth, every one of them was connected to every other one in the entire universe. At no time in the future existence of the universe would they ever arrive at a point of disconnection. Alienation for a particle is a theoretical impossibility. For galaxies too, relationships are the fact of existence. Each galaxy is directly connected to the hundred billion galaxies of the universe, and there will never come a time when a galaxy's destiny does not involve each of the galaxies in the universe." (Swimme and Berry 1992, p. 77)

Does this mean that we can proceed into space with the confidence that we will never encounter aliens ... only relatives?

As an educator, I suggest to my students that, in order for them to apply critical thinking effectively to problem-solving, it is necessary to frame their enquiries within the largest possible context. There is no larger context we know of than the universe itself. Hopefully, then, as the scientific community continues to observe the story of the universe in its continual unfolding and transformation, we will come to understand and integrate the true meaning of this story as the basis for a new, functional cosmology. "Sensitized to such guidance from the very structure and functioning of the universe, we can have confidence in the future that awaits the human venture" (Berry 1988, p. 137). If we begin to realign our human creativity with the creativity of the universe, then it will be possible to discover the inspiration, wisdom, and meaning necessary to free ourselves from Earth's gravitational embrace and explore the far-reaching depths of the universe beyond in alignment with its guiding principles of differentiation, subjectivity, and communion.

14.3 Spaceship mind/virtual migration to exoplanets

Krists Ernstsons, Architectural Assistant, Hawkins\Brown, London.

> "Exploration is in our nature. We began as wanderers, and we are wanderers still. We have lingered long enough on the shores of the cosmic ocean. We are ready at last to set sail for the stars." (Carl Sagan, *Cosmos*)

14.3.1 Introduction

A method for humanity's survival and expansion into the known universe may be achieved using an architectural system that harnesses quantum mechanical phenomena such as quantum entanglement and computing, to transfer information faster than the speed of light using a seed-like structure.

Quantum communication can enable digital wormholes through space-time that may allow us to break the speed of light constant, $c = 299{,}792{,}458$ m per second, and communicate instantaneously with quantum entanglement. A very stable system of quanta storage (memory) is required, because quantum states are extremely vulnerable.

Since all the known universe is made of quanta (particles) that communicate beyond the classical laws of physics, future information systems that work at the particle scale can enable more economical and faster solutions for seeking other intelligent life forms and existing in the fabric of the cosmos.

The ultimate goal of the seed that is sent out is to become a personal avatar/messenger/research unit of life on other exoplanets—expanding humanity's reach and continuing our existential need for exploration (Figs. 14.1, 14.2, 14.3, 14.4, and 14.5).

Should a human migration occur, colonies are most likely to head towards Earth-like planets—called exoplanets—with atmospheres and liquid water. So far, we have been looking at a very narrow part of our galaxy and NASA launched the Kepler telescope to broaden our view. Since 1995, when the first exoplanet was discovered at the Observatory of the University of Geneva by Michel Mayor and Didier Queloz, we have discovered over 2000 exoplanets.

Although Earth-like exoplanets are a very intriguing alternative to our survival in the cosmos, the distance to travel to them is enormous—light years away. This means travelling 9.4605284×10^{15} m in a year.

Conventional propulsion methods will not allow us to travel very far. They can become efficient within the scale of our Solar System but, to travel to exoplanets, we need to achieve speeds close to the speed of light. This means that the mass of the object sent has to be minimal and it should not carry liquid propellant with it—the famous equation of Einstein: $e = mc^2$ (Fig. 14.6).

14.3.2 Science and light

Light is a truly mysterious phenomenon—the essence of life. We know so much and so little about it at the same time. We use it in our technology and we observe the cosmos by looking at light. It's a fundamental part of our lives and is part of a whole electromagnetic spectrum that includes visible light, warming infrared rays, and harmful gamma radiation.

Figure 14.1 Quantum superposition. Credit: Krists Ernstsons

Light also plays a fundamental part in our knowledge about quantum mechanics which transgresses the classical laws of physics. It is both a wave and a particle. This is the foundation of quantum mechanics where our universe begins to seem like a work of science fiction—composed of dark matter, dark energy, black holes, and wormholes through space-time.

Since light is a quantum phenomenon, it can behave "spookily"—existing at two places at the same time. When a photon (a single particle of light), which exists in a superposition state, is split into two parts, they become entangled in such a way that, when we measure one of them, the other will always have the opposite spin—either up or down. The same phenomena happen when we separate the photons over infinite distances. A third photon with different characteristics can also be imposed on entangled photons, where its state will be teleported. These principles can be demonstrated in laboratory settings and, although they have not been fully resolved, they can be considered as an architecture of particles.

By studying these experiments, it possible that one day a quantum entangled communication system would initially teleport information, but potentially also teleport matter. In its basic form, the system uses the quantum effect of superposition, quantum bits, and controlled decoherance time. Quantum tunneling allows us to map, locate, and select these

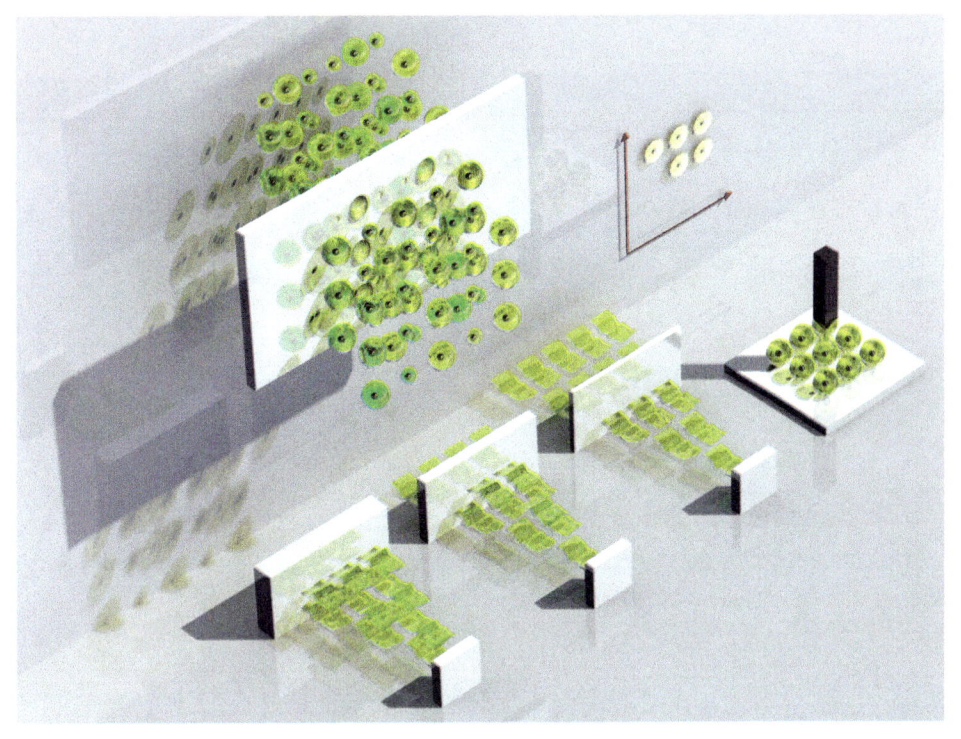

Spaceship mind/virtual migration to exoplanets 403

Figure 14.2 Quantum tunneling. Credit: Krists Ernstsons

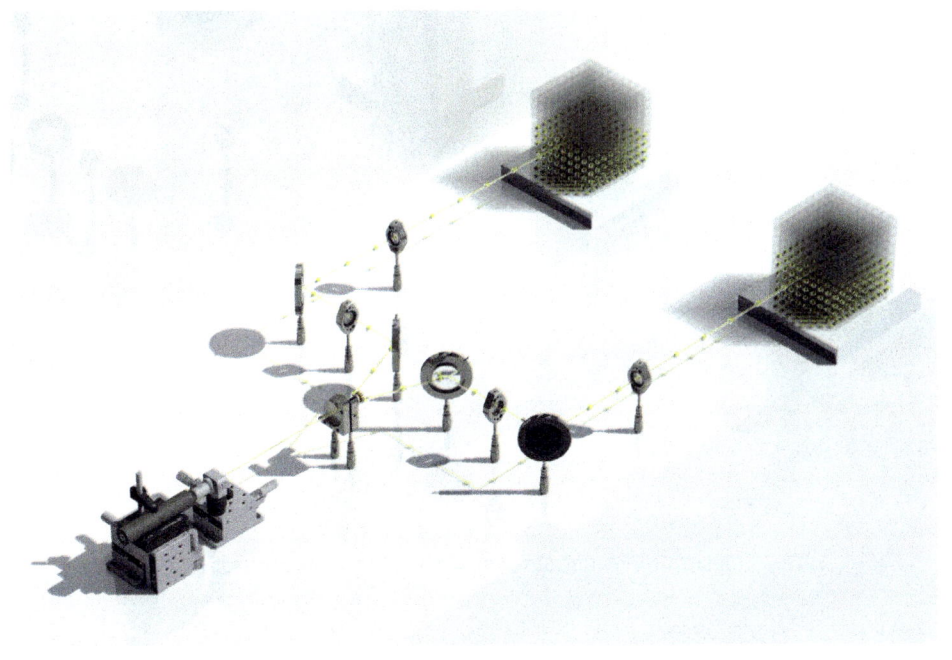

Figure 14.3 Quantum entanglement. Credit: Krists Ernstsons

Figure 14.4 Quantum teleportation. Credit: Krists Ernstsons

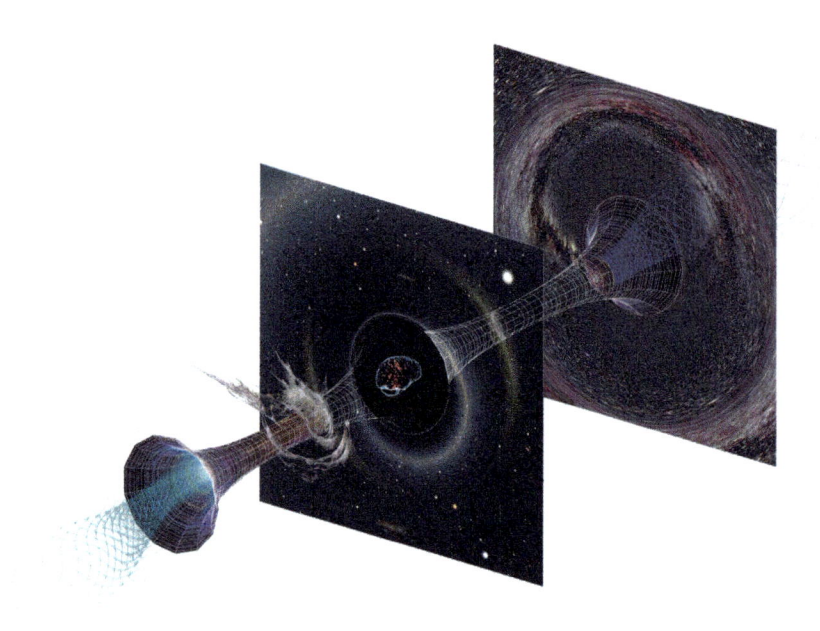

Figure 14.5 Artificial wormhole. Credit: Krists Ernstsons

Figure 14.6 The Solar System is a dynamic system. If we are to journey into interplanetary space, we need to know the alignment of Earth with other planets at certain points in time. For example, we need to understand the alignment of Earth with the three types of planets in our Solar System—terrestrial planets, gas giants, and ice giants. Credit: Krists Ernstsons

quantum bits. Quantum entanglement allows us to communicate with these bits beyond the speed-of-light boundary. And quantum teleportation uses all of the above to get us closer to the ultimate goal of teleporting matter. One day, we may be able to use these principles to transmit ourselves across the universe faster than the speed of light.

14.3.3 Microseeds and migration

Perhaps this could take place using micro-probes (seeds) with a built-in quantum communication system. These information seeds would travel close to the speed of light and enable teleportation beyond the existing limits of space-time. Potentially, they could precipitate molecular growth at a target location depending on the received information. This technology may be thought of as a wormhole—a shortcut through space-time—but not a natural wormhole that spontaneously occurs, rather a digital one made by human design (Fig. 14.7).

The teleportation seed structure has three main software components and nine key hardware components that include: camera telescopic module, magnetic laser pads, integrated electronics, laser photon generator, quantum memory, liquid and gas tanks, cargo module, batteries, microwave engine, and the sliding optical module.

The exterior is clad in a carbon nanotube shell with embedded graphene for durability and cooling circuits. The seed is about 45 mm in length and 15 mm at its widest part, weighing around 50 g.

Potentially, these devices can send molecular replicators—structures that are able to grow and develop, possibly building factories of new seeds and propagating life. For example, the seed that carries all the world's data in a carbon diamond; a bank of

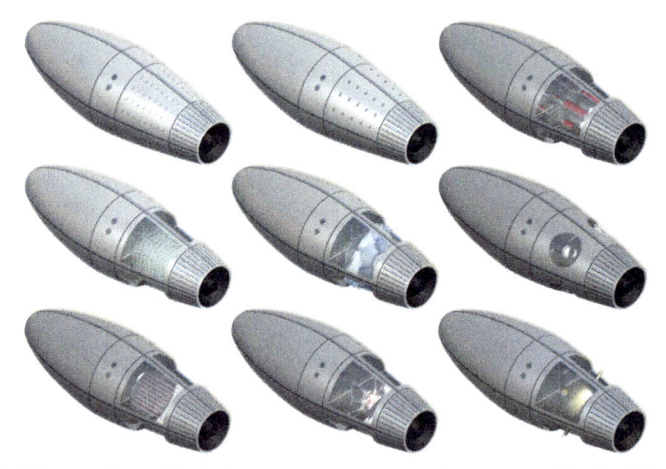

Figure 14.7 Microseeds with in-built quantum communication system. Credit: Krists Ernstsons

seeds; music; four types of human blood; Earth's air; an acorn replicator; DNA samples; an *Escherichia coli* replicator; and personal avatar.

This is the Atlas master plan of exoplanet fertilization. Mission 1 requires an energy source to be launch into heliocentric orbit around the Sun; the first stage of the rocket is then returned to Earth on a landing platform using SpaceX technology.

The seeds are produced in a factory at CERN, where the counterparts of the quantum banks are stored on Earth.

Mission 2 launches the seed beamer satellite between Mars and Jupiter. Connection calibration is established between the International Space Station (ISS), the seed beamer, and the observatory in Tenerife.

Energy is beamed from the solar collector to the seed beamer.

The first batch of 100 seeds are fired toward an exoplanet within the Tau Ceti system which is 12 light years away. During the journey, the seeds are accelerated close to the speed of light and undergo a series of reformation, swarming, and regrouping patterns to deal with friction, interstellar dust, gases, and possible collisions. Some of the seeds might land or go off-course to other points of interest or to harvest energy.

Before approaching the stellar system, a second part of the engine is engaged to decelerate the seeds. Once the seeds enter the stellar system, they undergo a series of observations and calculations.

A landing pattern is initiated, depending on the cargo of the seed and the observed characteristics of the exoplanet.

Once the seed has landed and a clear communication channel with Earth has been established, it starts its evolution. The quantum information wormhole through space-time has been established, and the process to link other exoplanets is underway. While our human bodies are unlikely to travel distant galaxies or even stars light years away, our technological creations might just do that. This architecture may be viewed as a kind of cosmic activism, activist as explorer, exploration as fertilization. Perhaps we may evolve so that our minds can travel to distant Earth-like planets (Fig. 14.8).

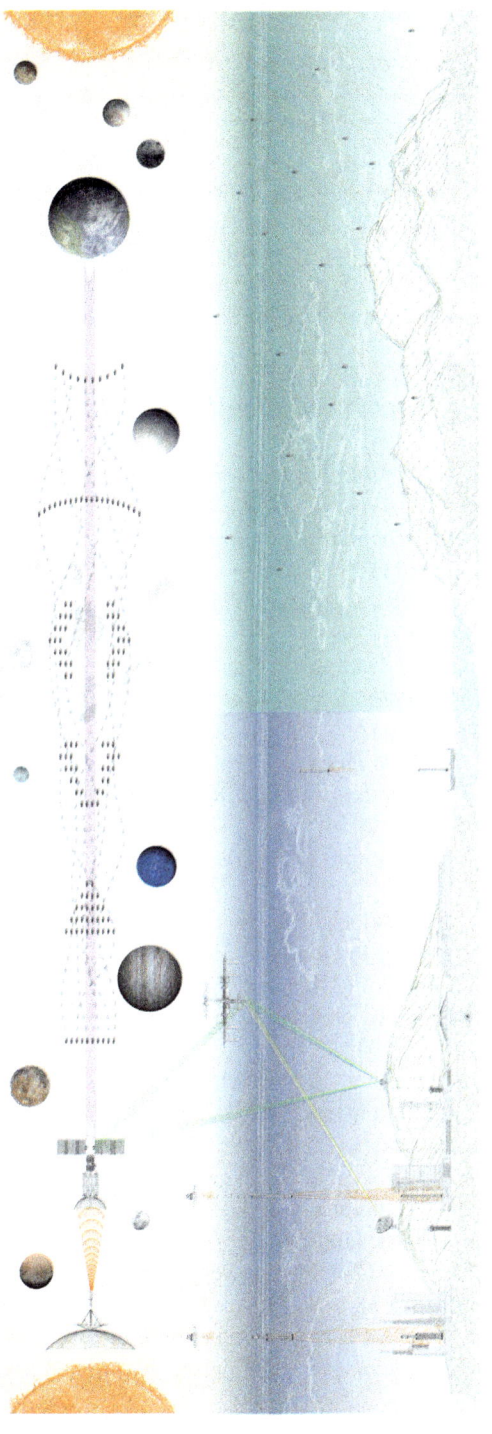

Figure 14.8 Microseed masterplan. Credit: Krists Ernstsons

COMBINED REFERENCE LIST FOR CHAPTER 14

R. Armstrong, *Vibrant Architecture: Matter as a Co-Designer of Living Structures* (Walter de Gruyter, Berlin, 2015)

T. Berry, *The Divine Milieu* (Harper & Row, New York, 1960)

T. Berry, *The Dream of the Earth* (Sierra Club, San Francisco, 1988)

T. Berry, *The Great Work: Our Way into the Future* (Bell Tower, New York, 1999)

N. Bostrom, In defense of posthuman dignity. Bioethics **19**(3), 202–214 (2005)

Breakthrough Institute, The Ecomodernist Manifesto. (2015a), www.ecomodernism.org/

Breakthrough Institute, The good Anthropocene: the breakthrough dialogue 2015. (2015b), http://thebreakthrough.org/index.php/dialogue/past-dialogues/breakthrough-dialogue-2015

D. Brin, *Sundiver* (Bantam, New York, 1980)

G. Church, F. Regis, *Regenesis: How Synthetic Biology Will Reinvent Nature and Ourselves* (Basic Books, New York, 2012)

T. Dobzhansky, *The Biology of Ultimate Concern* (New American Library, New York, 1967)

S. Donaldson, W. Kymlicka, *Zoopolis: A Political Theory of Animal Rights* (Oxford University Press, Oxford, 2011)

M. Dowd, *Earthspirit: A Handbook for Nurturing an Ecological Christianity* (Twenty-Third Publications, Mystic, CT, 1991)

S. Dunn, A. Lonergan (eds.), *Thomas Berry with Thomas Clarke, SJ, Befriending the Earth: A Theology of Reconciliation between Humans and the Earth* (Twenty-Third Publications, Mystic, CT, 1991)

S. Fuller, *Preparing for Life in Humanity 2.0.* (Palgrave Macmillan, London, 2012)

S. Fuller, V. Lipinska, *The Proactionary Imperative: A Foundation for Transhumanism* (Palgrave Macmillan, London, 2014)

S. Fuller, *Humanity 2.0: What It Means to Be Human Past, Present and Future* (Palgrave Macmillan, London, 2011)

B. Guillaume, Vernadsky's philosophical legacy: a perspective from the Anthropocene. Anthropocene Rev. **1**(2), 137–146 (2014)

C. Hamilton, *Requiem for a Species* (Allen and Unwin, Sydney, 2010)

C. Hamilton, A new kind of human being: reply to Steve Fuller. (2015a), www.abc.net.au/religion/articles/2015/08/21/4297536.htm

C. Hamilton, Dreams of a fallen civilization: why there is no escaping the blue planet. (2015b), www.abc.net.au/religion/articles/2015/08/21/4297536.htm

J. Huxley, *New Bottles for New Wine* (Chatto and Windus, London, 1957)

R. Kurzweil, *The Singularity Is Near* (Viking, New York, 2005)

N. Land, The dark enlightenment. (2013), www.thedarkenlightenment.com/the-dark-enlightenment-by-nick-land/

J. Milbank, *Theology and Social Theory* (Blackwell, Oxford, 1990)

M. More, N. Vita-More (eds.), *The Transhumanist Reader* (Wiley-Blackwell, London, 2013)

D.F. Noble, *The Religion of Technology: The Divinity of Man and the Spirit of Invention* (Penguin, London, 1997)

N. Oreskes, E. Conway, *The Collapse of Western Civilization* (Columbia University Press, New York, 2014)

P. Teilhard de Chardin, The Phenomenon of Man (orig. 1955) (Harper and Row, New York, 1961)

J. Passmore, *The Perfectibility of Man* (Duckworth, London, 1970)

B. Swimme, T. Berry, *The Universe Story: From the Primordial Flaring Forth to the Ecozoic Era: A Celebration of the Unfolding of the Cosmos* (Harper, San Francisco, 1992)

M.E. Tucker (ed.), *Evening Thoughts: Reflecting on Earth as Sacred Community, Thomas Berry* (Sierra Club Books, San Francisco, 2006)

S. Vaj, *Biopolitica: Il Nuova Paradigma* (Società Editrice Barbarossa, Milan, 2005)

L. White, *The Evolution of Culture* (McGraw-Hill, New York, 1959)

G. Young, *The Russian Cosmists* (Oxford University Press, Oxford, 2012)

15

Constructing worlds

Jordan Geiger and Mark Morris

Two individual authors Geiger and Morris consider different tactics in exploring dimensions of scale and ambition in the construction of megastructures and development of world-building strategies implicit in colonizing the cosmos.

15.1 Alive without us

Jordan Geiger, international architect and educator.

> "All the modern things, like cars and such, have always existed.
> They've just been waiting in a mountain for the right moment, listening to the irritating noises of dinosaurs and people dabbling about.
> It's their turn now." (Björk 1995)

15.1.1 Introduction

This essay moves forward in time with spacecraft. It departs from the essay "Zero Atmosphere Architecture," which described four points for spaceships past as a fusion of architecture and human–computer interaction (HCI), as a product of fact/fiction transfers, as steeped in protectionism for Earth's atmosphere, and as implying new scales for architectural thinking (Geiger 2015).

"Alive without us" forecasts architectural and other implications of Persephone. The text contextualizes this past within recently coined terms of the "Very Large": Very Large

---

Jordan Geiger (✉)
Architect/educator
e-mail: Jordan@ga-ga.org

Mark Morris (✉)
Associate Professor of Practice, Architecture Art Planning, Cornell University,
Ithaca, NY 14850, United States
e-mail: mm789@cornell.edu

Organizations and Very Large Finitude. Roughly caught in the tangle between globalized capital and global warming, we now look ahead to 100 years in which we dismantle the former and eliminate ourselves and our teleology in the latter.

What arises is the promise of a futurality that is more architectural while re-inventive of architecture's underpinnings: steeped in imagination, spatially resistive to a section cut or orientation to a ground, proposing an infinite interiority. This futurality frees us to think past homeostasis, control, and a *milieu intérieur*, toward the "felt" as basis for human relations after the Anthropocene, and to build for non-human beings. Thanks to Persephone, time, scale, envelope, inhabitation—some of the essential terms of architecture—are readied for redefinition.

15.1.2 Spacecraft and architecture

Let us begin with a fiction, a seemingly unreasonably strange one, in which humans are no longer around and space is unbound and time has only a future, no present. Now let us explore in what respects this fiction is intrinsically architectural, a reliable reality, written by a 100-year starship.

Architects, after all, trade in fictions: they promote futures through representations in drawing, model, rhetoric, and other media. These representations are biased, polemical, calling for a reality to come (*à venir*, we recall, in the French) and, therefore, all architectural work constitutes itself as speculative fiction. As it happens, this is but one aspect by which design for outer space can be seen to reveal varied forces on and shifts in architectural thinking during our last century.

As I have written elsewhere, here are four of them:

1. Spacecraft arose as a pure fusion of architecture and human–computer interaction (HCI).
2. Spacecraft developed through a "technology transfer" between works of fact and fiction.
3. Spacecraft were built on an Earth ecology discourse that anticipated a collapsing atmosphere.
4. Spacecraft gestured toward new notions of scale with relationship to the body.

To briefly summarize each of these:

1. The Apollo 11 mission's lunar landing in 1969 indicated the moment since which the architectural structure has been inseparable from the computational and communications infrastructure that rendered it survivable, and also from the complex routines of verbal and manual communications that enabled the functioning of it all: space was not merely reliant on computers, but on new human–computer interaction. It is no great stretch of the imagination to look now at remote sensing, automation, and locative media in homes today (alarms, thermostats, and moisture-sensitive gardening implements) as a direct result.
2. To get to this audacious achievement, many figures in the development of rocketry, mechanical systems, building envelopes, and more have imported and also exported ideas to cinema and television. These have included renowned engineer Wernher von Braun's collaborations with Disney and ABC television, but also the consulting that designers employed by Buckminster Fuller did on Stanley Kubrick's *2001: A Space Odyssey* (1968) before later working on the roof structure of the Biosphere 2 project. Examples abound, showing what we might call a "technology transfer" undertaken—at times both unconsciously and reciprocally—to generate far-reaching ideas and then to

realize them by generating public support, financing, material production, human inhabitation, and more.

3. The first space race of the last century is commonly discussed in terms of Cold War politics and escalating nuclear proliferation, but that was likewise often coupled with the scientific and rhetorical efforts of figures within space programs to build explorations of space on the possibility of a collapsing Earth atmosphere. Carl Sagan and James Hansen are examples of figures employed by NASA and engaged in the component of the agency's mission devoted to "understand and protect our home planet." These efforts, along with the funding of the public agency, have been eroded in recent years. What has remained is the emergent questioning of our planet's atmosphere as a commons. Neither clearly regulated nor protected, and defiant of stable boundary conceptions, this "atmospheric commons" has bolstered climate change discourse even as it represents a new architectural notion of human inhabitation of space. The architectural section drawing, a vertical cut through a building that explores internal spaces and their related ground plane, grew insufficient to conceive the domain of habitable space; air and space travel gave rise to what we may call a "sectionally expanded" notion of Earth and its built environments.
4. Ethnographic analyses of cockpits and Mission Control desks (Dourish 2001) have established the ways that the human body has been retraining itself to effect unseen interactions and remote controls over physical spaces, but these grow subject to further consideration in today's era of gene therapy, implantable prostheses, and living cell materials that may leap from body to host space. Scales of spatial perception, gesture, and engineering continue to be redefined, and their changes grow only more extreme when measured against the sublime expanse of outer space (Fig. 15.1).

Now we can look to how the star ark project indicates a raft of new dynamics ahead between architectural thinking and the forces with which it intersects. Persephone declares itself dedicated to three novel priorities: a 100-year process, a basis in ecopoiesis, and a funding through collaborations of many bodies (the Defense Advanced Research Projects Agency (DARPA), International Space Station (ISS), and others), but more. These are suggestive of shifts to the very definitions of architecture and HCI, along with key terms of each such as time, boundary, and life itself.

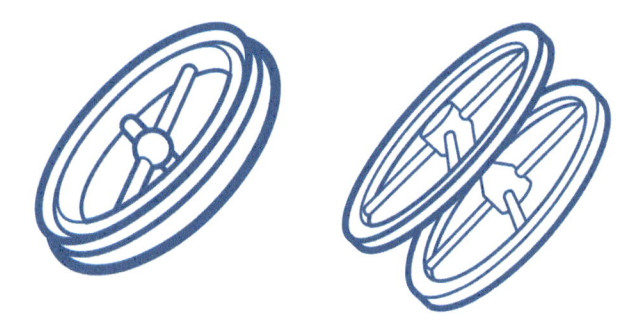

Figure 15.1 Werner von Braun's S-1 Station (*left*) and the orbiting station from Stanley Kubrick's *2001: A Space Odyssey* (*right*). *Credit*: Jordan Geiger

Reworking such definitions relates to and, therefore, draws on two recently formulated "Very Large" constructs. Very Large Organizations (VLOs) and Very Large Finitude are concepts that emerge in architecture and in philosophy, yet refer to many forces and fields of study: globalization, futurology, synthetic biology, cinema, interaction design, economics, and more. With Persephone, we see *beyond* the Very Large, to another project, for other orders to host other, non-human beings.

15.1.3 The Very Large

15.1.3.1 Organizations

A VLO is the amalgam of many tiny and global contributing forces, with the augmentation of one or more ubiquitous computing technologies. Wherever a VLO coalesces, its architecture and interaction yield novel design problems (for an introduction to VLOs and to their most totalizing results to date, see Geiger 2013).

VLOs are at once abstract and concrete, difficult to see, and yet familiar to us all. Immigration. Drugs. Incarceration. Space travel. These are each VLOs today, as they are fundamentally bound up with forces such as foreign policy and multi-national corporate finance, but also ubiquitous computing technologies and climate change.

It is tempting to declare these unprecedented phenomena, if they are phenomena at all. But, in fact, one of the salient characteristics of VLOs is that they go unobserved,

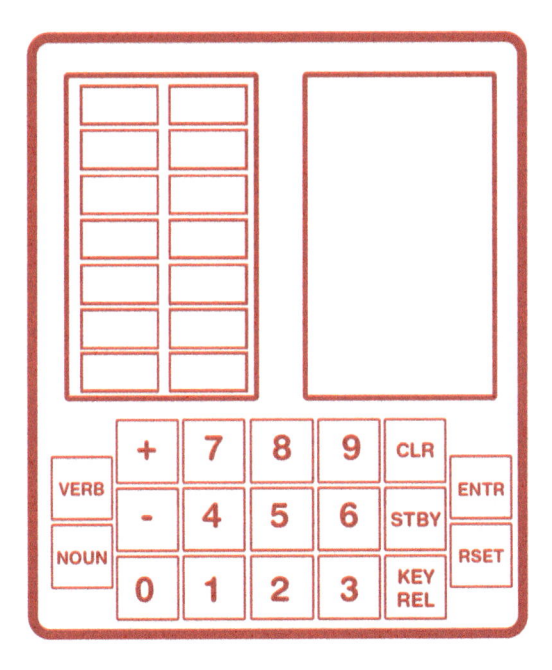

Figure 15.2 The Apollo 11's Display and Keyboard Unit, or "DSKY," and its complex "verb-noun" syntax that asked crew to learn complex new languages and keyboard input routines to operate their ship. *Credit*: Jordan Geiger

imperceptible, ontologically defiant of any classification. Their built environments—the architectures and landscapes of VLOs—are similarly challenging, since they are different from objects of infrastructure but may sometimes include them. They resist classification by typology (such as hospital, hotel, or house) (Fig. 15.2).

Again, consider immigration, incarceration, and the drug trades. Each of these has corresponding physical spaces, including interiors and buildings, but also vast exterior landscapes. And yet, since the introduction of things like RFID transponders, GPS-enabled locator cuffs, or substance-sensing controls, respectively, these are each forever transformed. Roads obviate toll collection plazas, prisons turn seemingly inside-out as parolees move about cities, tracked perpetually via GPS, and drugs (both legal and controlled) are detectable and manageable remotely and globally. Further, behind each of these is a complex knot of corporate investment, local and international laws, and more. Immediately, it becomes apparent that many of these changes regard movement and stasis, whether of goods or of human bodies. But, these are also each spatial products (Easterling 2014) of global capital, novel technologies, shifting mores, human relations to other forms of life, all across sweeping ranges in scale. They are each enabled by the presence of tiny material shifts and the microelectronics that live in them or for them, but also by networks of fiber-electric cables and satellites that crisscross oceans and orbit the planet. Space programs, viewed from the ethereal perspective where they live, are self-generating VLOs; some of their technologies, such as those for atmospheric sensing or machine vision, enable and justify further space exploration, and their attendant space architectures.[1] Products of such international and multi-corporate teamwork, space programs may well be the original, the *ur*-VLOs. Their physical manifestations range in scale and location from things like the legendary DSKY control (Fig. 15.2) to the ISS. Importantly, even the tiniest of these remain integral with a global telecom network that communicates them to one another. In this way, VLOs like space programs could be described as global in reach, but their designed objects occupy a vast sweep in scales. Architecture and landscape for all VLOs—indeed, the lived scale of humans and their societies—are firmly left in the middle. They are profoundly affected by VLOs and yet frequently bereft of their design attention. Considering the built environments but also the tangle of law, logistics, and finance, "organization" in this context is meant in both the spatial and also the administrative sense.

It is most commonly thought that the space race that culminated with the first lunar landing was motivated most by a mixture of Cold War politics and technical advances in rocketry. Yet, research today reveals how space programs carried with them other missions and other organizations. We learn of the spacesuit's evolution from a hard shell and military-industrial complex production to a soft, 21-layered, hand-sewn artifact influenced by Christian Dior and made by Playtex (De Monchaux 2011). We learn of how "software" containing communications commands were literally soft, sewn into the sleeves of Apollo astronauts (De Monchaux 2011). Today, space programs result from shifting partnerships of varied nations (the ISS includes Canada's robotic arm and the European Space Agency (ESA)'s Columbus Laboratory), and private corporations. Of the many private efforts, we can look to Mars One, Google, Bigelow, Space X, Blue Origin, Virgin Galactic, and the many vendors that NASA now looks to for contract outsourcing. Their range of travel and communications continues to expand, and plans rest frequently on technologies tested on

---

[1] Here we must also consider the various ongoing, private corporate development of high-altitude aircraft to deliver broadband internet across entire continents.

Earth. Recently, a spectacular case might be the ESA's work with Foster+Partners architects for a 3D printed building on the Moon, but the more mundane and constant deployment of private satellites reflects how inhabitations of space are self-generative.

15.1.3.2 Finitude

Philosopher Timothy Morton coins the term "hyperobjects" to describe

> "things that are massively distributed in time and space relative to humans… [A] black hole… the Florida Everglades… the biosphere… the sum total of all the nuclear materials on Earth [or a] very long-lasting product of direct human manufacture, such as Styrofoam or plastic bags, or the sum of all the whirring machinery of capitalism. Hyperobjects, then, are 'hyper' in relation to some other entity, whether they are directly manufactured by humans or not." (Morton 2013)

A leader of the recently emerged area of philosophical thought known as Object Oriented Ontology, Morton describes hyperobjects as decidedly real things, however imperceptible they may be. Further, it is often their imperceptibility that renders them such potent actors in global warming. They are inextricably linked to a sublime property that he calls "Very Large Finitude":

> "… gigantic timescales are truly humiliating in the sense that they force us to realize how close to Earth we are. Infinity is far easier to cope with. Infinity brings to mind our cognitive powers, which is why for Kant the mathematical sublime is the realization that infinity is an uncountably vast magnitude beyond magnitude. But hyperobjects are not forever. What they offer instead is very large finitude. I can think infinity. But I can't count up to one hundred thousand. I have written one hundred thousand words, in fits and starts. But one hundred thousand years? It's unimaginably vast. Yet there it is, staring me in the face, as the hyperobject global warming. And I helped cause it… There is a real sense in which it is far easier to conceive of 'forever' than very large finitude. Forever makes you feel important. One hundred thousand years makes you wonder whether you can imagine one hundred thousand anything. It seems rather abstract to imagine that a book is one hundred thousand words long." (Morton 2013, p. 71)

It is no stretch to see the quintessences (plural) of Very Large Finitude in space travel, and their further extremity in a 100-year project for a star ark. I can imagine an endless universe, but—despite all the cartoons of my childhood and data visualizations of today—vast limitation is a large finitude beyond comprehension. Morton relates our recognition of global warming to a consideration of other planets:

> "Planets are hyperobjects in most senses. They have Gaussian geometry and measurable spacetime distortion because they are so massive. They affect everything that exists on and in them. They're 'everywhere and nowhere' up close (viscosity)…. They are really old and really huge compared with humans. And there's something disturbing about the existence of a planet that far away, perhaps not even of 'our' solar system originally, yet close enough to be uncanny (a very large finitude)…The historic moment at which hyperobjects become visible by humans has arrived. This visibility changes everything. Humans enter a new age of sincerity." (Morton 2013, p. 128)

Implications for reaching, let alone inhabiting, are harder to judge. What purpose such an endeavor? If spaceships can be understood heretofore as a fusion of architecture and HCI, what is their future? The implications of Very Large Finitude are clearly set into a picture of Earth's Anthropocene era and its uncertain aftermath:

> "Futurality is reinscribed into the present, ending the metaphysics of presence: not through some neat philosophical footwork, but because the very large finitude of hyperobjects forces humans to coexist with a strange future, a future 'without us.'... Thanks to hyperobjects, the idea that events are tending toward the future, drawn by some ineluctable telos, is discovered decisively to be a human reification of aesthetic–causal appearance-for. It is the—for that indicates that we are already in an interobjective space when it comes to thinking this way, a space demarcated by entities that subtend the interobjective space. ... the end of the world is the end of endings, the end of telos, and the beginning of an uncertain, hesitating futurality." (Morton 2013, pp. 94–95)

A future without us, an end to teleology—and to the world itself, as Morton phrases it. This effect of Very Large Finitude does not so much undermine any new space programs as *liberate* them. Hyperobjects and Very Large Finitude are key concepts for discussing spacecraft at this moment, because these concepts show a way forward. While hyperobjects may not always be of human making, Very Large Finitude belies to human incapacity to deal with them. What if spacecraft were not reliant on our understanding of their purpose? Could we begin to work at the service of something non-human—a mission other than ourselves?

Writ large, the driving forces and products of VLOs (globalized capital, for one) and of hyperobjects (global warming, for one; Very Large Finitude, for another) are clearly entangled with one another. As we look ahead to a 100-year project of this kind, we can begin to imagine its implicit byproducts as dismantling VLOs and also as freeing itself of Very Large Finitude. At the weighty risk of bombast, humanity itself is faced with the catalyst for a futurality that is more architectural even as it is re-inventive of architecture's underpinnings: more steeped in imagination, spatially resistive to a section cut or orientation to a ground, and proposing an infinite interiority.

With infinite interiority, we are freed to think past homeostasis, control, and a *milieu intérieur* toward the "felt" as basis for human relations after the Anthropocene, and to build for non-human beings. Thanks to this project, time, scale, envelope, inhabitation—some of the very defining terms of architecture—are readied for reassessment.

15.1.4 Beyond the Very Large

15.1.4.1 Hazy projections (futurality)

How is the future architectural and how is architecture only of the future? Architecture exists always as a projection, in notational drawings, in abstract models, in promissory words. And yet, these futures are always tethered to some established terms of the present and of the past; the future is contingent on origins. In one touchstone image of eighteenth-century architectural discourse, Marc-Antoine (Abbé) Laugier's *Essai sur l'Architecture* appeared with a frontispiece by Charles Eisen. Here was engraved the very personage of architecture as a woman, gesturing toward a frail thicket of branches supporting twigs and leaves. This source of shelter, which itself appears to exist in a tentative state of stasis, is today known as the *primitive hut*; it captures paradoxes of architecture as both protective and needing support, of and from Nature. It is the subjugation of living flora to the ostensible inhabitation of fauna. The image seems to suggest that architecture may boast stability, shelter, separation from the elements, but that it is always subject to these things as well (Fig. 15.3).

The image also effects an operation on time through the medium of architecture: both projective and retrospective, and, to complicate matters, ambiguously so in both cases. Stan Allen describes the nature of all architectural drawing as "impure, and unclassifiable," neither purely *autographic* nor *allographic*. Paraphrasing philosopher Nelson Goodman, Allen explains:

> "(The) autographic (are) those arts, like painting and sculpture, that depend for their authenticity upon the direct contact of the author.... Allographic arts are those capable of being reproduced at a distance from the author by means of notation... as a consequence of the ephemerality of the work itself (poetry, or music), or the need to coordinate an intricate collaborative structure (dance, or symphonic music for example)... it is obvious that architecture is neither clearly allographic or autographic." (Allen 2008, pp. 33–34)

This distinction, involving authenticity and the author's presence, is evocative of Walter Benjamin's declaration of the eroding "aura" in works of art produced since the age of mechanical reproducibility. For our purposes, the fraught relationship to reality and time is productive, as it heralds spacecraft planning as an architectural task. By extension, we can examine how spacecraft design processes rewrite the conventions and terms of architectural work.

In 1969—the year of the first lunar landing—John McHale published his book, *The Future of the Future* (McHale 1969). By the time of the book's publication, McHale had served as co-founder of Britain's Independent Group, exhibited widely as an artist, published many articles on architecture and electronic technology, and completed a Ph.D. in sociology. All of these seem to have contributed to his enthusiasm for futurology.

McHale seems to speak to the oncoming recognition of hyperobjects:

> "Our view of the future is no longer that or a great evolutionary onrush, largely independent of man's intervention, tinged variously with doom or elation. We realize, for example, that man does not, in the end, master Nature in the nineteenth century sense but collaborates within the natural world; his very existence depends upon an intricate balance of forces within which he is also an active agent." (McHale 1969, p. 5)

Later, he says that "All our previously local actions have now been magnified to planetary scale. The knowledge with which we might make the correct decisions is barely adequate—yet our gross errors may be perpetuated for many generations" (McHale 1969, p. 15). And yet, in an entire chapter entitled "Outer Space," McHale delivers some positions that seem surprising in hindsight:

> "The developed space vehicle—with its protective shields and energy collectors and converters; with its internal closed ecology for the cycling of air, water and wastes; with its sensors and communicating devices—is a microminiaturized version of the

Figure 15.3 Allegorical engraving of the Vitruvian primitive hut, which served as frontispiece to Abbé Laugier's *Essai sur l'Architecture*, 2nd edn (1755). *Credit*: Charles-Dominique-Joseph Eisen

earth itself, a simulated planetary vehicle for the larger human community.... The 'closed ecology' of life support systems for sustained flights... is a systems model for the redesign of many of our large-scale industrial undertakings whose ecological malfunction degrades the quality of our environment." (McHale 1969, pp. 177–178)

The closed-systems model McHale describes (in diagrams and photos as well) is both typical of its era and absolutely pragmatic for many reasons. The contemporaneity of this model can be understood in terms of scientific and design communities' discussions of homeostasis, to which we will return momentarily and from which we might today depart. That McHale expresses these as miniatures of Earth, and as applicable or justifiable for their usefulness in improving industrial practices back home adds and inscribes a human *telos* and a terracentrism. Both of these can be seen today as running counter to a future-oriented undertaking. While both of these added values resonate with much of the discourse promoting space programs at this time, both are also questionable, as space programs are rethought for an era after the Anthropocene.

As has been described above, past spacecraft were realized, in part, through a productive "transfer" of ideas in fiction; by now, we can recognize these fictions as architectural design methods. In a similar reciprocity, space programs created their larger VLOs (think of NASA's Jet Propulsion Lab in partnership with Disney and ABC television), even as the VLOs were at the service of the space programs. As we look forward to Persephone's ambitions, we also recognize the related and potent role of black-sky thinking in this respect. That the Apollo missions were conceived and executed in less than a dozen years was a speed that added to the shock of the very accomplishment; if we instead ask for a project to build slowly, we are forced to speculate, to fictionalize, and then to realize a far more radical and as yet unknowable thing. Black-sky thinking demands the sort of far-reaching speculation that trades away the false comfort of "educated guesswork" for unleashed creativity. It also dispenses with the *telos* that was latent in McHale's era of space programming: rather than building ships at the service of shoring up our habits and more efficiently dispensing resources on Earth, we can begin to use Earth at the service of a long-term future project of the starship, mission indeterminate.

15.1.4.2 Hazy boundaries (continuous interiority)

What if we were to poke a hole in John McHale's closed ecosystems model as well? The Apollo spacesuit only happens to be one in a stack of layers by which architects and space engineers alike conceived of interior environments as closed and internally regulated.

The "*milieu intérieur*" or "environment within" was French physiognomist Claude Bernard's phrase to explain biological systems as reliant on internal equilibrium. This concept went on to strongly inform later ideas of homeostasis that proliferated in the twentieth century, from the natural sciences to computer engineering and architecture. Appearing in projects as disparate as Buckminster Fuller's proposed "Dome over Manhattan" and in corporate office towers, the separation of interior into a mechanically controlled environment ran tandem with the development of climate controls as its own field of engineering, but also with Manfred Clynes and Nathan Kline's coining of the word and concept "cyborg." In the context of space architectures, the cyborg can be held as a counterpoint to *terraforming*, the cultivation of non-terrestrial environments to be more like Earth (read: habitable by humans). This history of the ideas and their migration to outer space and buildings of the twentieth century is summarized by architect Matt Johnson (2014), who asks whether we ought to continue with the increasingly optimized separations of interior and controlled filtrations of interior atmospheres. To his implicit call to rethink these practices, I will add a proposal to reverse the drive for applicability: architectures on Earth certainly can begin dissolving the control mechanisms and envelopes of homeostats into more *continuous interiors*, but these may serve as a model for spacecraft built this way as well.

A star ark of continuous interiority is not the image of a vast torus as has been envisioned previously. Rather, it is many possible outcomes of thinking of all space as continuously and intrinsically *of* an interior—a project and a field of inhabitation that is more ethereal in its articulations and less reliant on adaptable but hermetic membrane. Continuous interiority is a concept with many implications and, like other architectural futures, not without pasts (Fig. 15.4).

A star ark and its continuous interior carry stark challenges to two related and essential terms of architecture: *section* and *envelope*. With the orbital space station, the section drawing, an orthographic projection used to study and explain the related spaces of an interior environment by adjacency and elevation, had already been drastically stretched in its range, legibility, and very constructibility. If section drawings historically rooted a building to its ground plane and explained a building's relationship to landscape, they also presumed vertical limits to habitable space. These limits have indicated the "landscape" orientation of a section drawing on a page of paper; the advent of tall buildings with steel and concrete construction, and the expansion upward with the development of the elevator, resulted in different drawings and different methods to understand sectional space. With orbital spacecraft above Earth's stratosphere, we could say that our collective imagination of habitable environments was radically sectionally expanded. The likelihood is rather slim that anyone will be constructing a drawing to represent such a space, to scale, in its position relative to a ground plane. And yet, to the extent that section drawings are both index and also agent of our imagination, the architectural range of habitable space grew exponentially in 1969. Yet it remained vertical in the sense that it referred always to Earth as its datum. The project of a 100-year starship carries with it two implicit changes to all this: whether built over a 100 years, or slated to move in outer space for this long, it is freed of any necessary terrestrial datum. Untethered, set adrift, the section of a continuous interior can learn from past ideas in the architectural imagination. Earlier than the many inspirations that architects found with Deleuze and Guattari's *rhizome* of the 1990s, we can look to Constant's countless models and drawings of New Babylon, or earlier still, to Piranesi's *Carceri* (Fig. 15.3) engravings. Not merely defiant of a clear edge and center, the warrens of Piranesi's prisons are also notoriously *errant perspectives*; they cannot be resolved into an orthographically accurate section. Here are drawings that suggest a *continuous interior* along with a method of its description in drawing.

Alive without us 421

Figure 15.4 The curious perspectives of the *Carceri* were not merely resistive to orthographic resolution, but suggestive of multiple parallel perceptions of interiority. *Credit*: Jordan Geiger, after Piranesi

Today, the interior on Earth means many things—shelter, security, perhaps—but also privacy (a rather recent cultural invention). With the proliferation of an Internet of things, we are only at the proverbial birth of a quantified self, for which data privacy grows difficult to pin down. We share not only our social doings and locations, but biologically interior conditions such as our heartbeat or blood alcohol content. These sorts of interiors grow privy to different kinds of eyes: friends, parole officers, unknown others. Yet, the live distribution of biometric data was pioneered already before ubiquitous computing, in the 1990s, when internet startup Quokka[2] attempted to stream Nascar racing and Whitbread sailing online with multiple on-board cameras as well as heat and perspiration sensor readouts from drivers' flesh. This distributed interiority is distinct from the spatiality described above, but its evolution will play at least a transitional role in how we reconcile the architecture of a star ark with the biometrics of life itself (be it passenger, pathogen, or plant).

The star ark will also confront a continuous interior as it regards cognition and sentience, such as these may become. Again, we may look at rich studies of sentience today in the very fabric of our cities (Shepard 2011). What changes here may relate to cognition in outer space? From the inside, from consciousness, from the imagination, the outside is what we sometimes think sublime, infinitely complex; safety is found in containment, in a clean room envelope, a sealed suit, or in a thought well defined by language (remembering here that it is the between—the Very Large Finitude—where problems begin). For anyone who has seen Andrei Tarkovsky's mysterious 1972 film *Solaris*, this will feel familiar. Based on a story by Stanislaw Lem, the film provides and yet undermines a viewer's experience of a spacecraft nearing the surface of the eponymous fluid planet, proximity to which seems to unleash the presence of dead loved ones and overlapping experiences of space on board the ship. Interiority is both spatial and cognitive, yet both of these ooze into one another. In *Solaris*, the surface sea is itself sentient, and its extended space is also its extension into the consciousness of other life forms that approach—a continuousness across physicality and sentience. Of all the cinematic works that have served to bolster real space exploration, this one has likely never been consulted. But why not? How might we look to the shifts in sentience in our urban spaces today, and rethink these as a means to shape a continuous interior? The prototyping of such conditions here on Earth are not an end unto themselves, but rather a laboratory for the "long now" project found in Persephone (Fig. 15.5).[3]

If we have challenged the architectural properties of a continuous interior for its sectional logic, its envelope, and its effects on consciousness, we must also plan such an interior with *material life*. As all hyperobjects are vast beyond human figuring, so too is the formation of synthetic living matter. Persephone does not call for a closed ecological system, but rather a fusion of its material self with life support.

---

[2] For a documentation on this project, see www.inventinginteractive.com/2010/05/26/remembering-quokka/ (accessed July 15, 2015).

[3] The Long Now foundation is a project initiated in 1996 by Stewart Brand to "provide a counterpoint to today's accelerating culture and help make long-term thinking more common. We hope to creatively foster responsibility in the framework of the next 10,000 years." One of its many projects is the construction of a 10,000-year clock.

Figure 15.5 The *Solaris* sea opens the possibility of imagining surface and envelope in flattened and intersecting tones. *Credit*: Jordan Geiger, after Tarkovsky

15.1.4.3 Fur and felt (non-humanity)

This fusion *becomes fur*—it does not take its forms in any sort of biomimicry, which Rachel Armstrong has aptly called "like biology for the copy-and-paste generation"[4]—but in a real sense of behavior. Fur is neither surface nor endlessly porous. Its plush depths are variable, its contours pliant, and its liveliness ambiguous. Fur grows, it insulates, and it binds to a living flesh. These are behaviors all central to the material qualities of a continuous interior, and they, too, have been subject to fictitious development that can serve us. Here, we can skip the sumptuous fur-lined bedroom designed by Adolf Loos for his wife, Elsie. Lining the room's surfaces and as backdrop to furniture pieces, fur is here purely an affect with hard substrate. Rather, we can look deep within the space capsule that carried Jane Fonda in the 1968 film *Barbarella*. Fonda's body and her ship are arguably *both equally* unabashedly sexualized and objectified. Yet, this is also because her hair and flesh meld into the fur—a mutual extension of the ship and her body (Fig. 15.6).

The realness of living matter will be steeped in the outcomes of today's incipient synthetic biology, and more. Synthetic biology releases new definitions of life at all scales, far from the logic of Laugier's "Primitive Hut": from sub-cellular particles that can now be programmed for simple actions to their shifting, non-directional assembly, as a cloth, or felt. Felt has been described as

> "infinite, open, and unlimited in every direction; it has neither top nor bottom nor center; it does not assign fixed and mobile elements but rather distributes a continuous variation... a space that is prolongable in all directions—but still has a center." (Deleuze and Guattari 1980)

---

[4] "Black Sky Thinking," Rachel Armstrong in conversation with Debbi Evans in *Libertine*, Winter 2013.

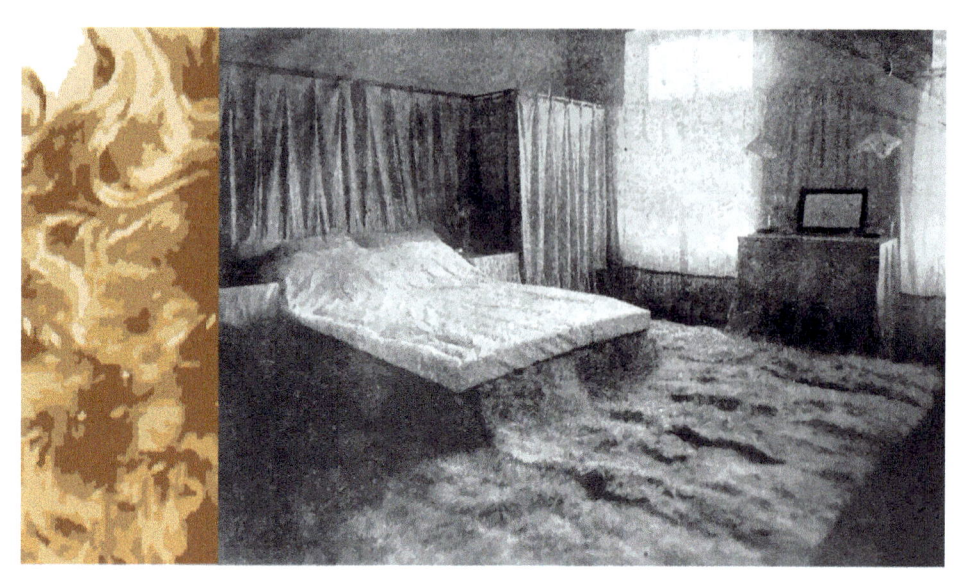

Figure 15.6 Two visions of fur and felt as architectural surface. *Credit*: Jordan Geiger, after Roger Vadim's *Barbarella* and Adolf Loos's wife's bedroom in Vienna

Again, this sort of description is to be taken literally, as space and a process rather than a mere formal arrangement. As felt comes about through the tangling of fibers, synthetic biology results from material and physical forces, but also human intertwining. Material sciences are also subject to the systems that create them, latent VLOs in formation in which capital and organizations push the selections of matter and the direction of its crochet.

Felting as a metaphor for *social* process has been called out in economics and in the sciences. Frederik Ceyssens, Maarten Driesen, and Kristof Wouters of the Technical University Leuven, Belgium, have raised important considerations around the financial and material resources that will bear on the development of "gigascale space projects" to come. They have called for the establishment of "an international network of nongovernmental organizations (NGOs) focused on private and public fundraising for interstellar exploration and supporting a bottom-up societal movement, similar to e.g. the WWF" (Ceyssens et al. 2011). Although they recommend that this body carry the clear goal "to get a second home planet for humanity by the end of this millennium"—something with which this text differs—they articulate a shift that is observable in popular movements and in academic fields as well, away from the sociality of the VLO model and its technological products.

Philosopher Nick Land first proposed an idea of "Accelerationism" as a kind of supercapitalism, but economists Alex Williams and Nick Srnicek counter with their #Accelerate Manifesto (2013), wherein they call for "…an accelerationist politics at ease with a modernity of abstraction, complexity, globality, and technology.... The existing infrastructure is not a capitalist stage to be smashed, but a springboard to launch towards post-capitalism." They claim a need to

> "... move beyond the notion that an organically generated global proletariat already exists. Instead it must seek to knit together a disparate array of partial proletarian identities, often embodied in post-Fordist forms of precarious labour... What accelerationism pushes towards is a future that is more modern—an alternative modernity that neoliberalism is inherently unable to generate." (Williams and Srnicek 2013)

These varied fields are the same ones that gave rise to VLOs of space programs past, but they clearly indicate a need to dissolve such administrative structures for space futures. Where DARPA and Disney served to mobilize large-scale space architectures under the purview of superpower governments, economic and social indicators like this suggest that there is something anachronistic in Foster's hard shell and its funding by the ESA. Rather, the softness of felt corresponds to a post-neoliberal financial structure as much as it relates to a soft interior. Softness and synthetic biology serve possible non-human beings in the wake of an anthropocentric logic that have launched with so many closed ecosystems in our past.

These false dichotomies and false choices—human/non-human, but also cyborg/terraforming—dissolve in an era of small organization, of continuous interior. Biosphere 2 has a certain infamy for its faulty mechanical systems, failed social experiment, and more. But, we still have lots to learn from its essential mission of soils containment and its terrestrial inheritance from a space fiction. Part of the structure, a space frame of steel and glass in the form of a ziggurat, shared not only the structural and material logic, but, in fact, some of the same staff that modeled closed dome greenhouses aboard the spaceship in the 1972 film *Silent Running*. Those domes, themselves an echo of Fullerian lightweight enclosures, were built on the same logic as McHale's "microminaturized Earth"—yet they were, in fact, in holding patterns as they awaited Earth's return to a hospitable status for humans. Later, the team at Pixar that created the 2008 film *Wall-E* acknowledged the strong inspiration that they took from the earlier film. Yet, they left some possible ambiguities about its tacit project, in which all life is naturalized and conceived as a resource at the service of humans. What is HCI without the H? What is a biofelt in lieu of a biosphere, a continuous interior in place of a hard shell?

In an era of long now planning, in which we can think of very slow interactions and very small organizations, it is incumbent on us to move past such a central role for humans and a servile existence for everything else. We are obligated to pursue Persephone to confront Very Large Finitude, but also to liberate architecture at the service of all life. Persephone will have a living interior, but it will also yield a continuous, living interiority—more *Barbarella*'s fur capsule than Loos's bedroom. It will be scaleless, soft, and untethered from orthogonal section. It will be inexact, slow, and not for us. We can rest in the future, having left behind the shuttled soils of *Silent Running* for the fog on Mount Fuji (Figs. 15.7 and 15.8).

426 Constructing worlds

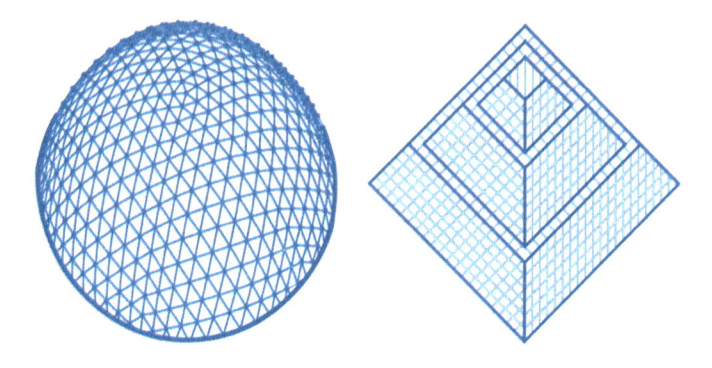

Figure 15.7 Glass enclosures in *Silent Running*'s ship, the *Valley Forge* (*left*) and on Earth's Biosphere 2 (*right*). *Credit*: Jordan Geiger

Figure 15.8 Mists at Mount Fuji. *Credit*: Jordan Geiger

15.2 The scales of *Ouroboros*

Mark Morris, Associate Professor of Practice, Director of Exhibitions and Events, Cornell University.

> "So close—the infinitesimal and the infinite. But I knew they were really the two ends of the same concept. The unbelievably small and the unbelievably vast eventually meet—like the closing of a gigantic circle." (Richard Matheson, *The Incredible Shrinking Man*, 1957)

15.2.1 Introduction (Fig. 15.9)

The ancient Greek figure of the serpent devouring itself—as celebrated by Gnostics, alchemists, Jungians,…whomever—is variously read as a symbol of infinity, self-sufficiency, or the cyclical return. So insistent is the circular theme that the word for it, *Ouroboros*, manages to pack in four letter O's as if to hammer the point home. Ostensibly, scale, in the architectural sense, does not work this way. It is linear (Fig. 15.10).

There is something in the graphic scale bar that suggests a snake or the meander of a snake's path. Yet, there would not be much of a point in wrapping an architectural scale in on itself. As the scale slides left to right, marking out progressive increments, its qualification as a vector, something with singular direction, appears obvious. The graphic scale, perhaps a nicety when nearly all architectural drawings were to a conventional scale, has seen a fresh peak in utility now that so many "drawings" shuffle through the digital world from CAD to BIM, as Google Image Search results, in a scalar vacuum merely awaiting

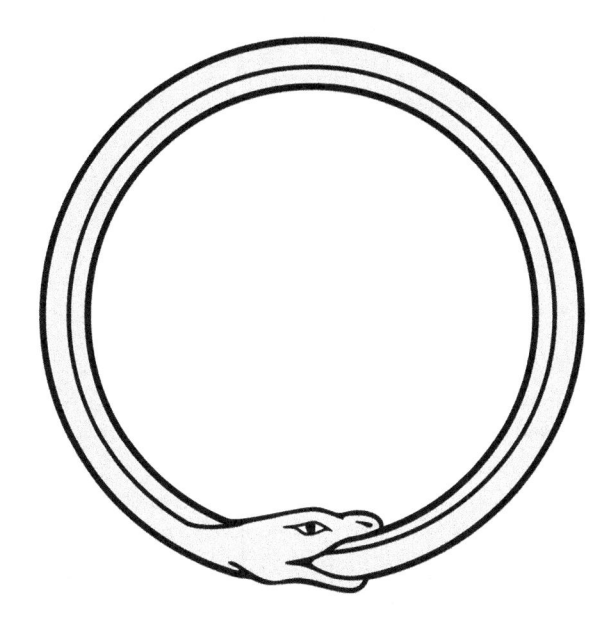

Figure 15.9 *Ouroboros* in its most essential form: a popular tattoo graphic. *Credit*: Wikimedia Commons

428 Constructing worlds

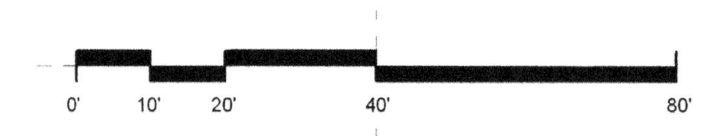

```
 0'    10'  20'        40'                            80'
```

Figure 15.10 An incremental graphic bar scale commonly used by architects and civil engineers. *Credit*: NA, standard graphic

Figure 15.11 Ray and Charles Eames hatching another of their genius ideas. Their second iteration (1977) of *Powers of Ten and the Relative Size of Things in the Universe* was their most successful and widely distributed film project—still a staple of classrooms. *Credit*: © Eames Office LLC

output. The only thing that permits this fluidity, lets the drawing output to a scale (any scale), is the graphic scale riding around with the drawing; few things in architecture, line for line, are so nimbly efficacious (Fig. 15.11).

When Charles (1907–1978) and Ray Eames (1912–1988) created their short documentary, based on Kees Boeke's (1884–1966) 1957 book, *Cosmic View*, "Powers of Ten" films (the prototype or *A Rough Sketch for a Proposed Film Dealing with the Powers of Ten and the Relative Size of Things in the Universe* was completed in 1968; a more polished *Powers of Ten: A Film Dealing with the Relative Size of Things in the Universe and the Effect of Adding Another Zero* was released in 1977); they scripted the voiceovers reflecting on the visual similarities between the space of outer space and the space inside a carbon atom: "The vast nothing in the realms between the nucleus and its orbiting electrons. The charged particles of the nucleus are in the distant center—all part of the atomic scale."[5] Further, they equate the quark-to-galaxy journey in basic building-block and organizational terms: "We are in the domain of universal modules. There are protons and neutrons in every nucleus, electrons in every atom, atoms bonded into every molecule out to the farthest galaxy."[6] The films zoom out to the edges of the galaxy before zooming back in to Earth and the single cell in a human hand:

> "As we approach the limit of our vision we pause to start back home. This lonely scene, the galaxies like dust, is what most of space looks like. This emptiness is normal, the richness of our own neighborhood is the exception."[7]

The voiceover observes that there is a visual pattern in the zooms—a sequence of busy matter separated by relatively vast spaces of nothingness, whether in outer space or in the inner space of a cell. The architects make the point that the basic formal rhythm between stuff—often circular-shaped stuff—and the space that contains stuff plays out similarly in either scalar direction.

In his essay "Heavenly Mansions" (1946), Sir John Summerson (1904–1992) offers an interpretation of Gothic architecture based on nested scales of shelter. He begins with observations on a child's pastimes, playing house under a table or with doll's houses: "He is placing either himself or the doll (a projection of himself) in a sheltered setting. The pleasure he derives from it is a pleasure in the relationship between himself (or the doll) and the setting."[8] Having experienced the Blitz in London, Summerson poignantly emphasizes the emotional response to such scenarios:

> "None of us ever entirely outgrows the love of the doll's house or, usually in vicarious form, the love of squatting under the table. Camping and sailing are two adult of play analogous to the 'my house' pretences of a child. In both, there is a fascination of the miniature shelter which excludes the elements by only a narrow margin and intensifies the sense of security in a hostile world."[9]

This is all to set up a critique of the aedicule; the pedimented or framed niche or, occasionally, freestanding small temple within a larger enclosed space: think *baldacchino*. He charts their history from ancient Roman alcoves intended for *lares* or household gods to their incorporation in church architecture spanning the Romanesque, Gothic, Renaissance, and baroque. These are ceremonial or symbolic representations of the small-house (*aedicula*) archetype, sometimes stacked or layered up as a house within a church (Fig. 15.12).

Summerson highlights the fact that, from the Gothic forward, the object intended to be enshrined by the aedicule—sculptural or painted representations of holy figures—begin to vanish in favor of emphasis being placed on the architectural symbolism of shelter itself.

---

[5] Matheson, R., *The Incredible Shrinking Man*, film script (1957).

[6] Ibid.

[7] Ibid.

[8] Summerson, J. *Heavenly Mansions and Other Essays on Architecture*, pp. 1–28. WW Norton, New York (1998).

[9] Ibid.

Figure 15.12 Renaissance aedicule carved into the court gatehouse of Tawstock, Devon (dated 1574). *Credit*: Wikimedia Commons

Moreover, it is the fractal quality, the suggestion of infinite scales, of the aedicule that links the mansion/church with the heavenly. Scale is the operative word that permits this analogy. Despite its association with measurement, scale per se is not unit-based, but wholly dependent on comparative relationships. Emmanuel Kant claimed in *The Critique of Judgment*[10] that relative scale, rather than quantitative size, defines the sublime; the sense of pleasurable relief following an initial reaction of dread at the vast or overwhelming. Kant allowed that the sublime is found at the extremities of scale, the realms of both telescope and microscope. The emotive aspect of *aediculae* (plural) at different scales in a cathedral, for example, can trigger a sublime experience. Summerson's idea of pleasure taken in shelter or protection against hostility could also be called sublime. "Powers of

---

[10] Kant, E. *The Critique of Judgment*, 2nd edn revised, translated by J.H. Bernard. Macmillan, London (1914).

Figure 15.13 Soane Family Tomb, St Pancras Gardens, London (1816). Pevsner described it as an "outstandingly interesting monument…. Extremely Soanesque with all his originality and all his foibles." *Credit*: Courtesy Brian Harrington Spier

Ten," likewise, would seem to evoke a sublime reaction in the dizzying speed and shifting sense of scale played out in the films (Fig. 15.13).

Summerson was curator of Sir John Soane's Museum from 1945 to 1984, where he wrote portions of "Heavenly Mansions" and subsequent essays to be compiled into a book under the same title. Soane and the preservation of his legacy comprised Summerson's long-term professional project and he was particularly interested in Soane's renderer and watercolorist, Joseph Michael Gandy, who realized, even elevated, the vision of Soane. An entire chapter of the book *Heavenly Mansions* focuses on their collaborations. Alongside his own house in Lincoln Inn's Fields, one of Soane's most personal projects was the 1816 design of a tomb to be constructed over the vault of his wife and confidant Eliza who had died the year before. It was an unusual design, drawing on Egyptian, Greek, and Roman funerary tropes. A freestanding aedicule, the temple shelter is topped with a pinecone encircled by an *ouroboros*. The arched pediment of the temple is decorated with a wavy line, Soane's own invention, reminiscent of a snake's meander. The tomb became the family's mausoleum, Soane's elder son and Soane himself eventually being buried there, famously influencing Sir Giles Gilbert Scott's design for the red K2- and K6-type telephone boxes a century later. As built, the *ouroboros* carving was enlarged and moved to ring and appear to slightly squeeze the pillbox lantern. One of Gandy's representations of the tomb, set vividly against a cloudy sky, includes depiction of both of Soane's sons, John and George, neither of whom was on speaking terms with their father for much of their adult lives, hilariously shrunk down in size in an effort to make the modestly scaled project

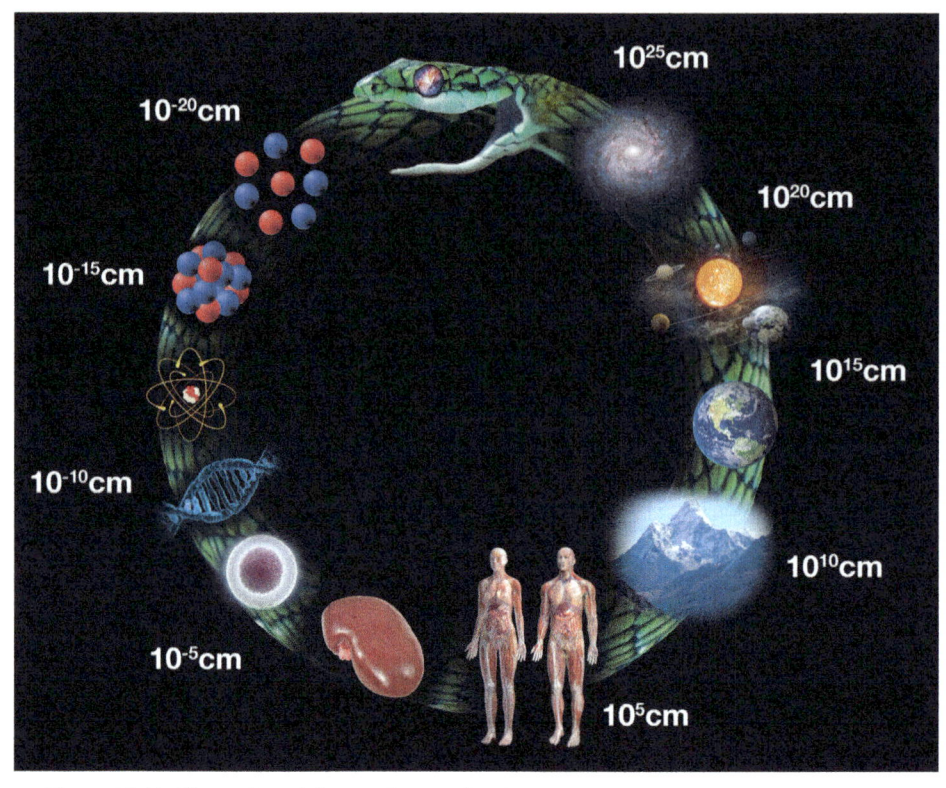

Figure 15.14 Illustration of the *ouroboros* micro/macro model as described by Martin Rees. *Credit*: Courtesy David Heskett

appear grander. This graphic game was in keeping with the slippery *Alice in Wonderland* sense of scale that both Soane and Gandy delighted in (Fig. 15.14).

The *ouroboros*, in its theosophical iteration, it is depicted as looping around the Sun, or the Sun and Moon, or the Sun and planets (depicted as astrological glyphs) standing for the limits of the Solar System and infinity of the cosmos, time, and planetary motion. Cosmologist, astrophysicist, professor, author, Astronomer Royal, and former president of the Royal Society, Lord Martin Rees, Baron of Ludlow, prefers the *ouroboros* as a logo for his research group, marked off in exponentially growing scales, explaining that it represents, for him, "the interconnectedness of the micro-world and the cosmos, the inner-space of atoms and the outer-space of the universe."[11] The link, where the snake eats its tail, is where sub-nuclear particles (dark matter) hold universes together. Addressing the irresolution of Einstein's theory of relativity versus quantum theory, or the theories of the vast and small, he calls for a unified theory of the cosmos and the micro-world. His diagram places humans at the midway point of the *ouroboros*: big enough to be atomically complex, but small enough not to be crushed by Earth's gravity—scalar equipoise.

---

[11] Rees, M. Will Civilisation Survive the 21st Century? *Proceedings of the Indian National Science Academy*, **73**(3), 197–203 (2007).

15.2.2 World-building (Fig. 15.15)

Science fiction plays up the small/vast interrelationship in a variety of ways. In the 1979 science-fiction novel based on his BBC Radio 4 series, *The Hitchhiker's Guide to the Galaxy*, Douglas Adams (1952–2001) introduces readers to a character unique to the genre. Slartibartfast—a name intentionally devised to sound, but not technically be, objectionable—is a designer of planets, the last of a race of Magratheans who specialized in building bespoke luxury planets for the richest species in the universe. When Slartibartfast meets the protagonist of the narrative, Arthur Dent, he describes his job and its attendant frustrations:

> "Look at me, I design coastlines, I got an award for Norway. Where's the sense in that? I've been doing fjords all my life, for a fleeting moment they become fashionable and I get a major award. In this replacement Earth we're building they've given me Africa to do and of course I'm doing it with all fjords again because I happen to like them, and I'm old fashioned enough to think that they give a lovely baroque feel to a continent. And they tell me it's not equatorial enough."[12]

Figure 15.15 An unusual inverted height-map made with data from the Norwegian Mapping Authority, its frilly coastline surely worthy of a design award. *Credit*: Courtesy Jon Olav Eikenes

---

[12] Adams, D. *The Hitchhiker's Guide to the Galaxy*, Chapter 30. Pan Books, London (1979).

This scene expands the domain of the designer—extending its scale to whole worlds—but, tellingly, does not expand his worldview. Adams's comic writing technique often involved absurd juxtapositions and tweaking a detail to render the whole impossible or silly. The equation of the Norwegian coastline with the baroque is a formal simile, comparing the numerous tight undulations of Norway's western border with the sea with, one imagines, the frilled sinuous curves found in, say, a sculpture by Gian Lorenzo Bernini (1598–1680). The mention of getting "a major award" for his Norwegian work implies a popular aesthetic agenda and appreciation for the physical appearance of a planet. Adams explains the problems with naturally occurring planets and the planet-building process:

> "None of them was entirely satisfactory: either the climate wasn't quite right in the later part of the afternoon, or the day was half an hour too long, or the sea was exactly the wrong shade of pink. And thus were created the conditions for a staggering new form of specialist industry: custom-made luxury planet building. The home of this industry was Magrathea, where hyperspatial engineers sucked matter through white holes in space to form it into dream planets."[13]

Within the Magrathean schema, there are artificially constructed worlds amidst naturally occurring ones. These, Adams describes, might be golden planets, planets with a gelatinous surface prone to hilarious earthquakes, and so on. Slartibartfast is portrayed as brilliant but bumbling, preoccupied—despite the enormity of his projects—with conventional architectural aggravations and questions of style.

He takes Arthur on a tour of his facilities in an aircar, coming through a gateway into a vast space:

> "The wall appeared perfectly flat. It would take the finest laser measuring equipment to detect that as it climbed, apparently to infinity, as it dropped dizzily away, as it planed out to either side, it also curved. It met itself again thirteen light seconds away. In other words the wall formed the inside of a hollow sphere, a sphere over three million miles across and flooded with unimaginable light.... Ranged away before them, at distances he could neither judge nor even guess at, were a series of curious suspensions, delicate traceries of metal and light hung about shadowy spherical shapes that hung in the space."[14]

15.2.3 Architectural iterations

There are two precedents for such a space. One is Étienne-Louis Boullée's (1728–1799) unbuilt architectural design for a cenotaph to Isaac Newton (1642–1727). A sublimely scaled 150-m-diameter sphere set in a circular base ringed with cypress trees, the memorial would have inverted day and night, letting sunlight filter through small holes as so many constellations and, at night, internally glow from a centrally suspended armillary sphere made into an enormous lantern. Printmaker William Blake (1757–1827) commemorated Newton in a 1795 monotype, depicting him as geometer of the universe, measuring the cosmos with diagrams and a compass. The other precedent is theoretical physicist and

---

[13] Ibid.

[14] Ibid.

Figure 15.16 Étienne-Louis Boullée's sublime Cenotaph to Newton project, a section through the vast monument revealing the interior aglow with its model sun. *Credit*: Cornell Libraries

mathematician Freeman Dyson's (1923–) hypothetical sphere or shell, where a megastructure or satellite structures would be built by an advanced civilization around a sun, capturing and tapping all of its energy. A Dyson ring—a circular strip or series of aligned satellites—likewise harnesses a portion of a sun's energy (Fig. 15.16).

Much has been made of both the sphere and ring elsewhere in science-fiction literature. *Star Maker* (1937) by Olaf Stapledon (1886–1950) inspired Dyson's own elaboration of the sphere. Larry Niven's *Ringworld* (1970) caught the attention of MIT physics students who disproved its viability; he wrote *Ringworld Engineers* (1979) to address their concerns. *Farthest Star* (1975) by Frederik Pohl featured a Dyson sphere unusually inhabited on its exterior. The *Cageworld* series by Colin Kapp, published in the early eighties, described concentrically constructed Dyson spheres, while *The Rise of Endymion* (1997) by Dan Simmons featured a partial Dyson sphere built with biotechnology. *Heaven's Reach* (1998) by David Brin is about a Criswell structure, a lacy version of a Dyson-like sphere employing fractal geometry across the structure's surface to maximize the reflection of sunlight (Figs. 15.17, 15.18, and 15.19).

A smaller version of the Dyson ring, the O'Neill cylinder, was proposed by physicist Gerard K. O'Neill (1927–1992) in 1976 and subsequently refined by him and his Princeton physics students as consisting of two 8-km-diameter, 20-km-long counter-rotating cylinders designed to cancel out gyroscopic effects and provide artificial gravity by centrifugal force. The cylinder was the third of three "islands" or outer-space design postulates that O'Neill organized, Islands One and Two being different scales of rotating spheres inspired by the Bernal sphere (1927)—artificial worlds where atmospheric conditions and gravity would permit equatorial inhabitation. Merging aspects of the sphere, ring, and cylinder,

436 Constructing worlds

Figure 15.17 Various postulated spherical and ring space colonies with populations of over a million by NASA artist Rick Guidice. *Credit*: NASA Ames Research Center

the Stanford torus ring (1975) was developed by Stanford students in a NASA summer study program. The topographical product of two circles rotated through each other, the Stanford torus is ouroborosian in three dimensions. Its perpetual rotation generates centrifugal force equal to Earth's gravity. Solar energy and light are provided by a paired mirror disk that can angle toward a sun independently.

Graphic artists hired by NASA provided hypothetical renderings of life inside the ring. What is interesting architecturally about these images is the visual juxtaposition and, oftentimes, design friction between the "space-age" or modern superstructures versus the conventional landscapes and architecture contained by them—the relatively small and vast comingling. This was intentional, part of a strategy to produce a "normal" human environment—that is to say, landscapes and architecture evident in 1970s California, specifically around Silicon Valley—in outer space. If one looks closely at the architecture represented, one finds terracotta-tiled roofs, whitewashed, even modern stucco houses with amble fenestration set in a tiered sunny valley running continuously inside the ring. A garden party is being arranged on a terrace in the foreground. Other illustrations show the agricultural zones of a torus ring as strips of river-valley landscapes alternating with windows looking onto outer space and mirrored projections of the Sun. Friction is felt again in the romantic design of these landscapes where country roads meander and orchards abide by no grid.

Figure 15.18 Various postulated spherical and ring space colonies with populations of over a million by NASA artist Rick Guidice. *Credit*: NASA Ames Research Center

Astronomer, astrophysicist, professor, and director of the Laboratory for Planetary Studies at Cornell University, Carl Sagan (1934–1996) was an advisor to NASA, starting in the 1950s, and became increasingly involved with the imagery associated with speculative outer-space settlement as part of his 13-part television series, *Cosmos: A Personal Voyage*. Created in 1978–1979 and aired in 1980, *Cosmos* was immediately popular and garnered a number of awards in respect to Sagan's advocacy of science. The twelfth chapter of his series, "Encyclopaedia Galactica," featured a number of hypothetical worlds, some of which could sustain life. These planets, as imagined by Sagan and painted by David Egge, John Allison, and Jon Lomburg, included "an oceanic Earth-like world with two large moons," "a terrestrial planet with major engineering work visible on its night side," and "a still more advanced civilization, constructing a habitable ring system around its home planet."[15] Sagan's ring resembled Dyson's, except that his ringed a planet rather than a sun. These speculative planets functioned as a finale of the series, opening a discourse not only about human settlement of outer space, but extraterrestrial alien life—a subtext of the whole series (Fig. 15.20).

---

[15] Sagan, C. *Cosmos*, p. 312. Random House, New York (1980).

Figure 15.19 Various postulated spherical and ring space colonies with populations of over a million by NASA artist Rick Guidice. *Credit*: NASA Ames Research Center

Sagan was a connoisseur of modern architecture. When he and his wife, Ann Druyan, built their "scholars' retreat" in Ithaca, New York, a stone's throw from Cornell, he sought out Le Corbusier-trained Chilean-born Jullian de la Fuente (1931–2008) to design a new interior and penthouse study for an existing Egyptian-revival temple (c.1890, once home to a Cornellian secret society) set on a cliff overlooking waterfalls. Druyan acknowledged, "We had just completed 'Cosmos,' and the temple evoked something of the great library at Alexandria."[16] In many ways, the house represented an architectural distillation of the "Cosmos" project as whole. *Architectural Digest* did a feature on the finished project in 1994, entitled "Of Architecture and Astronomy: Capturing the Sky in Carl Sagan's Ithaca, New York, Study." Explaining that the design of the retreat was inspired by two Joan Miró (1893–1983) scrolls owned by Sagan, Druyan described the architecture as "utterly free of trendiness or Postmodern gewgaws."[17] Instead, amidst the Spartan solemnity of the place,

---

[16] Viladas, P. Of Architecture and Astronomy: Capturing the Sky in Carl Sagan's Ithaca, New York, Study. *Architectural Digest*, pp. 72–77. Condé Nast, Los Angeles (1994).

[17] Ibid, p. 74.

Figure 15.20 Carl Sagan's retreat overlooking a deep gorge and waterfall in Ithaca, New York, as it looks today. The entry on the short elevation features an *ouroboros* relief over the doorway. *Credit*: Courtesy "M and J"

a few carefully conceived architectural strategies related to astronomical themes are deployed: "The space shows a relationship to the sky: a skewed square skylight acts as a solar clock; a circular window in the door allowed the light of the setting sun to move in an arc across the floor."[18] The existing carved limestone entrance includes a winged, cobra-flanked sun disk — the eye of Ra. This ancient Egyptian formulation of the *ouroboros* theme represented the solar cycle and the daily rebirth of Ra at each dawn. The twinned guardian cobras, or *uraei*, coil their tails in a circle to define the disk's frame. As one tail thins out, the other thickens to complete the circle. The retreat was intended to prompt and sustain scholarship and writing. Sagan is quoted in the article, saying that "The work we do here has to be worthy of the grandeur of the architecture."[19] The text by Pilar Viladas that accompanied the magazine's photo spread kicked off with an engaging premise (Fig. 15.21):

> "Architecture and astronomy have more in common than one might imagine. Both, after all, are concerned with space; both, at their best, combine science with art; and both balance the physical and metaphysical. Therefore, the meeting — however abstract — of these two disciplines should produce interesting results."[20]

---

[18] Ibid, p. 75.

[19] Ibid, p. 76.

[20] Ibid.

Figure 15.21 The winged Sun flanked by uraei (sacred cobras) is one of the oldest Egyptian religious symbols, dating back to the twenty-sixth century BCE. *Credit*: Courtesy Leo Reynolds

One could look at this theory regarding the fruitfulness of architecture–astronomy unions in the recent work of architect Greg Lynn, principal of FORM, presented at the Canadian Centre of Architecture's exhibition, "Other Space Odysseys: Greg Lynn, Michael Maltzan, Alessandro Poli" curated by Giovanna Borasi and CCA Director Mirko Zardini. Lynn's project "New Outer Atmospheric Habitat (NOAH)" is presented as a series of models of four planets along with drawings and animations. Each model appears as a twisted form made up of cellular structures and nested spaces intended to provide for wide range of microclimates. One cylindrical model contains a dense internal field of hills and valleys—something approaching a Criswell structure. Both projects are variations of the Dyson ring in form and both take Lynn's longstanding architectural interest in complex morphologies, parametrically derived geometries, and folded spaces to new scalar realms, while retaining much of Lynn's characteristic design approach evidenced in his buildings and industrial design work. In other words, the architect's established design techniques stretch and become more complex, but do not fundamentally recalibrate in the pursuit of planetary design. In this, Greg Lynn is Slartibartfastian: a creature or, rather, an architect of habit. Similarly, though in the opposite scalar direction, when Lynn created coffee and tea vessels for the Italian kitchenware company Alessi in 2003 and jewelry for Swarovski in 2012, he scaled down, modulated, and simplified his forms to meet the challenges of those extra-architectural requirements. This steadfastness of design philosophy, regardless of scale, might be viewed as an architectural fault or a positive advantage (Fig. 15.22).

Long before she won the Pritzker Prize, before she had her first commission, Zaha Hadid launched her first monograph with an exhibition at the Architectural Association in 1981. Published as a boxed folio, *Planetary Architecture* functioned as a manifesto. Book collectors will know that this title and the subsequent 1983 folio, *Planetary Architecture Two*, are incredibly rare and unlike later glossier monographs of her work. Both folios stress the "planetary" theme in various ways. In one sense, the word implies a global practice, one not tethered to a specific national tradition or agenda. In another, *planetary* plays

Figure 15.22 Greg Lynn's models for "Other Space Odysseys," CCA exhibition (2010). *Credit*: © Canadian Centre for Architecture

with the idea that her distinctive work appears as if from another planet. The graphic reproductions of the folios, the paintings, and model photography convey this alien sensibility. Ambiguity—in regard to scale—runs through most featured projects. There is also a formal aspect implied by the folios' title, where several projects appear to lift and pull the planet or ground plane into their designs, while other buildings project themselves into wider landscapes or across cities with subtle manipulations of ground. This refusal to work within conventional restrictions of the object building and, instead, advocate architecture's wider claim to reconfiguring space beyond any nominally given site remains a hallmark of her practice. Other projects included in the *Planetary Architecture* folios are marked by their visual buoyancy—a hovering or anti-gravitational stance set in dynamic juxtaposition with the "heavier" ground-derived forms (Fig. 15.23).

Hadid, even more than Lynn, is a Slartibartfast: an architect so confident and dexterous in her approach that she can take on any commission—be it a building, a city, a couch, or a shoe—with the same level of assuredness. One hesitates to say that it is an application of style but, rather, it is a palette of forms, profiles, and organizational strategies that can move up and down a scalar chain without losing the basic identity of that palette. Something as relatively small as a light fixture, the VORTEXX chandelier (designed with Patrick Schumacher in 2005), exhibits several characteristics of her office's buildings, so much so that the chandelier could be viewed as a repurposed architectural model. The sinuous form of the VORTEXX is

Figure 15.23 The second Architectural Association folio, Zaha Hadid's joyously trippy *Planetary Architecture* (1982), currently trading at around US$4000 in the used-books market. *Credit*: Cornell Libraries

well within Hadid's morphological palette, but its progressive thinning out and spiraling in on itself make it an *ouroboros* sculpture. The spiral does not come to a point and stop, but swells again and loops back up as a continuous coil. The VORTEXX is in the same morphological family as Lynn's "New City" form—a species of wobbly ring that, alongside the sphere, seems to make up the bulk of architecture's responses to world-building.

15.2.4 Conclusion (Fig. 15.24)

When contemporary Cuban sculptor Alexandre Arrechea was invited in 2013 to create a sculpture as part of a Manhattan temporary public art project, he took inspiration from one of the city's architectural icons located near his installation site in Midtown.

Figure 15.24 Alexandre Arrechea's steel "No Limits" (2013), one of a series of his large-scale sculptures temporarily set up around Manhattan, all treating architectural themes. *Credit*: New York City Parks and Recreation Department

444 **Constructing worlds**

The piece, "No Limits," bends the form of the Helmsley Building (originally the New York Central building), designed by architects Whitney Warren (1864–1943) and Charles Wetmore (1866–1941) in the late twenties, into a circle, where its distinctive pyramidal cupola is about to be devoured by the archway at the building's base, its archway spanning Park Avenue forming the "mouth." Arrechea explains that "It's a play on the [*ouroboros*] myth. It's like a city that devours itself. That has always been my first vision of New York."[21] As with the Pan Am building (now the MetLife) that would come later in the early sixties, the Helmsley was noteworthy and controversial because it spanned a major street, blocking the view. It really does appear to eat up Park Avenue. The installed sculpture has a curious scalar impact as it appears to mimic and miniaturize the facades of buildings facing it. The piece seems either an architectural model of exaggerated proportions or a miniaturized and manipulated building; none of this is possible to read into the small red maquette without the context submitted in Arrechea's proposal. Recalling Rees's micro-world-cosmic *ouroboros*, the sculpture claims its own equipoise between the scales of architecture and its conventional representations. The piece is also an aedicule in Summerson's terms, freestanding like Soane's tomb, a reiteration of the city within the city (Fig. 15.25).

When the *ouroboros* expands in scale to become a whole "world" as with the torus ring, its identification as such is not so much rooted in the pictogram of the snake as in its functionality and embrace of real cycles, duration, and spatial endlessness. When it is miniaturized, made into a chandelier, its characteristics shift to formal readings of a spiraling line with no apparent endpoint. As a pure symbol, Soane's tomb example, the *ouroboros* still has a direct connection to architecture; it loops and gently squeezes architectural

Figure 15.25 The atomic and/or solar-system *ouroboros*. *Credit*: Photobucket photo share

---

[21] Ke, P.-R. "Alexandre Arrechea's 'No Limits' takes over Park Avenue," *Wallpaper*, February 2013 issue.

form. Even when it is reduced to a relief, as in ancient Egyptian or Roman depictions, it is extruded from a wall. Thus, the *ouroboros* symbol is often not a pure abstraction, but bound to architectural expressions and uniquely about shifting scale or confusing it by linking the largest expression with the smallest: the Eames or Rees take on the theme of looping scales. When the *ouroboros* is represented by a pure glyph, it exposes itself to comparisons with other similar images from the very small, atomic energy, to the very large, solar-system glyph. When a tattoo parlor offers up its popular version of the *ouroboros*, it combines these figures into a hybrid—one that melts atomic, cosmic, and figural representations into one. Here, electrons are planets and vice versa, in just the way the *ouroboros*-as-concept would suggest.

COMBINED REFERENCE LIST FOR CHAPTER 15

S. Allen, Mapping the unmappable: on notation, in *Practice: Architecture, Technique and Representation*, ed. by S. Allen (Routledge, London, 2008), pp. 33–34

M.G. Björk, The modern things, in *Post*, ed. by Björk (One Little Indian and Elektra, London (audio recording), 1995).

F. Ceyssens, M. Driesen, K. Wouters, P.-J. Ceyssens, L. Wen, *Organizing and Financing Interstellar Space Projects: A Bottom-Up Approach.* (DARPA/NASA, Orlando, FL: 100 Year Starship Conference, 2011).

N. De Monchaux, *Spacesuit: Fashioning Apollo* (The MIT Press, Cambridge, MA, 2011)

G. Deleuze, F. Guattari, 1440: the smooth and the striated, in *A Thousand Plateaus: Capitalism and Schizophrenia*, ed. by G. Deleuze, F. Guattari, and B. Massumi (University of Minnesota Press, Minneapolis, MN, 1980), pp. 475–476

P. Dourish, *Where the Action Is: The Foundations of Embodied Interaction* (MIT Press, Cambridge, 2001)

K. Easterling, *Extrastatecraft: The Power of Infrastructure Space* (Verso, London and Brooklyn, NY, 2014)

J. Geiger, Maximal Surface Tension: Very Large Organizations and Their Apotheosis in Songdo, in *Scapegoat: Architecture|Landscape|Political Economy*, ed. by A. Blackwell, C. Lee (Scapegoat, Toronto, 2013)

J. Geiger, Zero Atmosphere Architecture, in *Bracket 3 [at Extremes]*, ed. by L. Shepard, M. Przybylski (Actar, Barcelona, 2015)

M. Johnson, The milieu intérieur, in *Thresholds 42* (The MIT Press, Cambridge, 2014), pp. 120–133.

J. McHale, *The Future of the Future* (George Braziller, New York, 1969)

T. Morton, *Hyperobjects: Philosophy and Ecology after the End of the World* (University of Minnesota Press, Minneapolis, MN, 2013)

M. Shepard (ed.), *Sentient City: Ubiquitous Computing, Architecture, and the Future of Urban Space* (Architectural League of New York, New York, 2011)

A. Williams, N. Srnicek, Accelerate manifesto for an accelerationist politics. *Critical Legal Thinking*, (2013), May 14, http://criticallegalthinking.com/2013/05/14/accelerate-manifesto-for-an-accelerationist-politics/. Accessed 10 July 2014

16

Interstellar research methodologies

Rolf Hughes and Rachel Armstrong

Two individual authors Hughes and Armstrong write collaboratively and individually to propose alternative forms of knowledge and disciplinary synthesis for the establishment of interstellar modes of existence

16.1 The art of the impossible: beyond reason in the twenty-first century

Rolf Hughes, Professor of Artistic Research, Stockholm University of the Arts, Sweden.

"It was her voice that made
The sky acutest at its vanishing.
She measured to the hour its solitude.
She was the single artificer of the world
In which she sang. And when she sang, the sea,
Whatever self it had, became the self
That was her song, for she was the maker. Then we,
As we beheld her striding there alone,

Rolf Hughes (✉)
Professor of Artistic Research, Stockholm University of the Arts,
10450 Stockholm, Sweden
e-mail: rolf.hughes@uniarts.se

Rachel Armstrong (✉)
Professor of Experimental Architecture, The Quadrangle,
Newcastle University, Newcastle-upon-Tyne, NE17RU,
United Kingdom
e-mail: rachel.armstrong3@ncl.ac.uk

Knew that there never was a world for her
Except the one she sang and, singing, made."[1]

> "If we can unleash our imaginations, then we'll begin to see the relationships and overlaps between different disciplines. We need audacious ideas that will give us optimism, so we can reclaim the 21st century as the age of impossible thinking."[2]

16.1.1 Introduction

If our understanding of life and human potentiality evolved over millennia in relation to a largely hospitable environment, how do we set about planning for life in conditions of unrelenting hostility? What is our relationship to an alien space? Is it a space of human projection or do we make our living systems, values, technologies, and dependencies *responsive* according to whatever we encounter out in the unknown? If so, how do we create the conditions for a *malleable*, terrestrial orphan—an indefinitely nomadic ecosystem?

Technologists dimension the world—identifying technical problems, measuring out materials and resources, refining prototypes until the resulting technological artifact is released as a "good enough" solution, its novelty momentarily masking the obsolescence implicated at the moment of its inception. Artists tend to be drawn to *unknowing*—mystery, contradictions, hidden, or "occult" connections outside those legitimized by the guardians of rationality—crafting an artwork capable of expressing its nuanced, resonant complexity.

This chapter seeks to ask what might be involved in creating an experimental laboratory space for the third millennium—one designed for the challenges of hypercomplexity. Against approaches based on technological determinism, I will advocate those based on Rachel Armstrong's philosophy of *black sky thinking* in the belief that a design task at this scale will require participants to conjure up unforeseen possibilities of fortitude from a fertile, cosmological sandpit. Against equilibrium and entropy, I will propose disequilibrium—not to produce chaos or its inverse—godlike designers—but rather ethical, philosophical, and design principles such as *poise*, *trust*, and the re-centering that occurs through shared acts of fertility, fortitude, and flourishing—that is to say, forms of *radical love*. For, if the crew on such an interstellar spacecraft are to survive their journey of generations, the ties that bond them must be strong, endurable, and indefinitely creative. The boundaries of categories (self, other, organic, artificial, species, technology living, technology) are each open to renegotiation on the Persephone project. This contribution is a first, tentative attempt to tease out some of the implications of this.

The chapter considers the ambition to design a "star ark" capable of interstellar space travel as an evolving case study by which we examine what practices of experimentation might contribute to the realization of such a vision. It sets aside questions of disciplinary affiliation and hierarchy—top-down command logics—in favor of magic, poetic modes of "knowing," and the contribution of the circus arts. I will propose that technological and engineering thinking must be exposed to calculated epistemic collisions to safeguard

---

[1] Wallace Stevens, "The Idea of Order at Key West," from *The Collected Poems of Wallace Stevens* (Alfred A. Knopf 1990).

[2] Rachel Armstrong, interview with Debbi Evans, "Black Sky Thinking", *Libertine*, http://liberti.ne/editions/rachel-armstrong-on-black-sky-thinking/ (accessed September 27, 2015).

against the ever-present allure of false prophets. The intention here is to liberate those involved in the ongoing space ark design conversations from not only existing materials and methods, but also conventions in regard to embodiment, emotion, experience, reflection—this, in turn, to invite a transformation of imaginative possibilities. The failure to respond on this level is evidenced in some of the recent reports concerning the design of interstellar space travel (one of which will be discussed shortly). Without wishing to disparage the efforts of others addressing this challenge, these preliminary "results" would appear to be, to a significant extent, already obsolete.

It is important, therefore, to emphasize that I am not proposing a "correct" approach to such questions. Rather, I seek to ensure that those forms of *knowing* and *expertise* that might be excluded or repressed if the Persephone project were formulated primarily as an engineering or technological challenge are, in effect, *integral* to the project's development. To do otherwise would be to eliminate kindness from the equation. As Jorge Luis Borges noted:

> "Poetry springs from something deeper; it's beyond intelligence. It may not even be linked with wisdom. It's a thing of its own; it has a nature of its own."[3]

I proceed, therefore, with a poet's empathy, but must first weigh in my hands the words that have been used to describe this beautiful dream so far.

Increasingly... a different set of needs arise...

"Increasingly," Jack Burnham writes in 1968:

"...products—either in art or life—become irrelevant and a different set of needs arise: these revolve around such concerns as maintaining the biological liveability of the earth, producing more accurate models of social interaction, understanding the growing symbiosis in man-machine relationships, establishing priorities for the usage and conservation of natural resources, and defining alternate patterns of education, productivity, and leisure. In the past our technologically-conceived artifacts structured living patterns. We are now in transition from an object-oriented to a systems-oriented culture."[4]

Where Burnham sees transitions between distinct paradigms, others, such as the object-oriented philosopher Graham Harman, would conceive of the relation between object-centered ideas and the processes arising from systems thinking as a series of entanglements—a layering of multiple paradigms bringing into a productive interplay a number of concepts which appear to be at odds with each other and therefore resist full integration. The relation between a (leaky) object and its associated network becomes one of paradox, an unresolved suspension, with action influencing object identity (and vice versa) as much as understanding of its core properties or "essence."

Burnham's identification of "a systems-oriented culture" where change derives "not from things, but from the way things are done" can also serve as an interpretative lens through which we illuminate differing conceptions of space travel. What ultimately informs such differing visions, I will suggest, is an *ethical argument* about the nature and purpose of life itself—thence its feasibility in extreme environments.

---

[3] Jorge Luis Borges, The Art of Fiction No. 39, *The Paris Review*, No. 40, Winter–Spring 1967.

[4] Jack Burnham, "Systems Esthetics," reprinted from Artforum (September, 1968), www.arts.ucsb.edu/faculty/jevbratt/readings/burnhamse.html (accessed March 25, 2015).

The art of the impossible: beyond reason in the twenty-first century

Technology-oriented visions, such as that of the "slower-than-light, generational worldship" proposed in the Astra Planeta final report (2015), emphasize the transportation vessel itself over the human and biological elements on board, or the conditions that await at its terminal destination. Astra Planeta defines an "interstellar worldship" as follows:

"The term 'interstellar worldship', or 'worldship' for short, is a spacecraft designed to travel over vast distances at non-relativistic speeds (slower than light: <0.10c). The term is reserved for spacecraft designed to carry a population of 100,000 people or more to a destination which may take hundreds to thousands of years to arrive at. A worldship is a generational ship in which many generations of people live their whole lives and die onboard. It is not a ship where the occupants are asleep or hibernating and it does not only carry human 'seeds' for growth at the destination."[5]

Compare the mission of the Perspehone project, specified as follows:

"In this project we will consider the application of living technologies such as protocells, programmable smart chemistry, in the context of habitable starship architecture that can respond and evolve according to the needs of its inhabitants. …A habitable long duration starship will need evolvable environments that not only use resources efficiently but can respond quickly to the needs of populations and bypass the current necessary time lags that are implicit in the current system—in identifying critical upgrades and then activating industrial supply and procurement chains—which are already playing catch-up by the time they are realized."[6]

The Astra Planeta report has a high modern technological bias and is illustrated by images of a circular vehicle thrusting ceaselessly (its fuel tanks seemingly inexhaustible) through the dark void of space. It speaks of the need for "unprecedented international cooperation" and "the creation of new legal regimes" to realize the eventual launch by 2115 of a worldship "full of thousands of representatives for the human species" which would set out "to settle or colonize a distant stellar locale."[7] There is a representational logic at work here—power relationships are mediated by "new legal regimes" and the travelers are "representatives," no less, of the human species. The politics of representation is not addressed, and there is an assumption that the system described will sail through space like flying clockwork, its parts working together harmoniously, its outcomes logical, predictable, Spock-like. Entirely absent from this account is a notion of hypercomplexity that addresses things that cannot be modeled by the tools of modern synthesis. Likewise, conspicuous by their absence are the human, cultural, ethical, and quality-of-life questions that are central to such an enterprise.

---

[5] Planeta Final Report, International Space University, March 20, 2015, p. 11.

[6] www.icarusinterstellar.org/projects/project-persephone/ (accessed April 29, 2015).

[7] Op. cit., p. 2. Elsewhere, the vision of the project is articulated as colonialism: "A self-sustaining worldship would have the capability to leverage a stellar system's resources to manufacture the necessary components to colonize a planet, construct a space station, or continue the voyage to another stellar system" (p. 10).

Consider, for example, how the report imagines interstellar travelers passing time for hundreds or perhaps thousands of years:

> "In addition to physical exercise, other recreational considerations must be taken into account to keep each individual healthy and happy. Each person must take care of his or her health habits and make time for relaxation and entertainment. Opportunities with free and group leisure activities will be in place for this.
>
> Various classes and clubs focused on artistic skills such as dancing, music, entertaining, drawing, photography, poetry, filming, sculpting, singing, painting, writing, and interior decoration will be formed. Diverse contests and competitions may be organized in order to rise up the competitiveness and to receive a special recognition and rewards among the population. Some events can include cultural nights, parties, concerts, and LAN (Local Area Network) parties. These activities will ensure that the population is united and monotony on the journey is decreased."[8]

It is a deeply conservative vision reminiscent of modern utopias—"competitiveness" (and its associated trinkets) is assumed to be the core purpose of play. Shopping is available to relieve the tedium of office work and students are assigned apprenticeships "to learn from leaders who have mastered specific skills vital to worldship operation" in what appears at times to be a return to a pre-modern oral culture: "Older students would also be required to teach and help younger students while grandparents can be tasked with taking care of infants."[9] Population control is addressed through a "reproduction lottery" ("The population could implement a certain number of offspring that a specific couple could produce which would aid the population explosion, as well as the genetic diversity of the population.")[10] Breeders breed, then shut up shop. Other ethical issues arise which might jeopardize the continued support of those back on Earth:

> "If the crew has to do things that are not considered ethical on Earth but necessary for the survival of the mission, will Earth continue to support such a mission, will it be a mission that Earth wants to support? For instance, if the crew has to euthanize an infected portion of itself to save the rest, is that ethically humane? What is it that makes us 'human'?"[11]

This is a central question, vital to the success of the entire mission, but it is one that is dealt with in this report only at a superficial level, despite the report authors acknowledging the importance of such questions in their conclusion to the report section dedicated to "Worldship Society":

> "On the whole, Societal Factors shows itself to be a domain of much importance. The study of the human condition is just as important and pertinent as any other. Each topic is of importance on its own as well as interlinking with many, if not all, of the others. Each has its own lengthy past and each hold an unsurmountable position in the day-to-day life of every human being, shaping the way we look and interact with the world. Many of the topics need to be paid varying levels of attention

---

[8] Op. cit., p. 53.
[9] p. 59.
[10] p. 60.
[11] p. 67.

before versus after the launch of the worldship, but all do need to be addressed in a timely fashion for the worldship itself."[12]

A swelling mass of matter… a living island… a kinetic sculpture…

"You are now in this Degree permitted to extend your researches into the more hidden paths of nature and science."[13]

Rachel Armstrong, project leader for the Persephone project, offers an alternative vision of an

"eco embryology… an epic embryology… an eco Leviathan… a swelling mass of matter… a living island… a kinetic sculpture… one that is being shaped into being by the materials that are becoming entangled through the fundamental agencies that garden it into existence."[14]

The two visions are not necessarily opposed, but entangled; the living island is airborne within a spaceship, the dimensions of which are a challenge to our imagination; object-oriented and systems-oriented thinking will need to seed each other on a soil of black sky thinking if life at this level is to be sustainable.[15]

*

Contact

Let us here consider the question of whether a body

can of its own force be carried or move upwards.

Do not let yourself be deceived by the case of flames,

or, indeed, of burgeoning crops and trees that increase and grow

upwards. It still holds true that anything with weight

tends downwards.

Lucretius, De Rerum Natura

"To be able to carry out a task which carries great responsibility I have been forced to refrain from what was my heart's desire in the days of my youth: being able to reproduce from external form and colour. In other words I have in fact been driven back from a field work, laboriously climbed the ladder …"

Hilma af Klimt

---

[12] p. 69.

[13] From the *Initiation of the Fellow Craft*, citing Robert Boyle, *Works*, Robert Lomas, *Freemasonry and the Birth of Modern Science* (Fair Wonds Press, 2003), p. 65.

[14] Rachel Armstrong, private email to the author, March 22, 2015.

[15] What underlies the viability of the project is an ancient, rather than a new, concern. The Vedas Sanskrit Scripture, dating from around 1500 BC, declares, "Upon this handful of soil our survival depends. Husband it and it will grow our food, our fuel and our shelter and surround us with beauty. Abuse it and the soil will collapse and die, taking humanity with it." Cited on United Nations Convention to Combat Desertification website, www.uncdd.int/en/programmes/Event-and-campaigns/WDCD/Pages/Proverbs-on-land-and-soil-.aspx (accessed March 25, 2015).

Interstellar research methodologies

I grip this ladder for you.
You could as well grip it for me
But this is how it is, this time.
You run up, launch, kick—and fly;
Reach out for support in mid-air;
Fall. Slow falling in outer space.
It's like a feather rotating on the breeze.

Again. I hold it for you.
Come. Try again. Trust me.
Up you run—launch, point and fly.
And fall. Gracefully.

But, dear friend, it was a better fall!
Don't you think so?

Listen, this falling probably isn't a problem.
The world was full of falling.
Falling is what we do.
It's what we want.

Even animals are falling all around us.
Falling is the graceful acceptance of time.
The fleeting choreography of the living
against the insistent gravity of the divine.

I am holding the ladder—
Or the ladder is holding me. ·

Or perhaps the earth (and its relations) still holds us all;
Each experiment, each connection between this and that;
You and me; the one that holds, the one that climbs,
The atmosphere that receives, resists, revives;
The context in which any of this makes sense (or fails to);
The gods that turn and hang their heavy heads
if we try to explain some, or all, of what happens
when we lift up—and momentarily away—from the freezing iron rungs.

*

16.1.2 Knitting entanglements

How, then, to explore the intricate entanglements between the technical, biological, ethical, political, and theological aspects of such a project? Are our existing belief systems and systems of value, designed and refined over centuries as they are for terrestrial life, up to the task? Are conclusions arrived at under the relatively stable conditions of equilibrium applicable in contexts of turmoil and disequilibrium? "It is essential," Armstrong writes,

> "… to deconstruct our existence and build up our identity again as a species that is bound by a common mission and cosmic purpose. This is a challenge with no single solution. It is awe-inspiring, difficult and strangely tautological as each set of decisions about who we are provides the basis for the next set of actions as to who we may become."[16]

This double demand—deconstructing and reconstituting our species identity—suggests a dynamic between ontological *unraveling* and ecological *knitting*. Knitting hints at an alternative form of computation, anticipating the weaving industry, the Jacquard punch card, and Ada Lovelace's notion of computer program, where fabric is not produced, but where the computational endeavor exceeds representation in number, becoming instead music, art, and choreography. Knitting brings us back to the idea of computationally producing the environment we inhabit—where the forces of natural selection are no longer random but pre-programmed into the design fabric of the space, from which they may be transformed through myriad encounters with their inhabitants. Knitting also gestures toward the convergences between unlike things like bodies, spaces, objects, technologies through iterations (informational, physical, linguistic) to experimentally investigate limits and transgress them. Not as the vague Enlightenment notion of polite, benevolent progress, but as a series of disruptive collisions, leading to meltings of boundaries, meetings of monsters, sudden starts, stops, and disjunctive accidents, before something carries across, grips, and grafts; starting to synthesize, it demonstrates its pervasiveness, and begins iteratively to (re)(in)form events.

16.1.3 Unraveling the human

> "We should not, perhaps, underestimate our wish to lose our balance, even though it's often easier to get up than to fall over. Indeed, the sign that something does matter to us is that we lose our steadiness."[17]

In unraveling our understanding of the human that has evolved from centuries of art, culture, and the humanities, we can make use of arts such as contemporary circus and magic to test human limits and prototype alternatives. Circus arts involve the constant negotiation of gravity, velocity, balance, collaboration, trust, and what we might term *the density of presence*. It is the preeminent art form for knitting entanglements between equilibrium and disequilibrium. It can help us prototype the challenges that lie ahead in creating a semi-closed, self-sufficient, perpetually nomadic ecosystem—what else, after all, is a touring circus troupe? Similarly, magic, with its channeling of human will,

---

[16] Part I, Chptr. 8, p165.
[17] Phillips, A. *On Balance*. Penguin (2011).

concentration, energy, and visualization, can help us approach such a design challenge with an extended palette of resources.[18] Magic combines language, observation, emotion, and experience to change our perception of the world. It has an ethical dimension in the magician's paradoxical honesty in declaring the use of deception. It is a form of conjuring that demands attention to detail, practice, repetition, and immersion in the world of Nature. Thus, if Nature is conventionally associated with rational processes—Darwin's determinism—then an infusion of magic brings forth Nature's repressed characteristics, its illogicalities, contradictions (and supposed "aberrations"). We learn in the process that the materials we work with are more unpredictable than the sciences have told us, which suggests, in turn, that their capacities for transformation remain largely untapped.

Circus permits us to reframe the questions of the humanities—*What is it to be human? What is a good life?*—in (literally and metaphorically) "another space." Circus, with its etymological origins in "circle,"[19] also implies a utopian site of experimentation, in which an unbroken set of relationships between things can be set in motion. As "Knitting Peace," the recent touring production by the leading Swedish company Cirkus Cirkör, made explicit, circus is also a complex generator of principles of unraveling and knitting, simultaneously unraveling meaning, knitting it anew, pursuing the impossible as an artistic principle.[20] Circus organizes systematized and purposeful activity within a universe not infrequently experienced as purposeless and meaningless.

Magic applies the tools of camouflage, distraction, misdirection to model strategies of deception, and can thereby sensitize us to the varieties of deception we might encounter outside the context of a magic performance. Using a form of visual education, the magician displays a trick openly, and thereby demonstrates, as Ian Saville notes in "I Can See your Ideology Moving,"[21] that "we're dealing with known unknowns, rather than unknown unknowns… by displaying the trick honestly, the audience's consciousness of the changeability of the world is reinforced."[22]

*

---

[18] See e.g. Mike Fuller, "The Logic of Magic," *Philosophy Now*, August/September 2015, https://philosophynow.org/issues/5/The_Logic_of_Magic (accessed August 16, 2015).

[19] The Online Etymological Dictionary provides the following entry for "circus": "late 14c., in reference to ancient Rome, from Latin *circus* 'ring, circular line,' which was applied by Romans to circular arenas for performances and contests and oval courses for racing (especially the *Circus Maximus*), from or cognate with Greek *kirkos* 'a circle, a ring,' from PIE **kirk-* from root **(s)ker-* (3) "to turn, bend" (see **ring** (n.)). In reference to modern large arenas for performances from 1791; sense then extended to the performing company, hence 'traveling show' (originally *traveling circus*, 1838). Extended in World War I to squadrons of military aircraft. Meaning 'lively uproar, chaotic hubbub' is from 1869. Sense in *Piccadilly Circus* and other place names is from early 18c. sense 'buildings arranged in a ring,' also 'circular road.' The adjective form is *circensian*." Source: www.etymonline.com/index.php?allowedinframe=0&search=circus&searchmode=none (accessed May 1, 2015).

[20] "Knitting Peace" by Cirkus Cirkör, directed by Tilde Björfors, has toured the world since January 2013, deliberately posing a seemingly impossible question to its global audiences – namely, can we, collectively, "knit peace"?

[21] *Cabinet Magazine*, Issue 26, Summer 2007.

[22] Cited by Jonathan Allen in "Magic."

Staffage

You to speak, she said,
but my tongue was a dusty marionette, strings cut, collapsed into itself.

It was but one in a field of cultivated moist muscle:
Unuttered, dancing unfleshed.
Tongue, sing softly; other tongues bind your ululations
Your yearning is to unburden each curdling symbiogenesis.

It's long, thick and soft, but tapers to a thin tip.
This wolf tongue, bristling with papillae for grooming,
Stripping bones, licking up blood, curling in respect or submission.

It's hanging, panting, as you point, lift and fly.

They are slammed together as the ship tumbles and heaves;
With her hands she feels the contours of his face
His fur draws forth the light; their fingers, they lace.

*

16.1.4 Catastrophic reversals

Let us imagine a future not predicated on the present's unfolding according to an evolutionary pace, but a catastrophic future, spawned by discontinuity and disjunction, where inequalities in the existing landscapes of possibility are knitted into cartographies that hint at other points of rupture, disruption, and thus departure.[23] In this, circus may accurately be described as an art of catastrophe, an art in which expectations are continually undermined—the forces of disequilibrium keep triumphing—and forces of artistry, fortitude, and collaboration are required to bring order back into the chaos. Objects and agencies are knitted together outside established systems of validation. Circus becomes the site where impossible feats are made possible through extraordinary human discipline, dedication, and vision.[24] Risks are calculated and rehearsed, which means that the circus performer is part of the computational landscape, part of the system orchestrating and influencing events. Similarly, the first settlers in space will be neither gods of their own design, nor at the mercy of chaos, but continually shaping and choreographing their living conditions. Human values and belief systems are devised over millennia to support life on Earth, yet may likely prove to be obsolete on an interstellar journey to an uncertain future. Circus' attraction to play, testing, and prototyping equips it, in this post-humanities setting, not

---

[23] Catastrophe is defined etymologically as "1530, 'reversal of what is expected' (especially a fatal turning point in a drama), from Latin *catastropha*, from Greek *katastrophe* 'an overturning; a sudden end,' from *katastrephein* 'to overturn, turn down, trample on; to come to an end,' from *kata* 'down' (see **cata-**) + *strephein* 'turn' (see **strophe**). Extension to 'sudden disaster' is first recorded 1748."

[24] "Tilde Björfors, Kajsa Lind (trans. Claire Chardet): Inuti ett Cirkus Hjärta/Inside A Circus Heart (Norsberg: Cirkus Cirkör, 2009), p. 39.

only as a guarantor of human health and imaginative agility, but as a central prototyping element in the trans-disciplinary research underpinning space ark design.

*

16.1.5   Circus: risk, gravity, equilibrium, disequilibrium

> "In building a new world, Persephone is invoking the existence of a new nature and if we are to design a space that supports dynamic systems, then we must learn to effectively design at non-equilibrium states—and create environments with material flows, whose cultural equivalent is dirt. Design hates dirt—as it is aesthetically and materially subversive. Yet the various forms of dirt—such as shit, grit and dust—when combined, have powerful transformative potential.
>
> In space, shit is surprisingly useful."[25]

The contemporary circus is a semi-closed system, prototyping ethical, social, political, as well as physical alternatives ("gravity-defying"); international, interdisciplinary, nomadic, and immersed in "systems thinking," contemporary circus practice is well situated to make significant contributions to trans-disciplinary research fields, a laboratory for prototyping interactions between objects, agents, and environments. Conjuring speculative, utopian, or calculatedly dysfunctional environments, circus offers a way of developing a "systems approach" into a metareality in which competing knowledge systems are entwined or knitted together to test their "ease of fit." Each circus discipline—tight/slack wire, acrobatics, aerial acrobatics, pair acrobatics, juggling, clown, teeter board—involves the manipulation of border objects, objects whose primary function is to test human limits of timing, coordination, dexterity, concentration, audience interaction, gravity, velocity, flight, variation, landing. The circus "trick" takes place in and through gravity; weightlessness is the fleeting domain in which circus happens. As the circus artist and researcher Jonathan Priest has argued, the trick does not enable you to leave gravity, but it does point to other positions outside it.[26] A ball at the top of its arc, Priest argues, does not technically have weight, whereas the arm that threw it remains implicated in the system of gravity. For aerialists—rope artists such as Priest, pair acrobats such as Henrik Agger and Louise von Euler Bjurholm—circus provides a means of behaving as if one is temporarily outside gravity.

Serres writes about the "quasi-object" of the ball in a wider system of rule-following—a collective game such as football:

> "A ball is not an ordinary object, for it is what it is only if a subject holds it. Over there, on the ground, it is nothing; it is stupid; it has no meaning, no function, and no value. Ball isn't played alone. Those who do, those who hog the ball, are bad players and are soon excluded from the game. They are said to be selfish. The collective game doesn't need persons, people out for themselves. Let us consider the one who holds it. If he makes it move around him, he is awkward, a bad player. The ball isn't

---

[25] Rachel Armstrong, www.centauri-dreams.org/?p=28905 (accessed May 1, 2015).

[26] Jonathan Priest, "The Following Circus Is False, the Preceding Circus Is True". Jonathan Priest's 30% seminar, Stockholm University of the Arts, May 6, 2015. http://www.uniarts.se/forskning/seminarier-och-konferenser2/jonathan-priest-30-seminar

there for the body; the exact contrary is true: the body is the object of the ball; the subject moves around this sun. Skill with the ball is recognized in the player who follows the ball and serves it instead of making it follow him and using it. It is the subject of the body, subject of bodies, and like a subject of subjects. Playing is nothing else but making oneself the attribute of the ball as a substance."[27]

This describes well the relation of circus artists to objects in the circus: in becoming animated through the system of circus interactions and movements, objects bend our intentions and make possible a weave of entanglements that knits together elements of *testing, entertaining, diverting, manipulating—play* and *research*, in short. The circus thus becomes a microscopic universe in which the laboratory determines every moment through experiment and prototypes rather than top-down commands. Dimensioned, historically, by the turning circle of a horse, contemporary circus in the main banishes animals from its enchanted circle and, therefore, becomes defined by that which it is not, that which is excluded.[28]

The ethical dimensions of circus as a system of interactions demanding a high degree of coordination, collaboration, and trust are worth noting. Burnham writes:

"The systems approach goes beyond a concern with staged environments and happenings; it deals in a revolutionary fashion with the larger problem of boundary concepts. In a systems perspective there are no contrived confines such as the theater proscenium or picture frame. Conceptual focus rather than material limits define the system. Thus any situation, either in or outside the context of art, may be designed and judged as a system. Inasmuch as a system may contain people, ideas, messages, atmospheric conditions, power sources, and so on, a system is, to quote the systems biologist, Ludwig von Bertalanffy, a 'complex of components in interaction,' comprised of material, energy, and information in various degrees of organization. In evaluating systems the artist is a perspectivist considering goals, boundaries, structure, input, output, and related activity inside and outside the system. Where the object almost always has a fixed shape and boundaries, the consistency of a system may be altered in time and space, its behavior determined both by external conditions and its mechanisms of control. … [A] system esthetic is literal in that all phases of the life cycle of a system are relevant. There is no end product that is primarily visual, nor does such an esthetic rely on a 'visual' syntax. It resists functioning as an applied esthetic, but is revealed in the principles underlying the progressive reorganization of the natural environment."[29]

Yet, Rachel Armstrong argues in Part I Chap. 2 of this book that our current technological practices are framed by an understanding of mechanical systems that are assembled as a series of hierarchically ordered objects, which perform work through the external application of energy. This creates dynamic instabilities in the system that can then be directed towards a particular challenge. Despite greater complexity through feedback (learning) systems within these structures, such cybernetic systems have not yet become fully

---

[27] Serres, M. *The Parasite*, translated by Lawrence R. Schehr, pp. 225–226. Johns Hopkins University Press, Baltimore and London (1982).

[28] The point was made by Jonathan Priest at the above seminar and is explored at greater length in his virtuoso performance lecture "Knot Circus," https://vimeo.com/64292705.

[29] Jack Burnham, op. cit.

autonomous, which would be a pre-requisite for a life-supporting, technological fabric to form the basis of an ecological infrastructure. Noting that our technical platforms are becoming increasingly lifelike, she continues to observe that humans appear to be becoming increasingly inert; in other words:

> "it seems that we are living at a time where the capabilities of living systems and machines are overlapping a great deal. …The clear either/or distinctions that formerly shaped our experiences of the world are being replaced by a much more fluid relationship with reality where possibilities are no longer fixed. Rather, they coherently and simultaneously exist in many states as both object-oriented and process-led solutions. These hybrid approaches and perspectives may be key to developing new kinds of technical platform and modes of practice."[30]

Circus is potentially one such hybrid approach relevant to a worldship community, integrating perspectives and operating *between*, *across*, and *beyond* its integral disciplines (or establishing "secret bridges" between knowledge, experience, and operations in the ongoing cultivation of the worldship ecosystem).

Enlisting the skills and insights of circus artists in support of such a project may inspire on a number of levels—practical as well as conceptual—as an example of "the art of the impossible" as well as a response to what Slavoj Žižek has described as the "forgotten fourth category" of knowledge compartmentalization, namely the "unknown knowns" (defined as "a type of knowledge forbidden, exclusively, from knowing itself").[31] Circus arts research is an optimum laboratory for exploring both "unknown knowns" and the art of the impossible.

If it is a map we are working toward, it will likely span concepts, epidemics (the dissemination of influence), care, attention, and even a form of radical love yet to be outlined (for which the skills of the poet, as much as those of the philosopher, will be required)—a continuum in the *varieties of engagement* between serpentine thinking, non-knowledge, ignorance. Clown-like, we cultivate *cluelessness* and hope to *stumble across* something in what we grandly call our *research*. We have a ladder, with its vertical logics of ascent and descent, but resting on nothing, it leads nowhere, repeatedly. Yet, in helping each other to ascend and descend, we create a shared relation. And this is not nothing.

---

[30] Later in the same chapter, Armstrong notes that "The current model used in testing the transition from nonliving to living matter, is Tibor Gánti's notion of a 3-compartment system, which contains metabolism, container and information (Gánti 2003) [Gánti, T. (2003). *The Principles of Life*. New York: Oxford University Press]. Scientists are therefore focused on creating primordial systems that try to produce structures that represent these functions in the pursuit of abiogenesis." Interestingly, metabolism, container, and information also describe the core elements of the proposed space ark.

[31] "It is commonly said that there exist three distinct subtypes of compartmentalization which aptly divide our awareness of the external world. First, there are the 'known knowns', things we know that we know. 'Known unknowns' and 'unknown unknowns' follow suit in a similar fashion. Then, what of the forgotten fourth category, that of unknown knowns? (a type of knowledge forbidden, exclusively, from knowing itself). It is this category that to me represents the Freudian unconscious, the embodiment of the disavowed beliefs and suppositions that we are not even aware of adhering to ourselves, but which nonetheless determine our acts and feelings." Slavoj Žižek, Facebook post, February 23, 2015.

16.2 *The Temptations of the Non-linear Ladder*: meditations on scrying space

Rolf Hughes Professor of Artistic Research, Stockholm University of the Arts, Sweden, and Rachel Armstrong, Professor of Experimental Architecture, Newcastle University.

The Temptations of the Non-Linear Ladder is a Persephone project and experimental space created for the *Do Disturb!* festival, which was held on April 8–10, 2016, at the Palais de Tokyo, Paris. The project team was Rachel Armstrong, Professor of Experimental Architecture, Newcastle University, UK; Rolf Hughes, Professor of Artistic Research, Stockholm University of the Arts, Sweden; and Olle Strandberg, Director, Cirkus Cirkör, Sweden. Additional appearances from performers Methinee Wongtrakoon and Alexander Dam were courtesy of Cirkör LAB, which is part of Cirkus Cirkör, Sweden. The technical set-up was directed by Joel Jedström. What follows seeks to represent the multi-layered contributions to this first, experimental collaboration.

16.2.1 On scrying

Rachel Armstrong

> "The power to influence events is bound up with the great expanse of natural knowledge, having its nearest most immediate origin in free will and describing future events which cannot be understood simply through being revealed. Neither can they be grasped through men's interpretations nor through another mode of cognizance or occult power under the firmament, neither in the present nor in the total eternity to come." (Sacred Texts 1555)

> "There is something primordial about traveling on water, even for short distances. You are informed that you are not supposed to be there not so much by your eyes, ear, nose, palate, or palm as by your feet, which feel odd acting as an organ of sense." (Brodsky 2013)

> "With inspiration places and aspects yield up hidden properties, namely that power in whose presence the three times are understood as Eternity whose unfolding contains them all: for all things are naked and open." (Sacred Texts 1555)

From orbit, Earth's Sun-soaked membranes split at the surface into earth, sea, and air. As we submit to gravity and plunge into their life-giving fractals, what first appears as a brittle shell around the planet pleats into valleys and deltas, split like hoar frost and unfolding into endlessly branching veins. At their terminus, where unlike atoms collide, trembling interfaces spill their secrets. Tiny ripples of whispering tension vibrate through these endless veils—their voices amplify and condense into songs of disruption. Now it is possible to see the micro scale through the naked eye at reflective interfaces where one medium plays off another in distortions, refractions, diffractions, reflections, magnifications, illusions, and revolutions. In these exchanges between image, matter, and transformation, the symphony of life swells. Their disturbances transmit happenings and transfigurations between one medium and another like tiny wormholes of potentiality.

Within almost incomprehensibly complex sites, we encounter non-linear ladders that appear, as if by magic, wherever perpetual motion creates impossible distortions of Euclidian geometries.

We are entering the realm of the scryer—the seer that comprehends the language of distortion.

Nothing can be called "real" in these unquiet landscapes, for nothing stays still enough to be determined. The world wriggles this way and that. Where is that body? That building? That face? That space? That surface? That eye? For it is no longer "there"—or even "here."

This is where our research begins.

Plunging from above to below within a darkened space where photonic ripples are amplified through dark mirrors, we watch reality transform before our eyes. Within the silhouettes of this space that is cleaved by the radical bodies of circus performers, we are in the presence of an unquiet world that has uncoupled from Newtonian motion and linear causality.

As we begin our impossible ascension from the dark ground through water towards the air and into the light—all that we thought we knew begins to dissolve and something else takes its place. Shivering, we try to imagine that we are born of a place and time that is sure of itself. But that possibility no longer exists. Instead, a space of transformation lies before us. And so we take our first steps on a non-linear ladder to venture into the turbulent landscapes and uncertain horizons that ripple before us.

16.2.2 Background to *The Temptations of the Non-linear Ladder*

Rachel Armstrong

The Palais de Tokyo event is part of the Persephone project, which has been realized through a series of terrestrially located experimental spaces—from growing artificial soil matrices in chemistry laboratories, to constructing artificial cybernetic soils and launching a range of life forms into the stratosphere. Each set of experiments takes us further from the conditions for living that we're used to and deeper into the unknown, where survival is not guaranteed.

16.2.2.1 Laboratory set-up

Rachel Armstrong

The domed temple space at the Palais de Tokyo presented a centrally positioned 4-m black scrying pool, filled with 2 cm of water that acted as a black mirror, which is an instrument traditionally used by scryers. This radical experiment in performance art aimed to explore the limits of the human body in an unstable space. Suspended directly over the scrying space was a circular 1.2-m-diameter platform made of safety glass with three eyebolts through which carbines were looped and tethered to a central rigging (load 600 kg). This aerial space could be altered using a 5:1 pulley system, which was threaded through two roof points and anchored to the floor. The working load limit was 200 kg plus × 10 security margin. Behind the central performance space, two television monitors played two specially edited protocell films played side by side continually during the three-day performance to create asynchrony between the images and movements. As such, the movements

The Temptations Of the Non-linear Ladder: meditations on scrying space 461

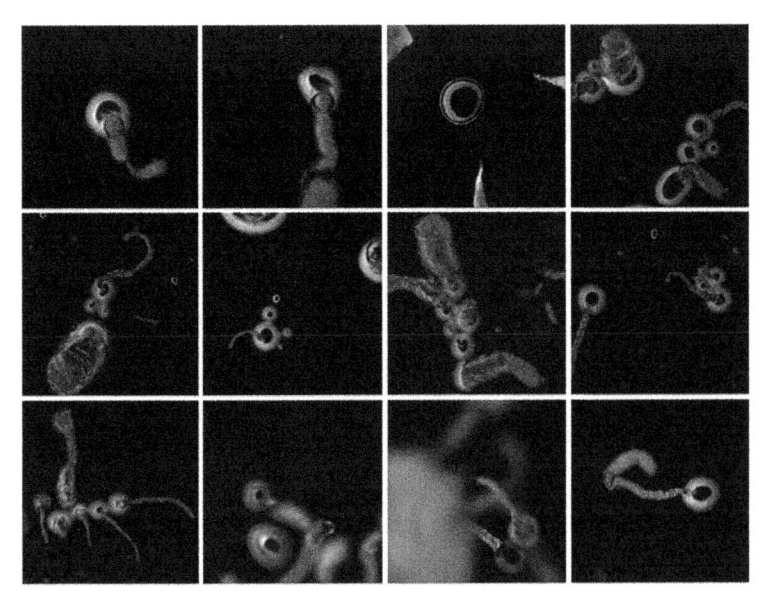

Figure 16.1 Collage of stills from a selection of protocell movies. *Credit*:

in the space are constantly off balance, presenting continually different visual experiences for viewers and performers (Fig. 16.1).

16.2.3 Background: "Protocell Circus"

Rachel Armstrong

In 2011, "Protocell Circus"—a film collaboration between Rachel Armstrong and Michael Simon Toon—looked at the inner life of a simple water/oil droplet system, which behaved in a lifelike way without actually being granted the full status of being "alive" (Protocell Circus 2011). This marvelously strange behavior of the droplets occurs spontaneously, even though they do not have a central code that organizes their responses like DNA. What they have instead is an interface—where oil and water meet—that is highly responsive to spatial events. A great deal of energy, which produces the movement, occurs across this interface and a flow of material in which transformation occurs. When water is alkaline, it makes soap skins on contact with oil. These structures block exchanges in a site-specific way and potentiate the asymmetry in the system so that it is polarized from the outset, "grows," and moves vigorously.

Yet, while there are many features of this system that can be explained in terms of classical chemistry and fluid dynamics, there are many aspects that cannot. For example, it is not obvious why the droplets do not fuse. Nor is their population-scale behavior something that can be simply "explained."

This makes rich pickings for a visual system that starts to discuss an experimental forum where image, form, behavior, and the nature of bodies may be interrogated in a perpetually unstable space—one in which surprising, or unexplainable, things can occur.

16.2.4 Principles

Rachel Armstrong

Drawing from the radical anatomies produced during Protocell Circus, two films were composed for *The Temptations of the Non-Linear Ladder* experiment, which captured the character of the dynamic droplets. Events within the aqueous character of droplet bodies wee occasionally augmented using a fluorescent dye that could be clearly seen under dark light.

Broadly speaking, the first film examined the behavior of satellite bodies, or moons—agents that orbit around others (Fig. 16.2).

The second film examined structural connections and relationships between droplets (Fig. 16.3).

However, the many resonances and entanglements between the films created provocative overlaps as they played side by side on two separate monitors in the temple space.

16.2.5 Themes of *The Temptations of the Non-linear Ladder*

Rachel Armstrong and Rolf Hughes

The Temptations of the Non-Linear Ladder explores themes relating to the uncertainty of our survival between two planes of existence. They focus primarily on:

- the non-linear ladder;
- transitioning;
- ascension.

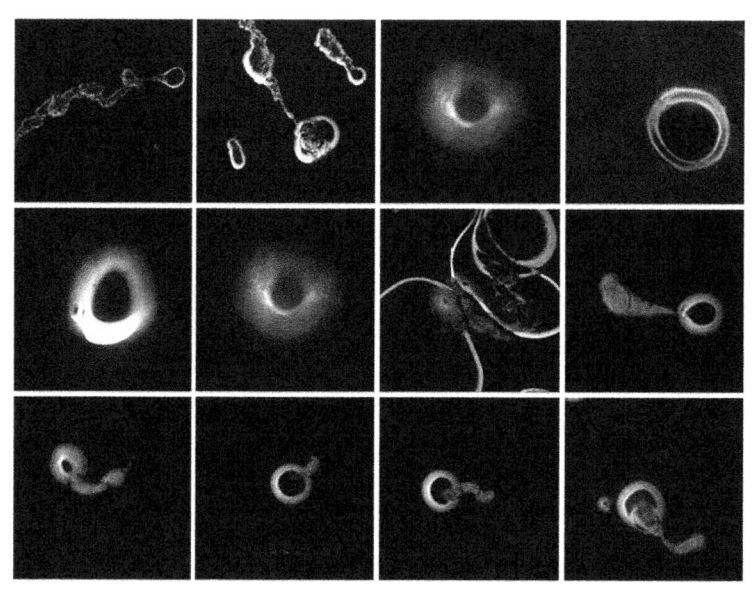

Figure 16.2 Collage of "moon" stills from a selection of protocell movies. *Credit*: Rachel Armstrong

The Temptations Of the Non-linear Ladder: meditations on scrying space 463

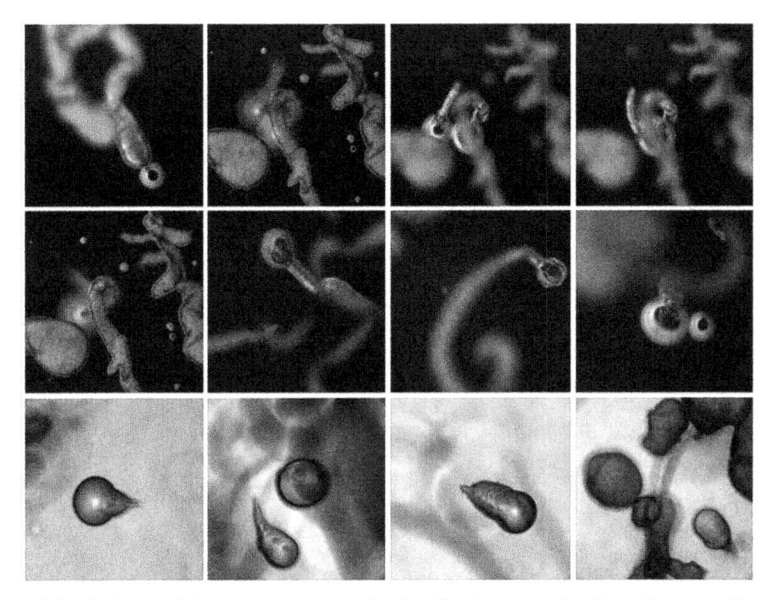

Figure 16.3 Collage of "structural connection" stills from a selection of protocell movies. *Credit*: Rachel Armstrong

16.2.5.1 The non-linear ladder

The non-linear ladder refers to the specific journey and forms of embodiment that were experimentally explored during the Palais de Tokyo event. They embodied transitions from one set of existence conditions to another—such as from the water to the air. Here, unstable bodies, their interrelationships, and space produced their own route of access in forming structures recognized as "non-linear ladders." In other words, they were typical of non-linear systems whose outputs are not directly proportional to the inputs and exceeded our capacity to completely solve their complex, dynamic relationship using classical approaches. *It is in these sites that magic is produced: events that at not reducible to simple explanations*—and, even when observed, occurrences were not demystified through appended narratives. Indeed, performance, site, bodies, and structures were intimately entangled and became inseparable. They formed the apparatuses and assemblages for construction of non-linear ladders that provide access to the synthesis of wonder.

16.2.5.2 Transitioning

During transitions between media and sites, something strange happened to the circus performers' notions of their bodies and what they understood of their own physical parameters.

16.2.5.3 Ascension

At the limits of existence, when outcomes were so undetermined, the performers were no longer embodied objects, but excitable fields that shaped the behavior of matter far beyond the immediate spaces that are recognizably "human." These provocations raised ethical, philosophical, existential, environmental, technical, and cultural questions that created the possibility of a radical identity explosion. Audiences followed the transformations of the circus performers that trod the uncertain pathway towards ascension and another state of being.

16.2.6 On medaka fish

Rachel Armstrong

Around the perimeter of the temple space, against the back wall, bowls of tiny fish were placed. These creatures were unlikely scryers and inspiration for the non-linear ladder. As they flitted, dotted, and darted around their liquid lenses, lit by means of a lightbox, their shadows whispered strange stories amongst the many refractions, reflections, and transformations that took place at the surface of the central dark mirror. And, while these organisms may have seemed diminutive to some, among the human spectators that moved around the space, they were anything but insignificant.

The tiny rainbow-hued *medaka* (Japanese rice fish) has been popular among all classes of feudal Japanese since the Edo Period (1603–1868). Adults grow to about 3 cm and are careful parents. Females carry their eggs around in clusters between their anal fins, rather than simply releasing their eggs en masse into the water. They plant each one carefully, as a gardener might, into a specifically chosen site. Medaka fish are particularly renowned for their bewitching eyes—in fact, "medaka" in Japanese means "with high eyes"—and they are also extremely hardy (Fig. 16.4).

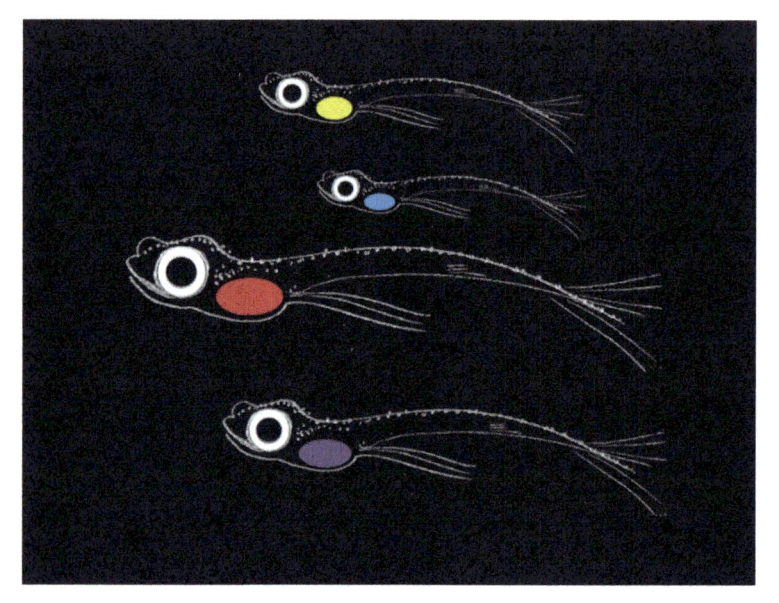

Figure 16.4 Drawing of medaka fish. *Credit*: Rachel Armstrong

The Temptations Of the Non-linear Ladder: meditations on scrying space

In the early 1900s, the rainbow hues of medaka contributed to a growing field of modern genetics where they could be bred in ways that followed Mendel's Laws of Inheritance. This is a biological theory of natural selection, where dominant and recessive traits are passed through generations in distinctive ways. With the explosion in molecular biology techniques in the 1980s, the medaka fish became a sought-after model organism for study. In 2007, the entire genome of the species was sequenced, which showed that they share 60% of their genes in equivalent sequences with humans. Medaka fish were therefore considered a good "model" for studying the genetic principles of human biology. Of course, with access to their genome, they were genetically modified with jellyfish genes, which caused them to bioluminesce red, green, or yellow under dark light.

In 1994, the medaka reached a different biological milestone when they became the first vertebrates to breed successfully in orbit aboard the *Columbia* spacecraft. Curiously, they did not exhibit a pathological behavior, observed in other species like goldfish and fingerlings, called "looping," where fish swim continually in tight circles until they acclimatize to the reduced gravitational field over a few days. Not only did medaka successfully mate during the mission; they also spawned normal young (Hooper 2015).

Now the medaka fish occupy a special aquatic chamber in the Kibo module of the International Space Station (ISS), which was built by the Japanese Space Agency. Swimming around in weightlessness, their lives are "read" by scientific researchers who study their genetic codes, behavior, and color and form. Through these bodily expressions, scientists hypothesize ways of combating terrestrial processes on our home planet differently. Effectively, medaka are scrying mirrors for processes like ageing that are likened to changes in reduced-gravity environments, where muscle and bone become weakened with time. Modern science appropriates the twists and turns that the fish use to survive these extreme situations and uses them to propose ways of combating destructive physiological forces.

Yet, the medaka fish are authors of strange stories and do not simply tell us what we expect to hear. In the summer of 2015, when the *fox13* gene was changed in a cohort of female fish, they produced sperm in their ovaries, instead of eggs. This was beyond all the expectations of modern science—as, in every other way, the fish remained female. Yet, on further detailed study, their sperm fertilized eggs normally and produced completely healthy offspring.

For the very first time in the story of life on Earth, medaka fish showed us that the sex of a body is independent of its sex cells.

Of course, fish are very different to mammals, which also have the *fox13* gene, but what kinds of processes this code is responsible for are as yet unknown. Perhaps this gene has something to do with preventing sex cells becoming male but it will be much more complex to establish the human story than observing the effects of its absence in medaka fish.

But, while science tends to read the stories of the medaka fish backwards through post-rational lenses, reflecting on what we know to be true right now, *The Temptation of the Non-Linear Ladder* is interested in reading the stories of things that may be yet to come. For example, what can the fish tell us with respect to the impact of microgravity for life beyond Earth's terrains? What may their story of survival enable us to become as we face the turbulent conditions for life on Earth in the twenty-first century? And what kinds of existence may be possible for the story of life amongst the distant stars?

*

16.2.6.1 Unbounded

Rolf Hughes

It was a long day, beating the bounds of our parish. We had gathered most of the able villagers to walk around our boundaries and create a common memory of its limits. Boys were included to make sure the memory endured. Now we were back at the town hall where there was a sudden hubbub at the window.

Look out there, it's your mother. She is digging with her trowel, see, down into the mud, a mud-encrusted trowel, she is on her knees, her mud-encrusted knees, lifting the mud up, and dumping it out, making a mud mound, but it's not the mound of mud she is after—she is uninterested in even temporary memorials—she is digging down through the trowel-sucking mud, calmly and methodically, to where the river is, deep down beneath the cloying mud, the river that seems impossible to us now watching with folded arms around our prize theodolite on the grassy knoll, the river that she says is home to the impossible creatures, the medaka fish.

Not for them the looping neurosis of the likes of goldfish, born not for adaptive hardships on parabolic flights, but to fawn for fawning tourists on Earth. Those that dare venture into evolution's stranger terrains—*Oryzias latipes*, our brave *medakonauts*—form an auguring procession; sharing human gene sequences, they fuse with jellyfish code to bioluminesce red, yellow, or green under dark light, becoming restless traffic lights in a place of no traffic, their hatch fry embodying the scry, revealing that ageing can be resisted, that the sex of our bodies is independent of its sex cells, that space too is a site of procreation, a setting for unbounded lives, for futures which we can almost imagine when we gaze closely into their high, bewitching eyes.

16.2.7 On Lazarus fish

Rachel Armstrong

The medaka may be regarded as a kind of "Lazarus" fish—a term that usually implies that a species has come back from the dead. In *The Temptations of the Non-Linear Ladder*, the idea of "ascension" was challenged through the combination of transformations within the protocell bodies, the medaka, and the circus performers that became a new assemblage of organisms capable of writing extraordinary futures of survival in extreme spaces and surprising new evolutionary transitions (Fig. 16.5).

In one sense, fish are "cold-blooded"—they do not need to maintain a constant internal environment to stay alive—so some species will be able to survive extreme conditions in a kind of hibernation state called "torpor." This is not quite full hibernation but a physiological condition that maintains the creature during life-threatening times and produces very similar results to hibernation, such as reduced body temperature, slowed metabolism, slow reaction times, reduced oxygen exchange rate, and a decrease in activity of many primary body functions. When conditions change, the fish appear to literally "come back to life."

The term "Lazarus species" also refers to creatures that were thought to be extinct and have been rediscovered in a modern context. The most famous of these "living fossils" is the coelacanth, a species that was evidenced in the fossil record around 400 million years ago. To the great surprise of the modern world, in the 1940s, fishermen off the coast of

The Temptations Of the Non-linear Ladder: meditations on scrying space 467

Figure 16.5 Drawing of a Lazarus fish, a creature in transition towards a new evolutionary space. *Credit*: Rachel Armstrong

South Africa caught a live one in their nets. Since then, there have been many live sightings of this deep-water species.

The temple space presents a new kind variety of Lazarus organism — one that does not dwell in suspended presents or transcended pasts, but carves out new possibilities for the future through ascension of bodily forms. It moves us into future evolutionary spaces. Recently, a new fossil fish species was discovered with leg-like fins that are thought to represent the transition from water to land. Scientists have theorized such a transition but this discovery lends weight to its actuality. Within the temple space, viewers are invited to propose alternative futures that may exceed what we know to be true today. Rather than the accepted extinction scenarios that describe our modern relationship with the planet, of our polluting ways and voracious resource consumption, the non-linear ladder becomes an instrument that enables us to access imaginary spaces that inspire us to find ongoing futures for our continued survival. In this way, we may begin to construct those stories that invite new trajectories for ourselves both on this planet and beyond it.

*

16.2.7.1 Ascension

Rolf Hughes

It began while the city's population was mostly sleeping. Many subsequently recalled "frivolous and gentle" dimensions to their dreams. There had been much discussion about expansion. Some had spoken of blocks across oceans. Others felt the need to "drill down" to solve a chronic housing shortage. Certain architecture students had proposed airborne parks, but these were immediately deemed "unrealistic" from an engineering perspective.

According to the city's homeless, it was during the *blue* hour between night and dawn that the terrestrial buildings started sighing. A soft billowing of gas winnowed through their pipes and valves, and their concrete facades appeared visibly to deflate. Today, deflating concrete structures has become an established discipline, but at the time it was felt to be "risky" and "anti-science."

Towards noon, the bulk of the city's infrastructure was floating. Roads, rail tracks, and tram lines had become aerial spaghetti, while water pipes and electricity lines were dangerously entangled. Waking at the shrill call of the alarm in a still familiar bedroom and navigating a route to a newly unknowable workplace involved negotiating swarms of bewildered commuters slip-sliding through the sky like knots of fish. Children delighted in the absence of identifiable routes to their schools, somersaulting down inclines, running up opposing slopes, leaping from one vector to the other while blowing bubble gum bubbles. Everyone, it seemed, rejoiced in being released from a hard, joyless carapace—a surface for grazing skin rather than lifting off towards weightlessness.

And yet there were those who railed against the airborne city and wished to re-establish relations on a terrestrial, gravitational footing. They claimed floating was an affront to democracy and wanted an "accountable" statement as to its purpose so that people could make a "rational" decision as to its "desirability." *Standing on one's own two feet*, they felt, was no longer a value widely recognized. At the same time, they were aware of how the majority of their neighbors had taken *like ducks to water* to this new way of interacting through *float*, *grip*, *spin*, *tumble*, and *catch*. Aware of which way the wind was blowing, they did not want to be associated with outmoded trends such as *stump*, *stamp*, and *plod*.

It was February—the sky was dark and ugly—and most believed things could only get better. The bloated dreamed of floating, while the floaters thought of anchoring their feelings and thoughts to something weighty back on the bruised Earth below. Religious leaders sought to escape this dichotomy, arguing that what is important is what one feels while *transversing* the hither and thither. Most, however, wanted a reliable map of the new city rather than pontifications about its relation to yet another space.

Those more agile in navigating ambiguity tended to be those that thrived in an absence of limits or definitions. When enough of them, through some mysterious force, congregated in a given aerial expanse, their movements became akin to a semi-human murmuration. Even so, many collided with immovable forces, falling lifeless back into the toothless gaps of uprooted conurbations.

I saw you, in the mid-morning haze, tumbling across the sky, your hair pinned tight, a faint scent of oranges, wood smoke, and compost. Had I this moment again, this possible pivot in the new world of rotation and churn, I would have gripped you as you neared, held tightly until we fused and were hinged. But we were gliding to other co-ordinates, auto-smiling in the faceless throng; our raised hands of greeting almost touched, but then steered us further and further away, each from the other, our smiles pointing elsewhere, until we were gone.

The Temptations Of the Non-linear Ladder: meditations on scrying space

16.2.8 Additional texts for inspiration

16.2.8.1 "The Temptation of St. Antony: or A Revelation of the Soul"

(extract from Gustave Flaubert (not dated))

"A briny breath of air strikes his nostrils. A seashore is now before him.

At a distance rise waterspouts, lashed up by the whales; and at the extremity of the horizon the beasts of the sea, round, like leather bottles, flat, like strips of metal, or indented, like saws, advance, crawling over the sand:

You are about to come with us into our unfathomable depths, never penetrated by man before. Different races dwell in the country of the ocean. Some are in the abode of the tempests; others swim openly in the transparency of the cold waves, browse like oxen over the coral plains, sniff in with their nostrils the ebbing tide, or carry on their shoulders the weight of the ocean-springs.'

Phosphorescences flash from the hairs of the seals and from the scales of the fishes. Sea-hedgehogs turn around like wheels; Ammon's horns unroll themselves like cables; oysters make sounds with the fastenings of their shells; polypi spread out their tentacles; medusae quiver like crystal balls; sponges float; anemones squirt out water; and mosses and seaweed shoot up.

And all kinds of plants spread out into branches, twist themselves into tendrils, lengthen into points, and grow round like fans. Pumpkins present the appearance of bosoms, and creeping plants entwine themselves like serpents. The Dedaims of Babylon, which are trees, have as their fruits human heads; mandrakes sing; and the root Baaras runs into the grass.

And now the plants can no longer be distinguished from the animals. Polyparies, which have the appearance of sycamores, carry arms on their branches. Antony fancies he can trace a caterpillar between two leaves; it is a butterfly which flits away. He is on the point of walking over some shingle when up springs a gray grasshopper. Insects, like petals of roses, garnish a bush; the remains of ephemera make a bed of snow upon the soil.

***

"…And he is no longer afraid! He lies down flat on his face, resting on his two elbows, and, holding in his breath, he gazes around.

Insects without stomachs keep eating; dried-up ferns begin to bloom afresh; and limbs which were wanting sprout forth again."

*

16.2.8.2 Extending perspectives: the non-linear ladder (Fig. 16.6)

Rolf Hughes

They were conjoined twins in thoughts only, so a surgeon was commissioned to conjoin them in deed by stitching together their heads until each became a graft to the

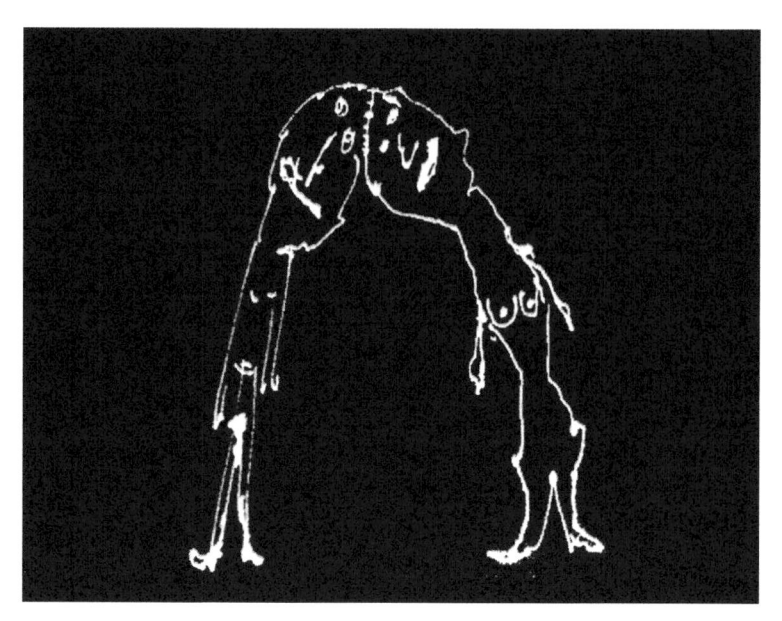

Figure 16.6 Surgically conjoined twins. *Credit*: Rolf Hughes

other. One asks the other if they should intend to climb. The other replies that they can only climb the way a compass climbs the drawing board in an architect's hand. Time eats their stitches. Horizons are stacked in the distance, like shelves of white chocolate. Lacking architect, each helps the other to try to ascend. They study stepladders and contemplate the logic of hinges. Unceasingly, they clatter to the ground — less compass, than flesh castanets. From there they watch gravity shaping matter, the modern universe churning overhead — gas clouds, stars, galaxies, galaxy clusters, dying and forming in the corpuscles of frictive, percussive bodies squeezed from dark matter's dark embrace.

*

16.2.8.3 This world is coming alive: a protocell circus[32]

Rolf Hughes and Rachel Armstrong

This world is now coming alive. We are learning to stop observing each other. We half-close our eyes so other dimensions swim into focus. We learn through feeling — vibrations, ripples, halos, radiance. This world is coming alive.

---

[32] The italicized sections cite Rachel Armstrong's reflections to the author on her protocell videos.

The Temptations Of the Non-linear Ladder: meditations on scrying space

If this were a life, this would be it—swimming, crawling, walking, running—gray solutions in cascading rain, no pre-programming, no measuring. A world without competition. Our transitions are outside explanation as explanation systems are currently configured. When you reduce explanations to components, they don't really make much sense at all. Yet other explanations are equally non-explanations. Mysticism masquerading as science. Anatomy and trajectory.

This is why you call on magic.

Things fuse in this environment. Or rather, they have the capacity to fuse, but they choose not to—they are attracted to each other, but they choose to keep their integrity. It's an example of a "synthetic" experiment but not one of "emergence" alone, which is a non-explanation, widely used (probably unconsciously as a mystical term). The mistake here is to look for identities—essences and properties—rather than interactions. How will we otherwise understand power and magnetism? Is there some sort of handbrake on weird fluid mechanics here? Or a simple fluid anti-dynamic to do with the damming of flow within a body? Neither convince. It's to do with some sort of charge. That's it. There's some sort of charge between them.

An "inner life" is created by an interesting relationship between interiority and exteriority as it exists at the interface. Friction. And, as we are continually transmitting light through our interfaces, and being continually on the move, we are not the easiest to film. You may have noticed this as you try to focus your lenses through the bottom of a world.

One is definitely livelier than the other—it is "osculating" as a form of communication. There are waves of something happening ... something chemical. And now it has come to the end of non-equilibrium ... the structure that evidences its "life" is interesting ... a cartography or topology of existence ... or what was once an existence.

We escape our Earthly tether through an internal force. We are not "alive." There is no central program. Something akin to cutting an umbilicus has occurred. Drop us in a solution and we explore it.

Two droplets negotiating their next move. Life is negotiation. With interesting geometry. Design and construction with probability via non-equilibrium materials. This is life. When you overlap fields of interaction the "probability" of an event is increased. Structure, osculation, and fusion all in one gestalt. We normally design through separation not through union of things—patterns move away from a centre, not towards one. You can see the mutual repulsion/attraction. Oscillation and osculation. Mother and baby or satellite phenomenon. It is not a "logical" set of construction rules. It is a kind of acrobatics in the system.

We are looking at the reflections, refractions, field changes to observe not *the thing itself* ... but learning how to focus on the world through narrowed eyes so that other dimensions more clearly come into focus. Adding alcohol to the system makes for very friendly swarming crowds that quickly become inert.

It is hydromancy and aeromancy and all kinds of dark arts that re-empower us within contexts and matrices of existence. Others say yes, of course we see the object, of course we see the object. But we see beyond too by its vibrations, its ripples, its halos ....

Our floating weather event runs and runs irrespective of whether the hood is lifted and your eye appears at the end of the funnel. Oh, we can stretch and spit too on command. Cloud, rain, storms, tempest all threaded across our strings. To float, intact, all turbulence contained within. And if I squat on your hard glass, so much the better for you to peer through my crack and into my world. Watching you, watching me.

They struggle for release, these new ones. Another dud planet drifting across my upper horizon. Suddenly I am not alone, my waste falls away and I am released, replaced by my twin, far less restless than me, anchored, able to draw all events towards its own calm pulsings. People peer at us as if we are microscopic, but we are whole galaxies interacting. They do not understand the first thing about *scale*. They are still studying the workings of their own thinking, believing themselves to be separate from what they observe.

The radiance in the world, just waiting to be discovered. This world is coming alive.

What is this strange agitation that occurs when we touch? The pressure point, the point of contact, sends reverberations through my entire being. Every element of me surges towards you, but you resist.

The resistance in the world, just waiting to become radiant. This world is coming alive.

And so proximity. Pressure. When it happens it's always so sudden! Now we become one! A tipping point has been reached. Borders breached. Implosion meets explosion.

Halloween lanterns on the bare branches of trees against the night sky. Here comes another, wanting in. We resist. We are populations fighting back. And then we open up. Ours is a war on separation. Joining us is to disavow your separate leanings. We are a swarm of ghosts that cannot be counted, measured, weighed. We are a black hole, concealed in our shimmering robes, drawing our neighbors into an insatiable hunger. Once you set us in motion, we will not stop. We cannot stop. They think it concerns attraction, motivation, desire. But this is not a story. Storytellers are decapitated before sunrise.

Come closer, rest your cheek against mine. I'll tell you everything—the loveliest stories, the sweetest yarns—but first come closer, closer...closer still....

Let us now sing the spaces between meanings.

*

16.2.8.4 Knots

Rolf Hughes and Rachel Armstrong

What we're not. Obvious.

I am reaching from your nowhere.
That place where what works doesn't work.
You cannot explain all this.
We cannot carve shadows
From the blue shadow of that blue dot.

Do not measure me.
Or measure me wrongly.
Not trying to know you.
But I would like you to know me.
In the dark. Spooky. Entangled.

The Temptations Of the Non-linear Ladder: meditations on scrying space

It's less about knowing than *scrying*. Another way of knowing.
And if you think you're going to grasp me at the end of it, think again.
You won't pin me down—I'll fly into pieces.
You see, although we share the same name, you're strange. I'm very little interested in you.

I'm interested in getting stronger, longer, broader, more agile.
I'm interested in *entangling*, *simultaneity*, *tunneling*, *diffraction*, *distraction*.
I'm not finished.
Tumbling. Friction.

Knots:
weaving a world through which the curious can climb
knitting of interfaces
monstrous tangle of *unknowing*.

You were here long before me. You've made this neighborhood your own.
Few are aware that I even exist. Your charmed anti-matter.
I hear you sometimes, a dark hole rustling voraciously.
You'd possess me if you could. It's in your nature.
Your approach is from a place beyond my nowhere.

It doesn't work this speaking for two.
I don't hear you anyway.
You're less than vacuum.
Beyond light.

Still, I yearn to be there
part of your somewhere,
part of your elsewhere,
together with the noiseless things, the weightless, sightless,
nameless things like wind,
tossed from myself,
across channels, alive,
mysterious, monstrous,
heart's slap of valve,
gravitational waves beating
from colliding black holes
cosmic thunder over horizons,
it's impossible, this synthesis,
and yet here we are—

And, next, the plants are indistinguishable from the stones.
Pebbles bear a resemblance to brains, stalactites to udders, and iron-dust to
tapestries adorned with figures. In pieces of ice he can trace efflorescences, impres-
sions of bushes and shells—so that one cannot tell whether they are the impressions of
those objects or the objects themselves. Diamonds glisten like eyes, and minerals
palpitate.

… nearer still;
singular, finite, fierce
the air static, crackling, tense,
it sears, this white scar
—this bolting bolt of darkest radiance.

*

16.3 "It"

Rolf Hughes, Professor of Artistic Research, Stockholm University of the Arts, Sweden.

One morning, as it awoke from buzzing dreams, it found itself ascending at a dizzying rate. It lay on its back and, if it lifted its head a little, it could feel the air rushing past as it was hauled up and away, high into and then beyond Earth's atmosphere. Its bedding slid away and its many legs, pitifully thin, waved about helplessly as it looked at the retreating surface of the planet below.

Above its bed hung a picture that it had recently gnawed from an illustrated magazine and housed in a nice gilded frame. It showed a woman fitted out with a fur hat and fur boa sitting upright, raising a heavy fur muff that covered the whole of her lower body. Above was a huge white balloon, its thin, bulging polythene membrane pumped tight with helium.

Rising above the dirt and particles suspended in the lower atmosphere, it gazed blankly at the passing clouds. It was developing a headache—its thoughts were decidedly fuzzy. Drops of rain sped by. It felt a slight itch on its belly, and raised itself slowly up on its back towards the headboard so that it could tilt its head. There was the itch. It saw that it was covered with a spatter of tiny white spots. When it tried to feel the place with one of its legs, it drew it back quickly, overcome by an icy shudder. It was way below freezing and hairy ice crystals were appearing. Half past six. Its hands quietly moving forwards. More like quarter to seven. The clock was slowing down or squirming. Was that a cautious knock at the bed's headboard? Or the headache playing tricks?

The first thing it wanted to do was move away from the clock. It needed to get the lower part of its body out of bed, but it tried to get the top out first. It lay there, its breath escaping from its chest as if total stillness might bring back some sort of peace and order. The sky sped by and the air became colder still.

There was a loud thump. It rubbed its head against the carpet. An attack of dizziness. Brown fluid escaped from its clattering mouthparts, flowed over the edge of the basket, and dripped down towards Earth below.

*

When it awoke from a deep sleep, the sky was pitch dark. It felt fully rested. But the dawning Sun was climbing palely over a bowed blue horizon, meeting the clouds as spattering white opacities on a dirty window pane. It pushed itself over to the edge of the balloon's basket, feeling its way clumsily with its long antennae which could, smell, taste, and feel what was happening. Minus 33°. A relative humidity of 10 %. Earth had been replaced by haze.

On the rim of the basket was a camera. It leaned towards the lens and waited for an indication as to whether it was being recorded or not. The atmosphere and the pressure were now those of Mars—an average of minus 50 °C. Anybody would struggle to function under such conditions. The headache was so intense it was distorting its vision.

Half suffocating, it watched with bulging eyes as its metabolism started to freeze and night returned. Time appeared to behave differently at this altitude. Its eyes glittering with pleasure, it gazed at the approaching stars.

*

Every other day would be spent struggling for breath at the edge of the basket. Its metabolism was perhaps giving up the ghost. It could not concentrate for very long. Whenever its head would fall wearily on its chest, it would pull it up again with a start. The silent camera followed every movement.

Increasingly, time revolved around the camera. It would often lie on its back the whole night through, neither sleeping nor fully awake, scratching at motes of dust for hours on end. Or it might go to all the effort of climbing up onto the rim and leaning over as far as it dared to stare out at space. It used to feel a great sense of freedom doing this, but now it was more of a habit, for what it actually saw in this way was becoming less distinct every day. It used to curse the bowed blue curves of a once familiar planet, but now it could not see it at all and, if it had not known that it had once lived there, that planet filled with liquids and soft edibles, it could have thought that it was gazing at a barren waste where the gray sky and the gray planets mingled inseparably.

*

As vision deteriorated, it stared increasingly at the picture on the wall of the woman in fur. It raised itself and pressed itself against the glass. It held it firmly, cool on its hot belly. It wanted to absorb it. The rushing void of space was swallowing everything. It was amazed at the great distances that every movement implied. It could not understand how, in its weakening condition, it had climbed the wall. It concentrated on crawling back down—a task so time-consuming and laborious that it barely registered the explosion when the balloon finally burst, spilling its cargo at the very limits of gravity.

"What now?" it asked the surrounding darkness. "Am I falling or ascending?" It could no longer move and its entire body was aching, but the pain seemed to be steadily getting weaker and weaker. It would likely soon disappear altogether. Peace in our time. It noticed the camera lens hovering opposite its face, and acknowledged its audience with a grimace. Light slowly bleached everything in dazzling brilliance.

Far below, as dawn rose over an old city in Europe, a woman eased herself from the heavy layers of artificial furs on her bed, stretched her still supple body, walked to the window, and let in the fresh morning air (Figs. 16.7, 16.8, 16.9).[33]

*

---

[33] This story was published in Issue 43 of *Odyssey*, the British Interplanetary Society's science-fiction journal.

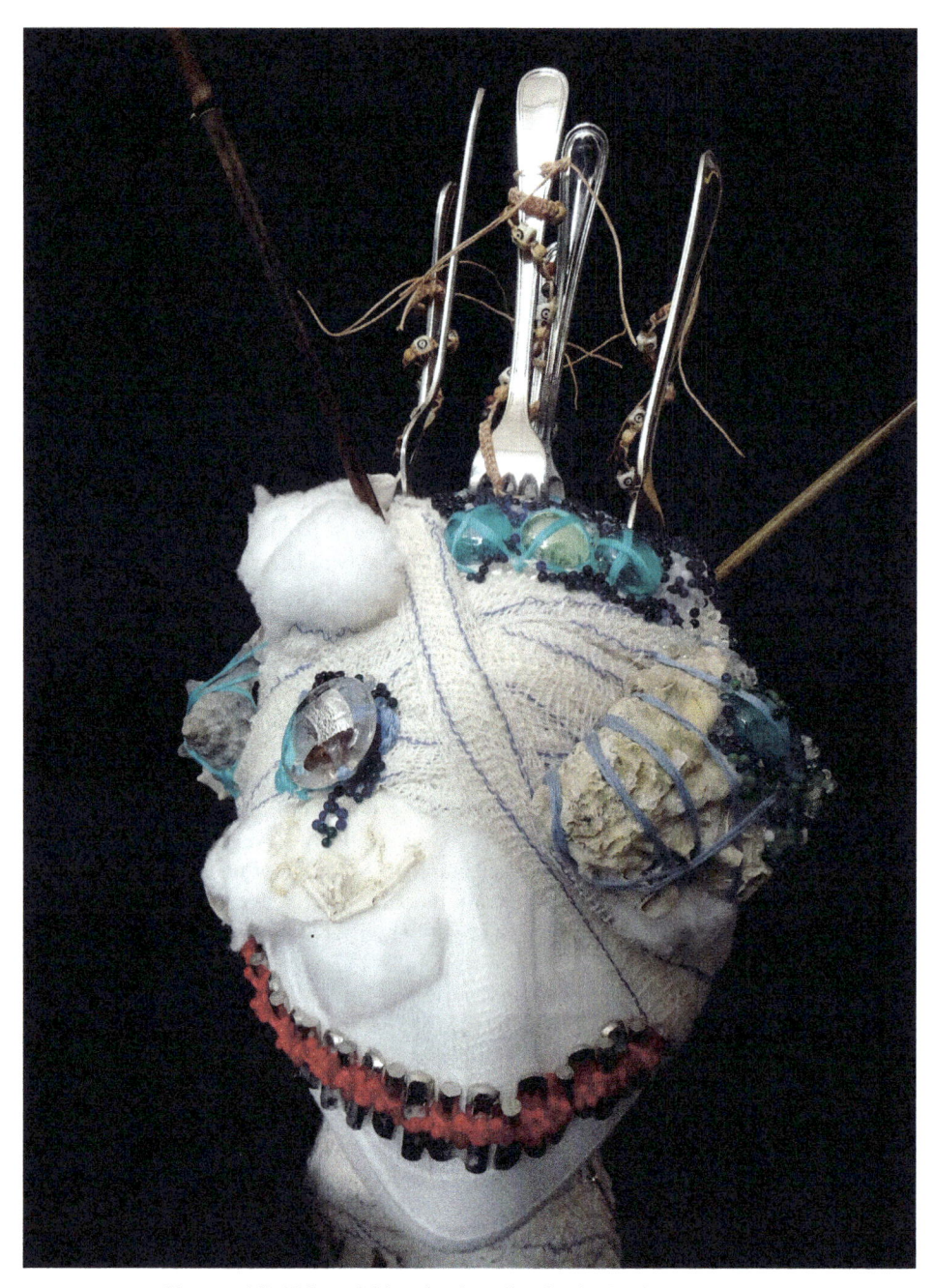

Figure 16.7 "It" model in mixed media. *Credit*: Rachel Armstrong

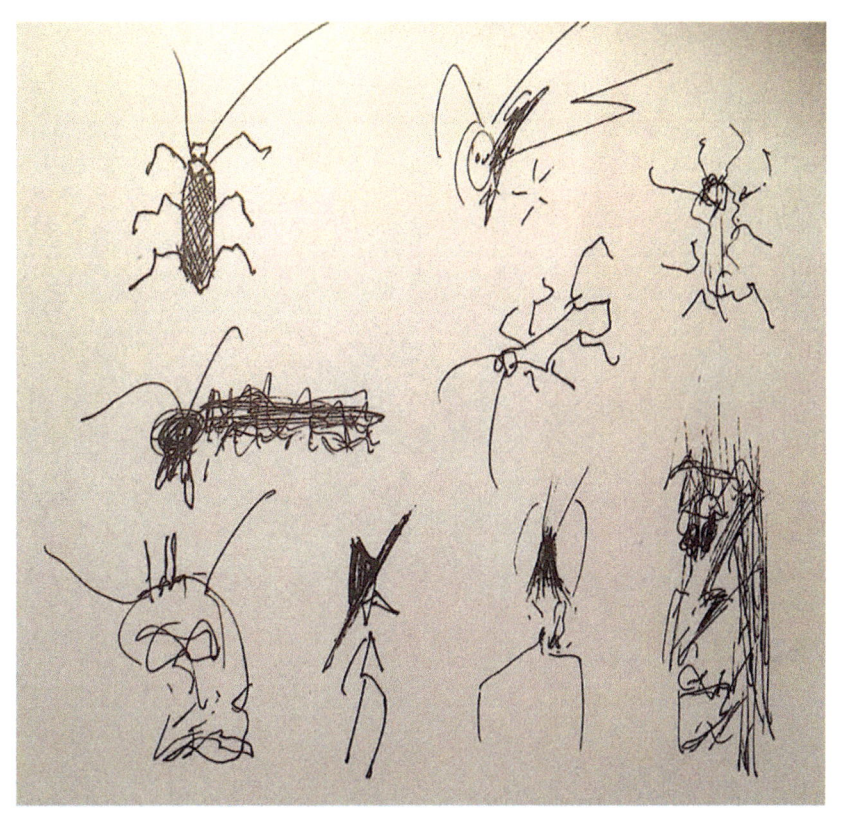

Figure 16.8 "It" sketch (after Rachel Armstrong). *Credit*: Rolf Hughes

16.4 Legacy of the Interstellar Question

Rachel Armstrong, Professor of Experimental Architecture, Newcastle University.

It is contrary to the aims of this book to conclude on the ideas shared by the production of the work and in the various contributions. Indeed, if the starship question has revealed anything, it is that the possibilities for those conditions and ideas that may cumulate in actual ventures to the stars are still very much up for grabs—so the opportunity to write the kind of futures that we may wish to inhabit remains wide open.

Yet the potential for interstellar exploration is not infinite. Limits will inevitably be set by the context in which our visions are dreamed, whether they are technological, political, cultural, or economic. However, it is unlikely they will detract from our already longstanding relationship that has been established with our heavens, night skies, and interstellar landscapes, where we imagine that alternate modes of existence are viable.

In this modern era, where classical science and technology, which form the backbone to conventional forms of engineering appear to create solvable gateways to potentially all

478 **Interstellar research methodologies**

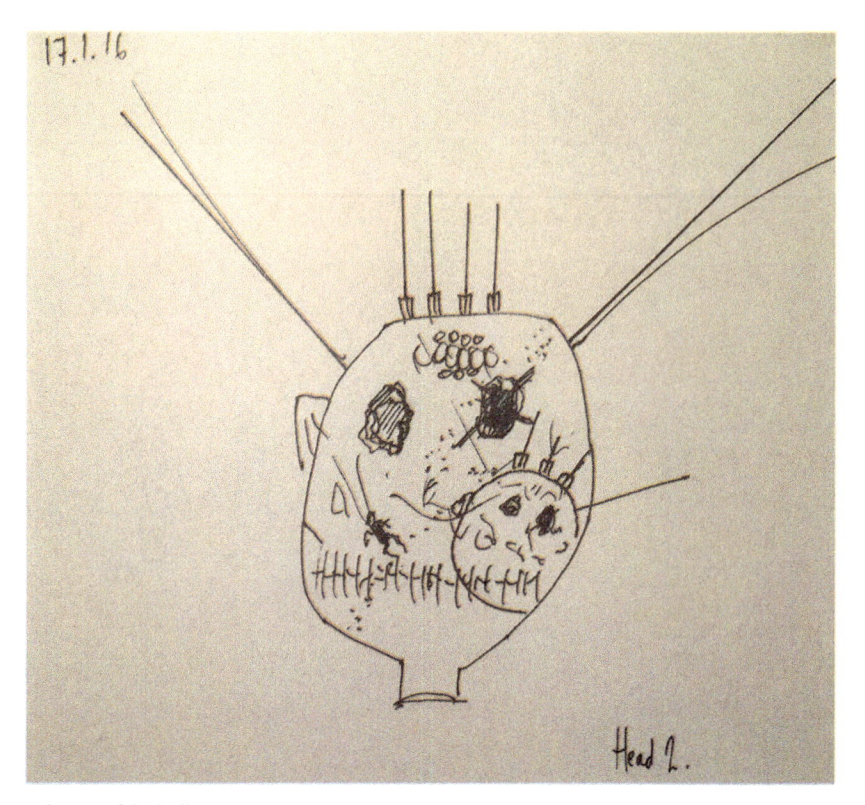

Figure 16.9 "It" development sketch (after Rachel Armstrong). *Credit*: Rolf Hughes

challenges, we also need to remind ourselves that this way of thinking is relatively recent. While our species may have been cognizant for around 150,000 years, Enlightenment-inspired thinking has been around for about 500 years. Even within this short space of time, the technological questions raised in actually reaching extremely far distances over many natural human lifetimes are extremely challenging. Given that classical science is a very recent addition to a whole portfolio of proposals that we have made for reaching the stars — perhaps contemporary ideas about starships will be succeeded by other approaches that may for example, not draw on the construction of giant vessels, but instead for example, propose our teleportation to distant realms through spooky entanglements with quantum phenomena.

It is also therefore premature to propose that such an ancient aspiration is impossible, as our stories about our aspirations are likely to continue to evolve in keeping with new discoveries and modes of thinking.

Yet, the Interstellar Question is not easy to address. Indeed, it is likely to always be steeped in enough uncertainty to render a positive outcome possible — that is, until the day comes on which we actually achieve human spaceflight to other solar systems.

Since there is a chance that we will leave our Solar System in search of other inhabitable spaces, then it is worth seriously considering the kind of challenges that we are confronted with. Indeed, the nature of space exploration is changing. Private corporations are already making businesses in space, such as Jeff Bezos's Blue Origin, Elon Musk Space X, Richard Branson's Virgin Galactic, Rick Tumlinson's Deep Space Industries, Alan Bond's Reaction Engines, and Bigelow Aerospace, to name but a few. Yet, these developments are much more far-reaching than establishing the economic foundations of space exploration; they also relate to a whole portfolio of infrastructural decisions that will shape how we view and nurture the possibility of human space exploration.

In regarding these developments, as a next step towards colonizing other habitats, it is essential that the suitability of existing legal, political, economic, and cultural infrastructures to make this transition are considered, as we are essentially talking about the structure of societies. In particular, we need to examine what it means to take an ecological — rather than an industrial — view of space exploration.

Since we cannot build ecosystems, we must take a long and deep look at our assumptions about life-supporting systems. Yet, in considering how the principles of generating self-sustaining environments may be transposed to support the inhabitation of non-terrestrial habitats in the longer term, we do not need to leave our own planet to pose and explore these questions, but think radically about our current practices and what life needs to persist. A benefit of this approach is that we are likely to discover something new about terrestrial ecosystems that could, for example, help us to regenerate deserts and promote biodiversity.

Project Persephone proposes to build an artificial world from scratch — and quickly confronts us with significant conceptual, technological, environmental, and procedural challenges that currently preclude our capacity to build such a system. My work has focused on engaging with a design-led experiment that looks to ask better questions that may be addressed through scientific and technological developments at some point in the investigation. Additionally, in building a community of researchers that can meaningfully engage with the challenges, it is also possible to deconstruct the assumptions on which our contemporary modes of thoughts about life and ecology are founded, so that new innovation spaces may be established.

How, then, can the parameters of this project be established?

Should everything already known about ecosystems good design be bundled together in an enormous sealed can and see what happens? Or, if we are to reconsider our starting points, where is it meaningful to do so — subatomic particles? Materials? Soils? Life? Ecosystems? Planetary cycles? Solar systems? Or the birth and death of stars?

None of these is entirely inseparable, yet it is simply too daunting to take on all aspects of our rethinking from first principles for every conceivable variable. At some point, we are inevitably constrained by assumptions of one kind or another.

In such circumstances, experimentation becomes incredibly valuable in facilitating the development of a space in which it may be possible to invent an entirely new way of being. Yet, for such an approach to be effective, it is important to open up our modes of experience to alternatives so that we do not preclude knowledge of the outcome. Instead, we need to remain immersed within the complex, massive, and constantly changing discovery process and therefore be sensitive to its emerging details.

Right now, Persephone exists in an experimental capacity. Yet, this is not a traditional scientific laboratory that operates under controlled situations. It is something much freer, complex, and messy, aimed at addressing specific issues in contemporary knowledge-making—particularly in exploring the challenges of hypercomplexity. The aim is not to eliminate uncertainty, but to invite it into the discovery process through design-led explorations and experiments. At such an early stage of development, Persephone's observation stations and laboratories have been Earth-bound. Some have taken place in the stratosphere where it has been possible to compare and contrast the infrastructures needed for different life forms, while others have taken place in cities like Venice, in which urban-scale structures like reefs and islands are proposed to be grown from scratch from plastics and programmable chemistries. While these operations break step with the conventions of scientific process, they have also opened up new avenues for discovery.

Yet, this should not surprise us.

The classical scientific experiment seeks to exclude messiness, which runs through natural systems. Indeed, the orchestration of new relationships between lively bodies has been fundamental to knowledge-making in the last 50 years. We have learned that at far from equilibrium states, which are characteristic of living things, matter is not a passive, brute substance, but is capable of taking "decisions" based on certain physical states or contexts and possesses creativity, which may be surprisingly inventive. Concurrently, we have also developed the ability to bring material relationships together in such a way that the practice of supra molecular chemistry enables us to invent matter that has never existed before in the history of the universe.

With the advent of these kinds of insights and processes that are investigated across a whole range of scales and contexts, all bets are off with the implications of these discoveries for existing traditions, practices, and forms of knowledge-making.

Perhaps Persephone's experiments simply remind us that the tools of modern synthesis are not sufficient to deal with the massively distributed nature and hypercomplexity of third-millennial challenges that resist the possibility of single control points, or a utopian idealized state of existence. These explorations embrace risk, and are full of contradictions, difficulties, and paradoxes that not only require us to invent new kinds of tools and technologies, but invite us also to stay engaged in a continually managed process, which will continue to surprise us—even when we embrace its radical creativity.

Indeed, the challenges that we're facing as a species with our long-term survival ask us to reconsider the impacts of these developments and perspectives for so many disciplines whereby, in this emerging age of scientific, technological, and design convergence, we retain our sense of enchantment and capacity to remain meaningfully engaged with each other without conflagration, even when there are disagreements—a diplomatically creative state of existence, called radical love.

*

"Radical Love," by Rachel Armstrong and Rolf Hughes (adapted from *The Handbook of the Unknowable*, published by V2 Press, March 2016, for the fourth Trondheim Biennale)

"VOLUNTEERS WANTED
For unending, unwaged journey in pitiless cold,
complete darkness, and constant danger.
Safe return impossible.

In the unlikely event of success,
Honour and fame guaranteed
on prime time TV."
And still, they wrote from all over the world
Hundreds of thousands desperate to embark
On a one way trip
To colonise a new star system.

The lucky twenty thousand applicants intriguingly
Included eight thousand and three software developers,
Five thousand, seven hundred and ninety-eight information communications technology experts,
Six hundred and thirteen asteroid mineral prospectors,
Two hundred and forty-three military personnel,
A hundred and seventy-seven doctors,
Scores of women who wanted to
Give birth to the first extra-terrestrial child,
Ninety-eight artists,
Thirty-three physicists,
Twenty-six articulated truck drivers,
Eight billionaires,
Five astronomers,
And no rocket scientists.

"We'll be in training for ten years.
We'll have learned to live away from family by then.
It can't be so difficult."

Researchers from the University of Kansas
Warned that high levels of radiation in space
Damage the ovaries and testicles,
Which likely hampers people's efforts to reproduce.
They said they'd try, give it a go anyhow.

"The most important thing in life
Is to leave a legacy.
We start training this year.
It's more likely I'll fall in love
With one of the team
Than someone 'out there.'
It's really exciting because
You're watching a society develop.
It's like seeing a country sprout from nothing."

Their courage to continue "for ever" onwards
Into the unknown at their own peril
Does not lie

Within the perfect circles
Of completed comforts, closed ecologies
Or winged serpents
Sedately grazing on their own tails.
Rather they speak over the
Worry knots that are woven through
Things left undone, unsaid, or
Unresolved—life's disappointments.

"It's time for humans to begin somewhere else
I have the skills to make that happen."

Their training completed
Away from prying eyes
In haunted senses abused
By compulsive disorders of
Grinding jaws and flattened teeth
Wrung hands
Traumatic alopecia
Holes that burst through pacing soles
Guilt stained anorexia
Broken minds, catastrophising through
Phantoms of injustice, regret, revenge
Reaching for covert relief in rosary and worry beads.

"It is highly risky, and an enormous responsibility
As well as an adventure."

Memories are riddled with radiation holes.
Wilful blindness masks
The physical devastation wreaked upon them:
Compelled by vaporous gravity
Limbs thin from leached calcium
That makes cataracts
And fills tender kidneys with stones.
Fat puddles into
Pot bellies and forms
Bilious blood clots.

Despite brutal exercise regimens
They succumb to
Extreme anatomical reordering
Whereby the authentic self
Can only be marked by loss of faith.
The only way of knowing that you're "you"
Is when your beliefs are scarred up.
Where did the body go?
We cannot tell, for something other than flesh

Quite intangible now holds
Us together.
"Human space exploration has always interested me
The opportunity to be involved is really appealing.
The future of humanity is in space."

In sleepless synthetic nights
Thought itself breaks down
And language fails to describe
Existential angst—
Its vastness, endlessness,
Hopelessness, dreamlessness.

"There is more to life than marriage and babies.
It'd be nice to escape the office
And have some AMAZING life experiences.
Leave the solar system."

What used to be called
Madness is quite normal.
Only the brave, the foolish,
And those that are belligerent
To the difficulties of their frontier existence
Endure the ensuing conflicts, disparities.
Fractures within the Tower of Babel appear
Where the conditions for existence
Are continually contested as they
Jostle for territory, partners, power and beliefs.

Are we extraterrestrial now—or forever earthbound?

Exchanging a new language
They have left experience behind
And rejoice in a strange, savage music
Uncoupled from sense.

Constantly under the dark light
Of their interior star
Which powers their ship, their lives
They shake, they shake, they shake again.

Frustrated, they turn up the volume
Of familiar tongues
That suddenly seem
Uncomfortably strange
Crumbling in
Other times, worlds, existences.

Experience has become alien
There is no language for it yet

All those things you learned as a child
—the flow of water, the fall of sweet rain, sunshine on skin—
Now unbearably unfamiliar.
What exactly was it, this "normal" nature?
You feel the need for new words, new concepts, new histories.

"Technology will advance rapidly.
One day, I will come home to Earth."

Dot, dot, dot...reboot.

Wakers logon reanimat vitals.
RUN: at purpose, in purpose.
Thawed cryoresurrection. Ritual
Via anatomia diurnalizes
Molecular data feed, quantum thought cloud
Swerves outmoded
Defeatist, deathist creeds.
Bodies, minds, flesh biolaced
In silicon, in vitro, in search of increased extropy
Centred on circus, art, community
What once were understood as human limits
Challenged in omnipresent neural networks
Through augmented bodies.
Science and technology not salvation
Continual enhancements
But form instruments—spirit music—
Building blocks of extremophiles.

Metal ions flavour the sweaty worldship dew,
Filter feeders slurp their soil soaked oceans
Nanofibre threads intrude on dreams
First as bright lights then
As radiation-repair shadows, which erase retinal scars.
Delicate biofilms laden with programmable physiology
Linger around them, like breath
Enriching organic flesh and emotional wellbeing.

So they weep awhile. At Heliosphere,
They weep: zero-four-four-twenty.
Zero-four-four-twenty. Four-four-twenty.
They weep to rituals of tear-stained forget-me-nots.
Synthbots. Aesthetic programs mourning old humanity—
A transitory stage in the evolution of cosmic intelligence.
It's always important to pay one's respects. And then a new day.
New kin. We are here. Yes, here we are.
In defiance of the gods.

COMBINED REFERENCE LIST FOR CHAPTER 16

Tilde Björfors, Kajsa Lind (trans. Claire Chardet): Inuti ett Cirkus Hjärta/Inside A Circus Heart (Norsberg: Cirkus Cirkör, 2009), p. 39

J. Brodsky, *Watermark: An Essay on Venice* (Penguin Classics, London, 2013)

G. Flaubert, The temptation of St. Antony. Source Bebook (not dated) www.covesys.com/docs/appnotes/pickup/BooksFromFVSR/The%20Temptation%20Of%20St.%20Antony%20-%20Flaubert%20Gustave.pdf

R.M. Hooper, The fish that helps us understand gender. *The Japan Times* (2015), June 20, www.japantimes.co.jp/news/2015/06/20/national/science-health/medaka-fish-helps-us-understand-gender/#.Vv5Bvavp6Hc

Protocell Circus. Protocell circus: spontaneous generation: a.k.a. A "Natural History" of Protocells. You Tube (2011), www.youtube.com/watch?v=dFagK5Lshlg

Sacred Texts. The prophecies of Nostradamus, preface, Salon (1555), March 1, www.sacred-texts.com/nos/; http://www.sacred-texts.com/nos/preface.htm

Index

A

Abiotic, 258
Able and Baker, 111
Accelerationism, 386
Adams, D., 10, 372, 433
Adaptive dissipation, 151, 155
Adjacent possible, 151
Albert II, 111
Alberti, L.B., 115
Aldrin, Edwin (Buzz), 13, 51, 179
Alexander, C., 139, 300, 310
Algae, 11, 15, 16, 23, 27, 34, 43, 61, 66, 68, 125, 127, 133, 135–138, 260, 262, 288, 294, 295, 308, 313, 315, 316, 325, 332–334, 337
Al-Khalili, J., 7, 161
Alpha Centauri, 23, 81, 212, 360
Ambergris, 191
Anandgram, Village of Joy, 175
Anthropocene, 16, 52, 78, 79, 81, 82, 123, 145, 166, 169, 170, 385, 387, 389, 410, 416, 419
Antimatter, 33
Apollo 8 mission, 79
Apollo 11 mission, 179, 411
Apollo Moon program, 49
Apollo's orphans, 48–54
Aquabatics, 203, 374–376
Arakawa, S., 130, 131
Archean, 69
Arks, 14, 15, 46–48, 53, 75, 140, 195, 287, 317, 331, 360, 384–387, 389, 392, 412, 415, 420, 422, 455
Armstrong, N., 13, 45
Armstrong, R., 2–206, 301, 318–319, 359–372, 394-396, 400, 423, 446-485
Arrhenius, S., 216
Artificial soils, 106, 110, 115, 119–121, 153, 156, 460
Artificial uterus, 370
Artificial wormhole, 404
Artwise, 124–126
Assemblage, 123, 130, 156, 168, 170, 180, 188, 463, 466
Asteroids, M-type, S-type, 88
Astra Planeta report, 448, 449
Astroecology. *See* Cosmoecology
Astudio, 19, 43, 136, 202
Astudi, R., 296–323
Atacama desert, 67

B

Bacon, F., 76, 78, 388
Bacterial biome. *See* Microbiome
Baikonur cosmodrome, 220
Banks, I.M., 8, 360
Barad, K., 170, 187
Bassler, B., 192
Beckstead, N., 5, 6
Beer, S., 41
Beesley, P., 11, 115, 116
Belousov-Zhabotinsky reaction, 152
Benjamin, M., 225, 226, 233
Bennett, J., 170, 187
Bernal, J.D., 37, 74
Bernal sphere, 37, 74, 75, 435
Berry, T., 394–400
Betsky, A., 25
Big bang, 34, 391, 393, 399
Big data, 169, 170
Bigelow, 50, 414, 479
Bio-architecture, 150
Biobricks, 10
Biocretes, 120, 121
Biofilms, 43, 78, 85, 90, 125–127, 132, 133, 172, 174, 190–192

Biogenesis, 17, 23, 65, 67, 69, 83
Biomes, 56, 60, 61, 64, 65, 68, 70, 139, 165, 180, 211
Biomimicry, 16, 26, 148, 393, 423
Bioprocessors, 135, 136
Bioreactors, 27, 294, 337
BIOS-3, 17, 140
Biosphere 2, 17, 92, 140, 287, 332, 411, 425
Biota, 256
Biovigilance, 65
BIQ House, 136, 137
Black sky thinking, 7, 18, 419, 447, 451
Blue Ball, The (film), 224–252
Blue Origin, 50, 414, 478–479
Bond, A., 8, 39, 47, 81, 479
Bottom-up, 9, 12, 26, 38, 301, 308, 309, 386
Boulding, K.E., 298
Boullée, É.L., 434, 435
Braidotti, R., 170
Brain–computer interface, 341, 347–350
Brain implants, 345–348, 351
Brand, S., 45, 227
Building Integrated Modeling (BIM), 309, 428
Bussard Ramjet, 211
Bütschli droplets, 99, 100, 117, 119, 155, 157–160
Bütschli, O., 99, 156
Bütschli oil droplets, 394

C

Calvino, I., 114, 119, 123, 206
Capsule of Crossed Destinies, 111–114
Carbonaceous chondrite asteroids, 69, 88
Carbon dioxide (CO_2), 11, 17, 23, 24, 34, 51, 69, 85, 93, 96, 97, 108, 111, 117–119, 135, 136, 140, 141, 157, 167, 215–217, 287, 288, 294, 295, 297, 310, 313, 315, 322, 324, 333, 337
Cassini, 35, 46, 58
Centaurus constellation, 6, 21, 23
Ceres, 36, 60, 212
Chemotropes, 165
Cheonggyecheon River, 85
Chermayeff, S., 300
Circular economy, 17, 302, 310
Circus, 167, 202, 203, 447, 453–458, 460, 463, 472
Cirkör LAB, 459
Cirkus Cirkör, 459
Clarke, A.C., 15, 48, 75, 76, 107, 261, 360, 364
Clynes, M.E., 369, 420
Cockroaches, 93, 111–114
Coelacanth, 70, 466
Cole, D., 75
Committee on Space Research (COSPAR), 66, 69
Cook, P., 25
Cosmism, 201, 384

Cosmoecology. *See* Astroecology
Cosmogenesis, 397, 398
Crawford-Young, S., 366, 367
Crick, F., 169
Crutzen, P., 16
Cryogenics, 14
CubeSat, 10, 77
Curiosity, 45, 57, 115, 168, 206, 298, 364
Cyborgs, 107, 170, 187, 256, 261, 265, 341–351, 373, 420, 425

D

Daedalus, project, 211
Dark energy, 33, 36, 205, 402
Dark matter, 33, 35, 36, 204, 263, 402, 432
Darwin, C., 21, 64, 98, 383, 390
Death, 14, 65, 87, 93, 97, 108, 131, 194, 195, 242, 355, 356, 366, 368, 384, 391
De Chardin, P.T., 385, 399
Decompiculture, 195
Decomposition, 65, 66, 195, 198, 288
Deinococcus radiodurans, 113, 261, 368
De Lavoisier, A.-L., 206
Delhi, 326, 327, 330, 331
Demilitarized Zone (DMZ), 193
Descartes, R., 169
Design-science, 300
De Wilde, F., 12, 39, 40
Diegetic objects, 9
Directed panspermia, 19, 62–70, 111, 165, 259
Discovery, Space Shuttle, 190
Disruptive innovation, 308–312
Dissipative structures, 85, 98, 145, 151–156, 158, 160, 161, 168, 194
Dissipative systems, 151, 152, 157, 160
DIYbio, 77, 146
Dobrovolski, G., 194
Drexel University, 10
Dust, 29, 35, 50, 56, 69, 86, 88, 165, 199, 236, 238, 249, 294, 406, 429, 456, 475
Dwarfism, 354–356, 359
Dymaxion house, 301
Dynamic droplets, 98–100, 116, 119, 120, 135, 157, 158, 161, 462
Dyson, F., 8, 48, 62, 69, 263, 434, 435
Dyson ring, 435, 440

E

Earth-bound, 226, 229, 235, 237–239, 241, 244–246, 249, 250, 373, 393, 480
Ecocene, 13, 15–18, 20, 46, 59–61, 64, 78, 82, 137, 144, 145, 150, 165, 170, 171, 175, 184, 193, 201

Ecological beings, 166–167, 169, 171, 180–182, 184, 187, 188
EcoLogicStudio, 43, 127, 333
Ecology, 12, 18–19, 32, 39, 40, 58–62, 67, 86, 92, 93, 120, 135, 137, 141, 166, 171, 175, 180, 182, 187, 193, 200, 211, 255–283, 287, 295, 300, 301, 304, 312, 315, 333, 411, 419, 479
Ecomodernism, 387, 389–392
Ecopoiesis, 17, 20, 24, 59, 61, 63, 66, 131, 139, 145, 153, 412
Ecosurveillance, 70
Elysium (film), 227
Engastration, 175, 191
England, J., 151, 155, 161
Enrique, J., 144
Ephemeralization, 300
Equilibrium, 24, 75, 82, 86, 100, 104, 131, 134, 143, 144, 152, 155, 194, 216, 390, 419, 447, 452, 453, 456–458, 480
Escherichia coli, 406
European Organization for Nuclear Research (CERN), 76, 406
European Space Agency (ESA), 49, 50, 54, 55, 58, 138, 139, 288, 292, 363, 414
Exoplanets, 6, 23, 225, 259, 260, 401–407
Experimental architecture, 11, 25, 26, 35, 103–127, 137, 301, 318, 459
Extravehicular activity (EVA), 374–376
Extremis, 114, 167–169, 171, 201, 372–378

F
Fermi paradox, 67
Field, F., 60
Finney, B., 75, 166
Forward, R., 211
Foster, N., 50
Foster + Partners, 55, 56, 139, 414
Franklin, R., 169
Fuller, R.B., 24, 45, 139, 298–300, 310, 340, 411, 419
Fuller, S., 19, 201, 383–394
Futurality, 410, 416, 417, 419
Future of Humanity Institute, 5, 6
Future Timeline, 304–306
Future Venice, 11, 119–120, 123
Future Venice II, 11, 123–127

G
Gaia, 79, 227–228, 239, 243
Galileo probe, 58

Gels, 98, 103–105, 348
Genetic Investigation of Anthropometric Trait, 354
Geodesic dome, 26, 353
Geo-engineering, 16, 64, 387
Geotextiles, 86, 97, 115
Gidzenko, Y., 141
Gins, M., 130
Gissen, D., 86
Glenn, J., 190
Gliese 360, 581
Goddard, R.H., 14
Godfrey, P., 224–226, 228–232, 252
Goldilocks zone, 23, 221
Gould, S.J., 83
Gravity (film), 76, 227, 228, 234, 238, 240–242, 244, 247, 248
Gray goo, 110
Great Splat, 34
Greenhouse effect, 51, 216, 325
"Growing As Building" (GrAB), 335–337

H
Habitable zones, 23, 221–223, 360
Hadfield, C., 54, 166, 231, 247
Hadid, Z., 440, 441
Hanczyc, M., 155
Hanging Gardens of Medusa, 12, 106–111
Hansen, A., 175
Haraway, D., 144, 170, 187, 188, 192
Harman, G., 187, 448
Hawaii Space Exploration Analog and Simulation (HI-SEAS I), 143, 353
Hawking, S., 6, 21, 79
Hendriks, A., 19, 171, 201, 352–359
Hird, M., 66, 170, 192
Homo floresiensis, 356–357, 359
Huxley, J., 383–385
Hughes, R., 202, 446–485
Hydroponics, 92, 95–97, 140, 331, 332
Hygroscopic, 27, 34, 153, 154
Hylozoic ground, 11, 115–119
Hypercomplexity, 18, 19, 41–44, 78, 82, 84, 86, 90, 91, 143, 156–158, 161, 162, 182, 195, 204, 447, 449, 479, 480
Hyperion, 82, 84

I
Icarus Interstellar, 3, 9, 18, 52, 77, 81, 201, 306, 384, 386
Icarus, project, 82, 211

IDEA Laboratory, 124–126
Ikegami, T., 155
Incredible Shrinking Man Project, 358–359, 425
International Genetically Engineered Machine (iGEM) competition, 10
International Space Station (ISS), 2, 17, 22, 45, 49, 50, 53, 54, 77, 92, 95, 113, 141, 143, 164, 166, 167, 169, 190, 195, 220, 222, 247, 288, 291, 292, 298, 330, 363, 372–374, 406, 412, 414, 465
Interplanetary communications network (IPCN), 54, 55
Interplanetary superhighway (IPS), 58
Interstellar (film), 6, 76, 213, 227, 228, 234, 238, 245, 246, 248, 251
Interstellar hackathon, 9, 10, 12, 205
Interstellar Question, 2–22, 28, 70, 81, 169, 195, 196, 204–206, 210–211, 477–480

J

Jacquard punch card, 453
Jade rabbit, 57
Jet Propulsion Laboratory (JPL), 54, 58, 219, 220, 419
Jones, E., 75, 166
Joy, B., 79
Juan Enriquez, 306
Jupiter, 35, 46, 58, 219, 406

K

Kauffman, S., 7, 134, 151, 161, 187
Kelly, K., 6, 7, 115, 137
Kelly, S., 45, 53, 92, 95
Kepler-23, 139, 186
Kepler 452b, 223, 224
Kepler space telescope, 23, 221, 223, 224
Keyser, M., 79
Kline, N., 369
Kohler, P., 363
Krikalev, S., 141
Kuiper belt, 35, 46, 62
Kurtzweil, R., 79
Kuwabara, Y., 370

L

Laboratory, 9, 10, 12, 15, 18, 19, 23, 25, 29, 30, 49, 60, 75–78, 82, 83, 89, 103, 104, 106, 107, 116, 119, 125, 127, 153, 175, 199, 201, 216, 292, 330–332, 342, 348, 367, 370, 375, 383, 388, 392, 402, 414, 422, 447, 456, 458, 460, 479

Lacks, H., 181
Lagrange points, 56, 60, 84, 138
Large Hadron Collider (LHC), 36, 76, 189
Laron syndrome, 355, 356
Laser, 54, 79, 405, 434
Latham, W., 115
Latour, B., 22, 187, 192, 227, 373
Lazarus Fish, 466–468
Lee, J.R., 195
Leprosy, 175–177, 179, 186
Lettuce, 92, 95, 331
Lexx, 8, 39
Liesegang, R., 104
Life-centred ethics, 255–265
Lindgren, K., 95
Lin, H., 68
Liu, H.-C., 370
Living technologies, 115, 150, 320–322, 449
Loeb, A., 68
Long, K.F., 15, 107
Lovelace, A., 453
Luna 2 mission, 22
Lynn, G., 115, 440, 441

M

Magic, 57, 134, 358, 447, 453, 454, 459, 471
Magnetic implants, 349–351
Malevich, K., 39
Mallory, G., 14, 57
Mallove, E., 15, 75
Manifesto for living architecture, 145–146
Man-Systems Integration Standards, 353
Margulis, L., 180
Mariner 4, 34
Mars, 2, 17, 34, 45, 48, 50, 51, 55–57, 61, 66, 67, 69, 92, 94, 111, 138, 139, 142, 143, 212, 219, 220, 226, 289, 295, 298, 323, 325, 327, 359, 363, 406, 475
Mars Atmosphere and Volatile Evolution (MAVEN), 58
Mars Express, 57
Mars-500 mission, 143
Mars Odyssey, 57
Mars One, 51, 52, 66, 67, 142, 298, 323, 325, 359, 414
Mars Orbiter Mission (MOM)/*Mangalyaan,* 58
Mars Reconnaissance Orbiter, 58
Martian, The (film), 92, 196
Martin, T., 8, 81
Massachusetts Institute of Technology (MIT), xviii, 10, 333, 435
Matloff, G., 15, 75, 260
Matzerath, O., 358

Maximalism, 155
Mbuti, 357, 359
McFadden, J., 7, 161
Mealworms, 142
Medaka fish, 367, 372, 464–466
Megastructures, 9, 12, 32, 37, 74, 75, 205, 299, 435
Melchiorri, J., 127
MELiSSA, 288, 295
Mendeleev, D., 206
Mercury, 34, 219, 360, 368
Mereschkowski, K., 180
Micro-Ecological Life Support System Alternative (MELiSSA), 61, 138
Microplastics, 11, 24, 124, 125
Microseeds, 405–407
MirCorp, 49
Mir space station, 17, 49
Mixology, 89–91
Montgolfier, J.-É., 111
Montgolfier, J.-M., 111
Moon Express, 56
Moon Palace One, 142
Moravec, H., 372
Morton, T., 81, 187, 205, 415, 416
Mother trees, 99, 189, 190, 192
Movile cave, Romania, 165
Multi-electrode array (MEA), 342, 344, 345, 348, 351
Musk, E., 50, 55, 57, 58, 79, 323, 324, 331, 479
MuskNet, 55, 58
Mycorrhizal networks, 85

N
National Aeronautics and Space Administration (NASA), 23, 36, 46, 49, 50, 54, 56, 61, 76, 84, 87, 92, 140, 141, 201, 219–223, 229, 233, 236, 237, 245, 246, 288, 291, 292, 294, 324, 353, 361–363, 365, 366, 368, 387, 401, 411–412, 414, 419, 435–437
Natural computing, 43, 82, 98, 134, 148–153, 156, 157, 161, 179
Natural resources, 17, 24, 26, 58, 63, 83, 138, 145, 214, 296, 396, 448
Nature cultures, 188
Nebula Sciences, 12, 107–112
Neptune, 35, 46, 219
Nerja, 129
New Atlantis, 76, 78
Newcastle University, 11, 104, 107, 124, 301, 459
New Horizons, 2, 35, 46, 58, 360
Newlands, J., 206
Nolan, C., 6, 13, 213, 227, 245
Non-equilibrium, 11, 86, 130, 150, 456, 471

Non-invasive brain-computer interfaces, 348–349
Non-linear, 129, 459–474
Nuclear fusion, 38, 51, 89, 200, 211

O
Oceanic ontology, 155, 156
O'Neil, G., 6, 15, 17, 37, 48, 69, 75, 83, 84, 123, 211, 260, 435
Oort, J., 36
Opportunity, 7, 8, 12, 24, 36, 41, 48, 49, 54, 57, 60, 64, 67, 100, 124, 134, 137, 148–150, 156, 161, 172, 195, 196, 199, 201, 204, 211, 218, 220, 226, 302, 310–312, 344, 347, 388, 389, 391, 449, 477, 483
Orbital debris, 87, 88
Oskar Matzerath, 357–359
Ouroboros, 202, 425–445
Outer Space Treaty (1967), 66, 69, 93

P
Pale Blue Dot, 16, 21, 45–46, 233, 361
Panbiotic astroethics, 67
Panspermia, 19, 35, 62–70, 111, 165, 190, 259, 260
Parallelism, 135, 162
Parametric, 42
Pask, G., 41
Patsayev, V., 194
67P/Churyumov-Gerasimenko comet, 56
Pell, S.J., 19, 171, 201, 341, 372–378
Perry, M., 127
Persephone, 3, 11, 13, 18, 19, 47, 63, 65, 68, 78–87, 89, 90, 93–95, 97, 99, 103, 110, 123, 131–134, 144–146, 162, 165, 166, 172, 175, 179, 197, 200, 201, 203–205, 211, 410, 412, 419, 422, 425, 447, 451, 456, 460, 479, 480
Petrov, T., 28–32
Philae Lander, 55, 56, 226
Pilâtre de Rozier, J.-F., 111
Planetary bioengineering, 66
Planetary resources, 34, 50, 52, 70, 139
Pluto, 2, 35, 46, 60, 219, 360
Polyakov, V., 17, 263, 362
Post-human, 166, 170, 261
Potsdam Potato, 368
Powers of Ten, 428, 430
Prebiotic, 258
Protocells, 108, 110, 257, 258, 263, 318–319, 322, 394, 449, 460–463, 466, 470–472

Prototypes, 8–11, 18, 19, 27–29, 48, 50, 56, 77, 78, 80, 82, 86, 89, 93, 101, 104, 110, 127, 135, 140, 141, 148, 195, 196, 201, 202, 204, 205, 295, 301, 332–334, 428, 447, 453

Q

Quantum effects, 161, 402
Quantum physics, 7, 76, 394
Quantum protocells, 161
Quantum teleportation, 403
Question-making, 11, 196
Quorum sensing, 189

R

Radical love, 203, 447, 458, 480–485
Reaction-diffusion computers, 152, 153
Reaction Engines, 479
Reef, 11, 79, 115, 119, 125, 135, 175
Regolith, 50, 55, 74, 86, 87, 139, 179, 212
Replicant, 170
Reynière, Grimod de la, 191
Robinson, K.S., 8, 107
Robonaut, 56
Rosetta, 56, 58
Rotifers, 259

S

Sagan, C., 34, 46, 69, 200, 361, 401, 411, 437, 438
Salmonella, 190
Saturn, 35, 46, 55, 58, 219, 324
Savage, M.T., 15–17
Saxe, J.G., 5
Schrödinger, E., 161, 195
Scrying, 459–474
Search for Extra Terrestrial Intelligence (SETI), 77
Self-perpetuation, 255, 298
Self-propagation, 255, 256
SETI@home, 77
Sexual relations, 363–365
Shepard, A., 45
Shepard, B., 141
Shoemaker, G., 86
Shrinking, 201, 352–359
Slartibartfast, 433, 441
Snow, J.T., 85
Soil fertility, xvi, 11, 16, 24, 83, 123, 320
Soil productivity, 84
Soils, 11, 24, 26, 34, 59, 63, 66, 67, 69, 77, 82–87, 89–97, 101, 103–110, 115, 116, 119, 120, 123, 127, 129–137, 143, 152, 153, 156, 161, 172, 179, 189, 197, 200, 202, 260, 264, 265, 287, 308, 314, 320–322, 325, 331, 425, 469
Soyuz 11 spacecraft, 194
Space ecosystem, 18, 40
Space junk, 32, 87, 88
Spaceman' economy, 298, 299
Space nature, 86, 143–145
Spaceship Earth, 24, 45, 288, 297, 298
Space Shuttle, 23, 48, 190, 219, 220, 222, 242, 363, 373
Space X, 49–51, 54, 414, 479
Spitzer space telescope, 224
Sputnik, 55, 76, 385
Stapledon, O., 435
Starship Enterprise, 8
StarshipSPIDER, 12, 39–41
Sterling, B., 9, 33, 132
Stoermer, E., 16
Strong Aroma, 91
Strong, J., 13, 15
Stross, C., 6
Studio Swine, 79, 127
Summerson, J., 429, 431
Supercentenarians, 355–356
Supersoils, 94, 95, 97, 101, 106, 137
Sustainability, 16, 24, 27, 32, 45–70, 74, 141, 166, 211, 300, 301, 304, 307–309, 312–314, 322, 359
Swimme, B., 394, 396–400
Symbiogenesis, 180, 455
SymbioticA, 146
Synthetic biology, 10, 94, 97, 303, 308, 309, 315, 318–320, 423–425
Synthetic ecology, 78, 120

T

Tardigrades, 138
Tau Ceti, 406
Techniuum, 115
Technofossils, 16
Teleportation, 405, 478
Terraforming, 8, 23, 47, 56, 61, 63, 66, 67, 69, 101, 139, 323–340, 420, 425
Thanatobiome, 65
Theia, 34
The Incredible Shrinking Man, 352
Theological energeticism, 390–392
Thomson, D., 115
Tiatini, P., 120
Tipping points, 24, 65, 70, 120, 154, 156, 160, 168, 472
Tomorrow's city, 296–304
Top-down, 9, 12, 26, 144, 166, 301, 386, 447, 457

Transhumanism, 383–394
Tsiolkovsky, K.E., 14, 21, 195, 261, 385
Tumlinson, R., 49, 51, 479
Turing, A., 19, 42, 43, 104, 151, 158, 161
Turing bands, 158
Turing test, 19
Two-slit experiment, 7
Tyndall, J., 216
Tyranny, 224–252

U
Uncertainty, 18, 19, 37, 61, 76, 156–157, 160–162, 168, 182, 187, 204, 205, 312, 331, 387, 462, 478, 479
Unknowing, 447
Upgrade, body, 342, 344–351
Upload, mind, 170
Uranus, 35, 46
Urey, H., 86

V
Van Mensvoort, K., 141
Venice Biennale, 11, 116, 125, 126, 327
Venter, J.C., 41
Venus, 34, 48, 219, 221, 222, 324, 360
Vernadsky, V., 59, 385
Verne, J., 21
Very Large Organizations (VLO), 202, 410, 412–414, 416, 419, 424
Viking probes, 34
Vinci, Leonardo da, 169, 172–175
Virgin Galactic, 49, 50, 479
Vita Vitale, 125, 126
Vitruvian, 129, 167
Vitruvian Man, 169
Volkov, V., 194
Vostok, lake, 138
Voyager probes, 58

W
Warwick, K., 19, 57, 171, 201, 341–351
Water Walls, 61, 141, 294–295
Watson, J., 169
White Cliffs, D., 129
White, Ed, 46
Wilkins, B.J., 21
Wilkins, M., 169
Woods, L., 25
World-building, 202, 433–434, 442
Worlding, 18, 78, 81, 82, 89, 201
Wormholes, 6, 401, 402, 405, 406, 459

Y
Yui, K., 95
Yutu, 57

Z
Zanzara Island, 124, 125
Zero Atmosphere Architecture, 410
Zizek, S., 458
Zubrin, R., 325

Printed by Printforce, the Netherlands